Water Governance, Stakeholder Engagement, and Sustainable Water Resources Management

Special Issue Editors
Sharon B. Megdal
Susanna Eden
Eylon Shamir

MDPI • Basel • Beijing • Wuhan • Barcelona • Belgrade

Special Issue Editors
Sharon B. Megdal
The University of Arizona
USA

Susanna Eden
The University of Arizona
USA

Eylon Shamir
Hydrologic Research Center
USA

Editorial Office
MDPI AG
St. Alban-Anlage 66
Basel, Switzerland

This edition is a reprint of the Special Issue published online in the open access journal *Water* (ISSN 2073-4441) from 2016–2017 (available at: http://www.mdpi.com/journal/water/special_issues/water-gov).

For citation purposes, cite each article independently as indicated on the article page online and as indicated below:

Author 1; Author 2. Article title. *Journal Name* **Year**, *Article number*, page range.

First Edition 2017

ISBN 978-3-03842-446-8 (Pbk)
ISBN 978-3-03842-447-5 (PDF)

Table of Contents

About the Special Issue Editors

Sharon B. Megdal is the Director of the University of Arizona Water Resources Research Center. Her work focuses on water policy and management, on which she writes and frequently speaks. The geographic scope of Dr. Megdal's work ranges from local to international. She served as the Lead Editor of Shared Borders, Shared Waters: Israeli–Palestinian and Colorado River Basin Water Challenges and MDPI's Policy and Economics of Managed Aquifer Recharge and Water Banking and has authored many journal articles. Professional service activities include present/past board and leadership roles with the Universities Council on Water Resources, International Arid Lands Consortium, American Water Resources Association, and National Institutes for Water Resources. Dr. Megdal also is an elected water official. Since January 2009, she has served on the Board of Directors for the Central Arizona Water Conservation District, also known as the Central Arizona Project.

Eylon Shamir received his Ph.D. in 2003 from the University of Arizona, Tucson, Arizona, specializing in hydrology and water resources sciences. Since his graduation, he has been with the Hydrologic Research Center, San Diego, California, a non-for-profit research, technology transfer, and training organization that aims to bridge gaps between scientific research in hydrology and applications for the solution of societal problems that involve water. Dr. Shamir conducts research in water resources planning and management, hydrologic risk assessment and is specialized in impact assessment of projected climatic change on water resources in arid environment. He has co-authored over thirty peer-reviewed articles and has been serving as an Assistant Editor for the Journal of Hydrology for the last ten years.

Susanna Eden is the Assistant Director at the Water Resources Research Center, University of Arizona. She holds a Ph.D. in Water Resources Administration from the University of Arizona and has been engaged with water resources research, outreach, and administration for more than 25 years. Her research centers on policy and decision-making in water management and the use of scientific information. Recent research focused on water harvesting as an adaptive strategy with multiple benefits to communities, stakeholder engagement, and the use of modeling information in water resources planning. She also writes for and edits the WRRC's quarterly newsletter, Arizona Water Resource, and the annual Arroyo publication.

Preface to "Water Governance, Stakeholder Engagement, and Sustainable Water Resources Management"

As editors of the journal Water, we embarked on effort project to produce a Special Issue because of the great interest worldwide in water governance and our knowledge of, and participation in, wide-ranging efforts to improve water governance and management. We write this Preface on World Water Day, a day for raising awareness of global water challenges and water management solutions. Improving how the framework of water laws, regulations, and management interfaces with processes for engaging stakeholders is crucial to sound water governance and management. It is well recognized that good water governance depends on transparency, accountability, and public participation, which are inherently related to the manner in which stakeholders are engaged in water policy and management processes. This volume is intended to expand our understanding of these relationships and processes. We entitled the Special Issue "Water Governance, Stakeholder Engagement and Sustainable Water Resources Management" and use the shorthand "GEM" for ready reference. We hope you agree that this Special Issue, which is now presented in book form, is a gem! The papers assembled in this volume reflect the breadth and depth of work designed to characterize good governance practices, develop effective stakeholder engagement processes and outcomes, and share efforts to improve water management. Along with the journal's announcement of an open call for papers, invitations were issued to experts working in this realm. The robust response is evidence of the excellent work being performed, as well as the value of focusing a volume on this general area. The resulting 20 papers, plus our editorial paper, were subject to peer review by at least two reviewers. We encourage you to read the papers, which, as explained in our editorial paper, we have grouped into three categories: stakeholder engagement; tools for building water management and governance capacity; and perspectives on water management and governance. We sincerely thank the excellent authors and peer reviewers, as well as the journal's staff, for their contributions to increased understanding of the critical importance of good water governance and stakeholder engagement to sustainable water management.

Sharon B. Megdal, Susanna Eden and Eylon Shamir
Special Issue Editors

water

MDPI

Editorial

Water Governance, Stakeholder Engagement, and Sustainable Water Resources Management

Sharon B. Megdal [1,*], Susanna Eden [1] and Eylon Shamir [2]

1 Water Resources Research Center, The University of Arizona, Tucson, AZ 85719, USA; seden@email.arizona.edu

2 Hydrologic Research Center, 11440 West Bernardo Court, Suite 375, San Diego, CA 92127, USA; eshamir@hrcwater.org

* Correspondence: smegdal@email.arizona.edu; Tel.: +1-520-621-9591; Fax: +1-520-892-8518

Academic Editor: Arjen Y. Hoekstra
Received: 3 February 2017; Accepted: 2 March 2017; Published: 6 March 2017

Abstract: Water governance and stakeholder engagement are receiving research attention for their role in formulating and implementing solutions to the world's critical water challenges. The inspiration for this Special Issue came from our desire to provide a platform for sharing results and informing the global water governance community about the wealth of excellent interdisciplinary and transdisciplinary research and projects being carried out around the world. The 20 peer-reviewed papers collected in this Special Issue have been grouped into three categories: stakeholder engagement, tools for building water management and governance capacity, and perspectives on water management and governance. Following a brief summary of the papers, concluding remarks that reflect on what the papers, taken as a whole, contribute to our understanding are provided.

Keywords: water governance; stakeholder engagement; water management; water resources; participation; collaboration; groundwater; water sustainability

---

1. Introduction

The role of water governance and stakeholder engagement in formulating and implementing solutions to the world's critical water challenges is receiving increasing attention. Over the past several years, United Nations organizations, the World Bank, the Global Environmental Facility, the Organisation for Economic Co-operation and Development (OECD), and others have joined forces with governmental, academic, professional association, Non-Governmental Organization (NGO), and other private sector representatives to explore efforts to improve water governance [1–3]. These efforts and water projects at multiple geographic scales point to the value of sharing experiences and lessons learned so that good practices can be transferred. They also point to the criticality of engaging stakeholders in deliberations on alternative approaches to addressing water scarcity, water quality, and water sanitation issues.

Because good water governance and effective stakeholder engagement are inextricably linked, we assembled this journal's Special Issue on water governance, stakeholder engagement, and sustainable water resources management to provide a platform for sharing results and informing the global water governance community. The robust response to our invitation to authors and the open call for papers is evidence of the wealth of excellent interdisciplinary and transdisciplinary research and projects from around the world. Twenty peer-reviewed papers were selected for this Special Issue. We have grouped these papers into three categories: stakeholder engagement, tools for building water management and governance capacity, and perspectives on water management and governance. Although there are overlaps among these three groups, papers were assigned to each category based on the nature of their primary focus: the development and impact of stakeholder participation, the development and use of tools for improving the water governance process, or specific aspects of the sustainable water management domain. The first group includes papers that present issues in stakeholder engagement through various case studies. The focus in the second group is on specific tools that were employed in order to raise the level of communication and understanding in the water governance context. The third group includes papers that examine water resources governance from different points of view.

After we summarize the contributions according to this categorization, we provide some concluding remarks.

2. Contributed Papers

2.1. Stakeholder Engagement

A substantial group of contributed papers addressed best practices for stakeholder engagement as a key element in good water governance. A great deal of innovative work has been directed at learning from the practice of stakeholder engagement to address challenges such as identifying the approaches and methods that should be used under different circumstances to produce the desired results. Guidance for engaging stakeholders is sorely needed as every day new efforts are initiated that would benefit from analysis of evidence from previous experience. The paper by Mott Lacroix and Megdal, "Explore, Synthesize, and Repeat: Unraveling Complex Water Management Issues through the Stakeholder Engagement Wheel", proposes a framework for understanding and developing effective stakeholder engagement [4]. Based on case studies at a range of scales in Arizona, USA, the authors developed the Stakeholder Engagement Wheel to visualize and highlight the need for flexibility and an iterative process for reaching common water goals. The Wheel demonstrates the value of a bridging organization to structure the process and a steering committee to guide it. The paper discusses four challenges that engagement processes are likely to experience, which are paucity of time, complexity of water resources management, difficulty of engaging diverse stakeholders, and lack of methods for engagement that are centered on empowerment, equity, trust, and learning.

Beyond lessons learned, general principles are being teased out of experiences with engaging stakeholders in water management and governance. For example, Akhmouch and Clavreul emphasize inclusiveness as a prerequisite to successful water governance along with process and structure requirements. In "Stakeholder Engagement for Inclusive Water Governance: 'Practicing What We Preach' with the OECD Water Governance Initiative", the authors refer to the experience of the OECD Water Governance Initiative to provide guidance on the practice of inclusive stakeholder engagement from an evidence-based assessment of its contributions to achieving water governance goals [5]. Their conclusions are consonant with the guidance provided by Mott Lacroix and Megdal on the importance of consistent ongoing communication among participants and institutional support in the form of facilitation, coordination, and continuity by organizations promoting good water governance processes [4].

The collaborative processes can make or break planning and implementation of water management, as illustrated by Dillon et al. in "Community Perspective on Consultation on Urban Stormwater Management: Lessons from Brownhill Creek, South Australia". The authors take the perspective of a community group to evaluate and draw lessons from two community consultation processes, separated by four years, addressing the same stormwater problem for a watershed in Adelaide, South Australia, Australia [6]. The first was a failure, due in large part to the mishandling of the consultative process, while the second was a success, which the authors attribute to the process of community engagement. Early and transparent consultation, inclusiveness, and responsiveness were found to be important characteristics of engaging communities in decisions.

Stakeholder engagement in water management is often fraught with conflict when interest differences lead to mistrust and noncooperation. A significant component of stakeholder engagement, therefore, may involve techniques for reducing conflict. In "Conflict Management in Participatory Approaches to Water Management: A Case Study of Lake Ontario and the St. Lawrence River Regulation", Furber et al. examine the ability of the participatory "Shared Vision Planning" approach to resolve conflicts in efforts to develop a new water regulation plan for the Lake Ontario and St. Lawrence River area [7]. They conclude that the approach succeeded when stakeholders perceived that they had something to gain from collaboration, but failed with stakeholders who perceived any change as a loss.

Even where interest conflicts are not a significant factor, stakeholder engagement can improve the efficiency of water management. This is especially true where unique contexts make the application of standard management practices problematic. An example provided in this collection, "Using a Participatory Stakeholder Process to Plan Water Development in Koraro, Ethiopia" (Alfredo et al.), demonstrates that the use of a one-day participatory workshop structured to learn from stakeholders in Koraro, Ethiopia, about their understanding, relations, and preferences with respect to their water supply provided the foundation for more effective interventions than a prescriptive one-size-fits-all approach [8]. Methods such as workshops were found successful in eliciting important stakeholder preferences that have an impact on water resources management [9].

The success of stakeholder engagement itself, however, may depend on prior knowledge of context in order to approach stakeholders with sensitivity. This is explicitly the case when working with Tribal Nations in the USA. In "Engaging Southwestern Tribes in Sustainable Water Resources Topics and Management", Chief et al. focus specifically on the requirements for effective stakeholder engagement with Native American tribes of the southwestern USA, evaluating five case studies of collaborations in terms of four simple rules for engaging tribes in research: "Ask about ethics, do more listening, follow tribal research protocols, and give back to the community" [10].

2.2. Tools for Building Water Management and Governance Capacity

The governance capacity of stakeholders can be enhanced through processes specifically designed to promote learning and to build competence in understanding the system they want to govern. The needed system learning does not always emerge from collaboration processes: the quality of the process is important. In "Public Participation in Water Planning in the Ebro River Basin (Spain) and Tucson Basin (Arizona, USA): Impact on Water Policy and Adaptive Capacity Building", Ballester and Mott Lacroix compare case studies on the Ebro Basin in Spain and the Tucson Active Management Area in Arizona, USA, to determine the factors with significant influence on outcomes of participatory processes [11]. They found that while policy impacts depended largely on contextual factors, an increase in adaptive capacity was related to the quality of the participatory process.

It is a given that sustainable water governance requires shared systems knowledge [12]. Failures of participation to increase capacity have led investigators to search for effective tools to create knowledge. Stakeholder engagement in modeling has been used in many instances and has been successful in creating a shared knowledge base [13]. King and Thornton explore the potential of collaborative modeling in "Staying the Course: Collaborative Modeling to Support Adaptive and Resilient Water Resource Governance in the Inland Northwest" [14]. As a method that encourages conversation, relationships, integrated systems thinking, and scientific understanding, it provides a foundation on which to develop adaptive and resilient water governance. The authors make this argument based on the examples of work done in the Spokane River Basin and the Palouse Basin in the northwestern USA.

Another example of model use in creating knowledge is examined in "Opening the Black Box: Using a Hydrological Model to Link Stakeholder Engagement with Groundwater Management" by Eden et al. [15]. The authors explore how governance of complex groundwater systems will benefit from the co-production of knowledge about the system. A case of hydrologic model refinement and use to inform groundwater management in the Upper Santa Cruz River Basin in Arizona, USA, demonstrated how knowledge was co-produced through two-way communication between scientists and stakeholders that focused on generating an understanding of the workings of the model, hydrologic system, and management information needs.

Other tools are being tested for their efficacy in achieving the desired learning goal. "Exploring the Potential Impact of Serious Games on Social Learning and Stakeholder Collaborations for Transboundary Watershed Management of the St. Lawrence River Basin" is one example [16]. Medema et al. interviewed stakeholders in the transboundary St. Lawrence River Basin to determine the extent of social learning that has occurred through stakeholder collaborations and reviewed the literature of serious games, proposing such games as a means for increasing social learning. Very little social learning was found and a number of barriers were identified, such as limited interaction between stakeholders and an uneven understanding of the issues, which the use of serious games may help to overcome.

Also in the category of capacity-building tools are systems that coordinate multiple levels of decision-making for integrated responses to demands. The development of these systems can require high levels of collaboration among stakeholders whose interactions have been constrained. A shared tool can overcome such constraints as is the case described in "International Severe Weather and Flash Flood Hazard Early Warning Systems—Leveraging Coordination, Cooperation, and Partnerships through a Hydrometeorological Project in Southern Africa" [17]. Jubach and Tokar describe a project in Southern Africa intended to coordinate severe weather and flood warning systems at multiple levels by multiple responsible organizations. Linkages and interfaces will be established and strengthened through development of a roadmap of activities and a concept of operation that lays out the requirements for routine forecasts and operations during unusual events.

2.3. Perspectives on Water Management and Governance

Water management and governance have many facets that shed light on the topic from different angles. The final section of this Special Issue contains papers written from a range of perspectives from law to economics, culture, and social justice.

How does public opinion affect the achievement of water management goals and what influences public opinion to support or reject those goals? These are questions addressed in "Undermining Demand Management with Supply Management: Moral Hazard in Israeli Water Policies" by Katz [18]. He argues that care must be taken when communicating with the public about the solutions to water scarcity: over-confidence in augmentation strategies can undermine public belief in the need for conservation. He concludes that desalination was oversold to the public as the solution to Israel's water problems, causing a general increase in water use although water scarcity continues to be an issue.

Cosens and Chaffin focus on the particular challenges for indigenous peoples in establishing rights to water. "Adaptive Governance of Water Resources Shared with Indigenous Peoples: The Role of Law" examines the influence of law on the emergence of adaptive governance that is capable of responding to new pressures, specifically the assertion of water rights by Native Americans and the Ngarrindjeri Nation in Australia [19]. They find that legal tools can provide the catalyst for the development of adaptive processes, build institutions that solidify these processes, and legitimize their results.

Focusing on the law with respect to groundwater, Mechlem takes a global perspective to her examination of its evolution, as the importance of groundwater is increasingly recognized. In "Groundwater Governance: The Role of Legal Frameworks at the Local and National Level—Established Practice and Emerging Trends", she notes that groundwater legislation is being developed that balances inherent uncertainty about availability and quality with the need for certainty about rights to water use [20]. These legal frameworks reflect context-based groundwater governance that accounts for the unique conditions of each country.

Groundwater management and governance face challenges specific to the nature of the resource as hidden from sight and difficult to assess in terms of amount, accessibility, quality, renewability, etc. [21]. As Mechlem and others have noted, contextual factors are the dominant influences on how groundwater is governed [20]. This finding was affirmed by Varady et al. from multiple, diverse case studies in "Modes and Approaches of Groundwater Governance: A Survey of Lessons Learned from Selected Cases across the Globe" [22]. The paper provides an integrated review of 10 carefully selected case studies of groundwater governance from across the globe, representing diverse local contexts. To a large extent, these contexts determined governance policy. Despite the contextual diversity, however, they synthesized a number of lessons across cases relating to four elements of groundwater governance: an institutional setting, access to science and information, the robustness of civil society, and economic and regulatory frameworks. Among other lessons, the authors note the need for research to support policy development.

The need for research is frequently highlighted as data and information are often lacking for groundwater systems. As a result, decision-makers are often unaware of all the factors affecting policy impacts. In a comprehensive review of what should be considered in order to effectively govern groundwater, van der Gun et al. make connections often overlooked with related subsurface resources including space for aquifer storage [23]. "Enhancing Groundwater Governance by Making the Linkage with Multiple Uses of the Subsurface Space and Other Subsurface Resources" describes the many uses of the subsurface and their impacts on groundwater resources to highlight the need for much greater awareness and integration of policies and practices relating to other subsurface uses.

Wastewater offers a context for examining the relation of centralized governance with efforts to achieve management goals at the local level. Flores et al. use their Governance Assessment Tool to analyze the implementation of wastewater treatment policies in the Tlaxcala Atoyac Sub-Basin in Mexico in "Water Governance Decentralisation and River Basin Management in Hierarchical Systems: Do They Work in Mexico's Tlaxcala Atoyac Sub-Basin?" [24]. Use of the tool allowed them to evaluate efforts to decentralize governance with respect to meeting wastewater treatment goals and find the failure of these efforts in lack of policy responsibility and resources by local authorities. The historical and political context of Mexico's centralized governing institutions had a decisive influence on attempts to decentralize water treatment, putting a damper on change. Institutional and cultural context also can determine the how policy is made and implemented, as shown by Golan in "The Fall and Rise of the Kishon River" [25]. He uses the history of the Kishon River in Israel to highlight the cultural and ideological influences on policy decisions and actions that allowed the river to be sacrificed to development and later drove its rehabilitation. Thus the change to a more sustainable regime was hampered and then promoted by cultural attitudes.

A key question in water policy-making relates to how benefits of a particular policy are counted. When economic benefits are paramount, other benefits are likely to be discounted with on-the-ground results on undervalued sectors. In "Water and Agriculture in a Mediterranean Region: The Search for a Sustainable Water Policy Strategy", Llop and Ponce-Alifonso used a computable general equilibrium model to examine the impacts of technological change and the principle of cost recovery for water service on economic efficiency, environmental sustainability, and political viability in Catalonia, Spain [26]. They conclude that there are tradeoffs among these goals that require agricultural and water policies, now aimed at increasing efficiency, to focus more broadly on strategies for reducing environmental and political costs.

Ignoring the consequences of a focus on economic efficiency can drive policies that undermine the condition of the poor. Sustainable water governance comprehends the justice of policies and actions and their impacts on all stakeholders [27]. In "Re-Linking Governance of Energy with Livelihoods and Irrigation in Uttarakhand, India", Buechler et al. investigated the impacts of hydropower projects on rural communities in order to derive recommendations to improve the social justice of energy development governance [28]. They conclude that governance instruments can be employed to compensate rural losses and share the benefits of hydropower projects more equitably.

3. Conclusions

Taken as a whole, the contributions to this Special Issue reinforce several messages from the rich study of collaboration in water management and governance. As stated repeatedly in these papers, context matters. Stakeholder engagement is an important tool in developing the common understanding of context that is necessary for making decisions that affect sustainable water management. Context also matters for effective stakeholder engagement, so it must occur over time, leaving space to develop and change. Research and lessons from the experience of others can help provide a foundation for engagement actions that promote mutual learning, but iterative interaction among the participants is needed for dealing with water resource management for the long term. The specific form this will take depends on context.

Another theme threaded throughout these papers is the importance of legal, regulatory, and institutional frameworks in supporting or hindering the development of governance for sustainable water management. Frameworks that lack clear engagement goals or represent a restricted set of interests can limit participation to the detriment of management and governance processes and outcomes. On the other hand, frameworks can be catalysts for action. They can support processes that bridge interest divides and resolve conflicts and can be a source of forward momentum for the planning and implementation of water management and governance strategies.

The need for holistic approaches was invoked several times to comprehend the complex and interlinked aspects of water governance and management. Water management and governance issues comprise many dynamically related components, which can only be adequately addressed by understanding their interconnections. Parties engaged in water resource problem solving are advised to examine the boundaries of the water governance issue being addressed. This examination can reveal interests, relationships, and physical, social, political, and cultural influences that may be excluded to the detriment of both the process and outcomes.

This collection of papers is a resource for insights and guidance on developing effective and sustainable water governance at a range of scales from global principles to practical local actions. It was assembled to display the breadth of research that has been inspired by the need to improve water management sustainability, promote effective water governance, and foster successful stakeholder engagement in vital decisions about water. As the importance of stakeholder engagement in water governance issues has been widely recognized, we believe that there is an essential need for scientific publication outlets that present varieties of case studies and discuss best practices. We hope that this Special Issue will serve as a template to encourage future research that improves stakeholder engagement and water governance and promotes sustainable water management.

Author Contributions: Sharon B. Megdal conceived and led development of the paper; Sharon B. Megdal, Susanna Eden and Eylon Shamir each contributed substantially to the writing of the paper.

Conflicts of Interest: The authors declare no conflict of interest.

References

1. Transboundary Water Assessment Programme. Available online: http://www.geftwap.org/ (accessed on 30 January 2017).
2. Groundwater Governance—A Global Framework for Action. Available online: http://www.groundwatergovernance.org/home/en/ (accessed on 30 January 2017).
3. OECD Water Governance Initiative. Available online: http://www.oecd.org/gov/regional-policy/water-governance-initiative.htm (accessed on 30 January 2017).
4. Mott Lacroix, K.E.; Megdal, S.B. Explore, Synthesize, and Repeat: Unraveling Complex Water Management Issues through the Stakeholder Engagement Wheel. *Water* **2016**, *8*, 118. [CrossRef]
5. Akhmouch, A.; Clavreul, D. Stakeholder Engagement for Inclusive Water Governance: "Practicing What We Preach" with the OECD Water Governance Initiative. *Water* **2016**, *8*, 204. [CrossRef]
6. Dillon, P.; Bellchambers, R.; Meyer, W.; Ellis, R. Community Perspective on Consultation on Urban Stormwater Management: Lessons from Brownhill Creek, South Australia. *Water* **2016**, *8*, 170. [CrossRef]
7. Furber, A.; Medema, W.; Adamowski, J.; Clamen, M.; Vijay, M. Conflict Management in Participatory Approaches to Water Management: A Case Study of Lake Ontario and the St. Lawrence River Regulation. *Water* **2016**, *8*, 280. [CrossRef]

8. Alfredo, K.; Montalto, F.A.; Bartrand, T.; Wolde-Georgis, T.; Lall, U. Using a Participatory Stakeholder Process to Plan Water Development in Koraro, Ethiopia. *Water* **2016**, *8*, 275. [CrossRef]
9. Phillipson, J.; Lowe, P.; Proctor, A.; Ruto, E. Stakeholder engagement and knowledge exchange in environmental research. *J. Environ. Manag.* **2012**, *95*, 56–65. [CrossRef] [PubMed]
10. Chief, K.; Meadow, A.; Whyte, K. Engaging Southwestern Tribes in Sustainable Water Resources Topics and Management. *Water* **2016**, *8*, 350. [CrossRef]
11. Ballester, A.; Mott Lacroix, K.E. Public Participation in Water Planning in the Ebro River Basin (Spain) and Tucson Basin (Arizona, USA): Impact on Water Policy and Adaptive Capacity Building. *Water* **2016**, *8*, 273. [CrossRef]
12. Van Rijswick, M.; Edelenbos, J.; Hellegers, P.; Kok, M.; Kuks, S. Ten building blocks for sustainable water governance: An integrated method to assess the governance of water. *Water Int.* **2014**, *39*, 725–742. [CrossRef]
13. Voinov, A.; Bousquet, F. Modelling with stakeholders. *Environ. Model. Soft* **2010**, *25*, 1268–1281. [CrossRef]
14. King, A.B.; Thornton, M. Staying the Course: Collaborative Modeling to Support Adaptive and Resilient Water Resource Governance in the Inland Northwest. *Water* **2016**, *8*, 232. [CrossRef]
15. Eden, S.; Megdal, S.B.; Shamir, E.; Chief, K.; Mott Lacroix, K. Opening the Black Box: Using a Hydrological Model to Link Stakeholder Engagement with Groundwater Management. *Water* **2016**, *8*, 216. [CrossRef]
16. Medema, W.; Furber, A.; Adamowski, J.; Zhou, Q.; Mayer, I. Exploring the Potential Impact of Serious Games on Social Learning and Stakeholder Collaborations for Transboundary Watershed Management of the St. Lawrence River Basin. *Water* **2016**, *8*, 175. [CrossRef]
17. Jubach, R.; Tokar, A.S. International Severe Weather and Flash Flood Hazard Early Warning Systems—Leveraging Coordination, Cooperation, and Partnerships through a Hydrometeorological Project in Southern Africa. *Water* **2016**, *8*, 258. [CrossRef]
18. Katz, D. Undermining Demand Management with Supply Management: Moral Hazard in Israeli Water Policies. *Water* **2016**, *8*, 159. [CrossRef]
19. Cosens, B.; Chaffin, B.C. Adaptive Governance of Water Resources Shared with Indigenous Peoples: The Role of Law. *Water* **2016**, *8*, 97. [CrossRef]
20. Mechlem, K. Groundwater Governance: The Role of Legal Frameworks at the Local and National Level—Established Practice and Emerging Trends. *Water* **2016**, *8*, 347. [CrossRef]
21. Alley, W.M.; Beutler, L.; Campana, M.E.; Megdal, S.B.; Tracy, J.C. Groundwater Visibility: The Missing Link. *Groundwater* **2016**, *54*, 758–761. [CrossRef] [PubMed]
22. Varady, R.G.; Zuniga-Teran, A.A.; Gerlak, A.K.; Megdal, S.B. Modes and Approaches of Groundwater Governance: A Survey of Lessons Learned from Selected Cases across the Globe. *Water* **2016**, *8*, 417. [CrossRef]
23. Van der Gun, J.; Aureli, A.; Merla, A. Enhancing Groundwater Governance by Making the Linkage with Multiple Uses of the Subsurface Space and Other Subsurface Resources. *Water* **2016**, *8*, 222. [CrossRef]
24. Flores, C.C.; Vikolainen, V.; Bressers, H. Water Governance Decentralisation and River Basin Management in Hierarchical Systems: Do They Work in Mexico's Tlaxcala Atoyac Sub-Basin? *Water* **2016**, *8*, 210. [CrossRef]
25. Golan, T. The Fall and Rise of the Kishon River. *Water* **2016**, *8*, 283. [CrossRef]
26. Llop, M.; Ponce-Alifonso, X. Water and Agriculture in a Mediterranean Region: The Search for a Sustainable Water Policy Strategy. *Water* **2016**, *8*, 66. [CrossRef]
27. Schneider, F.; Bonriposi, M.; Graefe, O.; Herweg, K.; Homewood, C.; Huss, M. Assessing the sustainability of water governance systems: The sustainability wheel. *J. Environ. Plan. Manag.* **2015**, *58*, 1577–1600. [CrossRef]
28. Buechler, S.; Sen, D.; Khandekar, N.; Scott, C.A. Re-Linking Governance of Energy with Livelihoods and Irrigation in Uttarakhand, India. *Water* **2016**, *8*, 437. [CrossRef]

water

MDPI

Article

Explore, Synthesize, and Repeat: Unraveling Complex Water Management Issues through the Stakeholder Engagement Wheel

Kelly E. Mott Lacroix * and Sharon B. Megdal

Water Resources Research Center, Arizona Cooperative Extension, College of Agriculture and Life Sciences, University of Arizona, 350 N. Campbell Avenue, Tucson, AZ 85719, USA; smegdal@email.arizona.edu

* Correspondance: klacroix@email.arizona.edu; Tel: +1-520-621-3826; Fax: +1-520-792-8518

Academic Editor: Davide Viaggi

Received: 15 February 2016; Accepted: 21 March 2016; Published: 28 March 2016

Abstract: Effective stakeholder engagement is fundamental to water management, yet there are as many approaches to consultation as there are efforts. This paper provides an evaluation of, and lessons learned from three water management engagement processes, and uses this assessment to offer a framework for stakeholder engagement. The Stakeholder Engagement Wheel framework is centered on a bridging organization that ensures that the process continues to move forward, and a steering committee that guides and changes activities according to stakeholder interests and concerns. Around the Stakeholder Engagement Wheel are four steps designed to examine iteratively the water management issue driving the engagement process and expand the sphere of interests involved. Many engagement processes have limited effectiveness because of: (1) paucity of time; (2) complexity of water resources management; (3) difficulty of engaging diverse stakeholders; and (4) lack of methods for engagement that are centered on empowerment, equity, trust, and learning. In this study, we have encountered all four of these issues and have addressed all but the first through a deliberate, iterative, and flexible approach. By cycling through activities and actions as proposed in the Stakeholder Engagement Wheel, we can build a community of practitioners with the nuanced and shared understanding needed for cohesive action and robust decisions in the face our considerable challenges.

Keywords: stakeholder engagement; water management; social learning

---

1. Introduction

Effective stakeholder engagement is fundamental to water management [1,2], yet there are as many approaches to consultation as there are efforts. These efforts are rarely examined for their effectiveness [3], and even less frequently analyzed to identify overarching elements of success. While general discussions of frameworks to support this engagement are limited [2,4,5], strategies are particularly lacking for engagement with stakeholders who have expertise in the subject based on their professional or personal interest. These stakeholders differ from the general public because engagement with them is ideally ongoing and often used to improve understanding of a critical issue through social learning. In contrast, engagement with the general public is not necessarily intended to be iterative, and typically is used to vet new policies or gain social license to operate [6].

Stakeholder engagement to promote social learning is distinct from engagement for policy making or obtaining social license because the overarching goal in social learning is understanding complex issues, and not necessarily resolution of conflict [6]. While engagement as a learning process may not solve the conflict per se, it is considered very important to overcome persistent norms and "difficult-to-change socio-technical systems" [7] (p. 398). Overcoming these systems and norms through a social-learning process can lead to improved water management outcomes [8].

While the methods and the definitions of social learning are debated [9], there is evidence that social learning, particularly in the adaptive natural resource management context, is enhanced by bridging organizations [10]. Bridging organizations are neutral third parties that serve to bring diverse groups together and include a "focus on trust building, sense making, learning, vertical and horizontal collaboration and conflict resolution" that is distinct from the organizations they connect [11]. Bridging organizations can also help balance power asymmetries in discussions [11], which is particularly important in engagement for water management [12]. They also play an important role in promoting social learning and provide opportunities for diverse perspectives to retain their identities while also forming productive relationships with others [13–15].

The success of an engagement effort can be related to who convenes the process, e.g., [16], but the approach to engagement is also important. Typically, stakeholder engagement has proceeded with a "tool-kit" approach, where engagement proceeds based on the selection of the relevant tools for the job [17]. According to Reed (2008) [17], the "tool-kit" approach is inadequate because it lacks an underpinning philosophy of how, why, and when to engage with stakeholders. Increasingly, the focus is shifting to the importance of engagement as a process that is centered on empowerment, equity, trust, and learning [17–19]. Antunes, *et al.* (2009) [2] suggest that instead of choosing a tool from the toolbox, the emphasis should be on using multiple tools within a framework.

Existing examples of process design for engagement generally follow a cycle that includes planning, identifying stakeholders, engaging with stakeholders, evaluation, and implementation, e.g., [20,21]. The existing frameworks, however, are focused on engagement with the general public to create or vet policy or to gain social license to operate. There remains a paucity of information on what processes (1) encourage social learning [22] and (2) are best for engagement with stakeholders with expertise on problem that is the focus of the engagement [3]. Notably, Mathur, *et al.* (2008) [6] states that "[e]xisting practices view stakeholder engagement: mostly from a management perspective; sometimes from an ethical perspective; less often as a combination of the two; and rarely have any element of the social learning perspective. There is a need for an approach that combines all the three perspectives if sustainability is to be pursued." Stakeholders with expertise can include lay stakeholders, who are people that have a deep interest in the subject, and professional stakeholders, who are paid representatives of organized interests [23]. While there are numerous recent handbooks for engagement with the general public to improve natural resources management or encourage corporate responsibility, e.g., [20,24,25], we did not discover any frameworks tailored to designing a social-learning process to engage with stakeholders who have subject expertise.

In this paper we use experiences from three engagement processes to build a framework for effective stakeholder engagement, where the goal was to unravel complex water management problems through promoting social learning. The engagement projects occurred at three geographic scales within the U.S. state of Arizona. Key to this framework is the role of a bridging organization as the convener and promoter of a process and a steering committee, who together explore and then synthesize disparate data and perspectives to increase understanding of water resource management issues.

2. Study Areas and Key Project Aspects

The University of Arizona Water Resources Research Center (WRRC) convened three engagement processes that occurred at three different geographic and governance scales (local, regional, and state) (Figure 1). The geographic scales included the U.S. state of Arizona, the Upper Gila River Watershed, and the Town of Clarkdale. The general approach for each engagement process was to establish a project problem and goal, create a steering committee, conduct engagement activities, and determine next steps or action items. A summary of activities and issues addressed in each activity for the three projects is included in Table 1.

Figure 1. Geographic context and project locations (shown in **green**). Engagement efforts examined in this paper occurred at the statewide level (Roadmap for Connecting the Environment to Arizona Water Planning (Roadmap), watershed level (Participatory Assessment for the Upper Gila Watershed (Gila)) and town level (Water Resources Management Program for the Town of Clarkdale (Clarkdale)).

Table 1. Engagement and outreach activities.

Project	Activity Name	Number Scope Duration	No. of People	Key Issues Addressed
Roadmap	Presentations on Water for Natural Areas	38 11 counties 1 hour	800+	What do we know about water for natural areas? How can the WRRC database on water needs of riparian and aquatic ecosystems be used for management and planning?
	Phase II Kick-off Meeting	2 Regional 3 hours	52	What are the goals of this project? What do we hope to accomplish?
	Survey	1 Statewide NA	171	What makes a community decide to consider water natural areas? And What makes water for natural areas vulnerable?
	Benefits and Challenges Workshop	1 Statewide ½ day	39	What are the benefits, challenges, and reasons for including natural areas in water management and planning?
	Focus Groups	43 10 Counties 1 hour	247	How should we consider water for natural areas in water management and planning and what incentives would make it more attractive to do so?
	Action Items Workshop	4 Regional ½ day	79	What strategies and actions should be taken to improve education, facilitate collaboration, set priorities, create incentives and encourage water conservation to include natural areas in water management and planning?
Gila	Science Coordination Workshop	1 Watershed 1.5 day	40	What data do you have and what data do you need? How can we better coordinate data collection and use in the watershed?
	Shared History Timeline	1 Watershed 3 hours	43	What is the history of the watershed? How can we learn from that history as we move forward?
	Key Informant Interviews	8 Watershed 1 hour	10	What do you think the watershed will look like in the future? What were the biggest surprises from the past?
	Scenario Drivers Workshop	1 Watershed 1 day	16	What are the key drivers of change in the watershed? What drivers of change are most uncertain?
	Survey	1 Watershed NA	27	What scenarios would be most useful for your management concerns? What questions do you have about the scenarios?
	Scenario Narrative Interviews	8 Watershed 1 hour	8	Questions about the future of the Gila Watershed based on scenario frameworks determined through the survey.
Clarkdale	Water 101 & 102 Public Meetings	2 Regional 2 hour	52	Information sessions on current water use and management in Clarkdale and regional groundwater modeling results. Attendees were "lay stakeholders".
	Open House	1 Town 2 hour	25	What concerns do you have about water in Clarkdale? What do you think about different management options that the Town is considering?
	Water Experts Workshop	1 Statewide 1 day	27	What should Clarkdale include or not include in its Water Resources Management Program (WRMP) and why? What management strategies will have the greatest impact?
	Water Expert Interviews	8 Multi-State 1 hour	8	What should Clarkdale include or not include in its Water Resources Management Program (WRMP) and why? What management strategies will have the greatest impact?
	Small Town Water Forum	1 Statewide 1.5 day	29	What should Clarkdale include in its WRMP? What challenges do small towns face in managing their water? How can towns in Arizona work together to improve water management?

2.1. *Statewide Geographic Scale: Arizona*

Arizona is a semi-arid state in the southwestern United States (Figure 1). Annual precipitation varies across the state from a high of 914 mm in the mountainous central region of the state to a low of 76 mm in the Sonoran Desert in the southwestern portion of the state. Water demand in Arizona was estimated to be 8,634,372 million cubic meters or 7 million acre-feet in 2014. In 2014, the Colorado River met 40% of this demand, 17% was met by other instate rivers, 40% by groundwater, and the remaining 3% by reclaimed water use. Water extractions and diversions in Arizona are dominated by agriculture (73%), followed by municipal (21%) and industrial (6%) uses [26].

Governance, defined as the overarching framework laws, regulations, and customs, which includes mechanisms of stakeholder engagement [27], varies across the state. For example, the Active Management Areas (AMAs) are within the central and most populous portion of Arizona. Within the AMAs groundwater pumping is tightly regulated. In general, beyond the AMAs only a notice of intent to drill is required and groundwater pumping is unregulated. There are four types of water recognized by Arizona water law: surface water (streams and springs), groundwater, Colorado River water, and effluent [28]. Particularly notable are the management difficulties created by varied restrictions on groundwater pumping throughout the state and by limited legal recognition of the physical connection between surface water and groundwater [29].

The project at a statewide scale, Roadmap for Connecting the Environment to Arizona Water Planning (Roadmap), took place over four years (January 2010–December 2014) and actively engaged over 1000 people across all Arizona counties. The goals of this project were to (1) synthesize available information about environmental water needs and identify knowledge gaps and (2) identify voluntary, stakeholder-driven options for addressing water needs of natural areas in the context of limited water supplies and existing water rights. These two goals were met in three phases. The first phase (January 2010–January 2011) focused on compiling and analyzing 121 studies on environmental flow needs for riparian and aquatic species in Arizona. The Arizona Environmental Water Needs Assessment and Methodology Guidebook summarized the results of this analysis. In addition, a database was created as a tool for land and water managers to identify available information on riparian and aquatic species flow needs [30]. The second phase of the project (January 2011–January 2012) focused on disseminating results from the environmental water needs assessment and preparing for the engagement process in phase three. Phase three (January 2012–December 2014) built the Roadmap for Considering Water for Arizona's Natural Areas. Key aspects of this last phase included formation of a 12 person steering committee that met every six weeks for the final two years of the project and engagement activities, including focus groups and workshops.

2.2. Watershed Scale: Upper Gila River Watershed

The Upper Gila River Watershed has an area of approximately 39,350 square kilometers that straddle the Arizona-New Mexico boundary. Forty-eight percent of the watershed is in Arizona, and the remaining 52% is in New Mexico. The engagement process described in this paper focused on the Arizona portion of the watershed. The Upper Gila River Watershed is rural. Total population of the Arizona portion is approximately 68,000. Approximately 70% of this population lives within five miles of the Gila River or its main tributary the San Francisco River. The principal industries are farming, ranching, and mining. Ninety-one percent of water use in the Arizona portion of the watershed is for agricultural purposes. Significant portions of the watershed are owned by the federal government and the San Carlos Apache Tribe. As a sovereign nation, the San Carlos Apache Tribe regulates and manages their water supply separate from any state laws [31]. The watershed is not within an AMA; however, the Arizona Water Rights Settlement Act of 2004 and the Globe Equity Decree of 1935 limit use of groundwater and surface water, respectively, near the Gila River, including portions that run through the San Carlos Apache lands [32]. Efforts to improve watershed health are led not by state or federal organizations, but by a watershed group (Gila Watershed Partnership, GWP), which does not have regulatory authority.

The Upper Gila Watershed project, Participatory Watershed Assessment for the Upper Gila (Gila), took place over two years (October 2012–September 2014) and actively engaged approximately 150 people. The goal of this project was to understand natural and social conditions in order to: (1) provide reliable, long-term water supplies for a resilient community; (2) preserve the rural, agricultural lifestyle; and (3) sustain and enhance the health of the Upper Gila Watershed in the face of uncertain physically and legally available water supplies. The Gila project included two phases, assessing baseline watershed conditions and creating stakeholder-driven scenarios for watershed planning. Both the baseline watershed assessment and scenario planning phases were designed to be stakeholder driven and were guided by a watershed planning working group. Project outputs included an Atlas of the Upper Gila River Watershed and Scenarios for the Future of the Upper Gila Watershed [33,34].

2.3. *Town Scale: Town of Clarkdale*

The Town of Clarkdale, with a population of 4087 in 2010, lies on high ground above the Verde River, at an elevation of 1085 meters. Several long-standing economic activities, such as farming and ranching, depend on sufficient Verde River flows. Activities tied to ecotourism have also become an ever-larger component of the area's economy. The river itself, as one of the few rivers with dependable surface flows in this semi-arid region, has garnered increased attention for kayaking and fishing. Verde Valley river tourism events, combined with ecological, recreational, historical, and/or agricultural elements, are also rising in popularity [35]. The Town of Clarkdale is not within an AMA, however it has adopted ordinances to limit new development if water supplies are inadequate and to set standards for low-water use plants in outdoor landscaping.

The project, Recommendations for a Water Resources Management Program for the Town of Clarkdale, Arizona (Clarkdale), took place over two years (January 2013–December 2014), and included approximately 130 stakeholders. The goal of this project was to develop recommendations for a water resource management program that equitably meets the needs of residents, businesses, and Clarkdale's natural environmental in order for it to be a robust and resilient community. These recommendations were created through research on existing water management approaches applicable to small towns, advice from a steering committee, and a series of workshops and meetings. The primary project output was the Recommendations for the Clarkdale Water Management Program publication [36].

3. The Stakeholder Engagement Wheel

Unraveling thorny water management issues requires us to move beyond the "tool-kit approach," and to instead employ a highly iterative and collaborative process that not only increases understanding of issues but also builds personal connection and trust among stakeholders. It is accepted practice to gather and integrate iteratively water resource data to build a comprehensive understanding of physical hydrologic and water management systems. Engagement for water resource management should be no different. Cycling through engagement efforts as activities and actions that are designed and then implemented allows us to build a community of practitioners with the nuanced and shared-understanding needed for cohesive action. This community of practice can then fully comprehend data and shape perspectives to make robust decisions. Because trust and collective knowledge are key outcomes of social learning, we look to the social learning literature for guidance. To encourage social learning an engagement process should be (1) inclusive [37,38]; (2) interactive [36,37,39]; and (3) flexible [40,41]. The Stakeholder Engagement Wheel (Figure 2) shows how our understanding of water management issues can be an aggregate of engaged learning opportunities with each step offering the prospect to encourage diverse and interactive participation.

The Stakeholder Engagement Wheel moves both in a larger circle as well as in smaller, iterative circles between steps. These smaller circles are a key aspect of the Wheel because within them information is collected, synthesized, and then used to help solve water management problems. Additional stakeholders, ideas, and understanding gained or changed through each iteration are shown by the dashed lines around the numbered steps. To summarize at the outset, at the center of the Wheel are a convener, which ensures that the process continues to move forward, and a steering committee, which guides and changes the process according to stakeholder interests and concerns. To promote inclusiveness and interactiveness, the problem is iteratively examined throughout steps 1–4 along the Wheel. Multiple iterations of problem-solving throughout the process ensure that the problem and goal are robust (between 1 and 2), that ample engagement to understand the problem and to reach the most diverse set of stakeholders possible occurs (2 and 3), and that actions are developed that meet stakeholder needs (3 and 4).

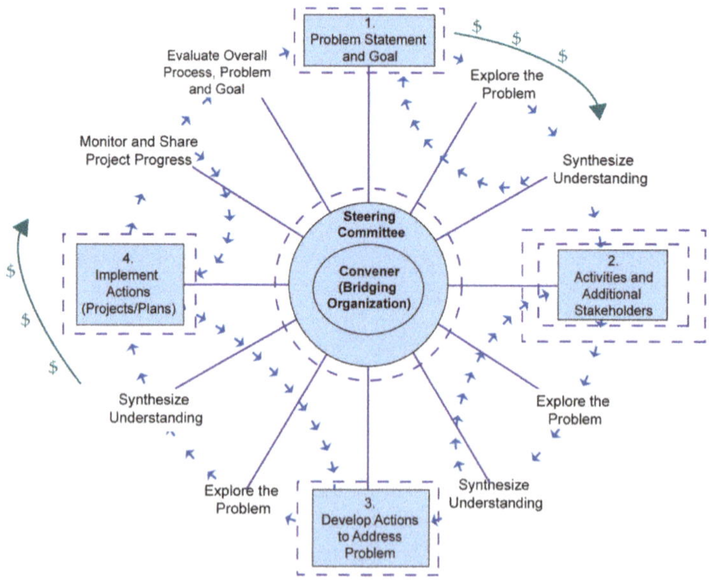

Figure 2. Stakeholder Engagement Wheel.

3.1. *Getting Started—The Importance of the Convener and the Steering Committee*

At the center of the Wheel are a bridging organization or another convener and the steering committee. The role of the bridging organization is to keep the Wheel turning through facilitating examination of the problem(s) and providing a neutral convening space that balances power asymmetries and ideas, which are well-noted functions of bridging organizations [11,14,15]. Because neutrality is key to building common ground among disparate ideologies [15] and expertise is important for facilitating the examination of the problem, the group that serves as a bridging organization should be carefully considered to ensure that it reflects these qualities. While this bridging organization role can be played by entities such as non-governmental organizations, university-affiliated organizations, such as Cooperative Extension and National Institutes for Water Resources in the United States, are uniquely suited for this task because not only do they not have water rights or a particular water interest to protect, they are also recognized by local communities as providing independent expertise [14].

Representing the bridging organization, we also found that we played a key role in moving each engagement process forward, which is a role beyond just bringing people together. For example, in the Clarkdale project the Town was originally designated as the project manager, and therefore, responsible for bringing the team together and designing engagement efforts. In initial team meetings, however, it was determined that the project was better served by the WRRC managing the effort because of our subject-matter expertise and the time constraints of Town staff.

Surrounding the convener on the Wheel is the steering committee. The role of the steering committee is to guide and change the process according to stakeholder interests and concerns, thereby promoting flexibility and responsiveness. For this reason, steering committees are important to an engagement process that seeks to reach the top rungs of Arnstein's ladder [42], or as redefined by Luyet, *et al.* [43], create a process that empowers stakeholders through "delegation of decision-making over project development and implementation." Although steering committees are mentioned in case studies and practitioner manuals on stakeholder engagement, e.g., [1,20,26,44], they are infrequently noted as key to successful outcomes. An exception is Slocum [45], who states that steering committees are key components of the process and suggests that a diverse and strong steering committee helps both to establish and complete the engagement process.

The steering committee is the linchpin to taking the process beyond a "tool-box" approach because they promote diverse participation, in-depth examination of the water management questions, and ensure the process is driven by the people who will be impacted by its outcomes. When acting as a microcosm for the spectrum of interests, the steering committee serves as an initial platform for exploration of the problem, goal setting, and identifying activities to broaden engagement and people to engage. For the steering committee act as a small-scale version of the broader array of interests, members should be selected based on their representativeness of the issues central to the engagement process and their willingness to sit at the table with people whose opinions may differ from their own [15]. Based on experiences with the three steering committees associated with the projects described here, we found that it is important consistently engage with this core group of stakeholders. The steering committee should be people who are either local to the engagement effort or are significantly vested in the project's outcomes.

Through each of the three projects, we found that the time we invested in forming and maintaining engagement with our steering committees aided our ability to reach out to additional stakeholders and develop successful engagement activities. During the Roadmap project, we made the most concerted effort to engage consistently with the steering committee. This interaction enabled us to connect agricultural and industrial water interests to the process who were initially not sufficiently represented. The Roadmap focus groups also provide evidence of the importance of a steering committee. Steering committee members recommended focus group participants through a stakeholder mapping exercise. This exercise identified 47 key organizations that should be included in the project because they would either support, be neutral to, or have reservations about allocating water to the environment. Of the 47 groups, 34% were identified by steering committee members as supporting environmental flows, 30% as neutral, and 36% as having reservations about allocating water for environmental flows. Over the course of only four months, we were able to conduct focus groups with 87% of the groups. Most importantly, 75% of the organizations identified as having reservations about water for the environment participated. The ability to have a relatively high success rate of engaging with potentially adversarial interests can be attributed to the advice and assistance we received from steering committee members in arranging focus group meetings.

3.2. Setting the Wheel in Motion—Creating a Problem Statement and Goal (Step 1)

Having created a steering committee, the Wheel can begin to move through refining a problem statement and goal. Establishing a clear goal has been noted by others as a key element of effective engagement, e.g., [23,46,47]. Without a clear problem to address and goals for the process, an otherwise effective approach will fall short [23]. Establishing a problem statement and goal are Step 1 and sit atop the Wheel because our experience confirmed the importance of goal setting. In the Gila and Clarkdale project the goals were established through workshops (Gila) and team meetings (Clarkdale). In the Roadmap project the goal was established by the objective of the grant. During the Clarkdale and Gila projects, the steering committees frequently referred to the goal when there were questions about which stakeholders to include or activities to conduct. The Gila project's desire to "preserve the rural, agricultural lifestyle" in the Gila project led to engagement activities targeted to farmers and ranchers. The need to "equitably meet the needs of residents, businesses, and Clarkdale's natural environment" as stated in the project goal, resulted in a Water 101 series where the Town could explain local hydrologic conditions and receive feedback from residents about their water concerns. Similarly, the WRRC used the Roadmap project goal of "identifying voluntary, stakeholder driven, options ... " to both shape workshop discussions and encourage participation of water interests that may be concerned with additional regulation and laws.

3.3. Maintaining the Momentum—Engagement Activities (Step 2)

Once the initial problem and goal(s) are established, the engagement process proceeds to Step 2 —engagement activities designed to improve understanding of the problem and stakeholder perspectives. Multiple iterations of exploring the problem through an engagement activity, synthesizing what was learned from the engagement, and then designing the next engagement activity based on this understanding help ensure the diverse participation, interactiveness, and process flexibility necessary for social learning. Based on the experience of others in stakeholder engagement, e.g., [18–20] and our own experience, we recommend that engagement rely on input from the steering committees, the goal(s) of the project, and integration of what has been learned from previous project activities.

Synthesizing results are particularly important in Step 2 because it enables the bridging organization and steering committee to adapt their approach to ensure iterative engagement with a diverse set of stakeholders. In synthesizing results of each engagement activity, and of the project as a whole, it is important to consider how information is displayed and how stakeholders will access the information. Effective communication about results from each activity facilitates steering committee decisions about next steps in the engagement process and can empower stakeholders to use the information assembled. The display of information was particularly important in the Gila and Roadmap projects. In the Gila project, graphics of the scenarios and the key drivers of change helped participants quickly understand the findings and assisted the project team in the dissemination of results to local councilmembers and county supervisors. For Roadmap project, graphical display of the qualitative analysis findings from focus group results led to the development of key questions for the workshops to develop action items (see [15]).

Examples of the evolving engagement approach as a result of exploring the problem and then integrating this understanding can be seen in all three projects. The original Clarkdale plan included three expert workshops focused on one of three building blocks for a water management plan: demand-based, supply-based, and augmentation-based options and recommendations. After our first meeting with the steering committee, however, it became clear that these categories contained too much potential overlap of subject matter and experts. Therefore, we decided to conduct one expert meeting and a series of follow-up phone interviews during which we discussed all three categories at one time. Because of this change in approach, we were able to convene one meeting with experts instead of three and then build upon the knowledge gained from that first meeting through a very well-received small town forum on water management. In this case, adjusting the approach based on examining who attended the first engagement exercise, as well as summarizing the results of that exercise, led to a more robust set of recommendations for the Town.

The Gila project was originally designed with a series of scenario building workshops to create the future scenarios. Discussion with our steering committee at the beginning of the scenario planning process made it clear that multiple, daylong workshops over the course of four months would be difficult for them to manage with their workloads. As a result, we decided to substitute the longer workshops with monthly steering committee meetings that were open to anyone interested. In this case, shorter meetings at regularly scheduled times interspersed with initial and final half-day day workshops made it possible for regular attendance necessary to build the scenarios. During the Roadmap project, the engagement approach was shifted from an initial set of workshops throughout the state that combined multiple water interests to focus groups with individual interest groups. This change in activities was suggested by the steering committee as a way to bring water interests into the discussion that otherwise might be reluctant to participate in a workshop. As previously discussed, as a result of this change we were able to reach 87% of the stakeholder interests identified as key to meeting the goals of engagement.

3.4. Getting Traction—Developing and Implementing Action Items (Steps 3 and 4)

Step 3 on the Wheel is to develop action items or next steps. While exploring and synthesizing a problem may help to better understand it and increase the diversity of participants, at some point it becomes necessary to develop concrete actions to address the problems identified in Steps 1 and 2. Similar to the makeup of the steering committee and engagement activities, the approach to developing action depends on the project's goals. For example, for the Clarkdale effort action items were developed through a synthesis of our research and the expert input gained throughout the project. These action items were then presented to the Town Council for adoption. The Gila project identified next steps through a synthesis of potential future conditions in the watershed based on a watershed baseline assessment and four future scenarios. Four regional workshops were used during the Roadmap project to develop action items. Using workshops as the mechanism to develop action items enabled all interested participants, not just the steering committee, to directly impact final recommendations. Roadmap recommendations (action items) were then distributed via presentations, webinars, and the report, Roadmap for Considering Water for Arizona's Natural Areas.

Having established action items or next steps, the engagement process proceeds to Step 4—implementation of selected action items. Of the seven action items identified as part of the Clarkdale project, three have been implemented, and the Town is actively seeking funding for the others. The recommendations report itself was accepted by the Town Council, but not adopted as funding is needed to proceed on each specific recommendation (Personal Communication, Jodie Filardo, Town of Clarkdale). Using the scenarios developed to create coordinated water and land management plans across the watershed's jurisdictions was the only action item from the Gila effort. This plan is proceeding through funding to the WRRC from the U.S. Bureau of Reclamation and has already resulted in water demand and supply estimates and identification of cross-jurisdictional opportunities for water management. Separate from the current WRRC convened process in the watershed, all of the local jurisdictions in one county have come together to discuss a standardized outdoor landscaping ordinance that could be adopted by each individual jurisdiction. Measuring implementation of the 10 action items from the Roadmap project is more difficult; however, we are aware of at least two efforts by environmental non-governmental organizations to implement action items. These early indications of implementation are encouraging, and although the projects described here have not been systematically examined for successful implementation, there are many examples of how failed engagement up to the point of implementation hampers successful implementation, e.g., [44,48,49] and some examples of how effective engagement can enable successful implementation, e.g., [18,50,51].

3.5. Coming Full Circle—Evaluation of Effectiveness

Evidence of social-learning as a result of engagement can take years to manifest, making a near-term evaluation of engagement processes difficult [11]. Near-term evaluation is, however, an important [42] and oft neglected [3] element of the study of engagement. Because the ultimate goal of this framework is to improve social learning outcomes, we evaluated the effectiveness of the Stakeholder Engagement Wheel approach against engagement elements that encourage social learning, *i.e.*, inclusiveness, interactiveness, and flexibility of approach [19,20,23,37]. For the projects that inspired the Stakeholder Engagement Wheel, inclusiveness was examined through documenting the diversity of water interests and calculating an index of stakeholder engagement. Interactiveness was evaluated through the variety of engagement techniques used and the participation level of the stakeholders over time. The flexibility of the engagement process was determined by comparing the original project plans with the actual project activities that were implemented.

The following steps enabled evaluation of the extent to which the process was inclusive of a diversity of interests: (1) key stakeholders were determined by stakeholder mapping (Roadmap and Clarkdale) or key informant interviews (Gila); (2) stakeholder type was documented at each activity; and (3) an involvement index was calculated based on the approach used by Walter, *et al.* 2007. To determine the involvement index, each interactive activity was assigned points based on its length, with twice the number of points given to interactive activities. For example, a two-hour workshop would receive four points and a one-hour presentation would receive one point. The points were then summed and divided by the maximum number of possible points for each project to calculate the index. An involvement index score of 1.0 would indicate that every participant in that water interest was involved in all project activities. The average for the projects includes steering committee participation in regular project activities, but the hours they contributed for steering committee activities were not used to calculate the average. All projects engaged at least once with each of the key stakeholder groups identified through stakeholder mapping or key informant interviews. Figure 3 shows the total number of participants and average of the involvement index for each water interest. It is notable that the Roadmap project, which had the most concerted effort to engage a diverse set of interests and the most engaged steering committee, the average index of involvement is similar across different water interests.

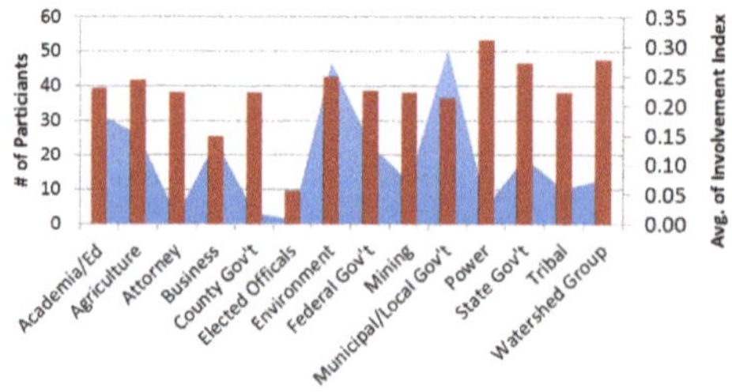

(a)

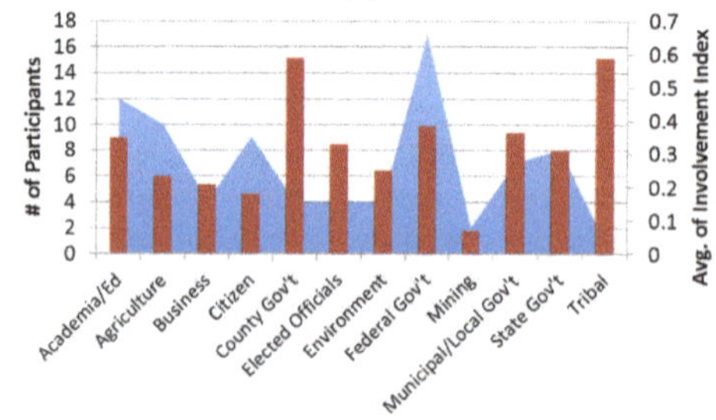

(b)

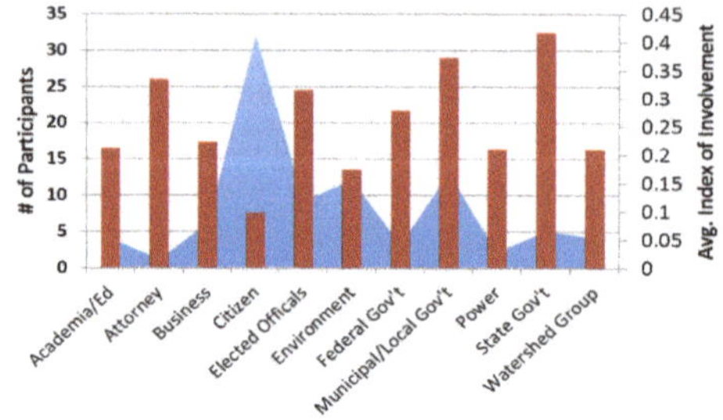

(c)

Figure 3. Evaluation of Participant Diversity and Involvement including number of participants (**blue**) and average of the involvement index (**red**) for each interest. (**a**) Roadmap; (**b**) Gila; (**c**) Clarkdale.

The following steps enabled evaluation of the extent to which the process was interactive: (1) repeat and increasing engagement was measured using the same data gathered for inclusiveness on participant involvement and (2) the variety of engagement techniques was described qualitatively.

Figure 4 indicates the growing reach of each of the projects and repeat engagement. We acknowledge that there could be some undercounting of repeat engagement and over counting of total participants because of anonymous participants in the Gila and Roadmap projects. Throughout the Roadmap and Gila projects, the number of people involved as well as those who participated more than once increased as the project progressed. Because the goal of the Clarkdale project was to gather as many different ideas a possible to create recommendations for the town's water resources management program, there were no attempts to engage the same stakeholders or interests more than once. As shown in Table 1, all projects included multiple engagement techniques recognized to encourage two-way interaction or engagement in order to improve opportunities for interactiveness. The variety of techniques also provided different types of opportunities for engagement. For both the Roadmap and the Gila projects used a survey. These surveys were able to reach more people for input than could have been engaged through meetings alone. Although this engagement is not equivalent to attending a workshop, it does provide a good opportunity to broaden the scope of those involved. The shortcoming of this method, however, is the need to allow for anonymous responses in order to increase participation, which makes it difficult to track individual participants.

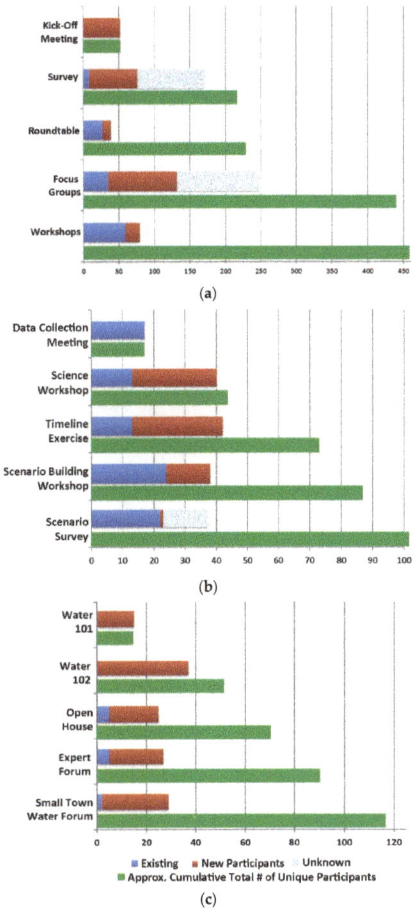

Figure 4. Number of participants in each activity and repeat and increasing engagement. (**a**) Roadmap; (**b**) Gila; (**c**) Clarkdale.

Finally, the flexibility of the engagement process was determined by comparing the original project plans with the actual project activities that were implemented. Process changes were made as a result of findings from formal and informal evaluations. For all projects, input from post-workshop surveys and steering committee advice changed the engagement approach. For example, suggestions from the evaluation after the science coordination workshop for the Gila project included concerns about the length of the meeting and the need for a better-defined purpose of the meeting. Because of these comments, subsequent workshops and meetings were shortened from 1.5 days to a single day, and the WRRC team worked to ensure that the purpose and outputs from each meeting were clear. Also during the workshop, participants stressed the importance of historical information on the watershed and expressed concern over the lack of data. As a result of this concern, the WRRC decided to add an activity not originally planned, and hosted shared history timeline exercise to gather historical perspectives on watershed management.

While engagement through a bridging organization provides a flexibility that may not be available to a state or federal agency, long-term engagement can be hampered by the need for continued grant funding. It is common for grants to last for only one or two years, which may hinder the ability of steering committees or broader stakeholder groups to continue their work. Engaging stakeholders on a project-by-project basis impedes the process of relationship building and reduces the likelihood of developing truly collaborative processes [45]. Issues with multi-year funding also highlight the importance of finding ways for bridging organizations to help "turn the wheel", but not necessarily be solely responsible for the forward motion of the process. In addition to other methods suggested for determining the long-term success of an approach, e.g., [3,38], another metric could be the ability for social learning and the project activities to continue without the bridging organization playing a role.

4. Discussion and Final Remarks

Effective stakeholder engagement through the process-design framework proposed here recognizes that in order to unravel complex problems we must repeatedly explore and synthesize information. Many engagement processes, particularly in water resources management, fall short of this goal [13,42,52]. The reasons for these shortcomings include: a paucity of time [20,53]; the complexity of water resources management [13,42]; the difficulty of engaging diverse stakeholders, especially those who feel disenfranchised [49,54]; and the lack of advice for an engagement process centered on empowerment, equity, trust, and learning that takes engagement beyond the "tool-kit" approach [17–19]. The projects that inspired and followed the Stakeholder Engagement Wheel encountered all of these challenges. Through the process described here they were able to address all but the first, namely a lack of time, through a deliberate, iterative, and flexible approach. This framework is particularly important for the many university bridging organizations, such as Regional Integrated Science Assessment Centers funded by the National Oceanic and Atmospheric Administration because they are increasingly relied upon to provide both unbiased research and stakeholder perspectives on key issues like water management and climate change.

Our findings are similar to others who have discovered that the convener of the process plays an important role in the efficacy of engagement [17,18]. Unlike most other approaches, however, we found that the people engaged through a steering committee also contribute to the efficacy of an engagement process. Because there are not many examples of an increased efficacy of engagement due to the use of steering committees, future research could focus on their role in promoting successful engagement. This research should examine best practices for informing, engaging, and maintaining these core groups of stakeholders and evaluating similar processes that include or omit a steering committee.

Water resource management is a complex endeavor, and sound management requires effective stakeholder engagement. The framework presented here provides an example of how a bridging organization can structure and evaluate the process of engagement. While the importance of bridging organizations in promoting social learning, and flexible, iterative approaches are recognized, there are few recommendations for a flexible process. As illustrated through the three projects presented here, the proposed framework provides for inclusivity, interactiveness, and flexibility in engagement through guidance by a steering committee and iteratively examining the problem. While further assessment is necessary, we suggest this framework provides a tangible approach that could aid other efforts engaging to solve complex water management problems across local, state, and regional scales. By cycling through activities and actions as proposed in the Stakeholder Engagement Wheel, we can build a community of practitioners with the nuanced and shared-understanding needed for cohesive action and robust decisions in the face our considerable challenges.

Acknowledgments: The authors would like to thank the many Arizona stakeholders who participated in these projects, in particular the almost 50 different people who served on the three steering committees for the projects, for their time and insights. Effective stakeholder engagement requires a dedicated team, and these projects would not have been possible without the assistance of current and former WRRC staff Brittany Xiu, Candice Rupprecht, Joanna Nadeau, and Aaron Lien, and graduate students Kathryn Bannister, Christopher Fullerton, and Ashley Hullinger. A sincere thanks to Susanna Eden, Shirley Papuga, Chris Scott, David Breshears, and Stuart Marsh for their valuable comments and ideas for how to present this research. Funding for the three projects described here, but not for this paper itself, came from the Nina Mason Pulliam Charitable Trust, the Desert Landscape Conservation Cooperative, and the Town of Clarkdale via the Walton Family Foundation. Partial funding for this research was also provided by the Arizona Technology and Research Initiative Fund.

Author Contributions: K.E.M.L. and S.B.M. conceived and designed the projects described here; K.E.M.L., under the supervision of S.B.M., convened the projects described here; K.E.M.L. and S.B.M. wrote the paper.

Conflicts of Interest: The authors declare no conflict of interest. The founding sponsors had no role in the design of the study, in the collection, analyses, or interpretation of data, in the writing of the manuscript, and in the decision to publish the results.

References

1. Lennox, J.; Proctor, W.; Russell, S. Structuring Stakeholder Participation in New Zealand's Water Resource Governance. *Ecol. Econ.* **2011**, *70*. [CrossRef]
2. Antunes, P.; Kallis, G.; Videira, N.; Santos, R. Participation and Evaluation for Sustainable River Basin Governance. *Ecol. Econ.* **2009**, *68*. [CrossRef]
3. Fischer, A.R.; Wentholt, M.T.; Rowe, G.; Frewer, L.J. Expert Involvement in Policy Development: A Systematic Review of Current Practice. *Sci. Public Policy* **2013**. [CrossRef]
4. Pahl-Wostl, C. A Conceptual Framework for Analysing Adaptive Capacity and Multi-Level Learning Processes in Resource Governance Regimes. *Glob. Environ. Chang.* **2009**, *19*. [CrossRef]
5. Turton, A.R.; Hattingh, J.; Claassen, M.; Roux, D.J.; Ashton, P.J. Towards a Model for Ecosystem Governance: An Integrated Water Resource Management Example. In *Governance as a Trialogue: Government-Society-Science in Transition*; Springer: Berlin/Heidelberg, Germany, 2007; pp. 1–28.
6. Mathur, V.N.; Price, A.D.; Austin, S. Conceptualizing Stakeholder Engagement in the Context of Sustainability and Its Assessment. *Constr. Manag. Econ.* **2008**, *26*. [CrossRef]
7. Bos, J.J.; Brown, R.R.; Farrelly, M.A. A Design Framework for Creating Social Learning Situations. *Glob. Environ. Chang.* **2013**, *23*. [CrossRef]
8. Kumler, L.M.; Lemos, M.C. Managing Waters of the Paraíba Do Sul River Basin, Brazil: A Case Study in Institutional Change and Social Learning. *Ecol. Soc.* **2008**, *13*, 22.
9. Reed, M.; Evely, A.C.; Cundill, G.; Fazey, I.R.A.; Glass, J.; Laing, A.; Newig, J.; Parrish, B.; Prell, C.; Raymond, C.; *et al.* What Is Social Learning? *Ecol. Soc.* **2010**, *15*, r1.
10. Berkes, F. Evolution of Co-Management: Role of Knowledge Generation, Bridging Organizations and Social Learning. *J. Environ. Manag.* **2009**, *90*. [CrossRef]
11. Sternlieb, F.; Bixler, R.P.; Huber-Stearns, H. A Question of Fit: Reflections on Boundaries, Organizations and Social–Ecological Systems. *J. Environ. Manag.* **2013**, *130*. [CrossRef]
12. Jacobs, K.; Lebel, L.; Buizer, J.; Addams, L.; Matson, P.; McCullough, E.; Garden, P.; Saliba, G.; Finan, T. Linking Knowledge with Action in the Pursuit of Sustainable Water-Resources Management. *Proc. Natl. Acad. Sci. USA* **2010**. [CrossRef]
13. Miller, C. Hybrid Management: Boundary Organizations, Science Policy, and Environmental Governance in the Climate Regime. *Sci. Technol. Hum. Values* **2001**, *26*. [CrossRef]

14. Carr, A.; Wilkinson, R. Beyond Participation: Boundary Organizations as a New Space for Farmers and Scientists to Interact. *Soc. Nat. Resour.* **2005**, *18*. [CrossRef]
15. Lacroix, K.E.M.; Xiu, B.C.; Megdal, S.B. Building Common Ground for Environmental Flows using Traditional Techniques and Novel Engagement Approaches. *Environ. Manag.* **2016**, *57*. [CrossRef]
16. Childs, C.; York, A.M.; White, D.; Schoon, M.L.; Bodner, G.S. Navigating a Murky Adaptive Comanagement Governance Network: Agua Fria Watershed, Arizona, USA. *Ecol. Soc.* **2013**, *18*. [CrossRef]
17. Reed, M.S. Stakeholder Participation for Environmental Management: A Literature Review. *Biol. Conserv.* **2008**, *141*. [CrossRef]
18. Areizaga, J.; Sano, M.; Medina, R.; Juanes, J. Improving Public Engagement in ICZM: A Practical Approach. *J. Environ. Manag.* **2012**, *109*. [CrossRef]
19. Jolibert, C.; Wesselink, A. Research Impacts and Impact on Research in Biodiversity Conservation: The Influence of Stakeholder Engagement. *Environ. Sci. Policy* **2012**, *22*. [CrossRef]
20. Thought Stream. Five Criteria for Effective Stakeholder Engagement in Education. 2013. Available online: http://www.thoughtstream.ca/downloads/5-criteria-for-effective-se-may2013.pdf (accessed on 16 December 2014).
21. Jeffery, N. Stakeholder Engagement: A Road Map to Meaningful Engagement. Doughty Centre, Cranfield University School of Management, 2009. Available online: http://www.som.cranfield.ac.uk/som/dinamic-content/research/doughty/stakeholder/Guide.pdf (accessed on 24 March 2016).
22. Muro, M.; Jeffrey, P. Time to Talk? How the Structure of Dialog Processes Shapes Stakeholder Learning in Participatory Water Resources Management. *Ecol. Soc.* **2012**, *17*. [CrossRef]
23. Fung, A. Varieties of Participation in Complex Governance. *Public Admin. Rev.* **2006**, *66*, 66–75. [CrossRef]
24. Durham, E.; Baker, H.; Smith, M.; Moore, E.; Morgan, V. *BiodivERsA Stakeholder Engagement Handbook*; BiddivERsA: Paris, France, 2014.
25. Stakeholder Research Associates Canada Inc. *From Words to Action: The Stakeholder Engagement Manual Volume 1*; Stakeholder Research Associates Canada Inc.: Toronto, ON, Canada, 2005.
26. Moreno, M.; Arizona Department of Water Resources, Phoenix, AZ, USA. Personal communication, 2015.
27. Megdal, S.B.; Gerlak, A.K.; Varady, R.G.; Huang, L.Y. Groundwater Governance in the United States: Common Priorities and Challenges. *Groundwater* **2015**, *53*. [CrossRef] [PubMed]
28. Pearce, K.L. Balancing Competing Interests: The History of State and Federal Water Laws. In *Arizona Water Policy: Management Innovations in an Urbanizing, Arid Region*, 1st ed.; Jacobs, K., Colby, B.G., Eds.; Resources for the Future: Washington, DC, USA, 2007; pp. 26–44.
29. Glennon, R. The Disconnect between Water Law and Hydrology. In *Arizona Water Policy: Management Innovations in an Urbanizing, Arid Region*, 1st ed.; Resources for the Future: Washington, DC, USA, 2007; pp. 106–120.
30. Mott Lacroix, K.E.; Xiu, B.C.; Nadeau, J.B.; Megdal, S.B. Synthesizing environmental flow needs data for water management in a water-scarce state: The Arizona Environmental Water Demands Database. *River Res. Appl.* **2016**, *32*. [CrossRef]
31. Colby, B.G.; Thorson, J.E.; Britton, S. *Negotiating Tribal Water Rights: Fulfilling Promises in the Arid West*; University of Arizona Press: Tucson, AZ, USA, 2005.
32. Arizona Department of Water Resources. *Arizona Water Atlas Volume 3: Southeastern Arizona Planning Area*; Arizona Department of Water Resources: Phoenix, AZ, USA, 2009; p. 619.
33. Bannister, K.; Chan, D.; Driscoll, J.M.; Fullerton, C.; Lien, A.; Lacroix, K.M. *Atlas of the Upper Gila River Watershed*; University of Arizona Water Resources Research Center: Tucson, AZ, USA, 2014.
34. Lacroix, K.M.; Hullinger, A.; Fullerton, C. *Scenarios for Upper Gila River Watershed*; University of Arizona Water Resources Research Center: Tucson, AZ, USA, 2014.
35. Von Gausig, D.; O'Banion, B.; Rooney, C. Verde River Economic Development Final Report. 2011. Available online: http://clarkdalesustainabilitypark.org/Verde%20River%20Reports%202011/VREDS%20Final%20Report-public-8-12-2011.pdf (accessed on 7 December 2015).
36. Lacroix, K.M.; Fullerton, C.; Rupprecht, C. *Town of Clarkdale Water Resources Management Program Recommendations Report*; University of Arizona Water Resources Research Center: Tucson, AZ, USA, 2014.
37. Saarikoski, H.; Tikkanen, J.; Leskinen, L.A. Public Participation in Practice—Assessing Public Participation in the Preparation of Regional Forest Programs in Northern Finland. *For. Policy Econ.* **2010**, *12*. [CrossRef]

38. Scott, C.A.; Varady, R.G.; Meza, F.; Montaña, E.; de Raga, G.B.; Luckman, B.; Martius, C. Science-Policy Dialogues for Water Security: Addressing Vulnerability and Adaptation to Global Change in the Arid Americas. *Environ. Sci. Policy Sustain. Dev.* **2012**, *54*. [CrossRef]
39. Walker, G.B.; Senecah, S.L.; Daniels, S.E. From the Forest to the River: Citizens' Views of Stakeholder Engagement. *Hum. Ecol. Rev.* **2006**, *13*, 193–202.
40. De Stefano, L. Facing the Water Framework Directive Challenges: A Baseline of Stakeholder Participation in the European Union. *J. Environ. Manag.* **2010**, *91*. [CrossRef] [PubMed]
41. Walter, A.I.; Helgenberger, S.; Wiek, A.; Scholz, R.W. Measuring Societal Effects of Transdisciplinary Research Projects: Design and Application of an Evaluation Method. *Eval. Program Plan.* **2007**, *30*. [CrossRef] [PubMed]
42. Arnstein, S.R. A Ladder of Citizen Participation. *J. Am. Inst. Plan.* **1969**, *35*. [CrossRef]
43. Luyet, V.; Schlaepfer, R.; Parlange, M.B.; Buttler, A. A framework to implement Stakeholder participation in environmental projects. *J. Environ. Manag.* **2012**, *111*. [CrossRef] [PubMed]
44. Larson, S.; Williams, L.J. Monitoring the Success of Stakeholder Engagement: Literature Review. In *People, Communities and Economies of the Lake Eyre Basin*; DKCRC Research Report 45; Measham, T.G., Brake, L., Eds.; Desert Knowledge Cooperative Research Centre: Alice Springs, Australia, 2009; pp. 251–298.
45. Slocum, N. Participatory Methods Toolkit: A Practitioner's Manual. King Baudouin Foundation and the Flemish Institute for Science and Technology Assessment, 2003. Available online: http://archive.unu.edu/hq/library/Collection/PDF_files/CRIS/PMT.pdf (accessed on 17 December 2014).
46. Burgin, S.; Webb, T.; Rae, D. Stakeholder Engagement in Water Policy: Lessons from Peri-Urban Irrigation. *Land Use Policy* **2013**, *31*. [CrossRef]
47. Susskind, L.; Camacho, A.E.; Schenk, T. A Critical Assessment of Collaborative Adaptive Management in Practice: Collaborative Adaptive Management. *J. Appl. Ecol.* **2012**, *49*. [CrossRef]
48. Bawole, J.N. Public Hearing or "Hearing Public"? An Evaluation of the Participation of Local Stakeholders in Environmental Impact Assessment of Ghana's Jubilee Oil Fields. *Environ. Manag.* **2013**, *52*. [CrossRef] [PubMed]
49. Innes, J.E.; Booher, D.E. Reframing Public Participation: Strategies for the 21st Century. *Plan. Theory Pract.* **2004**, *5*. [CrossRef]
50. Cundill, G. Monitoring Social Learning Processes in Adaptive Comanagement: Three Case Studies from South Africa. *Ecol. Soc.* **2010**, *15*, 28.
51. Gleason, M.; McCreary, S.; Miller-Henson, M.; Ugoretz, J.; Fox, E.; Merrifield, M.; McClintock, W.; Serpa, P.; Hoffman, K. Science-Based and Stakeholder-Driven Marine Protected Area Network Planning: A Successful Case Study from North Central California. *Ocean Coast. Manag.* **2010**, *53*. [CrossRef]
52. Jackson, S.; Tan, P.L.; Mooney, C.; Hoverman, S.; White, I. Principles and Guidelines for Good Practice in Indigenous Engagement in Water Planning. *J. Hydrol.* **2012**, *474*. [CrossRef]
53. Bartels, W.; Furman, C.A.; Royce, F.; Ortiz, B.; Zierden, D.; Fraisse, C. Developing a Learning Community: Lessons from a Climate Working Group for Agriculture in the Southeast USA. Available online: http://www.seclimate.org/wp-content/uploads/2014/07/2012-bartels-et-al-secc-technical-series-developing-a-learning-community.pdf (accessed on 22 march 2016).
54. Jacobs, M.H.; Buijs, A.E. Understanding Stakeholders' Attitudes toward Water Management Interventions: Role of Place Meanings: Stakeholders' attitudes toward water management. *Water Resour. Res.* **2011**, *47*. [CrossRef]

water

MDPI

Article

Stakeholder Engagement for Inclusive Water Governance: "Practicing What We Preach" with the OECD Water Governance Initiative

Aziza Akhmouch * and Delphine Clavreul

Water Governance Programme, Organisation for Economic Co-operation and Development (OECD), 2 rue André-Pascal, Paris 75775, France; Delphine.Clavreul@oecd.org

* Correspondence: Aziza.Akhmouch@oecd.org; Tel.: +33-1-45-24-79-30

Academic Editors: Sharon B. Megdal, Susanna Eden and Eylon Shamir

Received: 7 March 2016; Accepted: 20 April 2016; Published: 16 May 2016

Abstract: A cursory glance at the literature on water governance reveals that stakeholder engagement has long been considered an integral part of sound governance processes. However, a closer look at the literature reveals that, beyond this general assertion, there is a lack of evidence-based assessment on how engagement processes contribute to water governance objectives. This article addresses this research gap by presenting key findings and policy guidance from a study by the Organisation for Economic Co-operation and Development (OECD) on "Stakeholder Engagement for Inclusive Water Governance". This study employed comprehensive methods, including a survey administered to 215 stakeholder groups worldwide and separately, 69 case studies of specific stakeholder engagement initiatives on water management. This article also shares the experiences and lessons that have emerged from engaging stakeholders in the OECD Water Governance Initiative—an international multi-stakeholder policy forum created in 2013 to share policy and practical experiences on water governance at different levels. We hope this research will be used to stimulate and enrich discussions about the necessary conditions for results-oriented stakeholder engagement, and to guide decision makers accordingly.

Keywords: water governance; stakeholder engagement; inclusiveness; OECD; Water Governance Initiative

1. Introduction

There is much talk these days about involving citizens, and a broad range of actors from the private and non-profit sectors, more regularly and directly, in policy and decision-making processes. Traditional top-down approaches to policy design and implementation have increasingly lost political legitimacy and been replaced with more deliberative, inclusive and bottom-up processes. Town hall meetings are now commonplace during electoral campaigns. Citizen conferences have been used to elicit informed opinion and to probe for shared public interest in conjunction with major policy reform.

Water management has undergone this change as well: the traditional role of "governments" as the single decision-making authority has in many instances been replaced by multi-level, poly-centric governance. This transition acknowledges the important roles that stakeholders from different institutional settings can contribute to effective, efficient and inclusive water management. However, it would be naïve to think that stakeholder engagement has become globally institutionalized within water policy culture. Best practices from fully-fledged engagement initiatives can help demonstrate the effectiveness of stakeholder engagement and should be scaled-up. Notable examples in this regard include the Hurricane Sandy "Rebuild by Design" Initiative's approach to community-based solutions for redeveloping cities in environmentally and economically healthier ways [1], and Brazil's national

Pact for Water Management which has rekindled federal and state visions for water resources [2]. Governments now acknowledge that water policies, however well-intentioned, need stakeholder engagement for their implementation on the ground. In any case, people demand it: social protests rise against major infrastructure projects, such as the Sivens dam in France; new water charges in Ireland; or, most recently, high toxicity levels in drinking water in Flint, Michigan (United States).

The origins of this trend have been discussed and include arguments of declining citizens' trust in institutions following the economic crisis [3]; the open government agenda that transforms how decision-makers conduct their affairs and how accountable they are [4]; and reasons more specific to water management such as the move from the prevailing "technical" supply-driven and infrastructure-led solutions in the 1980s towards integrated water resources management [5].

Whether or not stakeholder engagement has become an institutional feature of water-related policy processes, the degree of attention being given to expanding stakeholders' role in water management underscores the need to consider how effective engagement processes have proven to be in reaching intended water governance objectives. What do we know about the contribution of stakeholder engagement to effective water governance? How does stakeholder engagement work in practice? What are its related costs and benefits? We address these threshold questions in two parts. First, by sharing the key findings from a study by the Organisation for Economic Co-operation and Development (OECD) [6], which relies on empirical data from a 2014 survey carried out across 215 stakeholder groups worldwide, within and outside the water sector, and separately, 69 case studies of specific stakeholder engagement initiatives on water management across a range of countries. Through this work we identify six necessary conditions addressed to decision makers and practitioners on how to set up the enabling environment for inclusive water governance. Second, we share the experiences of the OECD Water Governance Initiative [7], a full-scale exercise of multi-stakeholder engagement started in 2013 in the form of an international network of public, private and non-profit actors supporting better governance of water management. In drawing out these experiences we look back at the first three years of operation in order to flesh out some lessons learned.

For the purpose of this work, we have adopted the following definition of stakeholder engagement: the process by which any person or group who has an interest or stake in a water-related topic is involved in the related activities and decision-making and implementation processes. The person or group may be directly or indirectly affected by water policy and/or have the ability to influence the outcome positively or negatively [6]. Additionally, we define water governance as: encompassing political, institutional and administrative rules, practices, and processes (formal and informal) through which decisions are taken and implemented, stakeholders can articulate their interests and have their concerns considered, and decision-makers are held accountable in the management of water resources and the delivery of water services [8]. We wish to be clear from the outset of the article that the first section draws heavily on the OECD report "Stakeholder Engagement for Inclusive Water Governance", of which we are the main authors.

2. Part I: OECD Findings and Necessary Conditions on Stakeholder Engagement in Water Management

2.1. *Trends and Trajectories*

Conflicting and often competing water needs have resulted in water governance challenges that are both spatial and institutional. Increasing water scarcity and a territorial mismatch between water supply and demand are central challenges to several water crises. The intensifying competition for water resources is well documented [9–11]. Both demand and supply side pressures are on the rise, driven by economic development, population growth, deteriorating water quality and climate change. The *OECD Environmental Outlook to 2050* [9] highlights that water resources are already over-used or over-allocated in many places and that this will generate fierce competition between uses and users. In response, trade-offs will need to be managed, at the least cost for society. Engaging the broad range of stakeholders in water-related decision-making holds the promise of raising awareness about risks and related costs, building the social and political acceptability of related decisions, and reducing the potential for conflicts over water.

When tracing the history of public participation and stakeholder engagement, academic literature reveals a range of disciplines, fields of study and methodologies. Founding concepts have been forged by philosophers and socio-political theorists, such as Habermas [12], Mowday [13], and Ostrom, whose approaches to institutional analysis have demonstrated that solidarity-based economies are promising alternatives to traditional state-centred command-and-control economic solutions to pressing social and ecological problems [14]. More applied contributions have come from the fields of science, technology and environment policy. These contributions have investigated different typologies and processes of engagement [15–22]—moving from participation as an end in of itself, notably Arnstein's "ladder on citizen participation" [15], to considering it as a *means* to an *end* as in the case of Fung's "democracy cube" [21]. Structured around axes of authority and power, types of participants and communication and decision modes, the "democracy cube" aims to inform institutional design choices for public participation planning activities.

Three main distinguishing trends and trajectories of engagement processes have emerged over the past decades. First, there has been an evolution in water debates from the notion of "participation" to the concept of "engagement". While "participation" typically refers to the involvement of individuals and groups in the design, implementation and evaluation of a project or plan [17–19], "engagement" embraces a broader range of inclusive processes, with different intentions and different inputs to the decision-making process. Stakeholder engagement has also trended away from the sole consideration of civil society and end users, to reach out to other groups of actors with an influence on water-related planning, decision-making, implementation, and monitoring and evaluation such as property developers and long-term institutional investors. The OECD has developed a typology that distinguishes six levels of stakeholder engagement depending on the processes and the intentions they pursue (Figure 1).

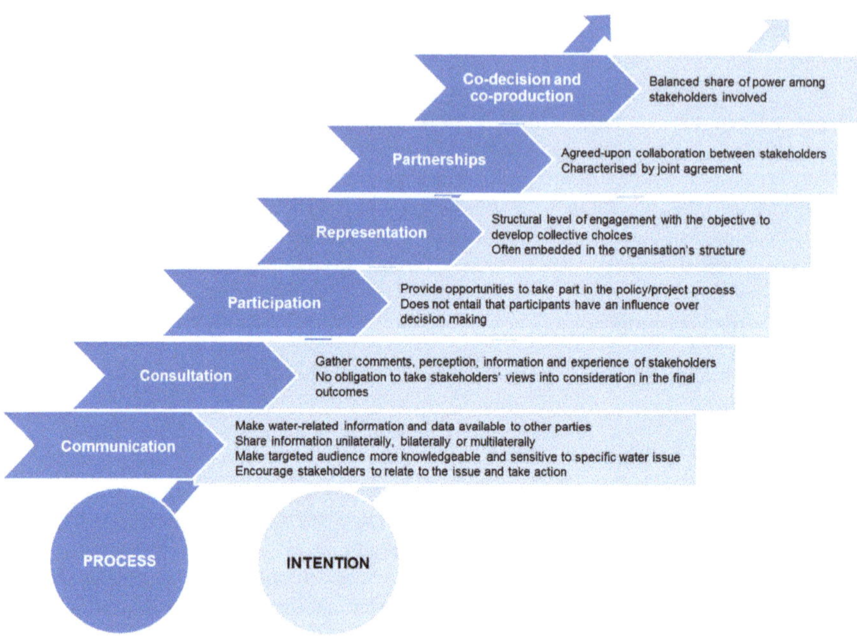

Figure 1. Organisation for Economic Co-operation and Development (OECD) typology of levels of stakeholder engagement [23].

The second notable trend is that, until recently, stakeholder engagement had remained mostly a "one-off" exercise. This can be explained by the flexibility often associated with issue-based stakeholder engagement, which decision-makers have preferred over systematic and more committed approaches. However, water management is now experiencing a move toward more structural forms of stakeholder engagement. New legislation, guidelines and standards at various levels have spurred the emergence of more formal forms of stakeholder engagement. The European Water Framework Directive [24] and the Aarhus Convention on Access to Information, Public Participation in Decision-Making and Access to Justice in Environmental Matters [25], or country-based initiatives such as the Dutch Delta Programme in The Netherlands [26] and the Chesapeake Bay Program in the United States [6] are noteworthy examples.

Lastly, stakeholder engagement has been more institutionalised for water *resources* management than water *service* delivery. This can be explained by the fact that legislation has primarily targeted integrated water resources management, such as the 1992 Dublin Principles [27], which has triggered important incentives and reforms of stakeholder engagement as a key pillar of water resources decision making and planning. Accompanying these changes, economic activities have grown increasingly dependent on water resources thus making stakeholder engagement all the more crucial. In comparison, engagement in water and sanitation services has been more heterogeneous and non-systematic, often consisting of handling customers' complaints. However, more recently, partnerships with users and citizens have emerged as a positive trend towards inclusive and innovative service delivery [28]. Examples of the latter include the Customer Challenge Groups in England and Wales [29] and Grenoble's Water Users' Committee in France [30].

2.2. Approach to the OECD Study: Survey and Methodology

Many international organisations and research institutions have carried out work on inclusive approaches to water management. Some have focused on specific types of stakeholders (citizens, women) while others have tackled aspects of the engagement processes (design, mechanisms). As a result, a plethora of good practices have been showcased around the world. However, as concluded during the session on "Stakeholders' engagement for effective water policy and management" at the 6th World Water Forum (Marseille, 2012), there has been a lack of evidenced-based analysis and policy tools to assess how these inclusive approaches ultimately contribute to the better performance of water policies and projects in terms of processes and outcomes.

Because more and more water players are seeking to influence certain trajectories, it is crucial to evaluate the actual weight of stakeholder engagement in water-related decision-making, and the extent to which this leads to greater effectiveness, efficiency and inclusiveness. In a high stakes context with political power plays and vested interests, an understanding of how common solutions are developed and hit the policy goals is crucial in order to ensure that no one is left behind. This was the purpose of the OECD study on "Stakeholder Engagement for Inclusive Water Governance", carried out in 2014 [6]. It proposes an analytical framework to assess stakeholder engagement (Figure 2) and defines six necessary conditions, along with a Checklist for public action and select indicators, in order to get it right while scaling-up success stories for more effective bench-marking.

This analytical framework is organised around five components: the *drivers* of engagement in water management; the *types of stakeholders* to be engaged; the *obstacles* to overcome; the *mechanisms* for engaging stakeholders; and the assessment of engagement processes, looking at *costs* and *benefits*. To support this analysis, an extensive online survey undertaken in 2014 helped collect qualitative and quantitative data on stakeholder engagement across a range of stakeholders, with various levels of interests and types of experience. Details on the survey's sample, methodology and results are provided in [6]. We wish to clarify that although the results of the survey provide valuable insights and feedback on the reality of stakeholder engagement practices, they do not intend to be statistically comprehensive or reflect the multitude of views, arrangements and players in the field. Drafts of the resulting report were discussed with multiple stakeholders and experts, notably as part of a multi-stakeholder workshop in Paris (France, 19 September 2014), at the International Water Association World Water Congress in Lisbon (Portugal, 24 September 2014), and at the 4th meeting of the OECD Water Governance Initiative in Paris (France, 24 November 2014). The final report was launched at the 7th World Water Forum (Republic of Korea, 13 April 2015) [6].

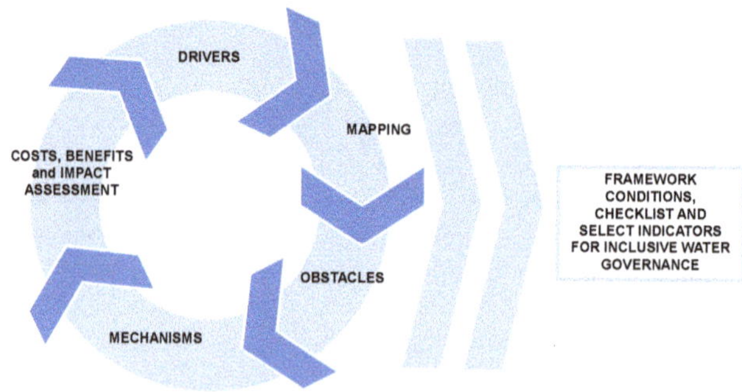

Figure 2. OECD Analytical framework of stakeholder engagement in water governance [23].

2.3. Main Findings

Taking a closer look at each of the five components of the analytical framework described in the previous paragraph, in this section we review the main findings from the OECD study before concluding with policy guidance in the form of six necessary conditions for successful stakeholder engagement.

To begin, we note that structural drivers—ranging from climate change; demographic trends of population and urbanisation growth; technological progress (e.g., rapid development of information and communication technologies); and socio-political changes such as the growing number and scale of citizen initiatives—have triggered a change in water governance paradigms to fit for the future. Simultaneously, conjunctural drivers, such as water-related disasters, policy reforms, large water-related projects such as dams, and democratic pressure have also pushed decision makers to explore new approaches that engage directly with stakeholders to solve water issues. Emergency-driven situations have an impact on stakeholder engagement as they shed light on the weaknesses of governments to properly address the risks, and are windows of opportunity for doing better through new innovative partnerships. Competition over water resources raises complex policy issues related to rivalry and excludability. Experiences in water allocation reforms in Canada, South Africa, the United States, as well as in England and Wales have shown that stakeholder engagement processes can be valuable to gain a deeper understanding of the preferences of different water users, reveal what a proposed allocation reform would mean for them and manage some trade-offs [31].

As policymakers seek to engage stakeholders, identifying *whom* to engage can be a daunting task. The OECD survey shed light on the categories of stakeholders involved in water governance. Evidence shows that beyond the "traditional" actors, newcomers are playing an increasing role in water management and need to be properly engaged. This is the case with property developers, who both generate future liabilities in terms of water management (and require proper coordination with flood protection, spatial planning and land use), but who also hold opportunities for innovative financing modes (e.g., property taxes). In addition, the corporate sector increasingly factors governance into their risk assessment frameworks and strategies. Long-term institutional investors, such as pension funds and insurance companies, are also now investing more in water infrastructure. Other players in the water landscape have yet to receive specific attention to ensure inclusiveness, particularly women, the youth, indigenous communities and the poor.

While stakeholder engagement processes vary greatly across places, common obstacles have been identified by the OECD survey, and the following barriers deserve particular attention:

- The lack of political will and leadership needed to shift the balance of power among stakeholders, including towards actors that may not share the same intentions, perspectives and interests;
- The lack of clarity on the use of stakeholder inputs (e.g., to build consensus, to take decisions, to share information, to raise awareness), which can result in mistrust and consultation "fatigue";
- Institutional fragmentation with overlapping mandates and conflicting goals, possibly leading to inadequate co-operation across authorities, water-related sectors and scales;
- The lack of funding to sustain the engagement process, logistical expenses related to meeting venues or support material and the lack of competent and dedicated staff;
- Conflicts of interest and consultation "capture", especially when certain groups of actors and lobbies are better organised to voice their concerns.

Results from the survey revealed a range of mechanisms to engage stakeholders, more or less formal, more or less costly, more or less timely and relevant. While stakeholders continue to use "tried-and-tested" instruments such as meetings, workshops and expert panels, innovative mechanisms are gaining traction because of technological advances and greater tools and openness. Information and communication technologies (ICTs), for instance, are a driving force of customised Internet applications [32], and the function of ICT platforms has taken new and varied dimensions as virtual meetings and vectors (e.g., social media, chat rooms, online forums) are used more frequently. For instance, the Water and Waste Services Regulation Authority of Portugal has developed a mobile application that provides relevant information to users in 278 municipalities on the quality of service provision.

Stakeholder engagement processes should be assessed. The OECD study finds that a compelling argument for evaluating stakeholder engagement is the need for accountability—to ensure the proper use of public or institutional resources, including stakeholders' time and efforts. However, there are other reasons. As with any process, evaluation assists in determining whether stakeholder engagement was successful and in drawing up an inventory of lessons learned towards future improvements. Undeniably, involving stakeholders in water-related decision-making processes raises costs—be they monetary or otherwise, but it also generates benefits. The costs relate to different phases of the engagement process and concern the production and disclosure of needed information, operational expenses (facilities, travel, staff, overtime, *etc.*), opposition to final decisions, as well as delays in decision-making or implementation (Figure 3).

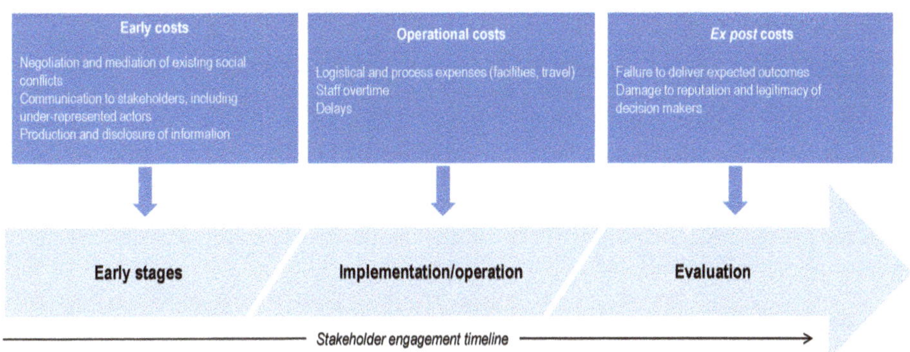

Figure 3. OECD typology of costs of stakeholder engagement in water governance [23].

The benefits were clustered into four categories (Figure 4): (i) acceptability and sustainability, in terms of effective implementation of water policy and projects, proper enforcement of regulation, political acceptability, and ownership of decisions and outcomes; (ii) social equity and cohesion, which is related to trust, confidence, customer satisfaction, as well as corporate social responsibility; (iii) capacity and knowledge development, which emanates from raising greater awareness, sharing information, and forming opinions; and (iv) economic efficiency, as it can assist in optimising cost-saving, value for money, time-saving, as well as broader economic benefits from policy coherence and synergies across sectors and projects. Costs and benefits accrue to different stakeholder groups at different times. Often, costs of stakeholder engagement are short term (e.g., operational costs), while benefits may arise during the engagement, immediately after on in the long run. Therefore, the sustainability of stakeholder engagement will not only depend on the net difference between aggregate costs and benefits, but also on how they are distributed between stakeholders, on stakeholders' willingness to bear them, and on adopting trade-offs to manage their dual temporality.

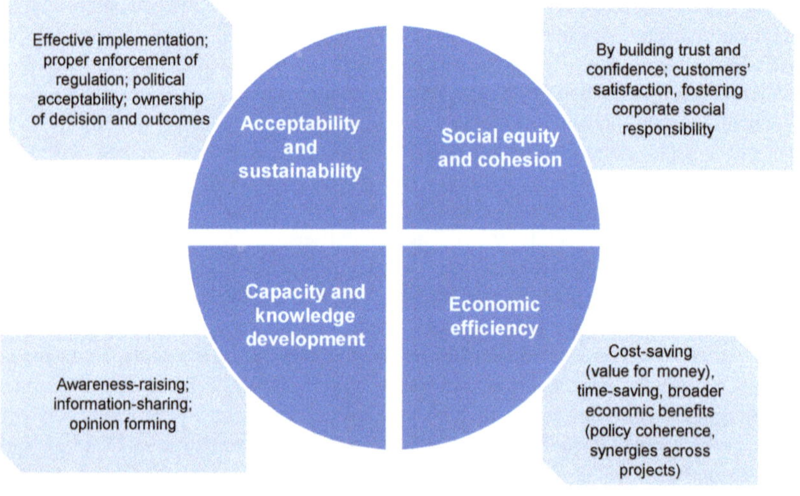

Figure 4. OECD typology of long-term benefits of stakeholder engagement in water governance [23].

2.4. Concluding Policy Guidance: Six Necessary Conditions for Inclusive Water Governance

The OECD study demonstrated that systemic, inclusive and foresighted approaches to water policy making are more likely to realise better outcomes, and returns on investment, in time and money. The case studies demonstrate how stakeholder engagement can lead to tangible outcomes. In France, a partnership between the energy supplier EDF and irrigators facilitates the optimisation of the water-energy nexus and led to the reduction of 90 million cubic meters in agricultural water consumption in the Durance Valley. In the Great Lakes region (Canada), multi-level engagement processes on water resources management between the province of Ontario, municipalities, local NGOs and First Nation communities have led to the formulation of common policy directions and long-term strategies for water protection.

For engagement processes to be relevant, a careful balance between what they try to achieve, the resources they require and whether they succeed in reaching the intended objectives is required. Although engagement processes cannot be easily replicated from one context to another, we conclude with the following proposed necessary conditions as policy guidance to decision-makers (Figure 5):

1. *Inclusiveness and equity*: Map all stakeholders who have a stake in the outcome or that are likely to be affected, as well as their responsibility, core motivations and interactions.
2. *Clarity of goals, transparency and accountability*: Define the ultimate line of decision making, the objectives of stakeholder engagement and the expected use of inputs.
3. *Capacity and information*: Allocate proper financial and human resources and share needed information for result-oriented stakeholder engagement.
4. *Efficiency and effectiveness*: Regularly assess the process and outcomes of stakeholder engagement to learn, adjust and improve accordingly.
5. *Institutionalisation, structuring and integration*: Embed engagement processes in clear legal and policy frameworks, organisational structures/principles and responsible authorities.
6. *Adaptiveness*: Customise the type and level of engagement as needed and keep the process flexible to changing circumstances.

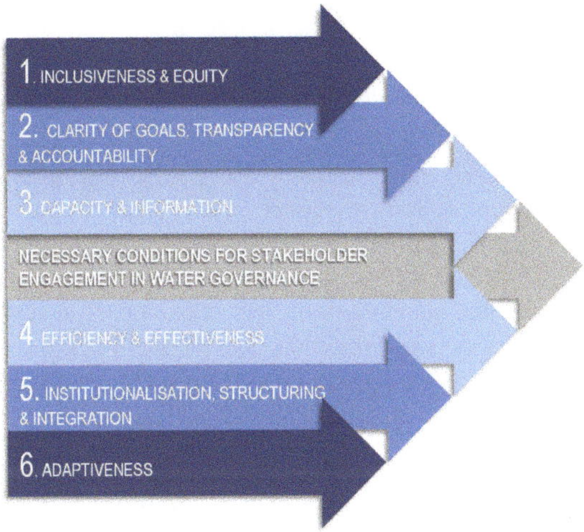

Figure 5. Necessary conditions for stakeholder engagement in water governance [23].

3. Part II: An Application of the OECD Necessary Conditions for Stakeholder Engagement to a Real-Life Experience: The Case of the OECD Water Governance Initiative

Part II presents the experience of the OECD Water Governance Initiative (WGI), a multi-stakeholder Policy Forum, and discusses its main characteristics through the lens of the six necessary conditions for inclusive water governance presented in Part I. This "6-condition prism" structures reflections on the first three years of operation of the Initiative that launched in 2013 [7] and provides a consultation platform to share knowledge and experience twice a year across over 100 stakeholders, within and outside the water community.

3.1. Why an OECD Initiative on Water Governance? Rationale and Inception

In 2011, the OECD was invited by the International Committee of the 6th World Water Forum to lead the governance discussions up to, and at the Forum, held on 12–17 March 2012 in Marseille (France). The responsibility entrusted to the OECD entailed bringing together a community of practitioners from around the world working on water governance to form a thematic taskforce, and devise the critical governance topics to be addressed at the forum [33]. At the time, this newly-created taskforce acknowledged that there had been no international platform or institution working solely and regularly on water governance issues. Rather, experts of water governance tended to gather on an ad hoc basis in the framework of major international rendezvous. A gap had to be filled to maintain greater continuity of water governance discussions in between events, and to move from good intentions to concrete actions. The OECD-led taskforce served as a basis to spur such a collective action.

One year after the 6th World Water Forum, the OECD launched the Water Governance Initiative (WGI) in March 2013, under the patronage of the OECD Secretary General, Angel Gurría. The first meeting gathered 75 major stakeholders from the public, private and non-profit sectors [33]. In all, 25 countries were represented, including spearheads of regional water networks in Asia, North America, Latin America, Africa and Europe. During two days, participants discussed their expectations for such a multi-stakeholder network, underlining the need to bridge the information gap on relevant case studies, good practices and expertise. They shared the ambition to see the Water Governance Initiative play a role of consultation platform to help governments take the needed steps for effective water governance reforms, and to discuss ways forward in this field. From March 2013 to November 2015, the WGI had met on six occasions, on a biannual basis, on: 27–28 March 2013 (OECD Headquarters, Paris) [34], 7–8 November 2013 (OECD Headquarters, Paris) [35], 28–29 April 2014 (Madrid, Spain) [36], 24–25 November 2014 (OECD Headquarters, Paris) [37], 26 May 2015 (Edinburgh, United Kingdom) [38] and 2–3 November 2015 (OECD Headquarters, Paris) [39].

3.2. The OECD Water Governance Initiative through the 6-Condition Prism

To highlight the main features of the WGI, we drew on the six necessary conditions for inclusive water governance described in Part I and used them as a "reading grid" to shed light on the key, and at times quite unique, characteristics and activities of the network.

3.2.1. Inclusiveness and Equity

The WGI rests on a sound understanding that water policy involves a range of public stakeholders across ministries, departments and public agencies, and between various levels of governments. Further, private actors, end users, investment banks, infrastructure, service providers and environmentalists have an important role to play in water policy design and implementation, and have a stake in the outcome too. This is why the WGI adopted a multi-stakeholder membership, acknowledging that managing water is a shared responsibility across policymakers and stakeholders. Institutions and actors that have joined the WGI since 2013 are representatives of national governments, local, regional and basin institutions, private sector, NGOs, public and private service providers, regulators, international organisations, donor agencies, academia as well as independent experts. The 100+ delegates of the WGI come from 32 countries of which 24 are OECD countries [7].

The diverse membership of the WGI has enabled cross-fertilisation of knowledge and has provided multiple checks to gauge what works, what does not work and what could work better in terms of water governance. This has proven instrumental to enrich the WGI analysis and key messages, going beyond what a single stakeholder group could achieve. Such a contribution was particularly relevant during the peer-review of national policy dialogues carried out by the OECD in The Netherlands [26], Jordan [40], Tunisia [41], and Brazil [42], where the views and experience from the broad range of stakeholders represented in the WGI helped shape better guidance to governments. In addition, members share the latest developments on water governance during the meetings (*i.e.*, on-going research and projects, as well as recent or upcoming events), which helps co-ordinate on-going initiatives and finding synergies and complementarities across institutions represented in the network.

3.2.2. Clarity of Goals, Transparency and Accountability

Two ingredients are essential to make multi-stakeholder processes work: clear goals and an unambiguous line of decision making.

The WGI was founded on a clear set of strategic objectives [43]. It has had the ambition to: (i) advise governments in taking the needed steps for effective water reforms through policy dialogue across decision makers at different levels; (ii) provide a technical platform to discuss analytical work on water governance through peer-to-peer exchanges and knowledge sharing; (iii) provide a consultation mechanisms to raise the profile of governance issues in the global water agenda; (iv) support the implementation of the governance targets designed for the 6th World Water Forum up to the 7th World Water Forum; and (v) contribute to the design of OECD Principles on Water Governance, and OECD Indicators on Water Governance to engage decision makers at all levels, within and outside the water community, to commit to action.

The WGI was also founded on a clear notion that it should not be considered as a formal body within the OECD structure, but rather as a technical network of experts, policymakers and practitioners gathering as part of the activities of the OECD Regional Development Policy Committee where the work on issues of multi-level governance, public investment, decentralisation, urban and rural policy, territorial indicators and spatial planning is carried out. This means, in practice, that no decisions are taken within the WGI, in which representatives have a consultative role.

3.2.3. Capacity and Information

The WGI has come at a cost, financially and in-kind. To sustain the network, financial and human resources have had to be deployed and have relied almost exclusively on voluntary contributions from the members.

The WGI did not establish a membership fee. Given the diversity of its membership, from large private companies to small NGOs, it would have been highly difficult to have a standard approach for all members. Therefore, costs related to running the Secretariat and organising the biannual meetings of the WGI were covered by select OECD countries and partner institutions that were convinced of the value added by the Initiative.

In addition, the functioning of the WGI has relied on the mobilisation of members' expertise on analytical content, the engagement of their respective networks, and the hosting of some meetings of the WGI in Madrid in 2014 [36] and Edinburgh in 2015 [38]. In addition, the Chair and members of the Steering Committee were actively engaged in the operation of the WGI (e.g., attending Steering Committee meetings and retreat, leading the thematic working groups, representing the WGI in international events, *etc.*) at their own costs.

Result-oriented engagement processes imply that stakeholders be aware and informed about the issues concerned. This is why the WGI' has had steady and timely exchanges of information and iterations with its members. Regularly, members would be asked to comment on draft reports, papers, scoping notes, and meeting proceedings. They also contributed to the development of policy guidance and standards related to water governance in a bottom-up fashion through working groups' activities, which resulted in the adoption of the Principles on Water Governance [8] and their backing at Ministerial level in June 2015.

3.2.4. Efficiency and Effectiveness

How appropriate is an experimental platform such as the WGI for achieving better water governance? Clearly, some activities or approaches may not work or new needs may arise which deserve careful consideration. This is why timely evaluation and monitoring are needed to figure out whether stakeholder engagement processes are fit to deliver expected outcomes; the WGI was no exception to this principle.

Originally, the WGI was set-up for a trial period of two years after which its outcomes and value added would be assessed to determine whether its activities should continue and if adjustments were necessary. For this purpose, a satisfaction survey was conducted in May 2015 across the 115 members of the WGI [44]. The purpose was two-fold: to collect members' feedback on the first two years of activities, both in terms of process and outcomes; and to identify areas for improvement that would help shape the strategic orientations of the WGI for the next programme of work in 2016–2018. In all, 82 members of the WGI responded to the survey, a 71.3% response rate [44].

The main results of the satisfaction survey can be clustered in two sets: what has worked well and what could be improved upon (for detailed quantitative and qualitative results of the satisfaction survey, see [7]). The first cluster of results showed a very high level of satisfaction for the WGI (above 90% of respondents). Survey respondents expressed a strong willingness to continue their involvement in the network, emphasising that the WGI is one of the few international forums that addresses water governance issues in a comprehensive and participative way, where mutual learning and experience-sharing can take place and international best practices can be identified and scaled up. The second cluster of results provided suggestions for improvement. Respondents called for refocusing the strategic objectives of the WGI onto new priority activities that include the collection of best practices and the development of indicators on water governance. This was recommended with a view to support the implementation of the OECD Principles on Water Governance in interested member and non-member countries; and the contribution to the global agenda, notably to raise the visibility of water governance in international processes such as the implementation of the Sustainable Development Goals, the preparation of the 8th World Water Forum, and climate change discussions. Survey respondents also suggested that the thematic scope of the WGI could be broadened to include new and rising water governance topics (e.g., drought and flood governance, the governance of the water-energy-food nexus, *etc.*). The conclusions drawn from the results of the satisfaction survey contributed to feed the second operating phase of the WGI over 2016–2018.

3.2.5. Institutionalisation, Structuring and Integration

There can be no stakeholder engagement without the proper incentives for bottom-up and inclusive decision-making. A clear set of rules and structure has been key to ensure the proactive engagement of members in the WGI.

The WGI was founded on a three-tier governance structure. First, a Secretariat, hosted by OECD's Public Governance and Territorial Development Directorate, is responsible for day-to-day activities. Second, a Steering Committee provides strategic guidance and ensures it is aligned with the OECD programme of work. It is composed of the co-founding institutions of the WGI who worked with OECD to organised the governance discussions at the 6th World Water Forum, *i.e., Suez Environnement,* the *Association Scientifique et Technique pour l'Eau et l'Environnement,* the International Network of Basin Organisations, United Nations Organisation for Education, Science and Culture (UNESCO)'s International Hydrological Programme, the Water Integrity Network, Stockholm International Water Institute and Transparency International. A Chair, elected by the aforementioned institutions following a call for applications, presides over the Steering Committee. Third, members participate in the activities of the WGI and related working groups. The broader water community benefits from WGI's outputs and results, which are disseminated online to a larger audience.

The WGI was also founded on clear rules of operation. Terms of reference were prepared and adopted to define the roles and responsibilities of the Chair, the Steering Committee, and the Working Groups within which the activities of the WGI have taken place. In each case, deliverables and timetables were devised to guide the work. In addition, each member signed an endorsement letter to both formalise their adhesion to the WGI and to commit to specific contributions within the WGI's strategic objectives. Together, the terms of reference and endorsement letters have formed a strong accountability framework that has kept involvement levels high and allowed regular monitoring and reporting on progress achieved.

The WGI's terms of reference [43] also set out a bottom-up process by which four thematic working groups were created to formulate and discuss key policy messages on select areas of governance (stakeholder engagement, integrity and transparency, performance of utilities, basin governance), which contributed to the development of the OECD Principles on Water Governance. This approach allowed strong buy-in of stakeholders who produced the Daegu Multi-stakeholder Declaration on the OECD Principles on Water Governance gathering 65 signatures from the institutions actively engaged in the WGI, to mainstream the Principles into their activities and practices and to work further with the OECD to contribute to their implementation.

3.2.6. Adaptiveness

Open memberships coupled with an adaptive structure have been two key factors that have sustained the WGI.

New actors gain prominence in water management, and increasingly weight on water-related decision-making processes. The open and evolving membership of the WGI has made it possible to reflect this growing diversity. New members have joined the WGI on a regular basis since its creation, and the network has grown from 88 in 2013 to 115 in 2015. Newcomers have included energy companies as well as countries looking to improve their water policies (e.g., People's Republic of China, Colombia). Other actors, who are often under-represented in water-related stakeholder initiatives, have also joined the WGI, such as the Water Youth Network.

The organisation of WGI activities has relied on a "fit-for-purpose" logic. Initially, four separate and parallel working groups were tasked with formulating key thematic messages to contribute to the development of the OECD Principles on Water Governance. Their life expectancy was conditional on the adoption of the Principles. Today, the WGI is shifting its focus to new activities on the implementation of the Principles (as explained in the above section "efficiency and effectiveness"). These activities are no longer clear-cut from a thematic point of view, and require some adjustment of the underlying working groups that can support future work. This is why it has adjusted its governance structure to evolve towards two working groups dedicated to identifying and sharing best practices and supporting the development of water governance indicators.

3.3. *Key Achievements and Lessons Learned*

In the previous section, we looked at the key aspects that characterise the WGI, from how it was set up, to how it has carried out its activities to how it was assessed after its first period. The 6-condition prism used shows that the WGI has strived to meet, or "comply", with the six necessary conditions for effective stakeholder engagement in water governance presented in Part I. The prism shed light on the key achievements of the WGI during 2013–2015, most notably its contribution to the development of the OECD Principles on Water Governance. It highlights some factors, beyond the six necessary conditions, which have proven instrumental in ensuring the WGI remains relevant and resilient. In this section, we primarily focus on three of these success factors with the intention to provide inspiration to similar stakeholder engagement processes.

First, achieving a balanced representation of all stakeholders from the water sector requires constant efforts and is a long-term endeavour. One cannot always expect to reach an optimal degree of inclusiveness from the start. Some actors may have different levels of interest, or they may not have the necessary resources to be actively engaged. Certainly, the WGI has faced this challenge. Engaging with certain stakeholder groups has proven difficult as in the case of local governments, non-OECD Members, consumer associations, trade unions, and other sectors with an influence on water. While some institutions from these categories of stakeholders have joined the WGI (as pointed out in the previous section "institutionalisation, structuring and integration"), more could be done to broaden the scope of the WGI's membership, e.g., through proactive communication, including web-based, to better disseminate the results and achievements of the WGI to catalyse further interest. This is currently a cross-cutting responsibility of WGI's Steering Committee to, among other objectives, bridge membership gaps.

Second, "flexible rigidity" and regular communication are paramount for resilient stakeholder engagement. Finding the right balance between clear and collectively agreed-upon rules of operation, and adaptiveness is important to manage expectations, avoid disappointment and endure in the long term. A mutual understanding on what the WGI was aiming to achieve, and what it was not, contributed to that. Good preparation ahead of meetings is key for quality contributions to substantive discussions, in line with the expectations. In the case of the WGI, regular communication (e.g., emails, news items), and intermediary milestones events (e.g., webinars and workshops on specific topics organised between the biannual plenary meetings) have proven instrumental to sustain the community of practice and ensure collective results.

Third, high-level support plays an important role to raise awareness and visibility. Leaders can set incentives and raise the profile of an engagement process. This has been the case for the WGI: the Secretary-General of the OECD, Angel Gurría, launched the network during the first WGI meeting through a video message and attended the 4th meeting. Four Ambassadors to the OECD from Korea, The Netherlands, Spain and France have also participated in WGI meetings. Moreover, the WGI welcomed high-level representatives from UNESCO, the World Water Council, the United Nations Secretary-General's Advisory Board on Water and Sanitation (UNSGAB), and the International Water Resources Association. The 3rd meeting of the WGI received the patronage of the Secretary General for International Co-operation and Development of Spain and Director of the Spanish Agency for International Co-operation for Development. Together, these leaders have reinforced the legitimacy of the WGI as a useful network for decision makers, and have helped to secure financial and in-kind support to sustain the WGI. Furthermore, they have contributed to raising the visibility of the network on the international stage. The WGI was at the forefront of major governance discussions in global events, including the Budapest Water Summit [45], the 2015 UN-Water Zaragoza conference [46] and the 7th World Water Forum [47]. The WGI also helped formulate key messages on governance that are featured in important international texts, such as the Ministerial Declaration of the 7th World Water Forum [48] and the Lisbon Charter on Guiding the Public Policy and Regulation of Drinking Water Supply, Sanitation and Wastewater Management Services [49].

4. Concluding Remarks: Capitalising on Promising Trends

Recent trends reflect the evolution of stakeholder engagement practices in water management. The 1992 Rio Declaration on Environment and Development introduced the emerging public involvement norms [50]. In the same year, the Dublin Statement on Water and Sustainable Development (adopted by the UN on 31 January 1992) included participation as one of its guiding principles [27], and the Agenda 21 envisaged public involvement in developing, implementing and enforcing environmental laws and policies, including management of freshwaters [51]. The Aarhus Convention on Access to Information, Public Participation in Decision-Making and Access to Justice in Environmental Matters focused specifically on participation [25]. Similarly, all World Water Fora highlighted the critical role of multi-actor partnerships (Marrakech, 1997); participatory approaches (The Hague, 2000); alliances, networks and dialogues (Kyoto, 2003); co-ordination across levels of government (Mexico, 2006); the critical role of vulnerable and marginalised groups (Istanbul, 2009); the need for multi-stakeholder platforms (Marseille, 2012); and the necessity to put stakeholder engagement into practice and fit to local realities (Daegu, 2015) [47].

The newly adopted Sustainable Development Goals [52] include several references to water, governance and stakeholder engagement through not only a dedicated goal on water (No. 6) and a target on local participation, but also other governance-related goals referring to inclusiveness, gender equality, capacity building, policy coherence, multi-stakeholder partnerships, data, monitoring and accountability. In line with this context, the OECD Principles on Water Governance adopted by the 34 Member countries and endorsed at ministerial level, contain a dedicated block on trust and engagement [8], which calls for promoting stakeholder engagement for informed and outcome-oriented contribution to water policy design and implementation (Figure 6).

Unarguably, the increasing attention at the global level to stakeholder engagement demonstrates a shift in how decisions should be made in water management. "Top-down hierarchical models" are no longer prevalent and more transparent and holistic models that involve public and non-state actors, including the private sector and not-for-profit organisations, have made their way to the forefront of water policy making.

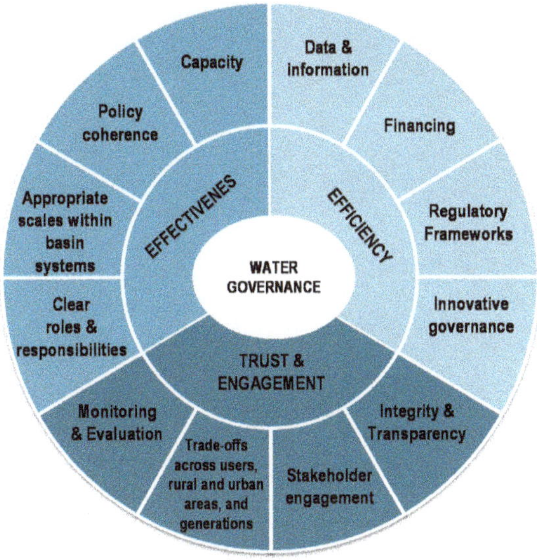

Figure 6. OECD Principles on Water Governance [8].

Looking forward, an apparent way to improve the implementation of engagement processes could be through greater knowledge on the distributional impacts of stakeholder engagement, particularly the danger of potentially inequitable distribution of benefits, and related policy guidance on whether, when and how to compensate the losers from a given water reform or project. Such knowledge could be derived from practical experiences within policy dialogues at different territorial (national, rural, urban, basin or cross border) levels. Public policies could, furthermore, benefit from improved tools for assessing the contribution of stakeholder engagement to better water quality and quantity management. The OECD is developing a set of water governance input, output and outcome indicators, including on stakeholder engagement, in the hope of providing a valuable contribution in this field.

Acknowledgments: The authors would like to thank the taskforce of 215 members that contributed to the OECD study "Stakeholder Engagement for Inclusive Water Governance" through participating in the survey, submitting the 69 case studies and peer-reviewing the findings and recommendations in several workshops and seminars. The authors would also like to thank all 100+ members of the OECD Water Governance Initiative and in particular its Chair and Steering Committee, for their active engagement in the network's activities and contribution to put stakeholder engagement in practice.

Author Contributions: Aziza Akhmouch and Delphine Clavreul jointly wrote the paper and the underlying OECD study on "Stakeholder Engagement for Inclusive Water Governance". The OECD Water Governance Initiative was spearheaded by Aziza Akhmouch in 2013 and has been coordinated together with Delphine Clavreul since then.

Conflicts of Interest: The authors declare no conflict of interest.

References

1. Rebuild By Design. Available online: http://www.rebuildbydesign.org/ (accessed on 12 April 2016).
2. ANA. Brazil's National Pact for Water Management. Available online: http://www2.ana.gov.br/Paginas/EN/programs.aspx (accessed on 12 April 2016).
3. Organisation for Economic Co-operation and Development (OECD). *Government at a Glance 2013*; OECD Publishing: Paris, France, 2013. Available online: http://dx.doi.org/10.1787/gov_glance-2013-en (accessed on 10 February 2016).

4. Organisation for Economic Co-operation and Development (OECD). Towards recovery and partnerships with citizens: A call for innovative and open government. In Proceedings of the Ministerial Meeting of the OECD, Paris, France, 7–8 October 2010; Public Governance Committee: Paris, France, 2010.
5. Organisation for Economic Co-operation and Development (OECD). *Water Governance in OECD Countries: A Multi-Level Approach*; OECD Studies on Water; OECD Publishing: Paris, France, 2011. Available online: http://dx.doi.org/10.1787/9789264119284-en (accessed on 10 February 2016).
6. Organisation for Economic Co-operation and Development (OECD). *Stakeholder Engagement for Inclusive Water Governance*; OECD Publishing: Paris, France, 2015. Available online: http://dx.doi.org/10.1787/9789264231122-en (accessed on 10 February 2016).
7. OECD. Water Governance Initiative. Available online: http://www.oecd.org/gov/regional-policy/water-governance-initiative.htm (accessed on 12 April 2016).
8. Organisation for Economic Co-operation and Development (OECD). *OECD Principles on Water Governance*; OECD Publishing: Paris, France, 2015. Available online: http://www.oecd.org/governance/oecd-principles-on-water-governance.htm (accessed on 12 April 2016).
9. Organisation for Economic Co-operation and Development (OECD). *OECD Environmental Outlook to 2050: The Consequences of Inaction*; OECD Publishing: Paris, France, 2012. Available online: http://dx.doi.org/10.1787/9789264122246-en (accessed on 10 February 2016).
10. World Resources Institute (WRI). *Aqueduct: Measuring and Mapping Water Risk*; World Resources Institute: Washington, DC, USA, 2015. Available online: http://www.wri.org/our-work/project/aqueduct (accessed on 26 January 2016).
11. United Nations Educational, Scientific and Cultural Organization (UNESCO). *Managing Water under Uncertainty and Risk*; The United Nations World Water Development Report 4; United Nations Educational, Scientific and Cultural Organization: Paris, France, 2012; Volume 1.
12. Habermas, J. *The Structural Transformation of the Public Sphere: An Inquiry into a Category of Bourgeois Society*; The MIT Press: Cambridge, MA, USA, 1989.
13. Mowday, R.T.; Steers, R.M. The measurement of organisational commitment. *J. Vocat. Behav.* **1979**, *14*, 224–247.
14. Ostrom, E. *Governing the Commons: The Evolution of Institutions for Collective Action*; Cambridge University Press: Cambridge, UK, 1990.
15. Arnstein, S.R. A ladder of citizen participation. *J. Am. Inst. Plan.* **1969**, *35*, 216–224.
16. Smith, D.H. Synanthrometrics: On progress in the development of a general theory of voluntary action and citizen participation. In *International Perspectives on Voluntary Action Research*; Smith, D.H., Van Til, J., Eds.; University Press of America: Washington, DC, USA, 1983.
17. Brown, M.; Wyckoff-Baird, B. *Designing Integrated Conservation and Development Projects*; Biodiversity Support Program: Washington, DC, USA, 1992.
18. Pretty, J. Participatory learning for sustainable agriculture. *World Dev.* **1995**, *23*, 1247–1263.
19. Yee, S. Stakeholder Engagement and Public Participation in Environmental Flows and River Health Assessment. Project Code P0018, Australia-China Environment Development Partnership, 2010. Available online: http://watercentre.org/portfolio/rhef/attachments/technical-reports/stakeholder-engagement-and-public-participation-in-eflows-and-river-health-assessments (accessed on 10 February 2016).
20. Rowe, G.; Frewer, L.J. A typology of public engagement mechanisms. *Sci. Technol. Hum. Values* **2005**, *30*, 251–290.
21. Fung, A. Varieties of participation in complex governance. *Public Adm. Rev.* **2006**, *66*, 66–75.
22. Mott Lacroix, K.E.; Megdal, S.B. Explore, synthesize, and repeat: Unraveling complex water management issues through the stakeholder engagement wheel. *Water* **2016**, *8*. [CrossRef]
23. OECD. *Stakeholder Engagement for Effective Water Governance*; OECD Publishing: Paris, France, 2015.
24. European Parliament and Council of the European Union. *Directive 2000/60/EC of the European Parliament and of the Council Establishing a Framework for Community Action in the Field of Water Policy*; Official Journal of the European Communities: Brussels, Belgium, 2000.
25. United Nations Economic Commission for Europe (UNECE). *Convention on Access to Information, Public Participation in Decision-Making and Access to Justice in Environmental Matters*; UNECE: Aarhus, Denmark, 1998.

26. Organisation for Economic Co-operation and Development (OECD). *Water Governance in the Netherlands: Fit for the Future*; OECD Studies on Water; OECD Publishing: Paris, France, 2014. Available online: http://dx.doi.org/10.1787/9789264102637-en (accessed on 10 February 2016).
27. Secretariat of the International Conference on Water and the Environment. *The Dublin Statement on Water and Sustainable Development*; Secretariat of the International Conference on Water and the Environment: Dublin, Ireland, 1992.
28. Organisation for Economic Co-operation and Development (OECD). *Together for Better Public Services: Partnering with Citizens and Civil Society*; OECD Public Governance Reviews; OECD Publishing: Paris, France, 2011. Available online: http://dx.doi.org/10.1787/9789264118843-en (accessed on 10 February 2016).
29. Consumer Council for Water. Customer Challenge Groups. Available online: http://www.ccwater.org.uk/waterissues/pr14/ccgpr14/ (accessed on 12 April 2016).
30. Eau de Grenoble. Available online: https://www.eaudegrenoble.fr/ (accessed on 12 April 2016).
31. Organisation for Economic Co-operation and Development (OECD). *Water Resources Allocation: Shared Water, Shared Risks*; OECD Study on Water; OECD Publishing: Paris, France, 2015.
32. Guimarães-Pereira, A.; Rinaudo, J.D.; Jeffrey, P.; Blasuqes, J.; Corral-Quintana, S.A.; Courtois, N.; Funtowicz, S.; Petit, V. ICT tools to support public participation in water resources governance and planning: Experiences from the design and testing of a multi-media platform. *J. Environ. Assess. Policy Manag.* **2003**, *5*, 395–420. [CrossRef]
33. Organisation for Economic Co-operation and Development (OECD). Condition for success 1 "Good governance": The Synthesis Report of Target 1 Stakeholders' Engagement for Effective Condition for success 1 "Good governance": The Synthesis Report of Target 1 Stakeholders' Engagement for Effective Water Policy and Management. In Proceedings of the 6th World Water Forum, Marseille, France, 12–17 March 2012. Available online: http://www.worldwaterforum6.org/uploads/tx_amswwf/CS1.1__Stakeholder__s_engagement_for_effective_water_policy_and_management_Report.pdf (accessed on 12 April 2016).
34. OECD. First Meeting of the OECD Initiative on Water Governance: Key Messages, Outcomes and Next Steps. 2013. Available online: http://www.oecd.org/gov/regional-policy/OECD-WGI-1st-Meeting-highlights.pdf (accessed on 12 April 2016).
35. OECD. Second Meeting of the OECD Initiative on Water Governance: Highlights. 2013. Available online: http://www.oecd.org/gov/regional-policy/OECD-WGI-2nd-Meeting-highlights.pdf (accessed on 12 April 2016).
36. OECD. Third Meeting of the OECD Initiative on Water Governance: Highlights. 2014. Available online: http://www.oecd.org/gov/regional-policy/OECD-WGI-3rd-Meeting-highlights.pdf (accessed on 12 April 2016).
37. OECD. Fourth Meeting of the OECD Initiative on Water Governance: Highlights. 2014. Available online: http://www.oecd.org/governance/regional-policy/water-governance-initiative-meeting-4.htm (accessed on 12 April 2016).
38. OECD. Fifth Meeting of the OECD Initiative on Water Governance: Highlights. 2015. Available online: http://www.oecd.org/gov/regional-policy/OECD-WGI-5th-Meeting-Highlights.pdf (accessed on 12 April 2016).
39. OECD. Sixth Meeting of the OECD Initiative on Water Governance: Highlights. 2015. Available online: http://www.oecd.org/gov/regional-policy/water-governance-initiative-6-highlights.pdf (accessed on 12 April 2016).
40. Organisation for Economic Co-operation and Development (OECD). *Water Governance in Jordan: Overcoming the Challenges to Private Sector Participation*; OECD Studies on Water; OECD Publishing: Paris, France, 2014.
41. Organisation for Economic Co-operation and Development (OECD). *Water Governance in Tunisia: Overcoming the Challenges to Private Sector Participation*; OECD Studies on Water; OECD Publishing: Paris, France, 2014.
42. Organisation for Economic Co-operation and Development (OECD). *Water Resources Governance in Brazil*; OECD Studies on Water; OECD Publishing: Paris, France, 2015.
43. OECD. Terms of Reference of the OECD Water Governance Initiative. 2013. Available online: http://www.oecd.org/gov/regional-policy/Terms-of-Reference%20-OECD-WGI.pdf (accessed on 12 April 2016).
44. OECD. OECD Water Governance Initiative: Satisfaction Survey Result. 2015. Available online: http://www.oecd.org/gov/regional-policy/WGI-Survey-synthesis.pdf (accessed on 12 April 2016).

45. Budapest Water Summit. Available online: http://www.budapestwatersummit.hu/ (accessed on 12 April 2016).
46. United Nations. 2015 UN Water Zaragoza Conference. Available online: http://www.un.org/waterforlifedecade/waterandsustainabledevelopment2015/ (accessed on 12 April 2016).
47. OECD. Outcomes of the governance sessions at the 7th World Water Forum—12–17 April 2015, Daegu-Gyeongbuk, Korea. Available online: http://www.oecd.org/gov/regional-policy/Outcomes-governance-7thForum.pdf (accessed on 12 April 2016).
48. 7th World Water Forum, World Water Council. Ministerial Declaration of the 7th World Water Forum, 13 April 2015, Gyeongju, Republic of Korea. Available online: http://www.worldwatercouncil.org/fileadmin/world_water_council/documents/press_releases/Ministerial_Declaration_7th_World_Water_Forum_1304_Final.pdf (accessed on 12 April 2016).
49. International Water Association. The Lisbon Charter—Guiding the Public Policy and Regulation of Drinking Water Supply, Sanitation and Wastewater Management Services. 2015. Available online: http://www.iwa-network.org/downloads/1444403418-Lisbon_Regulators_Charter_SCREEN.pdf (accessed on 12 April 2016).
50. United Nations. Rio declaration on environment and development. In Proceedings of the Annex I to the Report of the United Nations Conference on Environment and Development, Rio de Janeiro, Brazil, 3–14 June 1992.
51. UN. Agenda 21: Programme of Action for Sustainable Development; Rio Declaration on Environment and Development; Statement of Forest Principles: The Final Text of Agreements Negotiated by Governments. In Proceedings of the United Nations Conference on Environment and Development, Rio de Janeiro, Brazil, 3–14 June 1992; United Nations: New York, NY, USA, 1992.
52. United Nations. *Transforming our World: The 2030 Agenda for Sustainable Development*; Resolution adopted by the General Assembly on 25 September 2015; United Nations: New York, NY, USA, 2015.

water

MDPI

Article

Community Perspective on Consultation on Urban Stormwater Management: Lessons from Brownhill Creek, South Australia

Peter Dillon [1,2,3,4,*], Ron Bellchambers [1], Wayne Meyer [1,5] and Rod Ellis [1]

1 Brownhill Creek Association, Torrens Park, SA 5062, Australia; ron.bellchambers@icloud.com (R.B.); wayne.meyer@adelaide.edu.au (W.M.); rodaellis@bigpond.com (R.E.)
2 School of Civil, Environmental and Mining Engineering, University of Adelaide, SA 5005, Australia
3 School of the Environment, Flinders University, Bedford Park, SA 5042, Australia
4 CSIRO Land and Water, Glen Osmond, SA 5064, Australia
5 School of Earth and Environmental Sciences, University of Adelaide, SA 5005, Australia
* Correspondence: pdillon500@gmail.com; Tel.: +61-419-820-927

Academic Editors: Sharon B. Megdal, Susanna Eden and Eylon Shamir
Received: 21 February 2016; Accepted: 18 April 2016; Published: 25 April 2016

Abstract: There are salutary lessons from contrasting community consultation efforts in 2011 and 2015 to develop and gain support for an urban stormwater management plan for the Brownhill Creek catchment in Adelaide, South Australia. The 2011 process was a failure in the human dimension, precipitating loss of community confidence, unnecessarily entrained thousands of hours of time of residents who initiated a community action group for environmental conservation and caused a three-year delay to decision making. By contrast, the 2015 process was vastly improved, resulted in a landslide level of support for an obvious option not previously offered, achieved the required level of flood protection, saved Aus$5 million (14%) on the previously proposed option and protected a highly valued natural environment from an unnecessary dam. This paper presents a rarely heard perspective on these community consultation processes from a participating community environmental and heritage conservation action group (the Brownhill Creek Association) that was deeply engaged in reforming the Draft Brown Hill Keswick Creek Stormwater Management Plan. This reveals that the community needs to see that all options are considered and to have access to accurate information with which to assess them. It is also necessary that the proposed plan is consistent with existing agreed plans and policies developed through public consultation. Community concerns need to be heard, acknowledged and acted upon or responded to, and the consultation process needs to be transparently fair and democratic to win community support. A major contributor to success in the second consultation was that all community action groups were invited to meetings to discuss the purpose of the consultation and the methods to be used. Feedback was subsequently received before the process commenced to show what had changed and why any suggestions concerning the consultation process were not being adopted. This openness helped to mend the distrust of the first consultation process and is recommended as an essential early step in any public consultation process.

Keywords: public consultation; community engagement; urban water management; flood protection; conflict

---

1. Introduction

Flood management interventions had been under consideration for more than 10 years for the Brown Hill Creek catchment that transects the Adelaide metropolitan area. This was due to the range of alternatives and their combinations, their nested scales, and there being quite divided opinions in the community over elements of the evolving plan. Active community consultation on specific options was undertaken in 2011 and 2015. The primary issue was that the 100-year average return interval (ARI) flood was considered to inundate up to 7000 residential and commercial properties (across two catchments), predominantly in the downstream area [1] (Brown Hill Keswick Creek Stormwater Project (BHKCSP), August 2012). In the Brownhill Creek catchment, the consultants proposed channel works downstream and a flood control dam in the foothills upstream, that would bisect the 5 km long and 100 m wide Brownhill Creek Recreation Park (established 1841). Downstream channel works were generally accepted. The upstream dam seemed an obvious solution to interstate consultants looking at contour maps, unaware of the land use and the natural heritage, recreation, habitat and ecological services the dam site currently provides. There was an orchestrated, decisive and soundly based community backlash, and, subsequently, a quite different but obvious solution has been put forward with the same flood protection, savings of more than $5 M, protecting the upstream environment from a dam and with overwhelming community support. A three year delay and a significant cost in community volunteer time and anguish, could have been avoided had some basic principles been adhered to by the planners. This paper briefly describes the institutional and geographic setting that led to the case study presented, the technical and public consultation domains, and also discusses the local political processes that influenced the outcome.

2. Institutional and Geographical Setting

The institutional arrangements for decision making on urban stormwater management in Australia are currently shared across three levels of government. This influences the way in which local government bodies cooperate or compete, and, in the case of Brownhill Creek, this had a significant impact on the resulting stormwater plan. At a lower level, dissent among community action groups that favored different options fed upwards into the democratic processes of local government. The state and commonwealth governments each avoided involvement, (aside from elected local members declaring support for the no dam option), and used local government as a filter to produce consensus decisions representative of the five councils as a whole. Heterogeneity of community support for options among councils, therefore, became a consideration in reaching a final decision.

2.1. Responsibilities and Interactions between Levels of Government

In Australia, three levels of government are involved in flood mitigation in urban areas. Local government councils have responsibility for zoning land to prevent construction on flood prone land and to provide and maintain urban stormwater drainage infrastructure. State governments have ownership of all water resources and are responsible for land planning policies to which councils should adhere. The Commonwealth (National) Government formerly played a role in informing state water policies, through the National Water Commission, but, with its abolition in 2015, its primary national roles are serving as the Bureau of Meteorology's flood warning service and providing emergency funding to flood victims.

The South Australian (SA) Government formed catchment water management boards under the Catchment Water Management Act 1995 to produce catchment water management plans that addressed surface water and groundwater resources, flooding, riparian ecosystems, water quality protection and improvement and were funded by a levy on all land owners that was collected by local government councils on behalf of the boards. The Boards were to have community participation and to invest and help coordinate activities of councils where urban catchments involved multiple councils. The Brownhill Creek catchment fell within the responsibilities of the Patawolonga Catchment Water Management Board, which undertook a flood study that in 2006 produced a map of the urban area prone to flooding in the 100-year ARI flood and made recommendations on actions to prevent flooding and reduce damage.

The Catchment Water Management Boards (CWMBs) were seen to be effective in engaging communities in management of water and the environment, so, in 2004, the SA Government extended the concept by passing the Natural Resources Management Act which created eight regional Natural Resources Management Boards to cover SA. These were larger geographically and had a wider scope of functions including vegetation, wildlife, land and bushfire planning and subsumed the CWMBs. Brownhill Creek now fell under the responsibility of the Adelaide and Mount Lofty Ranges Natural Resources Management Board (AMLRNRMB). Levies were not increased so the amount of effort on water management declined significantly. Boards became known as places where water management plans were devised but under reduced funding were largely moribund unless executed by other parties, such as local government or through Commonwealth grants. State Government responded in part by establishing the Stormwater Management Authority (SMA) under the Local Government (Stormwater Management) Amendment Act 2007 with the support of the Local Government Association. By 2013, the two parties had signed a new nonbinding agreement over the role and scope of the Stormwater Management Authority, which appears to focus only on flood mitigation. It also commissioned a new study of flood management options for Brownhill and the adjacent Keswick Creek with costs to be shared by the five councils within the catchments of these streams.

The Brown Hill Keswick Creek Project Office was established and supported by councils and the SMA to develop a flood mitigation plan. The process of developing the plan has been carried out in accordance with the Stormwater Management Planning Guidelines of the SMA. These expect a level of community consultation which depends on factors such as the size of the catchment, the nature and complexity of stormwater management issues, the nature of the adopted management strategies and the impact of the strategies on the local community. The two formal community consultation processes (2011 and 2015) were carried out by consultants, hired by the Project Office, in accordance with relevant policies of the five project councils whose jurisdictions covered the catchment. The councils determined which options were to be considered and go to public consultation. Proposals for consultation were developed only to a concept level of detail. Subject to community support, the proposals would then form a stormwater management plan (SMP) for which approval would be sought from the Stormwater Management Authority (SMA). At the implementation stage of the approved SMP, further consultation would take place which would involve collaboration with directly affected stakeholders, including owners and residents of properties traversed by the watercourse in which works are proposed.

If councils failed to reach agreement on flood mitigation works, the SMA had the authority to direct works to be undertaken at the councils' expense. It was also implied that if councils failed to reach agreement, the state government may be unwilling to contribute to works from its own budget.

2.2. Brownhill Creek Geography and Flood Risk

The Adelaide Metropolitan Area is on a coastal plain between the Mt Lofty Ranges and the Gulf of St Vincent. It is traversed by three rivers, each of which has a water supply dam upstream that assists in mitigating floods from the upper rural catchment. Each river also has a linear park along its course through the urban area to act as a major recreational amenity, support natural habitat and to serve as a flood drainage channel to mitigate flooding from the urban catchment. The city also has 25 creeks and drains in the urban area but only four are considered to have a high risk of major floods, Brownhill Creek, Cobbler Creek, Dry Creek and Pedlar Creek. Of these, Brownhill Creek presents the most significant risk, as there is no flood control dam in its upper catchment, and councils have allowed residential development over its flood plain, and even over the channel itself. As recently as 2015, several meters depth of fill has been spread on the flood plain close to the edge of the creek channel and houses built with floor levels above flood level, substantially reducing the flow capacity of the river floodplain. Brownhill Creek and its smaller neighboring Keswick Creek, which meet near Adelaide airport, are largely canalized and used as the main stormwater drains by these councils, severely compromising the ecological value of the creek downstream. Where the natural channel has been retained it has been scoured by increasing peak flows due to urbanization. The urban catchment area has been marked for urban consolidation under the metropolitan area development plan, but the rural catchment is part of the hills face zone and new development there is prohibited, reflecting the value placed on the natural bushland visible from much of the city. It was clear that a stormwater management plan was needed, and that all councils would need to take responsibility to prevent excessive flooding in the downstream council areas, shown for a composite map of 100-year average return interval floods (Figure 1). Brownhill Creek flows from the south east through the Cities of Mitcham and Unley to the City of West Torrens in the west before discharging into the Gulf of St Vincent to the west of Figure 1. Keswick Creek flows from City of Unley in the east and is joined by a northern tributary passing through the Cities of Burnside and Adelaide before connecting with Brownhill Creek just east of the airport in the City of West Torrens.

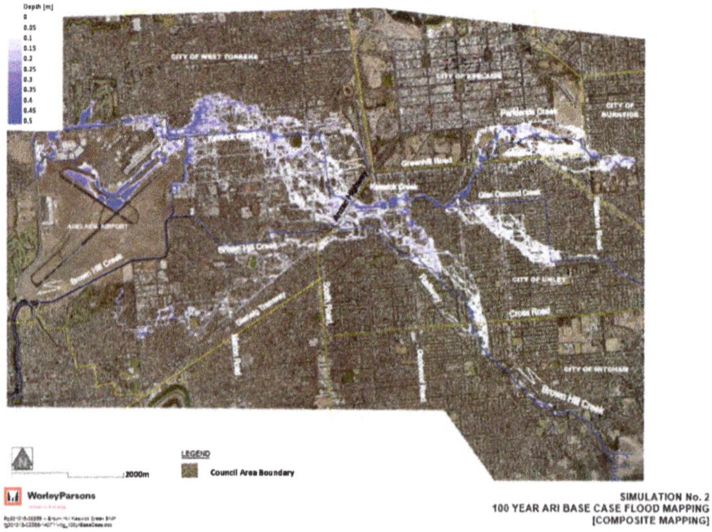

Figure 1. Composite flood map for 100-year average return interval floods of 1.5, 6 and 36 h duration events in the absence of a stormwater management plan (from BHKC, 2016 [2]). Flood depths for shown overflows from Brownhill and Keswick Creeks range from the shallowest 0 to 5cm (shown as white) through to the deepest at 45 to 50cm (shown as the darkest blue). Parts of unshaded areas may also be flooded to shallow depths by lack of capacity in local drainage systems that discharge to these creeks.

3. Brownhill Creek Consultation Case Study—Materials and Methods

3.1. First Consultation—Hydrological Aspects

The 100-year average return interval (ARI) flood was considered based on then-available Bureau of Meteorology (BOM) intensity-frequency-duration curves for 1.5, 6 and 36 h duration events and composite damage calculated [1]. Modelling showed that a flood mitigation dam would have no impact on flooding from short duration storm events that could produce localized flash flooding and stream bank overflow in the urban area but could reduce damage for the longer duration events. The evaluation averaged the damage from these three different duration events, as though they represented the full spectrum of 100-year ARI events. The expected downstream flooding in all events was found to be highly dependent on modelled channel capacity. The channel capacity data used in the initial flood modelling was 20 years old, and the channel had since become choked in places with exotic trees and debris. Furthermore, the BOM were in the process of releasing revised 100-year ARI rainfall data for use in all Australian flood studies.

3.2. First Consultation—Human Dimensions

The first public consultation program from 31 October to 12 December 2011 was to inform stakeholders and the community on the draft plan, and receive, collate and summarize feedback [3]. Materials were prepared to summarise the draft plan, including a summary report, brochure and fact sheets. A feedback form was mailed directly to 26,539 property owners and occupiers across the catchment, skewed towards the lower catchment where most potentially flood affected properties were located. Three open days were held in different council areas within the catchment and attracted 160 attendees. Although the draft plan relied on channel capacity expansion as the primary flood mitigating measure downstream, the only option given for flood management in the upstream part of Brownhill Creek was a dam. Effectively, the choice presented to residents was to have a dam or bear the risk of being flooded.

There was widespread community outrage in the City of Mitcham concerning the proposed dam in the Brownhill Creek Recreation Park and lack of consideration of alternatives by those responsible for the draft plan made evident in the consultation process. A group of concerned residents formed the No Dam in Brownhill Creek Community Action Group that subsequently became a subcommittee of Brownhill Creek Association. This group also pointed out that the environmental and social costs and benefits had not been included in the analysis, that the plan was in breach of the State Government's own plans for the Adelaide metropolitan area, and that it failed to address the Natural Resources Management Board's stated goals for improved urban coastal water. A petition was established, signs erected, brochures printed and distributed, a web site established to house factual information and to serve as a blog, a flagged rope slung across the creek at the elevation of the proposed dam crest to be a visual sign of the scale and impact of a dam at the project's preferred site, heritage status determined for significant trees at the proposed dam site, and consultations held that led to messages of support from Kaurna (custodial aboriginal) representatives, conservation and heritage groups, and the local council and local members of State and Commonwealth parliaments. The City of Mitcham requested that supplementary technical studies be undertaken on the basis of revised information and lack of evaluation of alternatives to a dam such as enlarging channel capacity in the upstream urban area or use of culverts to bypass constricted sections in this area.

3.3. Second Consultation—Hydrological Aspects

Following community dissent during the first public consultation phase, the plan was divided into two parts. The Keswick Creek Catchment and Lower Brownhill Creek Catchment work plan (about 80% of total budgeted costs) was agreed to by all five councils who were the stakeholders in the project, but, for the Upper Brownhill Creek Catchment, new investigations were commissioned to:

- use recently revised Bureau of Meteorology flood frequency-duration-intensity curves,
- re-survey the creek channel hydraulic capacity and status,
- evaluate culvert options to allow bypass flows in constricted sections,
- estimate the size and cost of expanding channel capacity (including land acquisition and easement costs) and also in making good overgrown sections of channel,
- consider locating the dam further upstream in a tributary, Ellisons Gully,
- remodel flood hydrographs and damage estimates for the three design storms for all Upper Creek intervention options assuming that the Lower Creek interventions were in place, and
- present the results to the constituent Councils and catchment community in a report and undertake a public consultation survey on a revised plan in Upper Brownhill Creek.

The councils, having recognised community opposition to a dam in the upper reaches of Brown Hill Creek, formed a preference to pursue a feasible and whole of catchment community supported "no dam" solution [2]. The subsequent revised plan [4] (BHKC, September 2014), which still did not evaluate water quality nor environmental externalities, found that a channel upgrade involving creek rehabilitation works and limited creek widening (not evaluated in the previous study), had the same flood mitigation benefits and was more than $5 M cheaper than the dam option. Culvert options in this area were found to be more expensive and technically difficult. Figure 2 is a map of the catchment showing key options.

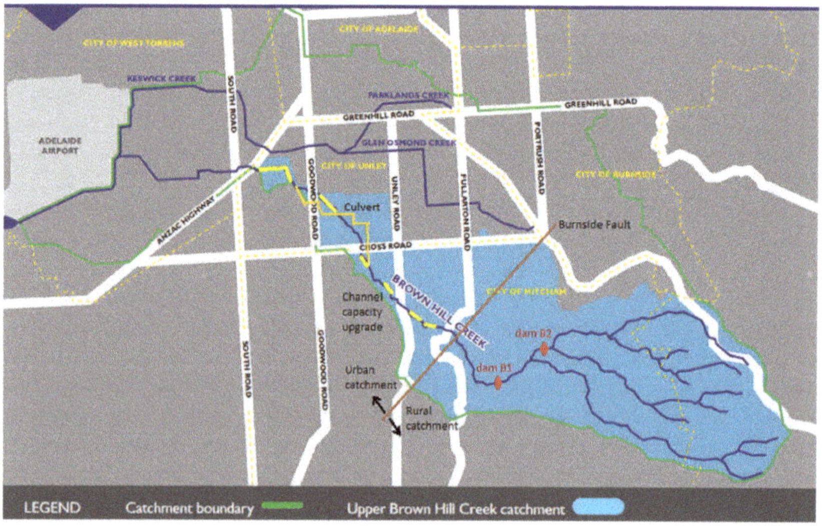

Figure 2. Map of Brownhill and Keswick Creek catchments showing council areas and locations of key options (adapted from BHKC 2016 [2]).

3.4. *Second Consultation—Human Dimensions*

The second community consultation process (May–June 2015) built on the previous consultation process, with direct contact with potentially affected property owners and ongoing liaison with community groups. However, unlike the first process, all community groups that favored or opposed various options were interviewed and invited to comment on the consultation process before it was formalized and undertaken. From a community perspective, this helped dismiss distrust and engender confidence in the impartiality of the process.

This public consultation phase included surveys direct mailed to creek-side owners and residents, with follow up letters inviting further contact. All other residents wanting to respond needed to contact the project office to request a survey form. Importantly, all community groups that favoured or opposed various options were invited to comment on the consultation process before it was formalised. Open days were also held in the two closest council areas, Mitcham and Unley, to allow residents to ask questions of technical staff working on the project. Information about forms and open days was placed in advertisements in a local newspaper that is delivered free to all households in the catchment. Consultants' reports were made available as reference material via the web and at council offices. Creek side land owners were given written assurance that they would be consulted on the specific design options for achieving the required channel capacity on their land, and they would have the option of council maintaining that treatment at no cost to them or to maintain it themselves.

Contrasts between the two consultation approaches included the second process was much narrower in scope and more targeted, good dialogue with community action groups before the second survey to discuss and seek input to the process and the feedback form, greater control over feedback forms by requiring them to be issued in paper form only from the BHKC Project Office, and reporting separately on responses on the forms mailed to creek owners and forms received from the general community upon individual request. Overall, the improvement in process rigor and in openness with all interested members of the community helped to achieve results for the second survey that were unambiguous, and all members of the community could see that their main concerns had been recorded and were addressed impartially.

The different geographic scale of the two public consultation processes and contrasting proximity of impact of options on different segments of the community would suggest that a range of tools be used within a broader framework, as advocated by Dean *et al.* [5] and synthesized by Lacroix and Megdal [6] as a Stakeholder Engagement Wheel. Lacroix and Megdal also identified the need for additional forms of engagement with stakeholders who have expertise in the subject based on their professional or personal interest, and the meetings with action groups in the second consultation met those needs. On the IAP2 Spectrum of Public Participation (IAP2 2015) [7], the first consultation was at the level of "consult" (to obtain feedback) but had deficiencies in the objective of listening to and acknowledging concerns. While the second consultation remained at the "consult" level in seeking feedback on the 2015 report, it was at the next level "involve" in working with key community groups and potentially affected property owners to ensure concerns were considered and understood within the process, and was successful in ensuring community aspirations were directly reflected in the report. Further along the spectrum are "collaborate" and "empower" and collaboration is expected during the detailed design of channel capacity restoration, when creek-side householders will be given the opportunity for input into how the required objective can be achieved (see Table 1). Note that the opportunity for public engagement on non-structural measures, such as a flood preparedness program, or flood insurance was not taken up. Furthermore, other elements reasonably expected in a stormwater management plan, including water quality improvement, stormwater harvesting for beneficial uses, urban amenity and green space were also not pursued, as the councils had previously agreed to limit the scope of the plan.

Table 1. The consultation programs for the Stormwater Plan development and implementation as determined by BHKC Project categorized in relation to the International Association for Public Participation (IAP2) spectrum of public engagement [8].

Inform	Consult	Involve	Collaborate	Empower
Consultation on Plan 2011	Options specified by BHKC project and feedback sought from community.			
Consultation on Plan 2015	Options specified by BHKC project and feedback sought from community.	Action groups given opportunity to comment on consultation process.		
Implementation of Plan		(not currently considered: Opportunities for participation in non-structural measures (e.g., flood preparedness programs), and measures to meet objectives other than flood mitigation.)	Channel capacity is specified by BHKC and designs to achieve this are developed and agreed with each riparian land holder, under either an easement or an agreement.	(not currently considered: However there are opportunities for household stormwater maangement options beyond council requirements.)

During this second public consultation phase, the Brownhill Creek Association [8]:

- updated its web site, Facebook and Twitter sites with the new information;
- letterboxed two council areas encouraging community members to respond;
- manned a booth at a shopping centre to spread the message and extend the petition;
- sent letters to the editors of local newspapers and were represented on radio and television;
- placed advertisements in local newspapers and sent out press releases;

- gained further statements of support from relevant credentialed aboriginal heritage and environmental conservation organisations;
- arranged expert evaluation of stone pines at the primary dam site which revealed they were Australia's largest specimens and were subsequently recorded on a national register of trees of significance;
- publicly displayed numerous banners and signs;
- made deputations to councils involved and lobbied for support;
- addressed community groups and visitors to the Brownhill Creek Recreation Park;
- conducted a heritage walk along the upper creek through Ellisons Gully;
- produced and circulated a children's book about yabbies in the creek; and
- produced two YouTube videos on children playing in Brownhill Creek.

4. Results

4.1. Outcome of First Community Consultation

A total of 2278 feedback forms were received, 8% of all those distributed [3] (Table 2). Response rates ranged from 5% in West Torrens (downstream) to 52% in Mitcham (upstream). The Consultation report [3] found strong support for all elements of the plan (at 71% overall) except for the dam in Brownhill Creek.

Table 2. Response rate from different community sectors (adapted from [3] and [9]).

Connunity Sector	Number of Forms Issued	Number of Responses Received	Percentage Responding	Distribution Method and Follow-up
1st Community Consultation, 31 October 2011–12 December 2011				
Owners of properties potentially affected by floods, creek-side owners, state govt. representatives and others determined by BHKC Project Office (criteria not specified) by council area . . .	26,539	2,278	8%	Direct mail to owner. Also community newspaper notice to advise that Feedback forms available from council web sites, from 3 open days in community centres. There was no differentiation in reporting by source of feedback form.
Mitcham	1,055	535	52%	
Unley	6,157	638	10%	
Burnside	2,290	90	4%	
Adelaide	176	73	42%	
West Torrens	16,861	942	5%	
Others		4,010	na	An unsolicited petition contained 4010 signatures of residents opposed to a dam in Brownhill Creek was submitted by the No Dam in Brownhill Creek Action Group to the City of Mitcham and forwarded into the consultation process.
2nd Community Consultation, 13 May 2015–23 June 2015				
Creek side property owners in Mitcham	90	49	54%	Direct mail to owner and followed up by a letter and phonecall
Creek side property owners in Unley	126	39	31%	Direct mail to owner and followed up by a letter and phonecall
All other residents	1,074	730	68%	Community newspaper notice to advise that forms are available on request. No follow up.
Total	1,290	818	63%	

The report stated that the flood control dam at Brownhill Creek Recreation Park was the least supported component of the Draft Plan overall with 33% of respondents opposed to a dam and 57% supporting a dam (Table 2). However, there was a large variation in level of support among council areas, ranging from 7% opposed downstream in West Torrens where residents had been led to believe that the dam was essential to prevent flooding there, to 74% and 76% opposed in Mitcham and Adelaide in closer proximity to the recreation park (Table 3). Hence, the extreme bias in distribution of mailed out feedback forms (Table 2) significantly inflated the reported aggregated support for the dam. Although reported, no weight was given to the petition against the dam signed by more than 4000 at that time (and subsequently closed in May 2015 with 11,617 signatures).

Table 3. Respondents support in the first public consultation program for no dam or dam in upper Brownhill Creek (adapted from [3]).

Council Area	Number of Respondents	% Support No Dam	% Support Dam	% No Opinion
Mitcham	535	74%	22%	4%
Unley	638	30%	61%	9%
Burnside	90	37%	11%	52%
Adelaide	73	76%	6%	18%
West Torrens	942	7%	82%	11%
Total	2,278			
Weighted Mean	100%	33%	57%	10%
Unweighted Mean		45%	36%	19%
Unweighted Standard Deviation		27%	30%	17%

The first public consultation report [3] summarized some of the objections raised at meetings with Friends of Brownhill Creek and residents living in close proximity to the proposed dam. A fuller description of these concerns not reported include impacts on stream ecology and water quality during and after dam construction, the extent of loss of recreational amenity, and specific impacts on local residents including noise, dust, quarrying and heavy haulage equipment on a small road which were an issue for local residents.

The plan also failed to meet the NRMB's stormwater management planning criteria [10], in addressing flooding to the exclusion of all other environmental and water resources considerations as specified in various SA Government plans (AMLRNRMB (2008)-integrated NRM Plan for Adelaide, now updated 2014-15 to 2023-24 [11] and Mt Lofty Ranges, Dept. Planning and Local Government (2010)-urban plan for Adelaide [12], Department for Water (2011)-stormwater management strategy [13] and Environment Protection Authority (2011)-coastal water quality improvement [14]. A full report of departures from these plans is found in public consultation submissions by BCA [15] but no mention is made in any of the BHKC reports.

The outcome of this consultation was that the Brown Hill Keswick Creek Project recommended and the five Councils agreed to proceed with the lower catchment flood management measures, and to undertake additional investigations to determine options for the upper catchment and undertake a separate community consultation based on recommendations arising from these studies.

4.2. Outcome of Second Community Consultation

Following the consultation process, a report was provided by the consultant [9], which revealed the breadth of views on the options for the upper creek management plan. Among Creekside land owners, there was a low rate of response, particularly in Unley where the most outspoken critics of the recommended option of channel rehabilitation (no dam) resided (Table 2). In spite of average easements being 240 sq. m. per property, 69% of residents in Unley and 46% in Mitcham had ignored appeals by the Project Office to respond to the feedback form. As outlined in the report, creek-side owners and residents were contacted by letter on several occasions, invited to attend meetings and encouraged to respond to the consultation process questionnaire because it was reasonable to assume that the proposed flood mitigation works were of more relevance to them than the rest of the community. It is possible that assurance by the project office that consultation would occur with individual land-owners on channel works was sufficient to satisfy non-responding land-owners. Photographs of example channel modifications that were provided in public consultation documents were aesthetically attractive and would be safer than existing steep and eroding banks and may have alleviated concerns.

In contrast, 89% of all responses came from members of the public who were not creek-side residents, and felt strongly enough on the issue that they requested and completed forms. The Brownhill Creek Association had also appealed to residents through a letterbox drop to all Mitcham residents and a stall in the Mitcham shopping centre to encourage them to request forms. Many who told stall volunteers that they would request forms were parents of young children, and middle-aged and elderly people who walked in the park regularly or who had in earlier years. They frequently mentioned their desire to preserve its natural beauty, particularly with urban consolidation and the park was so close to the city but hidden from it.

Results were convincing: 50% of respondents from creek-side properties who expressed support of any option supported the no dam option (Figure 3). The next highest level of support by creek-side owners was 26% in favor of a dam in Ellisons Gully. Thus, it is clear even among creek-side property owners alone, the strongest preference was for the channel rehabilitation with no dam. For all other respondents, the support was 90% for channel rehabilitation with no dam and 6% for a dam in Ellisons Gully. In aggregate, the no dam option had 85% support. In addition to the feedback form, individuals and organisations were invited to make written submissions to the Project Office concerning the Stormwater Management Plan. In total 76 submissions were received with 80% supporting the no dam option (Table 4). With this very clear outcome in favour of the no dam option, the five councils were informed and requested to support the proposal by 30 September 2015.

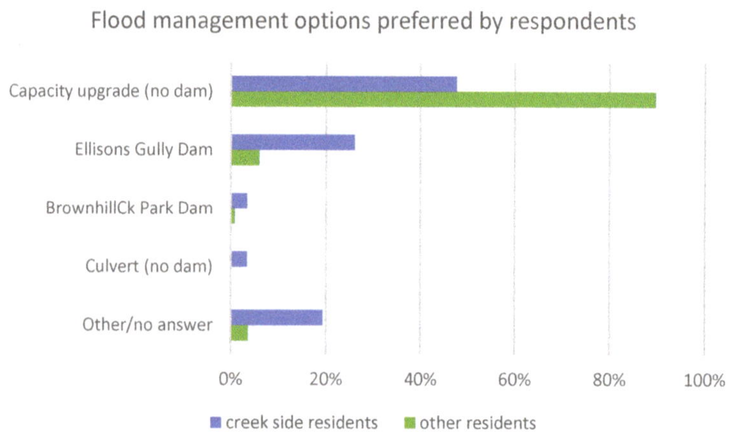

Figure 3. Preferences for flood management options from respondents in community consultation 2015 for creek-side residents and for other residents (adapted from [9]).

Four councils passed the motion to proceed with the no dam option including appropriate consultation with affected property owners concerning treatments to achieve the channel capacity at their location, and giving the option for easements (councils to maintain) or agreements (individual to maintain). In all cases, the motion was passed unanimously or with a significant majority. Unley Council could not reach agreement and called a special meeting to address the issue on 29 September. After several motions were lost, the original motion succeeded with some minor amendments, including that account be taken of initiatives to improve water quality and harvesting. The Project Office reported the agreement of the five councils on the preferred option to the Stormwater Management Authority for finalization of the plan (based on the no dam option) and its approval by the State Government. The plan was finalized in March 2016 (BHKC 2016) and submitted to SMA.

Table 4. Written submissions from the public during the second public consultation program 2015 (adapted from [9]).

Written Submissions	Number of Written Submissions	Supporting Option D (No Dam)	Opposing Option D (No Dam)	Support or Opposition Not Stated
Mitcham residents	26	24	0	2
Unley residents	13	7	5	1
Other residents	21	21	0	0
Interest groups	13	6	7	0
Businesses	2	2	0	0
Members of South Australian parliament	1	1	0	0
Total	76	61	12	3
% of submissions	100%	80%	16%	4%

5. Discussion

5.1. State and Local Government Interaction

Aside from covering the cost of an updated channel capacity assessment, there was minimal conspicuous participation from the NRMB. NRMB had expressed a desire not to interfere in local matters that were the subject of public consultation. However, the Stormwater Management Plan that was produced failed the planning criteria set within the NRMB's own catchment plans by focussing only on flood mitigation (and within this using only structural means). A second reason why NRMB or its parent, Department of Environment Water and Natural Resources, may have been expected to comment was on the potential loss of native vegetation, biodiversity and recreational amenity through the proposed construction of a dam midway along the 5 km long, 100 m wide Brownhill Creek Recreation Park. It was perceived by the community that the NRMB's lack of leadership in executing its own responsibilities was more related to deferring or potentially avoiding costs for state government rather than an unwillingness to play an active role while solutions for flood mitigation were being openly debated. State government also failed to intervene when one or more councils prevented the channel widening option from being included in the first public consultation exercise. This was in denial of the National Water Initiative to which all States are signatories that explicitly demands all water management options to be on the table and considered on their merits.

The councils also had a desire to reduce or defer costs. Unfortunately, councils had not been maintaining the channel capacity of the creek, nor had the NRMB required creek-side householders to do so. Hence, in the period since the CWMB flood study, channel capacity, particularly in the urban upstream area had in some places been halved by woody weeds including ash, elm and willow. Remarkably, none of the councils recognized the flood plain to be part of the stream, and there are many examples of fill being brought in and new houses established in the flood plain right up to and even over the main channel of the creek. Councils could see that should these need to be addressed they could accrue some liability for these blockages. No council was actively advocating the purchase of creek-side land as it came onto the market, nor a plan to prevent further creek-side development and fairly compensate owners for compulsory acquisition whenever the incumbent sought to change ownership. There are excellent local examples of linear parks along streams in urban areas, such as the Torrens and Little Para Rivers, concurrently serving recreation, wildlife habitat, urban green space, water quality improvement and flood mitigation needs while also providing increased opportunity for groundwater recharge and beneficial reuse. A larger scale example of flood management is in the Murray-Darling Basin where a wide diversity of water issues are addressed in a constraints-management-strategy [16], which enables creative use of water storage infrastructure to encourage minor flooding for the health of riverine and riparian ecosystems and mitigate against damage-causing major floods in a predominantly rural catchment.

5.2. Interactions among Councils

Initially, in response to the draft 2011 Stormwater Management Plan, the five councils were divided with four voting in favor of the plan presented. The most upstream council, the City of Mitcham dissented, objecting to a dam in Brownhill Creek and requesting that other options be explored. The Mayor of the most downstream council, the City of West Torrens, the council with the largest number of houses within the 100-year flood-prone area, was quoted in community newspapers as saying that wealthy (liberal-voting) Mitcham residents were holding a small part of the environment to be more precious than thousands of his middle-class (labor-voting) constituency who would inevitably be flooded without a dam. After the councils agreed to split the plan into upper and lower creek plans, the lower creek plan was unanimously supported by councils, and West Torrens made no further comment on upstream options that led to cost-effective achievement of satisfactory flood mitigation.

Conflict then ensued between Cities of Mitcham and its neighbor Unley, as to which of three options to pursue. These were:

- a dam in either the Brownhill Creek Recreation Park or on a tributary in Ellisons Gully further upstream;
- diversion culverts and pipes around constricted sections of creek; or
- widening of the creek channel (upgrading capacity) at critical sections.

In addition, for all options there was a requirement to also rehabilitate the capacity of the natural channel where it had been overgrown by exotic trees and reduced by unauthorized structures, at the expense of the Project. Creekside land owners were to be given the option of having an easement on their property title which would give council the right and responsibility for future maintenance of the channel capacity, or committing to maintain capacity at their own expense.

5.3. Local Action Groups

Three local action groups seized on different positions with respect to each of these options (Table 5):

- No dams action group of Brownhill Creek Association (BCA)—to oppose dams at both prospective sites, aimed at improving environmental and community outcomes at least cost;
- Save Our Streets Community Action Group (SOSCAG)—to protect street trees whose health and survival could be threatened by construction of major culverts under streets in Unley and Mitcham;
- Save our Creek Environs Trees (SOCKET)—to protect rights and autonomy of creek-side land owners or maximize compensation if easements are imposed.

Table 5. Membership of community action groups originated from different geographic locations that influenced their motivations. All groups attempted to influence the community and the Project Office with their preferred options.

Action Group	Geographic Spread	Primary Motivation	Other Motivations	Preferred Option
BCA	Dispersed widely but mostly in Mitcham where creek ecosystem values highest and nearest to recreation park where dam had been proposed	Environmental and heritage protection in recreation park and natural areas, Preserve recreational amenity	Improve creek environment in urban area, minimizing costs to the community, establish a green corridor, stream water quality improvement, enhanced stormwater harvesting.	D (selective channel widening with no dam)
SOSCAG	Predominantly residents of Unley streets where culverts had been considered an option	Protect street trees from culvert construction, avoid inconvenience of protracted major earthworks on residential streets	Preservation of existing trees in urban creek line, minimizing costs to the community.	Various including D (selective channel widening with no dam) and dam B2 (location in Ellisons Gully)

Table 5. *Cont.*

Action Group	Geographic Spread	Primary Motivation	Other Motivations	Preferred Option
SOCKET	Minority of residents owning creek-side properties in Mitcham and Unley	Retain control of their own land, fear of loss of land value, inconvenience and risks of earth works adjacent family home.	Preservation of existing trees in urban creek line, maximizing compensation to affected streamside landholders.	B2 dam (location in Ellisons Gully)

(BCA—Brownhill Creek Association; SOSCAG- Save Our Streets Community Action Group; SOCKET—Save our Creek Environs Trees)

These three groups continued public campaigns, with the latter two primarily with councils of the Cities of Mitcham and Unley, with creek-side residents and in community newspapers. SOCKET threatened legal action as it considered that the $3.2 M set aside in the budget for easements to be too low by an order of magnitude. If their desired allocation for easements was used the dam would then become the cheapest option. It was noted that the number of action group titles increased as the public consultation program progressed, in part due to the impression that each perceived separate group allowed a separate voice to be heard and could increase influence on the outcome.

6. Corollary

The unanimous agreement of the five councils to the no dam option has paved the way for a detailed flood management plan and cost sharing arrangements between the State and local government. The attention of the Brownhill Creek Association has refocused on restoring native riparian vegetation, removal of weeds, and expanding bushwalking and heritage trails connected to the walks already present through the Brownhill Creek Recreation Park. It is also actively facilitating environmental education opportunities and creek amenity enhancements and biofiltration of stormwater associated with a carpark expansion at a shopping centre built over the creek. Further work is planned on informing planning regulations for creek-side development, informing environmental flow requirements, informing improved stormwater management policies and practices including funding mechanisms for water quality improvement and improved amenity and stormwater harvesting via managed aquifer recharge, and exploring non-structural measures for flood damage mitigation, through for example the FloodSafe Program [17] for flood preparedness, while awaiting completion of structural measures. In December 2015, two members of BCA, David Wagner and Ron Bellchambers, received awards from the SA Conservation Council recognizing their role in environmental protection and Ron Bellchambers was awarded the Mitcham Citizen of the Year in January 2016.

7. Conclusions

Key messages from the Brownhill Creek Association to those engaging with the community on new water management projects are:

- Engage with the community early while draft plans and options are flexible and expandable, well before reaching a formal public consultation phase.
- Do your homework and ensure that data, which could change priorities of options, is correct before engaging in modelling work. Do sensitivity analyses on all assumptions that have potential to change the preferred option.
- Consider the full range of costs and benefits for each option and quantify or at least rank the impact of options on environmental and social externalities.
- Put all options on the table. Do not omit obvious solutions because of fear that they will be unpalatable with influential parties.

- Where options depart from existing soundly based policies and strategies that have developed with effective consultation, there should be a clear rationale for the departure and the consequences for the achievement of the objectives of those policies.
- Ensure those that may be affected have a clear understanding of how and how much they may be affected, and how this may be mitigated.
- Do not dismiss people's opinions because they have a different perspective from those of the project proponents or consultants. There are many dimensions to problems and solutions and a consultant cannot be expected to know them all, but can reasonably be expected to listen and broaden their perspective, and improve investigations and recommendations.
- The costs to a community to campaign for the ultimately most popular solution in this case amounted to more than $200,000 in in-kind labor as well as more than $10,000 in communications costs. Much of this could have been avoided by improved engagement with the community at an early stage. The costs to the community are generally neglected in decision making and achieving a suitable standard of rigor in investigations, and consulting the community in framing investigations could save on gross expenditure and speed up approval processes.

Participation in a community action group has revealed to the authors the anxiety and powerlessness experienced when factual information provided to officials is disregarded or at least not acknowledged because it is perceived to be selected to support a pre-determined position. The factual material posted on the action group's web site was written or audited by two water resources professionals of full professor standing and reviewed by peers with standing in the specific disciplines involved. We attempted to present balanced and accurate information. This contrasted with the original Draft Plan based on flawed investigations, and the biased design of the first community consultation process. The group observed that a much higher standard of technical work was achieved for the second consultation, and the second community consultation process was not only better designed and executed, but very importantly, used an inclusive process that enabled trust to be restored.

There is still room for improvement in a plan that surveys a 100-year time horizon for flooding, but does not yet also consider on the same time scale a plan for urban amenity, green space, recreation, and gives only token attention to water quality improvement, and stormwater harvesting via managed aquifer recharge in a consolidating urban area [2]. It was beyond the ability of our action group to widen the scope of the stormwater management plan beyond structural measures for flood mitigation, but ultimately this is our long term goal.

Acknowledgments: The authors are grateful to Natalie Fuller and Associates for reviewing the draft manuscript, and to two anonymous reviewers of MDPI Water Journal for constructive comments on the manuscript.

Author Contributions: Peter Dillon produced several articles to identify benefits of alternatives to the dam, suggested documentation of the consultation process, produced a draft of the paper and undertook revisions. Ron Bellchambers devoted himself to liaising with and making deputations to councils and many organizations in securing support for assessment of alternative options and in running the web site, petition and public communications reported. Wayne Meyer chairs BHA is a Professor of Natural Resources Management and helped construct a cogent case and was the main media spokesperson for the later stages of the campaign. Rod Ellis organized community meetings at Mitcham Village Uniting Church, liaised with Friends of Brownhill Creek, and with Wayne and Ron contributed ideas and reviews of drafts.

Conflicts of Interest: The authors are members of Brownhill Creek Association, and have summarized this information to give a community perspective while maintaining scientific impartiality to the best of our knowledge. A draft manuscript was provided to the public engagement consultant for review and revised to account for their comments. This work has not been sponsored or funded.

Abbreviations

The following abbreviations are used in this manuscript:

AMLRNRMB	Adelaide and Mount Lofty Ranges Natural Resources Management Board
BCA	Brownhill Creek Association

BHKCSP	Brown Hill Keswick Creek Stormwater Project
BOM	Bureau of Meteorology

References

1. Brown Hill Keswick Creek Stormwater Project. 2012. Available online: http://bhkcstormwater.com.au/ (accessed on 19 April 2016).
2. Brown Hill Keswick Creek Stormwater Project. Brown Hill Keswick Creek Catchment Stormwater Management Plan 2016. 2016. Available online: http://bhkcstormwater.com.au/ (accessed on 19 April 2016).
3. URPS in Association with Natalie Fuller and Associates Pty Ltd; Harlen Graphics. Brown Hill Keswick Creek Draft Stormwater Management Plan: Community Consultation Report; Prepared for the Brown Hill Keswick Creek Stormwater Project; 2012. http://bhkcstormwater.com.au/ (accessed on 19 April 2016).
4. Brown Hill Keswick Creek Stormwater Project. 2014. Available online: http://bhkcstormwater.com.au/ (accessed on 19 April 2016).
5. Dean, A.; Fielding, K.; Newton, F.; Ross, H. *Community Engagement in the Water Sector: An Outcome-Focused Review of Different Engagement Approaches*; Report Project A2.3; Cooperative Research Centre for Water Sensitive Cities: Clayton, Australia, 2016.
6. Lacroix, K.M.; Megdal, S.B. Explore, Synthesize, and Repeat: Unraveling Complex Water Management Issues through the Stakeholder Engagement Wheel. *Water* **2016**, *8*. [CrossRef]
7. IAP2 (The International Association for Public Participation). The IAP2 Public Participation Spectrum. 2015. Available online: http://www.iap2.org.au/resources/public-participation-spectrum (accessed on 19 April 2016).
8. Brownhill Creek Association. 2015. Available online: www.brownhillcreek.org (accessed on 19 April 2016).
9. Natalie Fuller; Associates Pty Ltd in Partnership with URPS. Consultation findings on the Brown Hill Keswick Creek Stormwater Project: Part B Report. Prepared for the Brown Hill Keswick Creek Stormwater Project. 2015. Available online: http://bhkcstormwater.com.au/ (accessed on 19 April 2016).
10. South Australia; Natural Resources Management Council. State Natural Resources Management Plan 2012-17. Available online: http://www.environment.sa.gov.au/about-us/our-plans (accessed on 19 April 2016).
11. South Australia; Adelaide and Mount Lofty Ranges Natural Resources Management Board. Creating a Sustainable Future: Adelaide and Mount Lofty Ranges Natural Resources Management Plan Volume 1 Strategic Plan 2014-15 to 2023-24. Available online: http://www.naturalresources.sa.gov.au/adelaidemtloftyranges/about-us/our-regions-plan (accessed on 19 April 2016).
12. South Australia; Department of Planning and Local Government. The 30-Year Plan for Greater Adelaide. 2010. Available online: http://www.dpti.sa.gov.au/planning/30_year_plan (accessed on 19 April 2016).
13. South Australia; Department for Water. Stormwater Strategy—The Future of Stormwater Management. 2011. Available online: http://www.environment.sa.gov.au/managing-natural-resources/water-use/water-resources/stormwater (accessed on 19 April 2016).
14. South Australia; Environment Protection Authority. Adelaide Coastal Water Quality Improvement Plan. 2011. Available online: www.epa.sa.gov.au/files/477449_acwqip_final.pdf (accessed on 19 April 2016).
15. Brownhill Creek Association. A stormwater plan that fails the planning criteria. 2012. Available online: http://brownhillck.org/submissions (accessed on 19 April 2016).
16. Murray-Darling Basin Authority. Constraints Management Strategy 2013 to 2024. Commonwealth of Australia, 2013. Available online: http://www.mdba.gov.au/publications/mdba-reports/constraints-management-strategy (accessed on 19 April 2016).
17. FloodSafe: Community FloodSafe Program in South Australia-Increasing Resilience in Urban Communities Facing Flood Risk. 2009. Available online: http://www.ses.sa.gov.au/site/community_safety/floodsafe.jsp (accessed on 19 April 2016).

water

MDPI

Article

Conflict Management in Participatory Approaches to Water Management: A Case Study of Lake Ontario and the St. Lawrence River Regulation

Alison Furber *, Wietske Medema, Jan Adamowski, Murray Clamen and Meetu Vijay

Bioresource Engineering, McGill University, 845 Rue Sherbrooke O, Montréal, QC H3A 0G4, Canada; wietske.medema@mcgill.ca (W.M.); jan.adamowski@mcgill.ca (J.A.); murrayclamen@rogers.com (M.C.); meetuvijay1@gmail.com (M.V.)

* Correspondence: Alison.Furber@mail.mcgill.ca

Academic Editors: Sharon B. Megdal, Susanna Eden and Eylon Shamir

Received: 19 February 2016; Accepted: 29 June 2016; Published: 8 July 2016

Abstract: The International Joint Commission (IJC) has been involved in a 14-year effort to formulate a new water regulation plan for the Lake Ontario St. Lawrence River ("LOSLR") area that balances the interests of a diverse group of stakeholders including shipping and navigation, hydropower, environment, recreational boating, municipal and domestic water supply, First Nations, and shoreline property owners. It has embraced the principles of collaborative and participatory management and, applying a Shared Visioning Planning (SVP) approach, has worked closely with stakeholders throughout all stages of this process; however, conflicts between competing stakeholders have delayed and complicated this effort. The overarching aim of this paper is to consider the extent to which the SVP approach employed by the IJC was effective in managing conflict in the LOSLR context. Audio recordings and transcriptions of public and technical hearings held by the IJC in 2013 have been systematically analysed using stakeholder mapping and content analysis methods, to gain insight into the stakeholder universe interacting with the IJC on Plan 2014. The principal conclusions of this paper are that (a) the Shared Vision Planning approach employed by the IJC had some significant successes in terms of conflict management—particularly notable is the success that has been achieved with regards to integration of First Nation concerns; (b) there is a distinct group of shoreline property owners, based in New York State, who remain opposed to Plan 2014—the IJC's public outreach and participation efforts have not been successful in reconciling their position with that of other stakeholders due to the fact that this stakeholder group perceive that they can only lose out from any regulation change and are therefore unlikely to be motivated to engage productively in any planning dialogue; and (c) a solution would require that the problem be reframed so that this stakeholder can see that they do in fact have something to gain from a successful resolution, which may necessitate bringing the prospect of compensation to the table.

Keywords: conflict; participation; Shared Vision Planning; Lake Ontario; St. Lawrence River

1. Conflict in Participatory Approaches

Participatory approaches to water resources planning are increasingly promoted as best practice [1,2]. For the purposes of this paper participation is defined broadly, following Carr [1], as involvement of the public, institutional decision makers, individuals, or representatives of groups with an interest in or ability to influence how a river is managed, in river management decision making processes. Participation can take many guises; the classic framework used to characterize participation is Arnstein's [3] ladder, which classifies participation processes based on the degree of power transferred from the process implementer to the participants. At one extreme, participation processes involve little transfer of decision making power to participants; these processes aim only to inform participants about, or perhaps even manipulate them into accepting, decisions that have already been taken. At the other extreme, participants are given real decision-making power, and are actively involved in the decision-making processes. In between these two extremes are participation processes in which participants are consulted on decisions but are not given final decision-making power.

Despite widespread acceptance of the concept, there has been some debate over the benefits participation brings and how it should be implemented in complex social-ecological settings [4]. To address the question of why participation is desirable, Carr [1] explored the mechanisms through which participation impacts river basin management. She concluded that participation can lead to better quality decisions being taken and, if conducted correctly, can increase the legitimacy of decisions, facilitating their implementation. She argued that participation both mobilizes and develops human and social capital, and provides space for deliberation and consensus building. Webler and Tuler [5] surveyed the opinions of watershed planners and activists from across Massachusetts and identified four prominent views of what a 'good' participation process should be. For some participants in their study, a good participation process is credible and legitimate, and maintains popular acceptance for outcomes decided, while for others it is one that is able to produce technically competent outcomes. A third set of participants emphasized the importance of fairness and procedural justice, while a fourth saw good participation as involving a process that educates and promotes constructive discourse.

Whichever of these views are held by planners, there is broad consensus that water resources problems are complex and multi-disciplinary; decisions are affected by, and affect, a broad range of stakeholders and actors, each of whom has their own knowledge and perspective of different aspects of the system about which decisions are being taken [6]. Participation is key to ensuring that the full breadth of existing knowledge is represented in the decision-making process. It is this amalgamation of different knowledge and perspectives that simultaneously makes participation so important but also so challenging [7]. When divergent world views and knowledge are brought into close proximity, the potential for conflict is great (ibid). A key challenge for participatory water resources planning, therefore, is finding ways to conduct participatory processes that manage this risk of conflict, so that the fruits of participation can be realized.

Shared Vision Planning (SVP) is a highly structured approach to planning that incorporates meaningful participation into each stage of traditional multi-objective planning [2]. First conceived for the National Drought Study [8] by the U.S. Army Corps of Engineers, it requires a team be involved in each step of the decision-making process, from problem identification to plan implementation. This team should be composed of stakeholders (those able to affect, and affected by decisions taken), decision-makers, and experts [2]. Collaborative modelling is used as a mechanism through which multiple understandings can be brought together to identify and resolve disputed causal effects and create consensus and transparency regarding the underlying system as a starting point for participative decision making (ibid). As such, it promises a tangible mechanism for effective conflict resolution in participatory approaches to water management. This paper explores the ability of SVP to manage conflict in the case of regulation planning for Lake Ontario and the St. Lawrence River.

2. Background to the Lake Ontario St. Lawrence River Regulation

Researchers and practitioners in the water sector increasingly recognise that the management of water resources is synonymous with the management of conflict [9–12]. Andrew [13] identified a number of reasons that natural resource management is particularly prone to conflict: (1) the widespread use of natural resources by large, diverse, and geographically dispersed groups creates complex networks of people and entities with differing power and influence; (2) the problem is compounded by the fact that the interconnectedness of the natural environment means that the actions of one group can have an impact on other groups a great distance away (i.e., upstream/downstream effects); (3) the use of the resource can also have different meaning to different people (i.e., economic livelihood for some; a way of life and cultural identity to others); (4) the diminishing supply of some natural resources may result in 'structural-scarcity' and unequal distribution.

Recognizing the potential for conflict, and the need to cooperate, in the management of waterways along their shared border, the United States and Great Britain (on behalf of Canada) entered into the Boundary Waters Treaty ("BWT") in 1909 and created the IJC. The joint Canada–USA staffed IJC and its various Boards has two basic responsibilities—to act on applications and issue orders if approved, and to conduct studies under formal references from the governments. The IJC is made up of six Commissioners, three appointed by the President of the USA and confirmed by the U.S. Senate, and three appointed by the Canadian Governor in Council, essentially the Prime Minister. The basic aim of the IJC and the BWT is to prevent and resolve disputes. Commissioners act not as representatives of their national government but in the common interest of the people of the basins the IJC works in.

In response to applications from both national governments, the IJC issued an Order of Approval in 1952 (and amended it in 1956) for the construction of the St. Lawrence River Hydropower Project. Figure 1 shows the location of the resulting Moses Saunders Dam in relation to Lake Ontario and the St. Lawrence River Drainage Basin.

Figure 1. Map of Lake Ontario and the St. Lawrence River.

Among the provisions contained in Article VIII of the BWT is a ranking of interests to be considered when devising regulation plans, set out as follows [14]:

1 Uses for domestic and sanitary purposes;
2 Uses for navigation, including the service of canals for the purposes of navigation;
3 Uses for power and for irrigation purposes.

Plan 1958-D was the management plan formulated for regulation of the dam; it respected this order of precedence and since 1963, regulation of Lake Ontario water levels and outflows have occurred under this plan. Notably absent is any mention of the environment and certain special interests such as shoreline property owners and boaters, although the BWT does require the IJC to give all interested parties opportunities to be heard. Plan 1958-D was based on hydrologic conditions experienced from 1860 to 1954. Since that time, there have been changes to water flow patterns, climate conditions, and the composition of interested stakeholders and, as a result, this plan has become outdated [15].

As a consequence, the plan is frequently deviated from, following an unofficial plan referred to as Plan 1958-D with Deviations (Plan 1958-DD).

By 1993, the IJC was receiving numerous complaints, especially from environmental groups and recreational boaters, that the regulation plan was not meeting their needs and as a result the IJC recommended that the Order of Approval be amended to better reflect the current needs of the users and interests of the system [15]. Embracing a new spirit of participatory management, the IJC created the International Lake Ontario St. Lawrence River Study Board (the "Study Board") in 1999 and entrusted it to perform a comprehensive scientific and environmental analysis of water levels and flow regimes in the LOSLR system and mandated that this effort include public input [16]. A novel feature of the study process was the creation of a special group of stakeholders called the Public Interest Advisory Group (PIAG). The PIAG was an independent advisory group, made up of volunteers, who created a link between the general public and the Study Board. The Canadian and U.S. PIAG chairs were also members of the Study Board; they were tasked to provide advice to the board, feedback to the public, and input at all stages of the process [15]. Between 2000 and 2006, an extensive study was undertaken to combine scientific knowledge, modelling, and a plurality of viewpoints in the development of a new regulation plan. The Study Board assembled numerous public interest and technical committees to model the lake and river systems and a Shared Vision Planning approach was implemented to steer and integrate the results [15,17].

A collaborative model was developed as part of the SVP approach. Technical work groups were tasked to undertake collaborative research and modelling of one aspect of the overall system. The task groups were: environmental; recreational boating and tourism; coastal processes; commercial navigation; hydroelectric power; and domestic, industrial, and municipal water uses [15]. Stakeholders were assigned to Work Groups based on their interests and concerns. These individual group models were then integrated so that regulation plan options could be simulated and the consequences in terms of multiple objectives (which were also defined by the Work Groups) assessed (ibid). Based on extensive ecosystem and environmental modelling, three alternative plans were developed that incorporated the preferences of interested stakeholders (ibid). This was followed by a period of public consultation that resulted in the IJC choosing to back one of the plans, which was subsequently referred to as Plan 2007. Having selected Plan 2007, the IJC held more extensive public hearings on this option. During the hearings it transpired that the Plan was widely opposed; environmentalists thought that the plan failed to offer the environmental protection they sought, while other stakeholders saw no benefit in the plan for their concerns.

Following the Plan 2007 public hearings and almost complete opposition to that plan, the IJC created one formal Work Group to find a solution. This Work Group was chaired by the IJC and made up of governmental representatives. They used all the work of the Study Board and some of the technical experts and came up with Plan Bv7 (Plan B being one of the three options proposed by the Study Board). Plan Bv7 provided a wider range of flow levels that more closely matched natural flow patterns than the more tightly regulated levels found in the originally proposed Plan 2007 and Plan 1958-D [18]. Facing strong opposition from New York State South Shore property owners concerned about flooding, shoreline erosion, and damage to build structures, Plan Bv7 was further modified to address their concerns. The current regulation proposal, termed Plan 2014, has undergone several iterations and has, at the time of writing, been submitted to the governments of Canada and the United States for approval. The plan comprises both a new approach to the management of water levels and an adaptive management plan, which aims to overcome any data and modelling uncertainties and to allow for continued improvement in the future [18]. Plan 2014 underwent an extensive period of public comment during the summer of 2013; dozens of stakeholder groups and hundreds of individuals participated in these occasionally contentious proceedings. In addition to these in-person hearings held in six cities in Canada and the USA, many individuals have made their positions known in a variety of other forums—to their elected representatives, to the media, at town hall meetings, in published articles, and online. A summary of the timeline of events can be found in Table 1.

Table 1. Timeline of events.

Timeline	Event
1909	Boundary Waters Treaty signed International Joint Commission Created
1952	Order of Approval for the construction of the St. Lawrence River Hydropower Project
1954–1958	Construction of the Moses-Saunders Dam
1999	The International Lake Ontario St. Lawrence River Study Board Created
2000–2006	Lake Ontario St. Lawrence River Study Conducted
2007	Plan 2007 Backed by the IJC as the Preferred Regulation Option
2008	Plan 2007 Widely Opposed during Public Hearings
2013	Hearings on Plan 2014

A full description and analysis of the public participation process up to and including the 2008 public hearings can be found in Carr, Loucks, and Blöschl [19]. At this mid-point in the whole process, they concluded that there were some considerable strengths of the IJC's approach including good access to information and meetings, commitment to involving all potentially affected communities and interest groups resulting in broad representation, impartial facilitation, and inclusion of a wide variety of knowledge. They specifically compliment the PIAG, describing them as, "dynamic, dedicated, and well supported". The present paper extends the work undertaken by Carr, Loucks, and Blöschl through an analysis of the 2013 hearings on Plan 2014, focusing on the extent to which the SVP approach employed by the IJC was effective in managing conflict during LOSLR regulation planning. To frame the research, the following research questions were posed: (a) To what extent was agreement reached by stakeholders in support of Plan 2014? (b) What evidence is there that the resolution process enabled stakeholders to overcome potential conflict and reach agreement? (c) What residual conflict persists and what is the root cause of this conflict? Finally, consideration is given to whether any opportunities exist to move the conflict further towards amicable resolution, and what lessons might be learnt more broadly.

3. Methodological Approach

The study of environmental conflict is, essentially, the study of stakeholder conflict. One definition of "stakeholder" that has been widely cited in the non-profit and natural resources management literature was first proposed by Freeman more than 30 years ago as: "any group or individual that can affect or is affected by the achievement of the organization's objectives" [11,12,20]. As such, this definition casts a broad net capturing actors, individuals, groups, and associations, whether formal or informal, holding different interests, perspectives, and viewpoints. In the current context, the IJC is confronted with the difficult task of balancing the competing demands of stakeholders and actors, some of whom hold strongly divergent views.

To determine the extent to which the IJC's planning process was effective at conflict management, a scheme was developed to identify, categorize and analyse the stakeholders interfacing with the IJC on the matter of Plan 2014. The IJC actively embraced the principles of collaborative and participatory management during the period of 14 years that a new water level regulation plan has been under consideration. A large number of stakeholder groups were engaged in various phases of the study. The IJC website provides copies of audio recordings and transcriptions of the public and technical hearings conducted during 2013. These data, which contain numerous statements made by a wide variety of actors and stakeholders, have been systematically analysed as part of the present work with the goal of gaining insight into the stakeholder universe interfacing with the IJC on Plan 2014.

The present analysis took the following steps, based on the methods and tools outlined by Freeman [20], Mendelow [21], Mitchell, Agle, and Wood [22], Elias, Cavana, and Jackson [23], and Elias [12].

To assess the extent that agreement was reached by stakeholders in support of Plan 2014:

i Stakeholders were identified and classified on a stakeholder chart;
ii A snapshot of stakeholder positions at the time of the technical hearings was visualised through a stakeholder mapping exercise;

To explore the effectiveness with which the IJC's planning process enabled stakeholders to overcome potential conflict and reach agreement, it was necessary to first determine the potential for conflict. This was achieved by:

iii. Undertaking a content analysis to identify where stakeholders held potentially conflicting needs, values, beliefs, or expectations relevant to the resolution process;

The potential conflict was then compared to the final positions achieved through the resolution process to see if potential conflict had been effectively avoided.

Finally, to assess whether residual conflict persists and identify the root causes of this conflict:

iv. Content analysis of the statements made by stakeholder classes who remain opposed to Plan 2014 was undertaken with the goal of elucidating the root causes of opposition.

Each of the methods is described in more detail, prior to results being presented, in the subsequent section.

4. Results

4.1. Categorisation of Stakeholders into Classes

The first task undertaken during the stakeholder analysis was categorisation of stakeholders into classes. The challenge presented by the process of categorisation is to find an appropriate level of aggregation of stakeholders that allows the stakeholder universe to be simplified to a manageable number of groups and perspectives without over-generalising and losing potentially important detail about stakeholder interaction. The goal was to split the stakeholder universe into a limited number of coherent and logical classes along important dividing lines. At the beginning of the stakeholder analysis process it was not clear exactly where these lines could most appropriately be drawn, however. An initial best guess was made and stakeholders were roughly classed as belonging to one of the following groups: environmental concerns, shipping concerns, hydroelectricity concerns, fishing concerns, First Nations, and the general public. As the stakeholder analysis progressed by undertaking the steps described below, it became apparent that these classes did not capture some of the most important attributes of the dispute. Classes were therefore continually re-shuffled in an iterative process throughout the analysis.

For example, during the stakeholder mapping exercise the initial class "general public" was found to be particularly incoherent and heterogeneous so additional effort was put into identifying dividing factors that diversify public opinion so that more homogenous sub-classes could be formed. During the first attempt at this process speakers were categorised geographically according to their state or province of origin. As the analysis proceeded, however, it became clear that geographic divisions could not adequately account for position differences. Opinion varied within the state of New York according to whether the speaker was a Lake Ontario riverside property owner or not. Conversely, the perspective held by those living adjacent to the St. Lawrence River did not appear to diverge depending upon whether the person lived in New York, Ontario, or Quebec. For this reason, the sub-categories used for the general public were "Riverine South Shore", "Non-riverine South Shore", and "St. Lawrence". The perspective of those holding political office was kept separate from individuals speaking on behalf of themselves or small community groups.

In general, if no discernible differentiation could be found between two groups in terms of either position or rationale for that position (see following sections), the groups were merged into one group.

If differences were identified in position or rationale within a group, but no apparent logical divisor could be identified, the group remained as one group (albeit one less homogenous group). Table 2 shows the final classification of stakeholders into classes that aim to remain logical but as homogenous as possible.

Table 2. Final classification of speakers participating in the 2013 public and technical consultations.

Environmental	Political South Shore	Riverine Public South Shore
• Large Environmental NGOs • The Nature Conservancy • World Wildlife Fund ("WWF") • Audubon Society • Ducks Unlimited • Nature Conservancy • Sierra Club • Local Environmental Groups • Upper St. Lawrence River Keeper • Save the River • Nature Quebec • Citizen's Campaign for the Environment ("CCE") • Royal Botanical Gardens • Hamilton Harbour Remedial Action • Thousand Island Land Trust • Goose Bay Reclamation Committee • Governmental Association • Conservation Authorities Ontario (CO) • Strategies Saint-Laurent • Zones d'Intervention Prioritaire (ZIP) committees	• Town/County Officials • Williamson, NY, USA • Wolcott, NY, USA • Sodus, NY, USA • Ontario, NY, USA • Huron, NY, USA • Greece, NY, USA • Monroe County, NY, USA • Niagara County, NY, USA • Orleans County, NY, USA • Wayne County, NY, USA • Oswego County, NY, USA • Cayuga County, NY, USA • Wilson, NY, USA • Olcott, NY, USA • Newfane, NY, USA • Somerset, NY, USA • Youngstown, NY, USA • Regional Alliances • Niagara/Orleans Regional Alliance • State/National representation • Congress men/women • Congressman Chris Collins • Senator Michael Nozzolio • Congresswoman Louise Slaughter • Senator Joe Roback • New York State Assembly	• Citizen Action Groups • Save Our Sodus • Lake Ontario Riparian Alliance • North Rainbow Shores Homeowner's Association • Grandview Beach Association (GBA) • Shore for All • Payne Beach Association • Baby Beach Association • Wilson Hill Association • Individuals who live in property adjacent to Lake Ontario's South Shore
Non-Riverine Public South Shore	**Public St. Lawrence**	**First Nations**
• Individuals who live in the counties along the South Shore of Lake Ontario but do not live directly on the shoreline	• Individuals living adjacent to the St. Lawrence in either Ontario, Quebec or New York	• St. Regis Mohawk Tribe (U.S.) • Mohawk Council of Akwesasne (Canadian)
Political St. Lawrence	**Shipping and Navigation**	**Fishing, Recreational Boating and Tourism**
• Officials from • Assembly (New York adjacent to St. Lawrence as far as Massena) • Jefferson County • St. Lawrence City • Clayton Town	• The St. Lawrence Seaway Management Corporation (RSMC) • Shipping Federation of Canada (SFC) • Canadian Shipowners Association • Association of Canadian Port Authorities • Montreal Port Authority	• Niagara Tourism and Convention Corporation • Tourism Department, Orleans County • Classic Island Cruises, Clayton • Schermerhorn Boat Sales and Marina • Arnie's Marina • Grenadier Island Country Club • Fédération Québécoise des Chasseurs et Pêcheurs • Antique Classic Boat Society • Alexandria Bay Fishing Guides Association • Lake Ontario Fishing Coalition
Scientific Community	**Hydropower**	**Municipality and Domestic Water Supply**
• State University of New York (SUNY) • St. Lawrence River Institute of Environmental Sciences	• Hydro Quebec • New York Power Authority • Ontario Power Generation	• Water and Wastewater Treatment Plants within the Basin • Shore well users • Industrial water users

4.2. Stakeholder Mapping

A key step in Freeman's analysis of stakeholders is the preparation of a high-level map of the universe of stakeholder classes [20]. In the LOSLR case study, stakeholders were analysed to identify their positioning in terms of the degree of support or opposition to the proposed Plan 2014. During the IJC's public and technical consultations in 2013, each speaker was permitted to speak for at least three minutes, longer if time permitted, and present their opinion on Plan 2014.

Each speaker's statement was examined for evidence to determine the speaker's position as either "strongly supportive", "supportive", "neutral", "opposed", or "strongly opposed". Many speakers explicitly stated their position as for or against Plan 2014, which simplified this task. The determination of whether a speaker was "supportive" or "strongly supportive" was made on the basis of whether the speaker also raised concerns about Plan 2014 during their speech. For example, one speaker said that their organization, "recognizes and supports the intent of Plan 2014" but also that they "have several concerns and recommendations". This speaker was therefore categorised as "supportive" rather than "strongly supportive". A distinction between "opposed" or "strongly opposed" was made on the basis of whether the speaker felt there was a need for a new plan but had strong enough concerns with the Plan to oppose it, or whether they felt that there was no need to make changes to the existing plan which they feel works adequately. A typical quote of a speaker who was determined to be "opposed" is, "we should not change the Plan until we have something that is more equal to all interests". This quote shows that the person is open to changing the regime but feels Plan 2014 is unfair and therefore their position is in opposition to it.

Despite efforts, made in the prior step, to classify stakeholders into coherent classes, some classes retained a degree of heterogeneity (when there was not further apparent logical basis on which to further divide a class) and therefore each aggregated stakeholder class was plotted relative to the range of positions that their sub-groups may hold. While there is undeniably a level of subjectivity to this approach, it is still a useful exercise as even a rough indication of position is useful to select stakeholders for more detailed exploration in the next step of the study.

Figure 2 shows results from the stakeholder mapping process, where stakeholder classes are mapped according to their position in terms of degree of support for Plan 2014. Evidence of the rationale behind the assignment of position of the stakeholders, excluding the general public, on the stakeholder map can be found in Table 3. Figure 3 looks specifically at the spread of opinion within the general public and their elected leaders, and the following section takes a closer look at the arguments of the Opposed South Shore public.

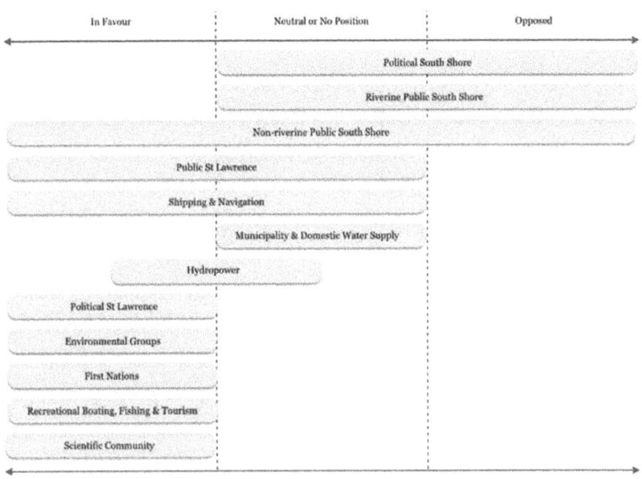

Figure 2. Final position of stakeholders on Plan 2014.

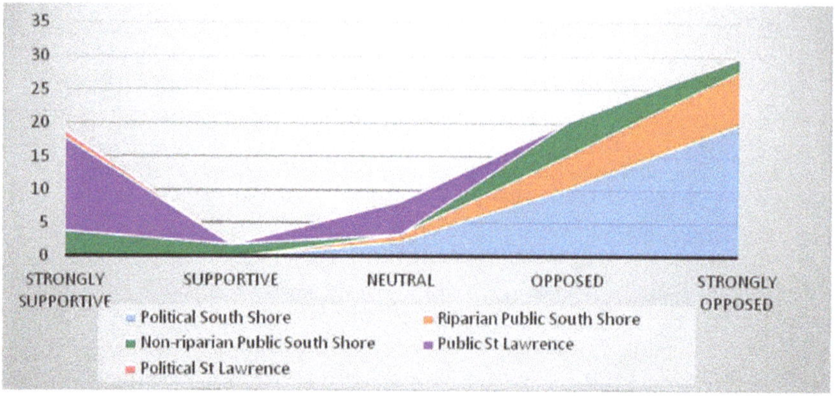

Figure 3. 2013 position of the public on Plan 2014.

Table 3. Evidence of assigned position of stakeholders (excluding the general public) towards Plan 2014.

Organisation	Category	Position	Evidence
Mohawks of Akwesasne	First Nations	Strongly Supportive	"We strongly support the effort to mimic the natural flows of the River versus the Lake environment model that we have ... I feel this Plan does give us the best approach forward ... we fully support this new regulation Plan 2014"
Shipping Federation of Canada	Shipping	Supportive/ Strongly Supportive	"We fully support the IJC's effort through Plan 2014 to protect and enhance the environment" but they have some lingering concerns, "we're concerned with the possibility of continued low water ... "
St. Lawrence Seaway Management Corporation (RSMC)	Shipping	Neutral	"The RSMC is not opposed to the prospect of refining the water management plan" but "given the above concerns, the implementation of the Plan in its current state will have significant negative impact to commercial navigation"
Canadian Shipowners Association	Shipping	Supportive	"The CSA recognizes and supports the intent of Plan 2014" ... "we believe that Plan 2014 can be achieved and support both increased benefits to the environment and the economy through the efficient and predictable control of water levels but it still requires certain specific amendments"
Montreal Port Authority	Shipping	Supportive/ Strongly Supportive	"the proposal generally seems good to us"
Hydro Quebec	Hydro	Supportive	"so this was done partly in the plan but not necessarily as we would have liked", "the present Plan is progressively good ... the Plan should be improved before it is implemented"
Conservation Ontario	Environment	Strongly supportive	"Conservation Authorities applaud the IJC for Plan 2014, proposing a more balanced approach to flow management that seeks to create more natural levels in the Lake and River while continuing to provide the basin's community with substantial benefits. It is a major step forward ... " "In conclusion, Conservation Ontario and its members support the implementation of Plan 2014"
WWF Canada	Environment	Strongly supportive	"We feel that Plan 2014 really does have the potential to have a national and global impact ... We're very pleased with plan 2014 ... "

Table 3. *Cont.*

Organisation	Category	Position	Evidence
Nature Conservancy	Environment	Strongly supportive	"I'm speaking today in support of Plan 2014"
Antique Classic Boat Society	Fishing, Recreational Boating & Tourism	Strongly Supportive	"I would definitely appreciate you accepting Plan 2014"
Lake Ontario Fishing Coalition	Fishing, Recreational Boating & Tourism	Strongly Supportive	"I really hope we can move forward with this"
Alexandria Bay Fishing Guides	Fishing, Recreational Boating & Tourism	Strongly Supportive	"I support Plan 2014 because I believe it will benefit the wetlands and the fish and the wildlife that live there"
Fédération Québécoise des Chasseurs et Pêcheurs	Fishing, Recreational Boating & Tourism	Strongly Supportive	"The Fédération Québécoise des Chasseurs et Pêcheurs is very happy with the main objectives of getting as close to a natural flow as possible. We strongly support this idea." If anything the speaker is suggesting that the plan should go further.
State University of New York	Scientific Community	Strongly Supportive	"I will point out every one of the damages that have been complained about happened under 1958DD. 1958DD never allows a low lake level to occur. I want to point out that low lake levels are really, really needed and critical"

4.3. *Content Analysis*

The content of the available transcripts was analysed with the goal of understanding the perspectives of each type of stakeholder, to assess the potential for conflict as a first step to assessing the success or failure of the IJC's process. For each stakeholder class, consideration was given to what the stakeholder needs or wants (i.e., what their stake is), the argument they use to support their position, the underlying values and beliefs that form the basis of their position, and any expectations they hold regarding the resolution process. A summary of the needs or wants of key stakeholders is provided in Table 4 below.

Table 4. Needs or wants of key stakeholders.

Stakeholder	Needs/Wants
First Nations	Inclusion in decision making To be able to live off the environment both now and in the future
Shipping and Navigation	Predictability in water level Minimum water levels
Environmental Groups	Increased biodiversity and ecological integrity through protection of environmentally significant areas and enhanced aquatic and terrestrial habitat
Scientific Community	Increased biodiversity and ecological integrity through protection of environmentally significant areas and enhanced aquatic and terrestrial habitat
Hydropower	Consistent flow Predictable flow
Recreational Fishing, Boating and Tourism	Abundant fish, extended boating season through higher water levels at end of season (which also makes removing boats from the water easier) and beautiful environment to attract tourists
Municipal and Domestic Water Supply	Problems arise for water treatment plants at very low water levels if water becomes difficult to draw or requires additional treatment due to more concentrated pollutants, and for water and wastewater treatment plants at high water levels due to potential flood damage to facilities and shoreline protection structures
Opposed South Shore	Property security and protection from water damage and erosion Protection from emotional devastation caused by disasters Water Access Building regulation changes Consistent water levels

An important limitation of the approach that was used was that the data on which the analysis is based are statements made at the end of the resolution process. Ideally a content analysis would have been undertaken both before and after the process to identify perspective changes. One of the significant attributes of the IJC's process was the effort put into developing a shared understanding of the environmental, technical, and social dimensions of the dam management regime through the collaborative modelling process. It would have been very interesting to see how lines of arguments, and the data upon which arguments were based, were changed by the shared visioning process. Despite this limitation, the content analysis revealed two features of the conflict with great potential to lead to conflict.

The potential for conflict between First Nations stakeholders and other stakeholders became evident through the content analysis process. The values of the Mohawks of Akwesasne stood in stark contrast to those of other stakeholders. The content analysis revealed a completely different worldview from that held by the other stakeholders. It was apparent that they had a structural concern that a process be employed that valued and included their way of knowing. In addition, previous relationship issues were referred to that highlighted the potential for conflict due to strained relations.

The potential for interest-based conflict was also particularly apparent. While the needs of some stakeholders were divergent but not necessarily mutually exclusive (for example, predictable water levels, high water levels during peak commercial times, and more variation in flow do not seem mutually exclusive), others simply seemed to conflict. The water regime cannot be simultaneously consistent in flow (as required by hydropower) and level (as required by shoreline property owners) and varied in flow and level (as required by environmental groups).

The final goal of this study was to identify the root causes of residual conflict. Having identified in preceding stages of the study that opposition to Plan 2014 persists among a subset of the public living along the South Shore and their local political leaders, content analysis of their statements was also used to identify complaints made by this group regarding Plan 2014. The outputs of this analysis are presented in Table 5.

Table 5. Complaints of the Opposed South Shore public.

There Is No Problem with the Current Management Regime:
"Why fix something that is not broken" (RK, Ontario NY)
Data Used in Shared Visioning Models Are Not Accurate:
"While this Plan reportedly reduces damages for home owners and recreational boaters, this remains an issue of great contention due to the fact that the data used remains old, outdated and inaccurate ... At no time has the assessment value decreased and since this was a critical feature for the development of coastal damage projections, we believe it clearly documents the cost for coastal shoreline protection is significantly undervalued in this Plan" (DE, Niagara County)
"We feel that this plan does not show improvement to the environment due to its assumptions and out-dated data" (LC, Huron)
No Environmental Review Took Place:
"I don't see any environmental review. In New York State there's a thing called a 'seeker'; it's anytime you do anything with wetlands; it's an extremely detailed process which is justified, that you evaluate the environmental impact. I don't see anything like that in this Plan that I am aware of" (JH, Williamson)
Review Comments Were Not Adequately Addressed:
"The study had as part of it a funding mechanism to bring in the National Resource Council on the American side and the Royal Society of Canada on the Canadian side to Peer Review the environmental science... The Peer Review did not come out well for the environmental science. As a matter of fact, page 65 of the Peer Review says: "Do not use those models as a decision-making mechanism" (TM, Niagara County)

Table 5. *Cont.*

Lack of Emergency Response Maps:
"Last year the town of Greece . . . asked how any plan can be considered when there are no FEMA (Federal Emergency Management Act) risk maps detailing the effects of these proposals on our lakefront properties" (JA, Greece)
Outcomes Are Biased Towards Hydropower Production and/or Shipping Concerns:
"What does make sense however, after six years of reviewing the data released by the IJC, the biggest winner with Plan 2014 just like with the other proposed plans, might not be the environment at all; it's hydro power" (SL, Sodus)
"I read an article in yesterday's Buffalo News that talked about the hydro-electric plants during the peak demand times, they can make more electricity when the water levels are higher. The difference there results in millions of dollars. I don't believe that it's only just about the wetlands and the muskrats; I also believe that it also has to do with hydro-electric plants, the money that is involved and the shipping industry as well" (TH, South Shore)
"Hydro power gets I calculated four times the benefit than the environment . . . I would submit that the Plan is non-balanced in terms of impact, in terms of damage" (JH, Williamson)
"What I'm hearing, what the real issue is here, is that hydro power and shipping take priority over infrastructure concerns" (MR, Bailey Beach)
The Cost Is Unfairly Distributed:
"This Plan, as did previous, concentrates large disproportionate loss on five South Shore counties including Niagara" (DE, Niagara County)
"The International Joint Commission prefers Montreal interests over the South Shore's problems with flooding" (CS, Hamlin)
"I see that what we are doing here is creating damages for a very small group of people in a very localised area. When we started the study, one of the basic tenants of the study was that there would be no disproportionate losses; no one would gain from someone else's loss. Well that has not happened" (TM, Niagara County)
It's Unfair to 'Change the Rules of the Game':
"The Moses-Saunders Dam was built with the South Shore residents permission and the agreement included a 4-foot level for lows and highs . . . now the IJC wants a Plan that will let the waters go 5 times higher than the original plan, they don't want to hear from the South Shore and the problems that would incur" (CS, Hamlin)
"So for the past 40 years, homeowners, municipalities and private businesses have engineered and built marina and water treatment infrastructure to that roughly 4 foot like variation" (LS, Wilson)
"Our marina, our structures, everything we have was built according to the existing plan. If we deviate from that, you're going to devastate our community" (TH, Newfane)
"much commercial, residential and municipal development along the lakeshore; it was done based on the belief that to the best of human ability the lake would remain as promised in plan 1958 DD" (RK, Ontario NY)
There Is an Alternative Agenda to the Process:
"I had the pleasure of observing muskrats this year in my pond because I got a wetland in there that I can't touch; it's about a quarter of an acre. Guess what muskrats eat? Cattails, yes. Well, when they were done eating every bit of cattails and completely defoliating the entire pond, they're gone... It's a dichotomy that makes me question the agendas" (DC, South Shore)
Meetings Are Being Held in Secret/behind Closed Doors:
"We fast forward to 2013, and by the way this is rumors, but the neighborhood, IJC and other groups have had closed door meetings. Is this true?" (SD, Crescent Beach)
"As required by the [Boundary Waters] Treaty, the IJC is to involve the public in all of its activities, yet after B+ the meetings have been behind closed doors with input by environmental groups only" (LC, Huron)
Not All Data Are Being Shared with the Public:
"Also a rumor is there is a map and it shows how many homes will be affected; possible up to 2000. Rumor is this is not going to be presented to the public" (SD, Crescent Beach)
The Canadian and U.S. Governments Are Unable to Work Together Effectively:
"My concern is your triggering points and how they get triggered. I have people here from the United States and people from Canada. I know our Congress can't work together because they're Republicans and Democrats" (RK, Grandview)

5. Discussion

5.1. Extent of Agreement Reached

Figure 2 shows that the vast majority of stakeholders were supportive of Plan 2014 at the time of the public and technical consultations. With the exception of a group of a few hundred shoreline property owners based in New York State and their local political leaders, consensus was reached across the majority of stakeholders in support of Plan 2014.

5.2. Effectiveness of the IJC's Resolution Process

The challenge presented to the IJC to manage conflict over regulation of water levels in LOSLR was significant. The IJC process has achieved many successes. Clear examples can be found of stakeholders changing their position to back the proposed regulation changes. One such example can be found in the statement of JH, speaking on behalf of the Hamilton Harbour Remedial Action Plan. In his 2013 statement to the IJC he recalls, "I appeared before the International Joint Commission in this same room I believe and I had suggested that at that time the Commission not approve Plan 2007 ... I'm very pleased to tell you that on behalf of our Remedial Action Plan Group ... that we're very pleased to see the work that's gone on, and you have our support for Plan 2014". The commission then asked JH what the difference is between the 2007 Plan and Plan 2014 that led to this change in position. JH responded that Plan 2014 goes further to mimic the natural variation in the water level fluctuation that is so important for ensuring biodiversity in the region. This example is typical of many in the environmental community who wanted improved/greater water level fluctuations, and whose position was opposed to Plan 2007 and supportive of Plan 2014.

A further significant success of the IJC was the effectiveness with which they were able to achieve consensus between First Nation communities and the majority of stakeholders in support of Plan 2014. In the technical hearing in which First Nations participated, two major ongoing issues were identified by Chief Brian David that could have impacted greatly on the conflict resolution process implemented by the IJC. The first is regarding land claims being made in New York State and the Province of Quebec, along with the North Shore of Cornwall Island. The second relates to longstanding problems over the rights of First Nations individuals to travel freely within their territory without being restricted passage by the presence of international borders. Against this backdrop, where relations must undoubtedly have been severely affected, obtaining the support of the First Nations communities was a victory.

The content analysis revealed that inclusion and influence over the decision-making process was important to the First Nations communities; "We have a concern that we know what's going on and we have some influence over the decision making". During the consultation process Henry Lickers, Director of the Mohawk Council of Akwesasne Department of the Environment and an important First Nation elder with great influence in the community who has been involved in the IJC's process for many years, is quoted as saying, "We know that there are many other teachers in this world and we sit here today and listen to our problems that we have, but we know that we have the knowledge that came to us down the corridors of time from elders and ancestors that have preceded us and of us have those trusted elders that we have listened to in the past and hear their knowledge today and we will build on that knowledge that this will be a better place. And so I ask you to bring together your minds and think about those teachers of the world, and can we agree that they are important to us?" indicating a concern that the knowledge of the Mohawks be treated as equal to that of other types of knowledge. A major success of the IJC's process is best explained by Mr. Lickers, who said that, "I think at that time a lot of the thinking from Akwesasne went forward in recommendations... I think we partly penned it, or actually had influence in the conception of it ... This is a really impressive effort and I think that you're trying to be sensitive and I really want to applaud you and thank you for this effort".

5.3. Root Causes of Residual Conflict

While the proportion of remaining opposition may be small, the impact of this group on the overall planning process has been significant to date. The content analysis of opposition statements identifies a number of unresolved concerns stated by the Opposed South Shore during the public hearings of 2013.

Comments made during the public hearings reveal issues in the relationship between the Opposed South Shore and the IJC. In particular, there seems to be a lack of trust by some in the IJC, which is apparent from concerns held by some of the public that the IJC had an alternative agenda, was holding secret meetings to which they were not invited or informed, and had data and graphs that they were not sharing. One speaker also questioned the ability of the U.S. and Canadian governments to work effectively together. The accuracy of the data used in the modelling process was refuted, calling into question the validity of the models and thereby the analysis of the likely impacts of management regime change. A particular concern was expressed as to the valuation given to South Shore properties, which was felt to be outdated.

The IJC's process was directly criticized by some, who argued that an environmental review was lacking, or that a review had taken place but no effort had been made to respond to criticisms made during the review, and finally that no emergency risk maps had been produced. Many questioned the fairness of any effort to update or change dam management policy at all, given that decisions had been taken and structures built on the basis of previous policy decisions. The statement made by RK (Ontario, NY), "Why fix something that is not broken" was the most extreme of a variety of statements that revealed a lack of either understanding of, or valuing of, the local environment. A common theme amongst the Opposed South Shore was the belief that 'muskrat' (or other indicator species) concerns were not as valid as their own. Some used the argument that muskrats do not pay tax. The values held by these individuals are therefore at odds with a core value of many of the supporters of Plan 2014, which led directly to the goal of increased biodiversity.

Many of the statements make it clear that the speaker felt too much emphasis was placed on the interests of hydropower and shipping concerns, or larger downstream cities, at the expense of South Shore property owners. They disputed Plan 2014 on the basis that their interests are were not being adequately safeguarded.

The above summary reveals that the Opposed South Shore presented the IJC with a broad array of criticism in 2013, which disputed Plan 2014 on multiple fronts. Is it really the case that this stakeholder class believes that the IJC are engaged in secret meetings and are prepared to put their reputations on the line by basing their arguments on dubious data, to manipulate the process because they have an alternative agenda to serve hydroelectric and commercial shipping companies? It is thought more likely that there is, in fact, an alternative root cause of the conflict that has led to a resolution within the public to refute Plan 2014 by any means necessary.

This seems particularly true given the length to which the IJC went to conduct a process that had all the hallmarks of a participative and democratic process; Carr, Loucks, and Blöschl [19] found little evidence of bias in the statements of the study board facilitators in their analysis of the participative process. The IJC put in place consultation and hearing processes that allowed all stakeholders to have a voice; the public had a direct link to the Study Board through the PIAG and individuals had ample opportunity to express their opinion through both three-minute speeches at the public consultations and via letters and online comment. By making recordings and transcripts of the public consultations freely available, the IJC sought to increase the transparency of the process. The Opposed South Shore, in particular, were also represented by citizen action groups who were invited to participate in the technical working groups.

Some claims made by the Opposed South Shore, for example that no effort had been made to respond to criticisms made during the environmental review process, seem to have little connection with reality. The evidence of extensive environmental sampling and analysis is available for all to see and more time and money was spent on this aspect of the study than any other. If the root cause of the Opposed South Shore's position can be identified, perhaps it will be possible to gain insight into how the conflict can finally be resolved.

Consideration was given to whether distance from the problem-solving process was a factor in determining an individual's position with regards to Plan 2014. It is noteworthy that those individuals who were involved in the Technical Working Groups through citizen organisations remained in opposition to Plan 2014 when final positions were stated, as well as those not directly involved. This suggests that even full integration with the IJC's resolution process was insufficient to bring the strongly Opposed South Shore on board with the Plan. While the SVP process succeeded to align the positions of the vast majority of stakeholders, it failed to align the position of the general public.

It is posited that the Opposed South Shore property owners are (or at least perceive themselves to be) 'playing' a 'game' of a slightly different nature from the other stakeholders. Every stakeholder, except for the Opposed South Shore, has something to gain through the introduction of new water level regulations. Therefore, each of the other stakeholders is playing a collaborative game in which they want to achieve final agreement; the resolution process is about negotiating how much they can benefit. The Opposed South Shore, conversely, believes that they have nothing to gain from any new plan. Their objective is to keep water levels constant, which is in direct conflict with the principal objective of any new plan (i.e., to return water levels to a more natural and varied pattern). The Opposed South Shore, therefore, is playing a zero-sum game in which there is no hope of benefitting from new plans. The IJC itself has reinforced this position in the past. They stated that "the current Regulation Plan 1958-D with Deviations comes close to minimizing damages for Lake Ontario shoreline property owners" [15]; this implies that finding alternatives that bring benefits to this stakeholder group is highly unlikely. They explicitly acknowledge the zero-sum nature of the problem when they say that, "Changes to the criteria and existing operation plan are not possible without harm to some interests" and go on to justify this harm by saying that the, "majority of Board members do not consider these damages a disproportionate loss" [15].

While the Opposed South Shore perceive themselves to be playing a zero-sum game, the benefits that can be brought by any collaborative process are arguably severely restricted. They have fairly limited power in comparison to the other stakeholders. Perhaps the only strategy which they feel to be available to them is to exert pressure on the IJC by standing in the way of any resolution process—a strategy which to date they have implemented highly successfully (as evidenced by the fact that the resolution process has now been going on for over 14 years).

5.4. Moving Forward

The positions of the stakeholders are now so well entrenched that reaching consensus appears an elusive goal. A solution to this impasse would require a significant re-framing of the problem such that the Opposed South Shore perceive themselves as having something to gain from committing to the resolution process, perhaps by bringing the possibility of compensation to the table. It should be noted, however, that taking a route involving compensation is not without difficulties of its own. For example, issues such as whether or not the owners of structures that have been developed during the time the studies and resolution process have been ongoing should be compensated are likely to be highly contentious. There is also the issue of whether compensation should be used to relocate those living in homes adjacent to Lake Ontario, or whether compensation would be for the rebuilding of damaged property following flood events. This could equally lead to a highly charged and politically contentious debate.

With every conflict there is a necessary compromise to be made between urgency and pressure to implement timely solutions, and a desire to ensure a democratic process is followed that encourages collective action. In the case of the LOSLR conflict, as with other conflicts, some may feel that the emphasis has been placed too firmly on achieving consensus at the expense of timeliness, whilst others may hold the opposite opinion. The fate of LOSLR now lies in the hands of the Governments of Canada and the United States of America, who must weigh up the increasing urgency and impatience to implement a new water regulation management plan against a desire to implement a solution that is supported by all stakeholders concerned. It is the authors' opinion that, as the environmental damage occurring due to the employment of Plan 1958-DD has now been demonstrated, the urgency to act outweighs the desirability of reaching full consensus. It is possible that the governments will concur with Plan 2014 for the greater good of both countries, and address shoreline property owners separately. It is not clear, at this time, exactly how much political power these few hundred individuals have to influence international boundary decisions.

5.5. Broader Lessons

The successes of the IJC's approach suggest that SVP can be effective at managing potential conflict between stakeholders who have something to gain from participating meaningfully in the process. It is probable that any participatory approach will fail to completely eliminate conflict where gains to one interest can only be realized at the expense of another. In such cases, other mechanisms, such as the possibility of compensation, might usefully be brought to the table to complement the SVP approach.

6. Conclusions

It is concluded that while the Shared Vision Planning process employed by the IJC had some significant successes, notably the success that has been achieved with regards to integration of First Nations and environmental concerns, the IJC's public outreach and participation efforts have not been successful in reconciling the positions of all stakeholders. There is a distinct group of shoreline property owners in New York State who remain opposed to Plan 2014 because they perceive that they can only lose out from any regulation change. They are therefore unlikely to be motivated to engage productively in any planning dialogue. A solution would require that the problem be reframed so that this group has something to gain from a successful regulation plan resolution, which may involve bringing the prospect of compensation to the table. The fate of Plan 2014 now lies with the governments of Canada and the United States of America, who may choose to concur with the Plan for the overall public good of both countries and address shoreline property owners separately.

Acknowledgments: The research team gives thanks to Bill Werick, who provided feedback to drafts of this paper. This study was funded by a Social Sciences and Humanities Research Council of Canada (SSHRC) Partnership Development Grant, held by Jan Adamowski, and the Brace Centre for Water Resources Management at McGill University.

Author Contributions: Research was conducted and the manuscript written by Alison Furber, building on an earlier paper drafted by Meetu Vijay. Murray Clamen contributed expert knowledge of the LOSLR context and the history and workings of the IJC, and contributed ideas and comment throughout development of the paper. The research was initiated, supervised and reviewed by Jan Adamowski and Wietske Medema.

Conflicts of Interest: The authors declare no conflict of interest.

Abbreviations

The following abbreviations are used in this manuscript:

BWT	Boundary Waters Treaty
IJC	International Joint Commission
LOSLR	Lake Ontario & St. Lawrence River
PIAG	Public Interest Advisory Group

References

1. Carr, C. Stakeholder and public participation in river basin management—An introduction. *WIREs Wiley Interdiscip. Rev. Water* **2015**, *2*, 393–405. [CrossRef]
2. Palmer, R.N.; Cardwell, H.E.; Lorie, M.A.; Werwick, W. Disciplined planning, structured participation, and collaborative modeling—Applying Shared Vision Planning to water resources. *J. Am. Water Res. Assoc.* **2013**, *49*, 614–628. [CrossRef]
3. Arnstein, S.R. A ladder in citizen participation. *J. Am. Inst. Plan.* **1969**, *35*, 216–224. [CrossRef]
4. Von Korff, Y.; Daniell, K.A.; Moellenkamp, S.; Bots, P.; Bijlsma, R.M. Implementing participation water management: Recent advances in theory, practice and evaluation. *Ecol. Soc.* **2012**, *17*. [CrossRef]
5. Webler, T.; Tuler, S. Public participation in watershed management planning: Views on process from people in the field. *Res. Hum. Ecol.* **2001**, *8*, 29–39.
6. Davies, K.K.; Fisher, K.T.; Dickson, M.E.; Thrush, S.F.; le Heron, R.L. Improving ecosystem service frameworks to address wicked problems. *Ecol. Soc.* **2015**, *20*. [CrossRef]
7. Lange, A.; Siebert, R.; Barkmann, T. Incrementality and regional bridging: Instruments for promoting stakeholder participation in land use management in Northern Germany. *Soc. Natl. Res.* **2016**, *29*, 868–879. [CrossRef]
8. Institute for Water Resources (IWR). *Managing Water for Drought*; IWR Report 94-NDS-8; Institute for Water Resources: Alexandria, VA, USA, 1994.
9. Mitchell, B. *Resource and Environmental Management*, 2nd ed.; Prentice Hall: Harlow, UK, 2001.
10. Peltonen, L.; Sairinen, R. Integrating impact assessment and conflict management in urban planning: Experiences from Finland. *Environ. Impact Assess. Rev.* **2010**, *30*, 328–337. [CrossRef]
11. Stanghellini, P. Stakeholder involvement in water management: The role of the stakeholder analysis within participatory processes. *Water Policy* **2010**, *12*, 675–694. [CrossRef]
12. Elias, A.A. A systems dynamics model for stakeholder analysis in environmental conflicts. *J. Environ. Plan. Manag.* **2012**, *55*, 387–406. [CrossRef]
13. Andrew, J.S. Potential application of mediation to land use conflicts in small-scale mining. *J. Clean. Prod.* **2003**, *11*, 117–130. [CrossRef]
14. International Joint Commission (IJC). Guidance in Seeking Approval for Uses, Obstructions or Diversions of Waters under the Boundary Waters Treaty of 1909. Available online: http://www.ijc.org/files/tinymce/uploaded/Guidance-in-Seeking-Approval-for-Uses_EN.pdf (accessed on 15 February 2016).
15. International Joint Commission (IJC). Options for Managing Lake Ontario and St. Lawrence River Water Levels and Flows. Final Report by the International Lake Ontario–St. Lawrence River Study Board to the International Joint Commission. 2006. Website. Available online: http://www.ijc.org/loslr/en/library/LOSLR%20Study%20Reports/report-main-e-6KB.pdf (accessed on 15 February 2016).
16. International Joint Commission (IJC). Plan of Study for Criteria Review in the Orders of Approval for the Regulation of Lake Ontario–St. Lawrence River Levels and Flows. Report by St. Lawrence River–Lake Ontario Plan of Study Team to the International Joint Commission. 1999. Available online: http://losl.org/PDF/PlanOfStudy_en.pdf (accessed on 15 February 2016).
17. Langsdale, S.; Allyson, B.K.; Elizabeth, B.; Erik, H.; Scott, K.; Rick, P. Collaborative modelling for decision support in water resources: Principles and best practices. *J. Am. Water Res. Assoc.* **2013**, *49*, 629–638. [CrossRef]
18. International Joint Commission (IJC). Lake Ontario–St. Lawrence River Plan 2014: Protecting against Extreme Water Levels, Restoring Wetlands and Preparing for Climate Change. A Report to the Governments of Canada and the United States by the International Joint Commission. IJC Website, 2014. Available online: http://www.ijc.org/files/tinymce/uploaded/LOSLR/IJC_LOSR_EN_Web.pdf (accessed on 15 February 2016).
19. Carr, G.; Loucks, D.P.; Blöschl, G. An analysis of public participation in the Lake Ontario–St. Lawrence River study, Chapter. In *Water Co-Management*; Grover, V.I., Krantzberg, G., Eds.; CRC Press: Boca Raton, FL, USA, 2013; pp. 48–77.
20. Freeman, R.E. *Strategic Management: A Stakeholder Approach*; Pitman: Boston, MA, USA, 1984.
21. Mendelow, A. Stakeholder mapping. In Proceedings of the Second International Conference on Information Systems, Cambridge, MA, USA, 7–9 December 1991.

22. Mitchell, R.; Agle, B.; Wood, D. Towards a theory of stakeholder identification and salience: Defining the principle of who and what really counts. *Acad. Manag. Rev.* **1997**, *22*, 853–886.
23. Elias, A.A.; Cavana, R.Y.; Jackson, L.S. Stakeholder analysis for R & D project management. *R D Manag.* **2002**, *32*, 301–310.

water

MDPI

Article

Using a Participatory Stakeholder Process to Plan Water Development in Koraro, Ethiopia

Katherine Alfredo [1,*], Franco A. Montalto [2], Timothy Bartrand [2,3], Tsegay Wolde-Georgis [4] and Upmanu Lall [1]

1 Columbia Water Center, Earth Institute, Columbia University, 500 W120 Street, New York, NY 10027, USA; ula2@columbia.edu
2 Department of Civil, Architectural, and Environmental Engineering, Drexel University, Philadelphia, PA 19104, USA; fmontalto@coe.drexel.edu (F.A.M.); tbartrand@coronaenv.com (T.B.)
3 Corona Environmental Consulting, 150 Monument Rd., Suite 207, Bala Cynwyd, PA 19004, USA
4 Consortium for Capacity Building at INSTAR, University of Colorado Boulder, Boulder, CO 80309, USA; tsegay@colorado.edu
* Correspondence: Katherine.alfredo@fulbrightmail.org; Tel.: +1-212-854-1695

Academic Editors: Sharon B. Megdal, Susanna Eden and Anna Scolobig
Received: 3 March 2016; Accepted: 20 June 2016; Published: 30 June 2016

Abstract: This article reports the results of a one-day participatory workshop in Koraro, Ethiopia conducted prior to major development interventions in the region. The methodology of the workshop, structured to generate data useful for understanding the physical and social systems integral to water resources planning, provides a framework for future water need explorations in similar settings in Ethiopia and elsewhere. The use of only improved water sources as a metric for access to water under-represents the situation in Koraro, as many rely on streambeds for water due to the perceived cleanliness and low salinity of this unimproved water source. The reliance on metrics common in the Millennium Development Goals and Sustainable Development Goals, such as a minimum distance to a water source and the categorization of potable water based on type of water source, using varying figures (from as many as 30 to as few as four) can lead to assessments regarding the amount of additional sources necessary to allow access to specific locales, that are not consistent with actual need. Since the workshop, the Millennium Village Project has constructed over 30 wells in the region, following the most commonly used distance and source type metrics with less than desirable results. The water access evaluations alone do not address the needs of Koraro residents.

Keywords: drinking water; Koraro; Millennium Village Project; participatory; stakeholder; water management

1. Introduction

In September 2000, eight Millennium Development Goals (MDGs) were outlined during the UN Millennium Summit in New York [1]. Essential to these goals were plans to achieve significant improvements in access to safe drinking water, basic sanitation, and adequate water resources for agricultural and income-generating activities in the world's poorest nations. Water was both directly and indirectly related to all eight of the MDGs, and is for this reason a critical component of global sustainability. These MDGs expired in 2015 and were replaced by the Sustainable Development Goals (SDGs), which strive to achieve the water related MDGs where they fell short and continue to increase the number of people with access to potable water [2,3]. The Millennium Village Project (MVP) was conceived in 2004 as a method of testing the ability to achieve the MDGs at the village level [4]. The MVP operates in 80 villages organized into 14 clusters throughout Africa.

Although Koraro, a Millennium Village (MV) located in semiarid highlands of Ethiopia, faces water development challenges common to many rural, arid areas including scarcity, poor access, long travel times due to low density settlement patterns, soil erosion, and resource degradation, specific local needs were undocumented at the start of the MVP project. Without a nuanced understanding of local need and opportunities, aid workers face difficulty in calibrating local interventions for maximum sustainability impact. This work was motivated by a goal of infusing local knowledge into the planning, design, and implementation of water interventions in Koraro. More generally, we posit that a poor understanding of local needs has hindered progress not only on the MDGs, but also on other efforts to promote sustainable development worldwide.

Approaches that feature singular solutions to complex water security problems like the one faced by Koraro have often proved ineffectual. Designed and implemented from afar, these plans often overlook the local constraints, opportunities, and needs of specific places and populations, leading to the failure on many fronts, not least the marginalization and demoralization of the intended beneficiaries [5,6]. For example, ignoring the concomitant ecological management of the water temple system in Bali led to an influx of pests during top-down implementation of the Green Revolution [7]. Likewise, the failure of soil and water conservation planning in the Chemoga watershed of rural Ethiopia was due in large part to unwilling and disinterested participation of local farmers in a top-down plan to construct structures widely perceived as ineffective [8].

In contrast, stakeholder participation, the information exchange between developers and those most likely to be affected by an intervention, is now widely recognized as an essential part of development for all nations [9–11]. Stakeholder participation was identified as a key component of integrated water resource management planning efforts in Uzbekistan [12], and was reported as crucial to successful and sustainable water resource management efforts in Tanzania [13]. Household involvement in decision-making was found to have had a positive impact on residents' satisfaction, equal access, and time savings in rural water supply projects in India [14]. In the Netherlands, the Netherlands Environmental Assessment Agency developed their *Guidance for Stakeholder Participation* as an attempt to standardize budding theory in the field and put them into practice [15].

In a similar fashion, though not technically specified as such, legislation in the European Union and the US requires public participation in water policy development and implementation [16,17]. In Arizona, a stakeholder engagement project included resident input into the development of a hydrologic model, linking participation with groundwater governance [18].

This paper demonstrates one methodology for rapidly assessing local needs in advance of the planning and design of water resource projects. The methodology seeks to uncover highly relevant, interconnected social and environmental factors, teasing out differences in needs and goals between different stakeholder groups [19]. At the outset, it appeared that such a model might reveal that current groundwater utilization in Koraro is completely reasonable, given the variability in regional precipitation and, further, indicate specific approaches to its use that would encourage safe and effective use of the resource.

The goals of the research were: (1) to assess the efficacy of a stakeholder workshop framework for the collection of stakeholder data in a rural setting to evaluate the distribution of "viable" water sources; (2) to evaluate the stakeholder data with regards to the MDG implemented water infrastructure to understand if needs could be met; and (3) to design a water intervention strategy that addresses the needs of the stakeholders. Ultimately, data from the workshop were used to evaluate the resultant physical and social changes that stemmed from implementation of water resource interventions.

The paper presents the workshop findings and subsequent development activities in Koraro and evaluates the degree to which implemented strategies met the stated needs of residents. Only now, with the conclusion of major interventions occurring as part of the MVP, can the importance of stakeholder engagement in water access planning undergo proper evaluation with respect to its role in development practices, especially as they relate to the expired MDGs and the new SDG agenda.

2. Methods

2.1. Workshop

The research centered on residents in Koraro, Ethiopia and participation in a workshop with participant solicitation carried out by Millennium Villages' (MV) staff as they often engaged in community mobilization. A representative cross-section of the community was requested through random participation recruitment, though a minimum degree of literacy was required to fully participate. The workshop was structured to generate data useful in understanding the physical and social systems integral to water resources in Koraro and took place on 27–30 September 2007. This workshop was conducted prior to the water-related interventions in Koraro, and sought to gather information listed in Table 1:

Table 1. Breakdown of workshop data collection goals.

Physical System Data Needs	Social System Data Needs
• Current land use practices • The availability of natural water sources • The availability of constructed water sources • What water resources are used by individuals and why? • Rain strength perceptions and the influence on agriculture and seasonality of available water sources	How water availability is related to: • Children's lives • Women's lives • Men's lives • Livestock

The workshop was divided into eight sequential activities, each designed to assist in meeting the physical and social data collection goals. The workshop followed a "participation by consultation" model (also referred to as participatory rural appraisal (PRA)), whereby external participants listened to the viewpoints expressed through the workshop activities, while constructing decision framework models representing the mental models held by the stakeholders [20]. Though this form of participation is more passive as described by Mikkelsen [21] in that self-mobilization and direct community organization were not within the scope of the workshop, it was an appropriate means of documenting the opinions, behavior, and cultural realities that influence how the local population accesses and uses water, a necessary first step to planning appropriate water interventions. Several common PRA techniques were incorporated into the design of the workshop activities and are described in detail below [20,22–25].

2.1.1. Activities

Introduction: Water-Related Anecdotes

This activity was an "ice-breaker" intended to set the tone of the workshop and reinforce the importance of sharing in PRA [26]. Each participant was asked to relate his or her worst water-related experience to the group. This activity was the only opportunity within the delineated bounds of the workshop proper where participants could express themselves outside the constraints of a programmed activity. All participants were given a chance to speak and all statements were recorded and translated into English after the workshop and coded to remove identities.

Activity 1: Enumeration of Water Uses

During the plenary session, participants were asked to name ways in which water is used in Koraro. The objective of this activity was to assess whether different uses of water could be satisfied with different water sources, consistent with the physical and social needs of the Koraro populace. Asking participants to identify uses explicitly prevented non-residents from merely presuming water uses, or approaching the task with preconceived notions of how water might be used. Each participant was asked to contribute and participants were encouraged to think of uses that had not yet been previously mentioned. Uses were listed in the local language on a flip chart in the front of the room. After all uses were listed, they were translated into English so that the research team (from North America) and workshop facilitators (not-villagers but from the region) could jointly group them into six use categories: household uses, hygiene, agriculture-related uses, construction, income-generating activities, and ceremonial uses.

Activity 2: Enumeration of Water Sources

The second activity asked participants to identify water sources used in the Koraro region. Participants named specific water sources by name and location. Each participant was given at least one opportunity to name a unique source and the activity continued until no participants could identify sources not yet mentioned. The sources were labeled and placed by the workshop facilitators on a map of Koraro.

Activity 3: Water Source Selection Criteria

The objective of the water source selection criteria activity was to document the criteria used by villagers to select specific water sources for specific uses. Participants were divided into three groups based on their administrative Kushet (Talla, Koraro, and Tensuka). Groups were asked to identify the single, most important factor in their selection of a water source for each of the six general categories of water use identified previously. Although participants were told that the objective of the activity was not to reach a consensus, the activity generated a lively discussion as individuals related their experiences to the group. All answers were recorded on flip charts mounted in front of the groups. Tally marks were added to the flip chart when group members identified the same criterion for source selection for a particular water use (e.g., proximity to house, etc.).

Activity 4: Satisfaction Criteria

The objective of the satisfaction criteria activity was to collect information regarding water use specific to household practices. During this activity, participants were again divided by Kushet and each individual was asked to answer questions related to the amount of water used for household purposes, the time required in water collection, per capita water consumption, land holdings, livestock holdings, and water availability. The facilitator posed the questions and recorded responses.

Activity 5: Labor and Traditional Roles

The objective of this activity was to quantify the time spent by women, boys, girls, and men on water-related activities. Data from this activity were used in assessing the impact potential water resource interventions had on the time available for women and children to pursue educational, income-generating, or other activities related to the MDGs. Organized by Kushet, participants were asked to identify the person in his or her household with the primary responsibility for securing water for each use category. Next, each participant was asked to estimate the time spent each day by that person to secure the required water. Uses considered were the general categories identified during Activity 1 Participants were not constrained in their choice of person with primary responsibility and in many cases, donkeys, not people, were selected as the primary responsible party. Individual answers were recorded on flip charts mounted in front of the groups.

Activity 6: Seasonal Diagramming

The objective of the seasonal diagramming exercise was to quantify the variability in water availability for household and agricultural uses. The activity was developed based on methodology described by Mikkelsen [20]. Grouped by Kushet, participants were asked to develop consensuses about the timing of seasonal activities and occurrences. Answers were recorded by facilitators as time lines on flip charts mounted on the wall.

Closing

At day's end, facilitators gave brief statements thanking participants for their assistance described how results from the workshop would be used. Representatives from the participant groups were also given the opportunity to provide brief statements.

2.2. Intervention Design

Using consumption and source selection preference data collected from the workshop, appropriate water interventions were designed and evaluated. Evaluations consisted of crop modeling, groundwater modeling, and seasonal forecasting. Predictions on how the MDGs would assess current access of Koraro residents were conducted using simple distance and source characterization criteria.

3. Results and Discussion

A general timeline of the work completed as part of this research is included in Table 2. The bulk of the discussion will focus on the participatory workshop findings as well as evaluations of implemented interventions.

Table 2. Timeline of work in Koraro.

Year	Activity
2005–2006	Water supply inventory
2007	In depth study of climate and agriculture; participatory workshop and baseline surveys
2008–2009	Development of microdam design; diversion and infiltration structures built at Tonseka
2011	Mai-Mageb and Tsegei-Koraro trench construction; irrigation well construction
2012	MVP self-evaluation
2012–2013	Water level monitoring to evaluate effectiveness of trench construction

Participant solicitation for the workshop was challenging. The population of the area is nearly equally divided between men and women (51% and 49%, respectively); yet, the project was unable to reflect the same gender spread in the workshop. The workshop, instead, was primarily male with only 29% female participants. This is most likely due to the minimum literacy requirement for participation. According to the Ethiopia Demographic and Health Study (EDHS), 38% of women in Ethiopia are literate, but in rural settings this number drops to less than 30%, whereas 65% of men are literate with 60% of rural population males literate [27]. Due to the literacy disparities, the workshop sample is more representative of the literacy rates than the gender ratio for the area. Despite these literacy requirements, many still could not read complete paragraphs.

In total, 45 residents participated, with 19, 14, and 12 participants from the Koraro, Talla, and Tenskua Kushet, respectively. The participants were comprised of the following: village administrators (5); priests and religious representatives (5); female heads of home (8); students (9); water, health, and development sector personnel (10); community/agricultural facilitators (2); community elders (3); farmers (2); and a non-specified community member (1). The participant pool represented major water users and community decision makers for the area.

Participants provided stirring and, in many cases, highly personal accounts of water difficulties. Participants often related water quality to water source and health outcomes. For example, a young female participant bathed, out of professed necessity, using water from a pool she suspected to be contaminated and the girl later developed a rash she attributed to use of the pool source. Other participants attributed dysentery or other illnesses to consumption of water from pools used by livestock and wildlife.

A second recurring theme in the anecdotes was extreme thirst leading to measures not usually taken for acquisition of drinking water. In many cases, the extreme thirst occurred during migration, such as that which took place during the 1983–1984 drought, during periods of military service, during illness, or under conditions such as being lost. In numerous anecdotes, participants were forced to drink their urine or consume discharge from a sewage system, as no other water sources were available.

Difficulties in acquiring water were cited by many participants as their worst water-related anecdote. Difficulties encountered were physical hazards (well collapse, attempted rape, and attack by wild animals) in addition to other difficulties such as long times allocated to water acquisition and conflicts that arose at water points during periods of scarcity. In several cases, clay pots used for transporting water to households broke in transit, resulting in no household water for that day. One respondent provided an overview of water resource availability and coping strategies used by Koraro villagers during the 1983–1984 drought. Because drought and its response are central to the well-being of people in Koraro, a translation of this participant's description is provided in full below.

> In 1984 were dark days. Our crops started better. In the beginning of August, there was a huge dark cloud. I watched hoping for rain but it did not pour. We migrated to Western Tigray. Our animals got water and pasture. We returned but the same happened the following year. There was no rain. The groundwater had dried up. All the springs dried except those under the cliffs. Animals were also thirsty. There was no water on the flood lands. We had to migrate. We had to go early to water one animal at a time. There was conflict and competition for water. Some people were scaring people to go away from the water. We used to drink dark brown water that the animals even rejected. There were cases when people went to other villages and borrowed or bartered water for grains. Sometimes it was easier to get food than water. Sometimes we could not get both. In the past water was really scarce.

The severity of the 1983–1984 drought and the compounding political issues that lead to widespread famine and the fact that most of the worst memories come from water scarcity issues related to that time period is important to consider in water resources planning moving forward. In an investigation of the spatial distribution of rainfall in Ethiopia, Cheung, Senay, and Singh [28] found little variability in the nation's rainfall between 1960 and 2002, even though the years of the drought were among the lowest rainfall in the study. Most of the North and Northeastern parts of the country saw higher variability in the Belg rainfall season, the lighter rainfall season lasting from March to May preceding the main rainfall season (Kiremt) from June through September [28]. The Kiremt season is susceptible to El-Nino Southern Oscillation influences, marked by suppressed rainfall [29], potentially making drought conditions and the ensuing economic impact for the area a regular occurrence. Implemented water development projects must, therefore, consider climate forecasts and predictions to sustainably supply water into the future.

3.1. Water Uses

The water uses identified by workshop participants are listed in Table 3. Additional information on the use of particular water sources was developed based on discussions among participants and with the facilitators. For example, hygiene-related water activities (bathing and clothes-washing) are generally performed concurrently by different individuals at surface water sources. The categories into which uses were grouped were developed by workshop facilitators (independent of participants) and were based on activity location and not the intended use; hence "dilution of pesticides" is listed under household use as most reported physically conducting this activity at the house and not out in the field. These categories were chosen to make subsequent activities practical.

Table 3. Water Uses Identified During the Workshop.

Activity or Use	Associated Uses
Household	drinking; cooking; local beer making; making injera; making tej (wine); extinguishing fires; cleaning floors; washing dishes; making coffee and tea; milk and butter separation; dilution of pesticides; dilution of anti-malaria medication.
Agricultural	irrigation; animal watering; planting (seeding); construction of threshing circle; animal bathing.
Hygienic	bathing; washing clothes.
Ceremonial	wedding celebrations; funeral services and wakes; church services; baptism.
Construction	house wall construction; house painting; roof mudding; brick making; grain storage construction.
Income-generating	pottery; fishery; development/irrigation water; grinding mill operation; hide and skin processing; attracting bee swarms.

3.2. *Water Sources*

Koraro residents generally refer to particular water sources by a specific name, often preceded by "mai," the Tigrina word for water. General water source types identified include developed springs, undeveloped springs, one river with year-round flow (the Agefet River), one river with flow only in the rainy season and part of the dry season (the Werei River), wells with hand pumps, wells without hand pumps, constructed rainwater storage, and beds of ephemeral streams. All of the water sources in the Koraro region are depicted in Figure 1. Streams depicted in Figure 1 include ephemeral and permanent streams extracted from available world stream datasets and created by manually tracing riverbeds using satellite imagery. Similarly, no GIS shape layer existed for the Koraro Kushets requiring the manual creation of all three using administrative maps obtained from Ethiopia. The only stream with year-round flow in the study area is the Agefet River, which forms the southern boundary of the region under review. The Werei River, forming the northwest boundary of Talla Kushet, typically flows eight or nine months of the year. All other streams flow only for a short period after rain events.

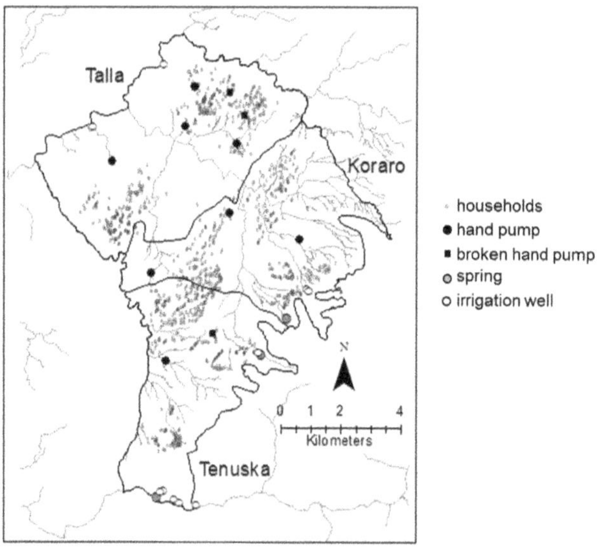

Figure 1. Water Sources in the Koraro Region including household locations (unpublished household data from the MVP baseline survey). Hand pumps are represented by the black icons (functioning hand pumps in black circles and broken pumps in squares), springs by grey circles, and irrigation wells in open circles. The smaller open circles represent the locations of households within the region.

Water points constructed immediately downstream from spring sources allow access and encourage spring protection. Undeveloped springs are generally found near the cliffs that form the eastern edge of the study area, and along the banks of the Werei and Agefet Rivers. The undeveloped springs are commonly located in a rapidly-eroding gulley approximately 50 m from the discharge of the gulley to the Agefet River. One such spring in the gully was reported to be used extensively by residents of the Tensuka Kushet.

Riverbeds of ephemeral streams were cited as an important source of water in the study region. The streams are filled with coarse sands and store runoff from the nearby cliffs. Water can be found in the streambed within 50 cm of the ground surface immediately at the end of the rainy season. Toward the end of the rainy season, the depth to water may be 2 m or greater below ground, based on information drawn from interviews with Koraro residents. Livestock were not mentioned to have restricted access to riverbeds.

3.2.1. Water Source Selection Criteria

The criteria used for selecting water for the six general use categories (from Table 3) are summarized in Figure 2. The criteria chosen for selection of household water indicates that the source selection process is more complex than evaluation of distance to the water point which the MDGs and other development standards often use to gauge access. Since "cleanliness" was the dominant reason presented for selection of household water in all locations, the perception of cleanliness must be evaluated. A hypothesis of this research was that villagers would associate water from constructed water points with higher water quality (cleanliness) especially given the surrogate use of "improved" water sources as a means to quantify access to potable water by the MDGs [30]. Discussion with workshop participants and observation of water use during field data collection indicated this assumption is not necessarily valid. Villagers were observed collecting water from shallow (<1 m) holes dug in sand in empty streambeds within 50 m of a constructed well that had a hygienic pad and hand pump. In light of the distribution of selection criteria for household uses, this choice of water source (between two alternatives with nearly identical proximity to houses) indicates that the people using the riverbed source consider the riverbed's water quality to be equivalent or better than that of the well's. While evaluating water coverage, this perception detail is often neglected when, from a practical standpoint, it should weigh more heavily in resultant discussions regarding access to safe water. In studies investigating well switching behavior of residents in Bangladesh choosing between an arsenic contaminated well and a well free of arsenic, the perceived taste and health of the water were often the determining factor influencing the act of switching to the new, safer source [31]. If the nearby hand pump is not used because the streambed water is favored, simply providing more hand pump wells might not increase access.

Another conclusion drawn from the source selection criteria activity is that certain taste aspects and the perceived presence of salinity play a larger role in source selection than was anticipated. Salinity is often a concern of farmers when discussing agricultural water; however, salinity was a concern not only for agriculture and income generating water uses, but also for ceremonial water uses in all Kushets except for Talla. When asked the reason for using the riverbed for water within a short distance of a developed source, the women drawing water attributed their choice to the "salty" taste of the water from the developed source. It is hypothesized that the water in the developed well and the riverbed come from the same shallow aquifer. If one considers cleanliness, salinity, hardness, and taste (other than "salty") all as indicators of "quality," then perceived water quality is the dominant concern in almost every use category except for hygiene, agriculture, and construction. When considering the entire Koraro region, collectors for hygienic uses of water are as equally concerned with water quality (40%) as proximity to the source (40%); however, considering the response of "flowing water" to be yet another indicator of water quality, quality concerns (47.5%) begin to dominate this decision as well.

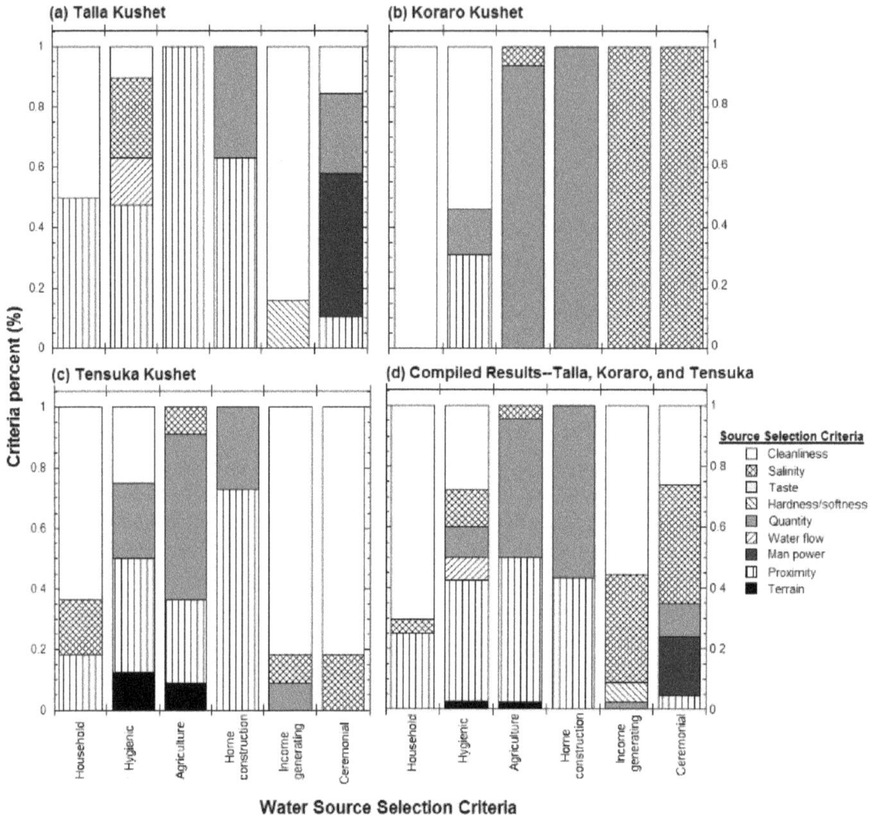

Figure 2. Water use criteria according to use and Kushet location: (**a**) Talla Kushet; (**b**) Koraro Kushet; (**c**) Tensuka Kushet; and (**d**) compiled results from all three Kushets.

Finally, proximity and quantity are the primary determinants of sources selected for agricultural uses and home construction. Agricultural uses include livestock uses and small-scale irrigation of kitchen gardens.

Satisfaction Criteria

Results from this activity allowed estimation of parameters such as average and range of per capita water consumption, time allocated for water fetching, family size, livestock holdings and land use. Insights into seasonal differences in water consumption and transportation were also documented.

Assuming jerrycans each hold 20 L, the median per capita water consumption in households of all participants is 10 L/capita/day. This value is at the lower range cited for daily use for sub-Saharan Africa of 10–20 L/person/day), [32], but, considering a more recent study by [33], wherein, families in Bongo, Ghana collected higher volumes of water per household at about 25–30 L/person/day, this value appears even more modest for the Koraro community.

The cumulative frequency distribution gathered from responses for household volume in terms of water fetched per person per day for each Kushet and for all Koraro is shown in Figure 3. Over 80% of residents collect one jerrycan or less per person per day. On average, households collect three jerrycans a day. If, perhaps, a single trip can satisfy all household activities, then there is little expectation that a separate trip to a hand pump will occur simply for drinking and cooking water, especially when taste factors are driving source selection for household needs.

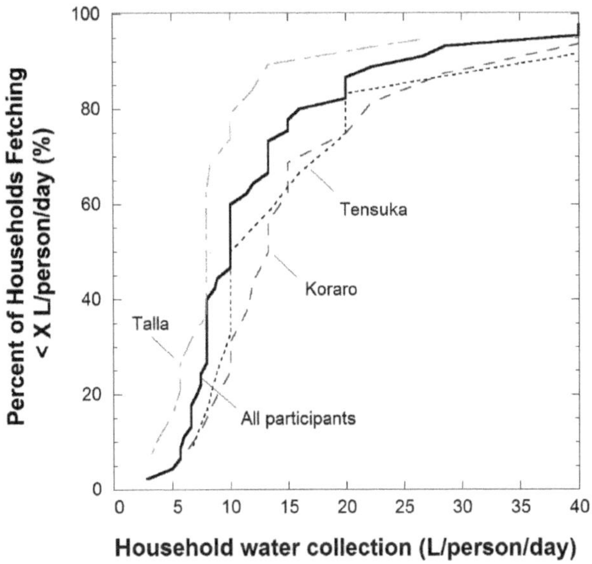

Figure 3. Cumulative distribution function for the volume of water fetched for each Kushet and for all participants.

Proximity to a source was a major contribution to source selection as discussed earlier and represented in Figure 2. As seen in Figure 2a, the Talla Kushet values proximity the most of the Kushets when considering water source selection. While Talla residents fetch the greatest amount of water per person per day, they also spend the most time, on average, traveling to collect water (Figure 4). Therefore, meeting the needs of residents in Talla will require properly differentiated interventions.

Collection times for fetching water ranged from five to 120 min. Per capita consumption (in L/person/day) is plotted against time to fetch water in Figure 5. Contrary to expectations, there does not appear to be a correlation between time to fetch water and per capita consumption. One interpretation of this result is that household water consumption is determined more by household use and conservation practices than by distance to the source and water-use activities.

Labor and Traditional Roles

All data collected during the activity are presented in Figure 6. In general, there was little variation in results between Kushets, potentially indicating that traditional roles are the primary determinants in the division of water acquisition labor. Of interest is the selection of animal as the party with primary responsibility for fetching water. This choice was not listed among responsible persons on the flip charts used to record answers, but the answer "donkey" appeared for all three Kushets. The person from the household who brings the donkey to collect the water was not recorded.

Approximately 75% of the household water collection responsibility was that of the adult females of the household, consistent with previous studies of sub-Saharan Africa [33,34]. This study, however, also highlights the importance of women in collecting water for hygienic and income generating activities. As interventions that alter the water landscape are introduced to Koraro, understanding the primary demographic that will interact with the source and soliciting their opinion is important. If, for example, a source that would provide many residents with income generating and household water is planned, understanding that women are the primary users of both but that source selection criteria are variously influenced will determine whether sources closest to the activity or the perceived salinity is most important.

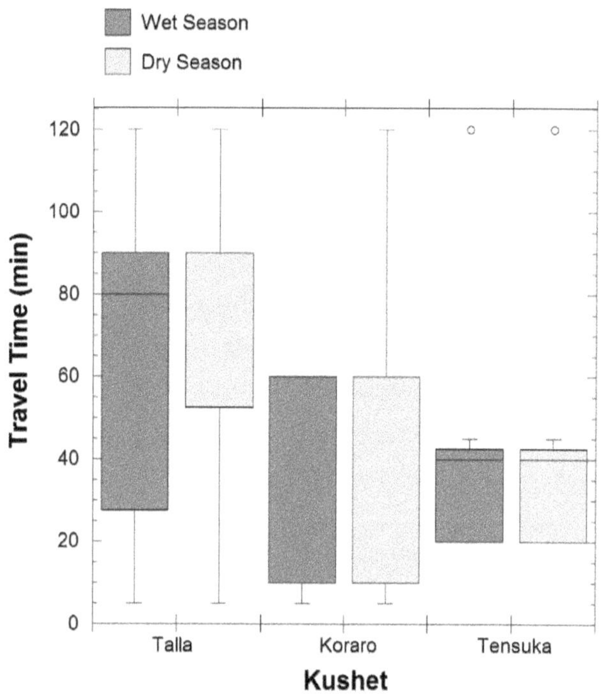

Figure 4. Water collection travel time by Kushet and season. The boxplots represent the 25th, 50th, and 75th percentiles with the upper whiskers marking 1.5 the interquartile range from the corresponding quartile for each dataset.

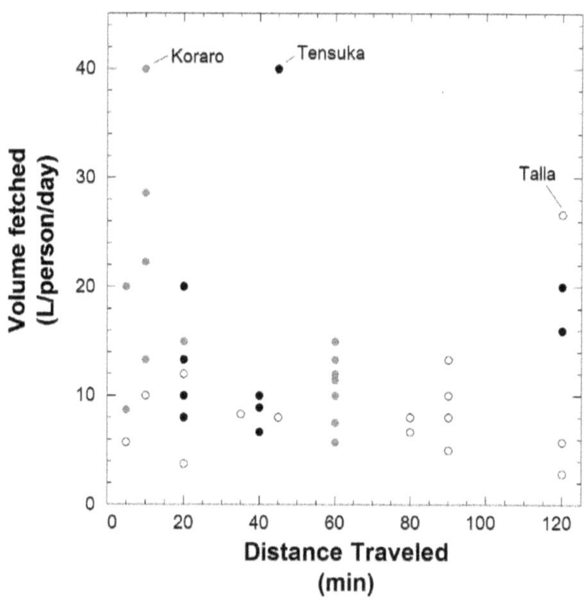

Figure 5. Volume of water fetched with respect to distance traveled.

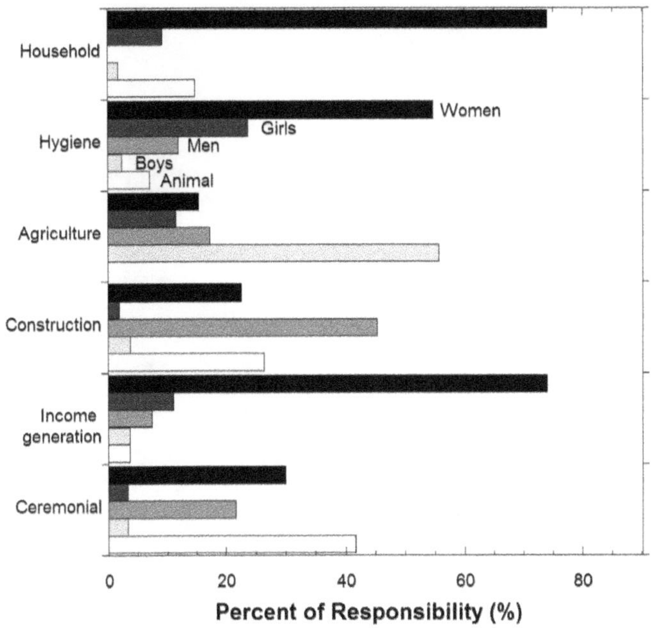

Figure 6. Division of water collection responsibilities by water use category.

The results from the workshop highlight the complex nature of source selection and use categories, even when the primary water users are exactly the same. Future research to identify the selection criteria of primary water users will further highlight the factors influencing choice of source for each activity.

3.3. Seasonal Diagramming

Figure 7 contains the participants' depictions of duration of an average rainy season and the distribution in the planting periods of four important crops: teff; barley; local beans (adengor); and corn. The Ethiopian calendar has 13 months (12 of which are 30 days long, and one, the thirteenth of the year, lasting five days) and those months are shown at the top of the figure, along with the corresponding day of the start of the month in the Julian calendar. Teff is a particularly important crop because it is a high-value crop and is the main ingredient for the local bread (injera) that accompanies meals. Barley is important as a relatively drought resistant crop and because it is used in production of local beer. The crop reported as having the earliest planting date, according to the resident participants, is the dark sorghum grown in Talla Kushet—planting occurs approximately three months before the onset of the average rainy season as reported by the Talla Kushet participants.

The estimates of the span of an average rainy season were surprisingly different for the three Kushets. Anecdotes from experts in the region and from workshop participants indicate that there are often interruptions in the rain during the rainy season. Prolonged interruptions are particularly detrimental to crop production.

Estimates of the range of planting times for the crops were similar from Kushet to Kushet. The timing of the crop plantings appears to coincide with the arrival of rains, corn with the Belg rains in March and the other crops with the Kiremt in June. Corn and millet (not shown) were reported as being planted at the earliest rains, with local beans planted later into the rainy season when there is more consensus among participants regarding consistent rainfall patterns.

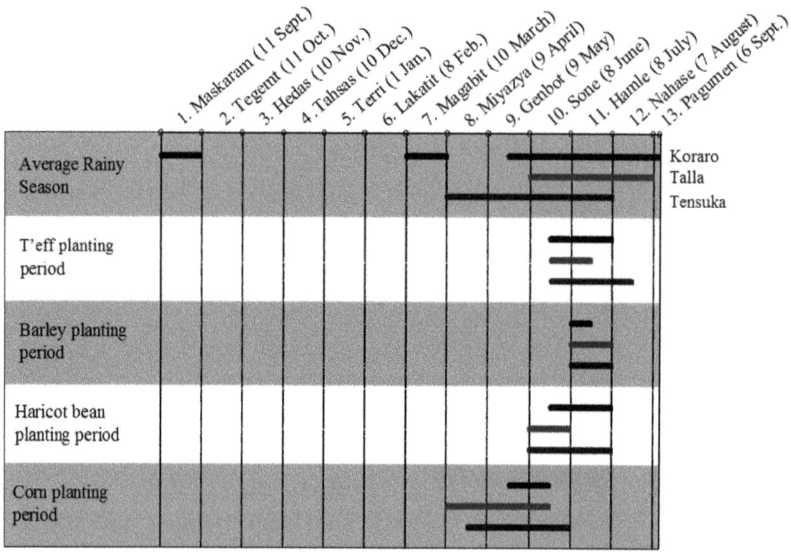

Figure 7. Planting Periods for main crops, by Kushet.

In general, Koraro residents indicated that water is available year-round in most sources, even in years of low rainfall. In riverbeds, water can be obtained in shallow holes (<1 m) at the end of the rainy season but only in deeper holes at the end of the dry season. Despite reports of failing pond water storage systems, the participants indicate that operational ponds yield water through much of the growing season, drying only before the most arid part of the year. Two water sources were identified as seasonal: the Werei River and the Koakih Riverbed. All springs (developed and undeveloped) were reported to flow year-round in normal years.

3.4. Merging Data Collected in the Workshop with Engineering Design and Decision

The purpose of the workshop was to evaluate water needs and preferences to enable the investigation of MDG interventions as well as those implemented by the Columbia Water Center. This screening process presents an alternative to specification-oriented engineering design, in which the goal of design is to meet criteria that were developed as general guidelines. If the accepted definitions of the MDGs for access to safe water are used, then many residents would be classified as not having access since sources currently being used are not all improved (MDG indicator 30: "*proportion of population with sustainable access to an improved water source*"). While the MDGs helped to increase the efforts related to water, broader and more context-specific frameworks were needed for the post-2015 water development agenda [30]. The language in the new SDGs is less prescriptive than the MDGs (SDG Goal 6.1: "*achieve universal and equitable access to safe and affordable drinking water for all*"), addresses the need to consider water quality (SDG Goal 6.3: "*improve water quality by reducing pollution, eliminating dumping and minimizing release of hazardous chemicals and materials*") and requires local participation (SDG Goal 6b: " *... support and strengthen the participation of local communities in improving water and sanitation management*") [30]. No longer is the standard (yet vague) metric of "improved source" being used as a marker of "safe" water since water quality and wastewater management are both primary target areas under the SDG [30]. Using this sustainability assessment approach to design may allow improved intervention selection and design through participatory community decision-making and integrated and scaled-up interventions that consider cross-sector impacts.

To contrast the use of specifications and sustainability in intervention selection and design, a relatively simple analysis based on the access criteria of distance and source type was performed.

Then, the determination of the number of wells that would have to be added to ensure "adequate" access to water for all Koraro residents was calculated. The process used was similar to that used by Hassan [35] in selection of deep well location for providing improved access to safe drinking water in Bangladesh. A geographic information system (GIS) model was used to generate service zones for existing water sources, where the service zones were buffers of 500 m, 1000 m, 1500 m and 2000 m around each of the water sources. The service zones for existing wells for a 1000 m service area around each water point are shown in Figure 8. Even this process overestimates the number of sources deemed acceptable by the MDG metrics as all constructed sources were used in the analysis. Expert judgment was used to determine the minimum number of additional wells that would be required to satisfy the service zone specifications that all houses be within 500 m, 1000 m, 1500 m and 2000 m of a water source. The resulting well field for a 1000 m service area is also depicted in Figure 8.

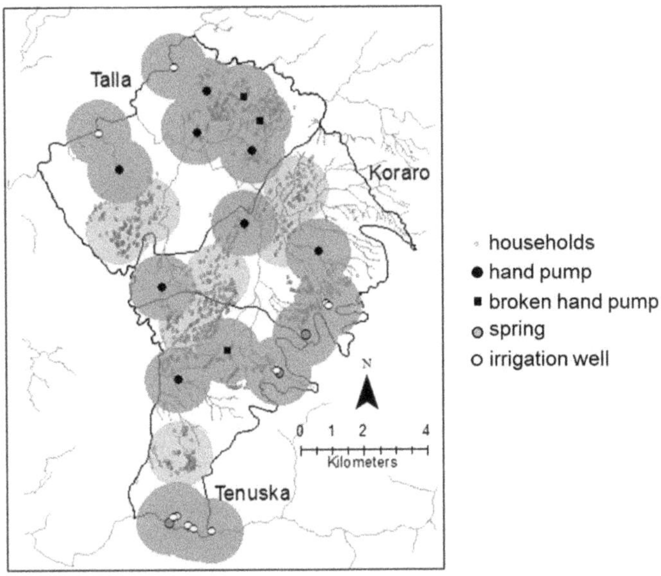

Figure 8. Well Field with well placement to satisfy the criteria of households being 1000 m of a water source (additional wells depicted by the light grey buffer).

Maps similar to those in Figure 8 were developed to illustrate service areas with 500 m, 1500 m and 2000 m radii from water points. The number of additional wells required to achieve the design specification corresponding to 500 m, 1000 m, and 1500 m are 30, 7, and 4 wells, respectively. If a 2000 m service area radius were adopted, no new wells are required. Of course, this approach ignores the fact that many of the current sources being used by Koraro residents are not "improved" sources.

In 2007, the typical cost for installation and drilling of individual shallow wells such as those used in the Koraro region was $38,000. Thus, for the well fields proposed, construction and development costs associated with a 500 m service radius are approximately $840,000, or around $170 per Koraro resident. As of the 2008 MVP report, 16 new wells, out of a planned 39, were drilled in the region [36]. From these numbers, sustainability of access aligns closely to the 500 m radius analysis and distance threshold and might not reflect the actual perceptions of residents. Moving forward, particular attention must be paid to these metrics and the community specified needs as the SDGs begin to implement development plans and design evaluation metrics.

To determine whether the expense of adding wells is "worth it," the anticipated impact of the wells on individuals' activities and contribution to sustainability in the region, an alternate map was created in which 1 km buffers were generated around current water sources and riverbeds (Figure 9). The justification of buffering riverbeds stems from the observed incidence of use of riverbeds as a household water source as reported in the workshop, even when a constructed water source is available at the same location. The reported presence of water in river sediments year-round (reported in the seasonal diagramming activity) and the association villagers appear to make between riverbed water, cleanliness, and agreeable taste makes the riverbed a source that must be considered in development activities in Koraro. As few as two additional wells would be needed to achieve a goal of each household being within 1 km of a water source when riverbeds are considered as viable water sources. This approach, while it contradicts the MDG format, might be more acceptable under the new SDG guidelines. It is apparent that the goals need more flexibility in determining adequate water sources and additional attention to actual water quality at the source, water quality from household containers, and household treatment practices is required to understand the quality of consumed water.

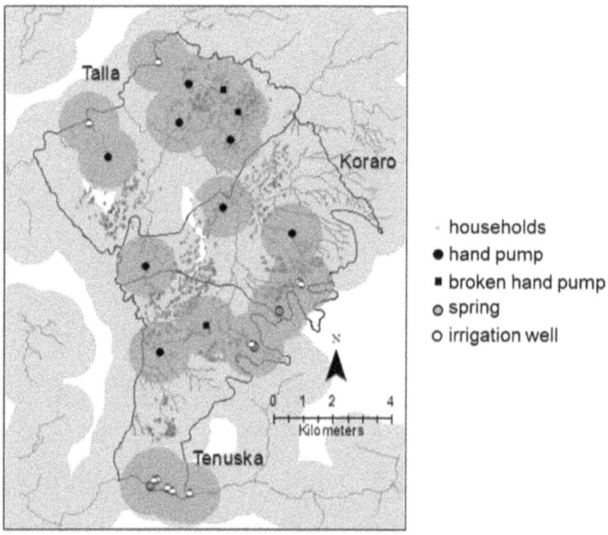

Figure 9. One thousand-meter buffer area on all sources and riverbeds in the research area (buffer area shown in grey).

It is realized that the preceding analysis is simplistic. Not all of the riverbeds provide easy access to water. For example, it is expected that in the regions where stream density is high (the headwater streams with intermittent flows in the higher elevations in the Eastern portion of Koraro) storage may be limited and water may not be available year-round. Regardless, the depiction of served areas better reflects use patterns observed and reported in the Koraro region and it illustrates that any addition of wells based on buffers around only constructed water points does not necessarily generate the greatest impact per expenditure in terms of contribution to the global sustainability of the region.

In light of the workshop results and the above analysis, the construction of several diversion and infiltration structures as well as irrigation wells (a total of eight) was completed between 2009 and 2013 as part of the MVP project. Gullies were situated along the eastern part of Koraro where the presence of several springs is coupled with high probability of enhanced recharge. Given the workshop discussions, water needs, and preferences expressed, it was hypothesized that properly planned recharge infrastructure would increase water availability to all water sources in the region (wells and streambeds) and thereby increase drinking water and irrigation potential. From the workshop it is known that agricultural water source selection is predominately based on quantity and proximity;

therefore, increasing the quantity to all sources might satisfy both criteria. The goal was to ensure water availability during the driest period before the rains return in March.

3.5. Implementing and Evaluating One Solution: Irrigation Diversion and Percolation Ponds

The first diversion and infiltration structures were constructed in 2009 in Tonseka. In 2011, the Mai-Megab and Tsegei-Koraro trench were completed along with eight irrigation wells. The success of the project was evaluated in 2011 through 2013 through harvest benefits, MVP self-evaluation reports, and visits by the Columbia Water Center team to verify assessments. Prior to the installation of irrigation measures, average yields were approximately 800 kg/ha barley, 1200 kg/ha maize, 700 kg/ha millet, 1000 kg/ha sorghum, and 600 kg/ha teff as reported by the MVP project staff. Additionally, in the 2008 Annual Report for the Koraro Millennium Village Project, 98% of the households surveyed had insufficient food [36].

Given the lack of baseline information regarding measured stream flows, it is not possible to determine the extent of the increase in water availability; however, increased flows in seasonal streams were reported by both farmers and the MVP team. Tracking water in the wells downgradient from the installed Mai Gaba trench was a more accurate measure of groundwater recharge from the constructed trenches and a more reliable marker of increased water viability in the region (see Figure 10). In Figure 10, location of the wells are presented in the Google Earth plot (Figure 10a), as well as time series well depth measurements (Figure 10b) and average monthly rainfall for 2012–2013 (Figure 10c) from the nearby meteorological station in Gondor (data downloaded from KNMI Data Explorer [37]). In 2013, six wells were tracked weekly from the last week in 2012 until the second week of April 2013. During the team visit in December 2012, several wells that were previously dry in 2008 contained water. In the first week of March, the well furthest downstream (1.07 km away aerially, approximately 1.33 km away from the trench via the streambed) was the first to go dry. The well closest to the trenches (0.35 km, both aerially and via the streambed, away from the trench) maintained at least 0.2 m of water until going dry by the third week of March. Furthermore, it appears that small amounts of rainfall in the area might directly result in groundwater recharge. More monitoring of the observation wells and local precipitation is needed to definitively assess recharge potential for the area. Similarly, the eight irrigation wells constructed in the area all held water during visits in December 2012.

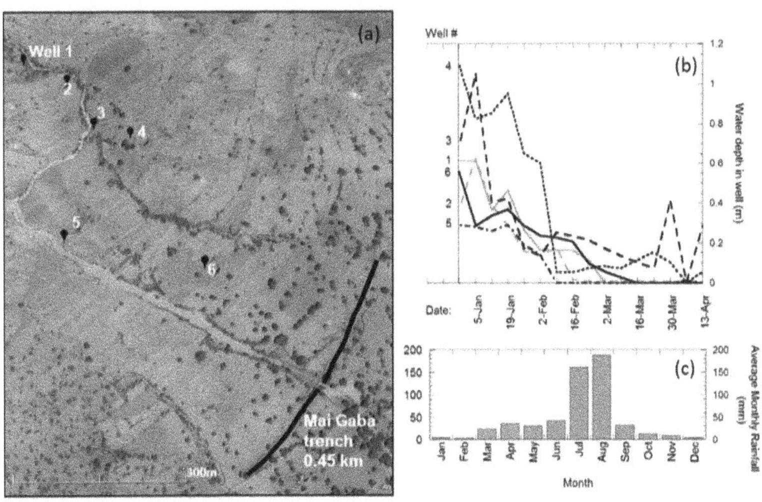

Figure 10. (**a**) Google Earth map of the observation wells and nearby Mai Gaba trench; (**b**) well depth observations for all six wells; and (**c**) monthly rainfall averages from 1992 to 2007 from nearby Hawzen meteorological station [38].

Direct farmer benefit was measured in increased hectare of productivity as well as availability of water. Irrigable area in Koraro has increased, however, this increase is a direct result of all the interventions and impossible to attribute to one intervention alone. One additional economic benefit as a consequence of increased water availability is the establishment of a sericulture (silkworm) cooperative.

3.6. Future Approaches for Development Interventions

The process of evaluating water needs and development interventions has resulted in a complex diagram of water needs for the area, as depicted in Figure 11. Representing all the dynamics in a single water assessment model is complex, yet is required for future interventions. As the SDGs continue into the future, models that attempt to understand the complexity surrounding water use and the impacts of implemented techniques are required. For a place like Koraro, the complex interactions depicted in Figure 11 can be subdivided into three main interaction systems: the community user, the hydrologic, and the agricultural production system. Future research is required to properly understand the interactions as development interventions are introduced.

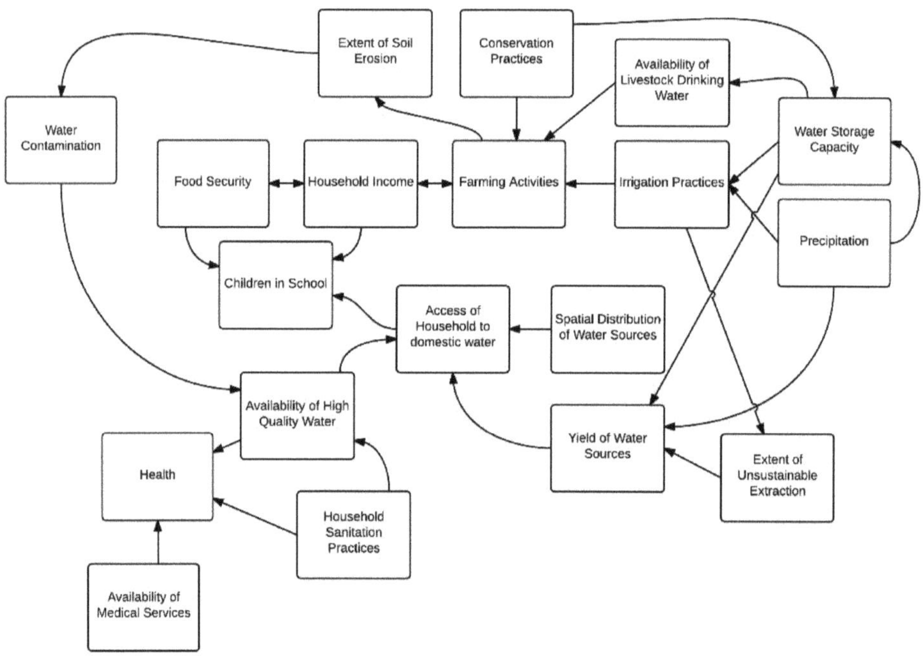

Figure 11. Interrelationships of various sectors related to water development.

The authors conclude that using the stakeholder meeting as outlined in this paper to inform development can help to address many of the inputs required by the SDGs, and are crucial to successful implementation of future waterworks, but more and prolonged research is needed to adequately assess the success of this approach to data gathering.

4. Conclusions

From the information collected by the Koraro stakeholder meeting, it is apparent that a one-size solution is not even applicable for the differing Kushets let alone the region. The reliance on streambeds for water and the perceived cleanliness and salinity requires further analysis, but does shift a development initiative's focus away from the normative, prescriptive approach of constructing boreholes. Interventions that improve or increase streambed water resources appear to be most desirable among residents of the Koraro region and those that were implemented are increasing water availability and potential earning power for many residents. If a 500 m buffer to an "improved" water source were the only criteria used to satisfy "reliable water provisions," the area would need to spend over $800,000, but doing so may not actually serve the needs of the community as determined by the selection criteria voiced in the workshop. The irrigation and recharge projects constructed as a direct result of the workshops helped to increase water availability in the available sources.

Finally, considering only the gender of the responsible individual to collect would not necessarily determine the criteria for source selection. By conducting the activities associated with the workshop that highlight the selection criteria, the gender roles, and the distance of sources, it is apparent that all impact the volume of water collected by households and which sources are favored. Even though the MDGs are no longer in use, access to clean water remains a prominent focus of the SDGs, indicating that there is a still considerable progress needed in solving the water access problems, especially in water scarce regions like Koraro. Development must consider the complex interactions of all systems using water within a location. This paper documents one method that can be used to assist in customizing interventions to maximize multi-domain sustainability benefits and understand the community user perspective. The only way in which the SDGs can meet the objective of achieving equitable access to safe and affordable drinking water for all is if stakeholders are involved in the process and all voices are considered and documented in the process.

Acknowledgments: The authors would like to acknowledge the Earth Institute Postdoctoral Fellowship, the Pulitzer Foundation, and the Columbia Water Center in support of this work. In addition, the authors would like to thank those associated with the Millennium Villages Project that helped with data collection, sharing, and local logistics for this research.

Author Contributions: Franco Montalto and Upmanu Lall conceived and designed the experiments; Franco Montalto, Timothy Bartrand, and Tsegay Wolde-Georgis performed the experiments; Katherine Alfredo led the data analysis with input from all the authors; and Katherine Alfredo, Franco Montalto, and Timothy Bartrand wrote the paper.

Conflicts of Interest: The authors declare no conflict of interest.

References

1. United Nations. United Nation Millennium Declaration. Availiabe online: http://www.un.org/en/ga/search/view_doc.asp?symbol=A/RES/55/2 (accessed on 29 June 2016).
2. Sachs, J.D. From millennium development goals to sustainable development goals. *Lancet* **2012**, *379*, 2206–2211. [CrossRef]
3. Suresh, P.; Johnson, L.R. The post-2015 agenda: From Millennium Development Goals (MDGs) to Sustainable Development Goals (SDGs). *Int. J. Curr. Res. Rev.* **2015**, *7*, 62–68.
4. Sanchez, P.; Palm, C.; Sachs, J.; Denning, G.; Flor, R.; Harawa, R.; Jama, B.; Kiflemariam, T.; Konecky, B.; Kozar, R.; et al. The African Millennium Villages. *Proc. Natl. Acad. Sci. USA* **2007**, *104*, 16775–16780. [CrossRef] [PubMed]
5. Scott, J.C. *Seeing Like a State: How Certain Schemes to Improve the Human Condition Have Failed;* Yale University Press: London, UK, 1998.
6. Hanna, N. *Analytical and Advisory Services for Comprehensive and Participatory Development;* OED Work. Pap. Ser. No. 12; The World Bank: Washington, DC, USA, 2000.
7. Lansing, J.S. *Perfect Order: Recognizing Complexity in Bali;* Princeton University Press: Princeton, NJ, USA, 2006.
8. Bewket, W.; Sterk, G. Farmers' participation in soil and water conservation activities in the Chemoga Watershed, Blue Nile basin, Ethiopia. *Land. Degrad. Dev.* **2002**, *13*, 189–200. [CrossRef]
9. Reed, M.S. Stakeholder participation for environmental management: A literature review. *Biol. Conserv.* **2008**, *141*, 2417–2431. [CrossRef]

10. Greenwood, D.J.; Levin, M. *Introduction to Action Research: Social Research for Social Change*, 2nd ed.; Sage Publications, Inc.: Thousand Oaks, CA, USA, 2007.
11. United Nations Development Program (UNDP). *Supporting Capacities for Integrated Local Development*; UNDP: New York, NY, USA, 2007.
12. Khasankhanova, G. Ublic participation to improve water resource management in Uzbekistan. *Water Sci. Technol.* **2005**, *51*, 365–372. [PubMed]
13. Dungumaro, E.W.; Madulu, N.F. Public participation in integrated water resources management: The case of Tanzania. *Phys. Chem. Earth Parts A/B/C* **2003**, *28*, 1009–1014. [CrossRef]
14. Prokopy, L.S. The relationship between participation and project outcomes: Evidence from rural water supply projects in India. *World Dev.* **2005**, *33*, 1801–1819. [CrossRef]
15. Hage, M.; Leroy, P.; Petersen, A.C. Stakeholder participation in environmental knowledge production. *Futures* **2010**, *42*, 254–264. [CrossRef]
16. United States Environmental Protcction Agency (USEPA). *Memorandum: Guidance for 2006 Assessment, Listing and Reporting Requirements Pursuant to Sections 303(d), 305(b) and 314 of the Clean Water Act; No. 1*; USEPA: Washington, DC, USA, 2006; pp. 1–89.
17. De Stefano, L. Facing the water framework directive challenges: A baseline of stakeholder participation in the European Union. *J. Environ. Manag.* **2010**, *91*, 1332–1340. [CrossRef] [PubMed]
18. Eden, S.; Megdal, S.; Shamir, E.; Chief, K.; Mott Lacroix, K. Opening the Black Box: Using a Hydrological Model to Link Stakeholder Engagement with Groundwater Management. *Water* **2016**, *8*. [CrossRef]
19. Cleaver, F. Paradoxes of participation: Questioning participatory approaches to development. *J. Int. Dev.* **1999**, *11*, 597–612. [CrossRef]
20. Mikkelsen, B. *Methods for Development Work and Research: A Guide for Practitioners;* Sage Publications, Inc.: New Delhi, India, 1995.
21. Mikkelsen, B. Participation in Development—The concept and critical perspectives. In *Methods for Development Work and Research: A New Guide for Practitioners;* Sage Publications, Inc.: New Delhi, India, 2005; pp. 53–87.
22. Campbell, J. A critical appraisal of participatory methods in development research. *Int. J. Soc. Res. Methodol.* **2002**, *5*, 19–29. [CrossRef]
23. Chambers, R. The origins and practice of participatory rural appraisal. *World Dev.* **1994**, *22*, 953–969. [CrossRef]
24. Chambers, R. Participatory rural appraisal (PRA): Analysis of experience. *World Dev.* **1994**, *22*, 1253–1268. [CrossRef]
25. Bell, S.; Morse, S.; Shah, R.A. Understanding stakeholder participation in research as part of sustainable development. *J. Environ. Manag.* **2012**, *101*, 13–22. [CrossRef] [PubMed]
26. Narayanasamy, N. The concept and principles of participatory rural appraisal. In *Participatory Rural Appraisal: Principles, Methods and Application*; Sage Publications, Inc.: New Delhi, India, 2009; pp. 25–41.
27. Central Statistical Agency and ICF International. *Ethiopia Demographic and Health Survey 2011*; Ethiopia Central Statistical Agency: Addis Ababa, Ethiopia; ICF International: Calverton, MD, USA, 2012.
28. Cheung, W.H.; Senay, B.; Singh, A. Trends and spatial distribution of annual and seasonal rainfall in Ethiopia. *Int. J. Climatol.* **2008**, *28*, 1723–1734. [CrossRef]
29. Korecha, D.; Barnston, A.G. Predictability of June–September Rainfall in Ethiopia. *Mon. Weather Rev.* **2007**, *135*, 628–650. [CrossRef]
30. Worldwide Asset Position (WWAP). *The United Nations World Water Development Report 2015: Water for a Sustainable World*; Worldwide Asset Position (WWAP): Paris, France, 2015.
31. Mosler, H.J.; Blochliger, O.R.; Inauen, J. Personal, social, and situational factors influencing the consumption of drinking water from arsenic-safe deep tubewells in Bangladesh. *J. Environ. Manag.* **2010**, *91*, 1316–1323. [CrossRef] [PubMed]
32. Rosen, S.; Vincent, J.R.; Esrey, S.A.; Gugerty, M.K.; Larson, B.A.; Malaney, P.; Simon, J.; Timmer, C.P.; White, G.F.; Whittington, D. *Household Water Resources and Rural Productivity in Sub-Saharan Africa: A Review of the Evidence*; Harvard Institute for International Development, Harvard University: Cambridge, MA, USA, 1999.
33. Alfredo, K.A.; Lawler, D.F.; Katz, L.E. Fluoride contamination in the Bongo District of Ghana, West Africa: Geogenic contamination and cultural complexities. *Water Int.* **2014**, *39*, 486–503. [CrossRef]

34. Bauchspies, W. The community water jar: Gender and technology in Guinea West Africa. *J. Asian Afr. Stud.* **2012**, *47*, 392–403. [CrossRef]
35. Hassan, M.M. Arsenic poisoning in Bangladesh: Spatial mitigation planning with GIS and public participation. *Health Policy N. Y.* **2005**, *74*, 247–260. [CrossRef] [PubMed]
36. The Earth Institute; Millenium Promise; United Nations Development Program. *The Millennium Villages Project: Annual Report 2008*; Columbia University: New York, NY, USA, 2008.
37. KNMI Climate Explorer. Monthly GONDAR Mean Monthly GHCN v2 Precipitation. Available online: climexp.knmi.nl (accessed on 17 May 2016).
38. Nedaw, D. Water Balance and Groundwater Quality of Koraro Area, Tigray, Northern Ethiopia. *CNCS Mekelle Univ.* **2010**, *2*, 110–127. [CrossRef]

water

MDPI

Article

Engaging Southwestern Tribes in Sustainable Water Resources Topics and Management

Karletta Chief [1,*], Alison Meadow [2] and Kyle Whyte [3]

1 Department of Soil, Water, and Environmental Sciences, University of Arizona, Tucson, AZ 85721, USA
2 Center for Climate Adapation and Science Solutions, University of Arizona, Tucson, AZ 85721, USA; meadow@email.arizona.edu
3 Department of Philosophy and Department of Community Sustainability, Michigan State University, East Lansing, MI 48824-1032, USA; kwhyte@msu.edu
* Correspondence: kchief@email.arizona.edu; Tel.: +1-520-626-5598

Academic Editors: Eylon Shamir, Sharon B. Megdal and Susanna Eden
Received: 5 April 2016; Accepted: 13 July 2016; Published: 18 August 2016

Abstract: Indigenous peoples in North America have a long history of understanding their societies as having an intimate relationship with their physical environments. Their cultures, traditions, and identities are based on the ecosystems and sacred places that shape their world. Their respect for their ancestors and 'Mother Earth' speaks of unique value and knowledge systems different than the value and knowledge systems of the dominant United States settler society. The value and knowledge systems of each indigenous and non-indigenous community are different but collide when water resources are endangered. One of the challenges that face indigenous people regarding the management of water relates to their opposition to the commodification of water for availability to select individuals. External researchers seeking to work with indigenous peoples on water research or management must learn how to design research or water management projects that respect indigenous cultural contexts, histories of interactions with settler governments and researchers, and the current socio-economic and political situations in which indigenous peoples are embedded. They should pay particular attention to the process of collaborating on water resource topics and management with and among indigenous communities while integrating Western and indigenous sciences in ways that are beneficial to both knowledge systems. The objectives of this paper are to (1) to provide an overview of the context of current indigenous water management issues, especially for the U.S. federally recognized tribes in the Southwestern United States; (2) to synthesize approaches to engage indigenous persons, communities, and governments on water resources topics and management; and (3) to compare the successes of engaging Southwestern tribes in five examples to highlight some significant activities for collaborating with tribes on water resources research and management. In discussing the engagement approaches of these five selected cases, we considered the four "simple rules" of tribal research, which are to ask about ethics, do more listening, follow tribal research protocols, and give back to the community. For the five select cases of collaboration involving Southwestern tribes, the success of external researchers with the tribes involved comprehensive engagement of diverse tribal audience from grassroots level to central tribal government, tribal oversight, on-going dialogue, transparency of data, and reporting back. There is a strong recognition of the importance of engaging tribal participants in water management discussions particularly with pressing impacts of drought, climate change, and mining and defining water rights.

Keywords: indigenous; tribes; Native Americans; stakeholder engagement

1. Introduction

Indigenous people often understand themselves as having an intimate relationship to the environment and cosmos in which they consider every element and species to have life and to be sacred. Some indigenous people believe that human and nonhuman individuals come from the earth and the ability to reach harmony among individuals is dependent on being a steward of the natural environment by giving back more than what is taken. To many indigenous people, water is life and water is sacred. Water is part of religious and cultural practices for purification, prayers, and various ceremonies. Water is also part of indigenous identities and origin stories; for a Diné-specific example, Diné deities include "Born-for-water" and Diné clans include "Big-water", "Near-the-water" and many other water-based clans. To indigenous people, the use of water is integrated with respect for the water as a living entity that gives life and supports the health, integrity, and character of an individual. Similar relations exist across North America. For Anishinaabe people (Ojibwe, Odawa, Potawatomi, among others), the value of water arises from the creation story in which water is considered to play the role of a source and supporter of life. Water mediates interactions among many living beings on the earth and is considered a relative with responsibilities to give and support life. Bodies of water are considered to have their own unique personalities. Humans have special responsibilities to respect and care for water and to encourage its life-giving force. Ceremonies are organized to ensure people renew their connections to water [1,2]. The sacredness of water represents a cross-cutting way Southwestern tribes approach their relationship to the environment.

One hundred and eighty-two federally recognized tribes have tribal lands in the six states of the Southwestern United States: Arizona, New Mexico, Colorado, Utah, Nevada, and California [3]. Tribal land holdings range from small rancherias, colonies, and reservations, such as the 20-acre reservation of the Augustine Band of Cahuilla Indians in California to the 27,413 square-mile reservation of the Navajo Nation located across Arizona, Utah, and New Mexico. Tribal lands are in rural and urban areas, for example, the Fort McDermitt Paiute-Shoshone Reservation lies within the rural Nevada-Oregon high deserts and the Las Vegas Tribe of Paiute Indians of the Las Vegas Indian Colony is located in the heart of the Las Vegas metropolitan area. Tribal economies vary widely including ranching, agriculture, mining, gaming, tourism, retail, and various other industries. Tribal lands span diverse ecosystems and climatic regions with highly varied precipitation and temperature patterns from arid deserts to fertile valleys as well as coastal and mountainous terrains. The Southwest is home to seven of the largest tribes and five of the largest Indian reservations. Many cities in the Southwest have large and diverse tribal communities that have organized centers and other institutions that provide health and other services, legal advocacy, and support of cultural continuance. California has the largest percentage of citizens (14%) who identify as American Indian and Alaska Native alone or in combination with another race (followed by Oklahoma, Arizona, Texas, and New York) [4] and 33% of the state of Arizona is tribal land. Tribes in the Southwest are diverse in their languages, traditions, beliefs, and geographic settings.

The legal, political, and cultural frameworks surrounding water in the U.S., which often treat water as a commodity that can be transferred and sold [5], can cause dissonance and challenges for indigenous communities who wish to assert their legal and political rights to water through their own cultural frameworks that treat water as sacred. There are at least four overarching challenges that face tribes regarding this dissonance, some of which are issues primarily for U.S. federally recognized tribes and others that are concerns for indigenous peoples more broadly. First, water laws governing the Western U.S. are based upon the concept of prior appropriation, or "first in time, first in right" and give priority to those water users presumed to have first put the water to beneficial use [5,6]. One specific barrier created by the idea of prior appropriation is that only federally recognized tribes are given this Westernized right [7]. For tribes that have occupied ancestral lands for thousands of years through complex cultural and political systems, which often emphasize kin, clan, and spiritual relationships, the concept of divvying up waters to tribal governments or to individual persons, settler or indigenous, is difficult to accept. Moreover, recognition of water rights for cultural and ceremonial practices is not always considered or part of the legal processes of water quantification [8]. Secondly, water contamination can mean two very different things to indigenous and non-indigenous communities. When water is contaminated, the contamination can be emotionally devastating and traumatic for indigenous peoples, particularly when the water is connected to sacred sites, religious concepts, and subsistence activities [9–13]. Third, partly as a result of the legal and political challenges discussed above, indigenous communities have often been excluded from negotiations and discussions about water management practices. Often, indigenous people are not at the table when water management decisions are made and data are not available in transparent or accessible ways [14,15]. Finally, the fourth overarching challenge—which certainly is influenced by and influences the other three—is that traditional knowledge(s) (TK) related to the management of water have often not been given equal standing or respect in U.S., Canadian, and other Western water management frameworks [14,15]. Traditional knowledge is local knowledge which have been gathered and refined over hundreds of years, passed down from generation to generation through particular cultural, economic, and spiritual practices and is part of the fabric of indigenous communities [16]. Traditional knowledges can be used to structure and guide scientific research, corroborate environmental data collected using instruments, and offer options for resilient management practices for indigenous people [17,18]. In order to effectively engage tribes in water management topics, it is important to recognize these challenges and ways to address these challenges using culturally appropriate and tribal specific engagement.

The objectives of this paper are (1) to provide an overview of the context of current tribal water management issues, especially for U.S. federally recognized tribes in the Southwestern United States; (2) to synthesize approaches to engage tribal individuals, communities, and governments on water resources topics and management; and (3) to compare the success of engaging Southwestern tribes in five examples to highlight some significant activities for collaborating with tribes on water resources research and management.

2. Context of Tribal Water Resources Management Issues

2.1. Indigenous Governance and Sovereignty

The legal, political, and historical frameworks that govern tribes within the U.S. strongly influence the ways in which tribal water rights are allocated and the ways in which water is and can be managed by tribes. In the following section, we outline the key principles of federal Indian law and tribal water rights, especially ones that are relevant to U.S. federally recognized Southwestern tribes. In the United States, there are 567 federally recognized tribes and at least 34 state recognized tribes [3]. The former are sovereign nations in the eyes of the U.S. who maintain government-to-government relationships with the U.S. government. Tribal sovereignty refers to the principle of self-governance; in the context of federal-tribal relations it means that any rights not explicitly ceded through treaty making are assumed to remain under tribal control—such as rights to water, natural resources, and internal self-governance. However, that sovereignty is limited because tribes now exist within the borders of the U.S. and there are limits on tribal capacity to regulate non-tribal members living in their jurisdictions. Tribes are legally described as "domestic dependent nations," in a guardian-ward relationship with the federal government [19]. The domestic dependent nation status has led to the concept of the trust doctrine, which requires the federal government to act in the best interests of tribes [19]. It is important to note that, while legal principles hold that tribes are on equal footing with the federal government and the government must support their best interests, neither the government-to-government relationship or the trust doctrine have been consistently applied throughout the history of federal-tribal relationships. Although the Department of Interior, Bureau of Indian Affairs (BIA) is assigned to enact this trust responsibility and oversee tribal assets, the BIA acting on behalf of federally recognized tribes has mismanaged programs, lands, and natural resources, and often failed to act according to the best interest of tribes [20–23]. In recent decades, progress has been made in terms of tribal control of tribal resources, such as the 1975 Indian Self-Determination and

Education Assistance Act and subsequent amendments (known informally as "638 contracts") which allow tribes to contract with the federal government to operate programs serving their tribal members and the Indian Gaming Regulatory Act of 1988, which can support natural resource management with gaming revenue.

As inherently sovereign nations, federally recognized tribes technically have the right, in the eyes of the U.S. federal government, to determine their own governing structures. Some tribes have maintained their traditional governments while other tribes have democratic Westernized tribal governments institutionalized by the U.S. government through the 1934 Indian Reorganization (Wheeler-Howard) Act [24]. For example, the Cochiti Pueblo in New Mexico maintains a traditional form of government, where the chief spiritual leader appoints the War Chief and Lieutenant War Chief, who are responsible for over-seeing and preserving the cultural and ceremonial calendar and the Governor and Lieutenant Governor, who carry out governmental affairs. On the other hand, the Ho-chunk Nation has a democratic election where candidates vying for President are elected based on popular vote. Tribes' abilities to control and manage natural resources are also affected by the ways in which land is held in their regions. Although reservation lands are controlled by tribal governments (and held in trust by the federal government), due to legislation such as the 1887 Dawes Act [25], some tribal lands have been allotted to individual owners, making them fee-simple lands controlled by individuals, not tribal governments. In some cases, reservations are "checkerboarded" with some tracts owned by tribal members or non-tribal members and others still under the control of tribal governments—making comprehensive natural resource management all the more challenging [26,27]. This is the case for the Cherokee Nation of Oklahoma where the tribal nation must manage their resources where there is significant non-tribal activity. It is also the case that within federally recognized, state recognized and unrecognized tribes, groups of tribal members continue their own "grassroots" water stewardship or management systems that are not recognized by any U.S. federal or state entity as self-determining entities. This initial use of the word "grassroots" is in quotes because sometimes these groups are exercising stewardship or management systems based on traditional knowledge that predates the U.S. or the era of federally recognized tribal governments. These groups, whether operating within federally recognized or unrecognized tribes, face the jurisdictional and governance challenges of not being respected as caretakers.

2.2. Tribal Water Rights

Federally recognized tribes have federally reserved water rights with a priority associated with the date their reservation was created. These rights were codified in the 1908 Winter's Doctrine [7]. Specifically, a tribe has rights to sufficient water to fulfill the purposes of its reservation, as defined by the U.S. government, such as agricultural production, and could not be lost due to non-use [28]. The McCarran Amendment [29] waived federal sovereign immunity in state water adjudications and allowed states to quantify federal reserved water rights including tribal reserved water rights in basin-wide adjudication, in which the courts assess and catalogue water rights contested in court. Tribes can also pursue negotiated water settlements bringing federal and non-tribal and tribal water users to the table to negotiate water rights [30].

In 1983, in Arizona v. California, tribal water rights were associated with the practicable irrigable acreage (PIA) policy, in which tribal water rights are quantified based on the amount of water needed to irrigate irrigable lands on the reservation [31,32]. PIA places limits on tribes insofar as only allocating water for agriculture, even though in some cases, tribes are not agriculturally based or not in a climate that necessarily supports agriculture on a large scale. In 2001, the Arizona Supreme Court ruled that PIA is not the only standard for determining water allocations. In Adjudication of All Rights to Use Water in the Gila River System and Source, the Court added the "homeland standard" for calculating the amount of water a tribe is entitled to, in recognition of the rights of tribal members to adequate water to live on reservations and not just for farming [33].

Tribes have been advocating for cultural uses through beneficial uses and water for the environment such as leveraging in-stream flow rights, minimum flow requirements, and Endangered Species Act. Rancier [8] analyzed 27 tribal water rights settlements and concluded that negotiated water rights allowed more flexibility to meet the needs of parties particularly for cultural water uses. Rancier [8] compared two tribal water settlements, The Snake River Water Rights Settlement ("Nez Perce Water Rights Settlements") and Truckee-Carson Pyramid Lake Water Rights Settlement Act and the Preliminary Settlement Agreement ("The Pyramid Lake Settlement") using a 28 point criteria to determine the success, strength, and weaknesses of the settlements [8,34]. Both of these tribes rely on fisheries for subsistence and cultural purposes and their settlements supported tribal fisheries listed under the Endangered Species Act. The Pyramid Lake Paiute Tribe (PLPT) used the Endangered Species Act to prevent further reduction of instream flows in the Truckee River System where overuse by agricultural activities resulted in fish floundering in a dry stream bed [8,34,35]. Two other water settlements, the Taos Pueblo Indian Water Rights Settlement and Agreement and the Zuni Indian Tribe Water Rights Settlement Act of 2003 and Agreement included significant provisions for cultural wetland restoration addressing groundwater and surface water use. For the Taos Pueblo Indian Water Rights Settlement, limits on groundwater withdrawals were set to limit impacts to tribal cultural and religious wetlands. For the Zuni Indian Tribe Water Rights Settlement, 5500 AFY of unappropriated flows were authorized to restore cultural and religious wetland Kolhu/wala:wa and Sacred Lake Hadin Kyaya with a 1984 priority date. Tribes have used other environmental protection legislations to advocate for the protection of their waters from overuse and contamination. These include the Clean Water Act, treaty rights, Endangered Species Act, as well as the ability of tribes to attain Water Quality Treatment as a State securing their authority to regulate water quality on their tribal lands.

3. Examples of Tribal Water Resources Challenges

A range of water resources challenges face tribes today. This section gives examples of water challenges facing tribes in the Southwestern United States. The first is the process of defining tribal water rights. The second is lack of access to water. The third is the effects of excessive use of water impacting water quantities on the reservation. The fourth is water contamination from mining, industrial, and agricultural activities. Finally, climate change and drought will amplify these existing water challenges.

The process of defining water rights can be locked in decades of litigation, adjudication, or negotiation due to the number of water users in the watershed and the legal complexities in court. If litigation is pursued, the tribe is typically the sole bearer of financial burden. In addition, the final rulings can have repercussions on other tribes in their water rights litigation. Even if a tribe wins the water right on paper, which is called a 'paper water right,' the tribe may not necessarily have the financial capital to extract and deliver their water [30]. This is the case in the Wind River Reservation where litigation resulted in water rights strictly for agricultural purposes and not for beneficial use, such as instream flows administered by the State Engineer [36]. Since the United States is trustee of federally recognized tribes, they are a party in tribal water rights cases, however, the U.S. may also potentially represent other competing interests such as that of federal agencies. Representing two parties in a case is normally a conflict of interest but current U.S. laws allows the U.S. to represent competing interests and thus are not held to the same standard as private fiduciary in tribal water rights cases. In sum, the U.S. Department of Justice may litigate on behalf of the tribe as an active partner or may play a passive role as a trustee for a federally recognized tribe.

Even if a water settlement is negotiated, it may receive opposition from the tribal community or grassroots because of discontent in the negotiated allocation such as in the Senate Bill 2109: Navajo-Hopi Little Colorado River Water Rights Settlement Act of 2012. In the final step, this settlement needed Congressional approval and Senator Jon Kyle (R-AZ, 1995–2012) introduced the settlement before Congress to garner support. In his opening remarks, Senator Kyle said, "Legally, the Navajo Nation and Hopi tribe may assert claims to larger quantities of water [than are outlined in the settlement] but they do not have the means to make use of those supplies in a safe and productive manner." This created uproar from the Navajo grassroots community because his statements implied the lack of the Navajo Nation's ability and resourcefulness to use the water. Yazzie identified this approach as a model of minimization of water rights not quantification [37]. In addition, grassroots communities were not involved in the negotiation at the community level as these negotiations were held behind closed doors. This widespread disapproval from both the Navajo and Hopi grassroots halted approval of the water settlement.

The second challenge is access to water. Approximately 9% of Native American homes in the United States lack safe and adequate water supplies and lack access to waste disposal facilities in comparison to 1% of U.S. homes [38]. Not only does this have implications for quality of health and life for tribal members, but it also dampens the economic growth on tribal lands subsequently impacting quality of life in terms of available jobs and infrastructure. In addition, some tribes may not be able to use groundwater sources due to lack of quality. For example on the Navajo Nation, in the southwestern portion of the reservation, groundwater is highly saline and can contain heavy metals, making it unsuitable for drinking or livestock purposes [39].

The third challenge is excessive water use and diversion impacting tribes. For the Pyramid Lake Paiute Tribe, excessive diversion at Derby Dam for agricultural use blocked access to upriver spawning grounds for fish and during a drought left dying fish for two miles downstream of the dam [35]. Since the Pyramid Lake Paiute Tribe is located at the terminal end of the river system, they are subject not only to excessive use but also by non-point pollution into the river by municipal, industrial, and agricultural practices. Another similar example is the Pueblo of Zia located 35 miles northwest of Albuquerque, NM near the confluence of the Jemez River and Rio Grande who are facing upstream impacts including recreational activity, agricultural diversion, and encroachment of the large cities of Rio Rancho and Albuquerque. The Pueblo of Zia encourages sustainability, self-sufficiency, and subsistence activities. Today, most of the tribal members hunt, gather, cultivate food crops, and raise livestock and Zia's economy is based on agriculture which is intertwined with their religion, government, social organizations, and livelihoods [40].

The fourth challenge is off-reservation and on-reservation contamination of water by mining, industrial, and agricultural uses. Regulating and minimizing impacts can be difficult when there are many non-tribal users who may be hard to identify until it is too late, such as in the case of abandoned mines. For the Laguna Pueblo, the abandoned Jackpile-Paguate Uranium Mine covers almost 8000 acres east of the village of Paguate and acid mine drainage leaks into the arroyos and streams flowing into the Rio Paguate and Rio Moquino, both of which are upstream of the village. In southern Arizona, thousands of gold and copper prospector abandoned mines exist that create both physical hazards and pools of acid mine drainage. Finally, hundreds of thousands of abandoned mines exist in Colorado upstream of the Mountain Ute, Southern Ute, and the Navajo Nation. Recently, on 5 August 2015, the Gold King Mine Spill, spilled three million gallons of acid mine drainage into the Animas and San Juan Rivers, which the tribes depend on [10,41].

Finally, short-term drought and climate change, which has impacts on natural systems affecting indigenous peoples everywhere, strains sometimes already stressed water infrastructure that may be lacking, inadequate, or poorly maintained. This increases the vulnerability of tribes to flooding, drought, and water-borne diseases [42]. Existing infrastructure can be damaged due to changes in water quality, water resources, and land subsidence [42]. A flood on the Fond du Lac Reservation in 2012 resulted in extensive crop loss of wild rice, which is a sacred food of great importance for many of the Midwestern tribes. On the Standing Rock Sioux Nation in North Dakota, silt and sludge closed a water supply intake pile when a 2003 drought caused water levels to drop to very low levels. In Alaska Native villages, beavers which can carry giardia are moving father and father north and are in northern rivers of Alaska which have not been occupied since the last ice age [43]. In light of these examples of the experience of diverse tribes, climate change and drought will likely amplify existing water challenges that face tribes in the Southwest today.

4. Collaborative and Participatory Approaches to Research and Resource Management

4.1. Tribal Sovereignty and Research Practices

The principles embedded in federal Indian law and policy (discussed above) also govern the ways in which research is conducted by and with tribes. As outlined above, the federal government has a set of legal principles that govern its relationships with tribes and its responsibility to act in the best interest of tribes. The rights of inherent sovereignty retained by tribes dictate that researchers and other professionals seeking to work with tribes adhere to the standards set by tribal governments. As sovereign nations, tribes have the authority to control when, where, how, and whether research is conducted on their lands or among their citizens. Unfortunately, this right was not always recognized or respected. Prior to the resurgence of tribal sovereignty in the area of research, researchers often felt they had "*the authority of the federal bureaucracy... behind them... [R]esearchers could set their own agenda and do as they pleased without having to consult with or defer to tribal polities. Research has always been deeply implicated in the colonial political context*" [44]. This lack of oversight often led to abuse, misrepresentation, and exploitation of Native peoples by researchers [44–47].

Since the mid-1990s, a growing number of tribes have established their own research protocols and review boards [44,47,48]. While each tribe's regulations vary, some general principles are evident across the protocols. First, researchers must gain permission to conduct the research on tribal lands or with tribal members. This process may start with speaking with tribal elected officials who have the authority to approve research and/or applying to the tribal research office for a research permit. During the process of gaining permission, researchers should have open discussions with the approving body about data ownership, use, and sharing [46]. Some tribes consider all data collected on their lands to be their property and do not allow it to be shared with the general public (often a requirement of federal research grants). Researchers should be clear about the extent to which they can publish or share data from their proposed research and whether additional permissions or review might be required prior to publication. Some tribes may simply request that the original data be archived with them for use and access by tribal members.

In addition to any tribal government oversight, researchers must have any research involving people reviewed by their Institutional Review Board (IRB) to ensure that it complies with federal regulations regarding human subjects research (i.e., The Common Rule, 45 CFR 46) [49]. However, a critique of the standards set by The Common Rule governing human subjects research is that it focuses too much on individual rights and protections and does not explicitly cover the concerns of tribal communities regarding, for example, protection of local or traditional knowledge [46]. There are also questions of whether the research will benefit tribal members who invest their time in the research process. To strengthen IRB processes to increase protection for tribes, some institutions, such as The University of Arizona, review all research proposed in or with tribal communities to ensure compliance with tribal protocols [44,48].

While these official review processes are fundamental to conducting ethical research with tribes, they are often just the first step in ensuring that tribal interests, needs, and knowledge are fully respected and braided into research and outcomes. It is important for researchers to take the time to learn the cultural context and sensitivities of the information being collected, such as how and when certain knowledge can be transmitted; the cultural significance of particular resources (see discussion above of water as a cultural resource), which imbues discussions with greater significance than the researcher may be aware of; and any prohibitions or proscriptions against discussing certain topics. It can sometimes be the case that certain communities within some tribal nations that are keepers of certain forms of knowledge may have their own expected protocols for researchers to respect that are stronger than those of their tribal governments. Given, as discussed earlier, that many tribal governments today do not correspond to the traditional governments, researchers should do the work to ensure they are paying respect to the overall research situation within a tribe, considering both the tribal government and the community [16,50]. Though in no circumstances should the complexity of tribes be used to play different governmental authorities and constituencies against one another or to serve as an excuse for avoiding formal tribal IRB processes or the formal involvement of tribal council. Lastly, researchers need to consider that certain members of tribes, especially elders or culture-bearers, are being constantly burdened with requests to work with researchers. These members often do not receive benefits from participation in research projects that are comparable to what the researchers gain in relation to their own careers and aspirations.

Scientists have a moral responsibility to respect indigenous knowledge systems as more than sources of data. Many indigenous peoples see themselves as sensitive to social, cultural, and political dimensions of knowledge systems that scientists have forgotten to pay attention to in relation to their own knowledge systems. Scientists may assume that it is normal to view plants, animals, elements (e.g., water) and ecosystems as mere resources with no cultural or spiritual value.

Yet many indigenous peoples see "water as a resource" as just one possible kind of relationship on which a knowledge system could be based. Many indigenous knowledge systems produce empirically valid claims using approaches to plants, animals, elements, and ecosystems as kin or clan relatives, who have agencies, spiritualities and personalities of their own [51–53]. Hypothetically, a researcher who treats water as a mere resource could accidentally insult or disrespect tribes or imply forms of handling or using water in research that are inappropriate. Thus, scientists should take pains to ensure that they do not privilege often unseen assumptions within their own knowledge systems that could lead them to sour their relationships with tribal collaborators.

External researchers and scientists are often not accustomed to or comfortable with the ways in which indigenous persons describe indigenous knowledge systems. For example, Native Americans often describe knowledge as knowledge in experience that is carried and embodied [52,53]. This is called 'lived knowledge' where knowledge cannot be separated from human experience and action. Burkhart gives an example of the Seneca people who planted corn, beans, and squash together because of traditional ceremonies and cultural stories about Three Sisters that told that planting these together would feed the people. This relationship is an illustration of the nitrogen cycle and nitrogen replenishment keeping the soil productive and fertile. Indigenous knowledge is based on patient observation and contemplation. Though just because indigenous persons describe their knowledge systems in this way, it should not be presupposed that this means that indigenous persons cannot understand and connect their knowledges to the approaches to knowledge assumed by others [54]. Understanding the differences between approaches to knowledge, even if just at the heuristic level, will arguably help researchers be better prepared to work with tribes and to have appropriate cultural sensitivities [54].

Tribal knowledge systems also involve different social institutions and relationships to politics than scientific knowledge systems do. As fields such as science studies and the philosophy of science have pointed out, any knowledge system works through social institutions, from political authorities to sources of support (e.g., funding), that set research priorities, assign powers and privileges to certain members or groups, create rituals and rites of passage, and so on. External researchers are often unaccustomed to seeing themselves as embedded within social institutions given their personal interest and passion in the pursuit of knowledge. Yet they are nonetheless embedded in social institutions that differ from those of tribal knowledge systems, creating the possibility for mismatch. For example, knowledge taken to be public by scientists may be tribal knowledge that should only be appropriately expressed in a certain ceremony. Another example is tribal knowledge about the location of a medicinal plant may not create political problems for scientists but it may for members of a particular tribe who could stand to lose access to harvesting those plants if their location is made public. Climate and Traditional Knowledges Workgroup (CTKW) calls this the "governance value" of knowledge, which scientists often times do not pay attention to with their own knowledge systems [16,50,55]. While tribal members often are open about the relationship of their knowledge systems with beliefs in the spirituality of nonhumans or the inherent ties between knowledge and social institutions and political agendas, scientists are often not comfortable with this. Scientists should approach collaborations with tribes initially more as diplomats than as fellow inquirers, given these differences.

Lomawaima summarizes the legal and political standards of research as well as the cultural and ethical requirements when working with tribes, into four "simple rules": (1) ask about the ethics of conducting research in each particular community; (2) do more listening than talking; (3) find and follow any and all tribal research rules or protocols; and (4) give something back to the community in exchange for their cooperation with your research [44]. Others have emphasized, in relation to these rules, that researchers do have to respect and follow tribal research protocols that emerge from cosmological frameworks for approaching the world that are very different from science [51,52,55]. Moreover, researchers should endeavor to consider their roles within larger U.S. settler colonial structures as professionals working for powerful universities, scientific organizations and agencies. Instead of understanding their research ethics as simply a matter of getting information from tribes without harming anybody; they can instead try to determine how their research will ultimately support the sovereignty, cultural revitalization, and well-being of tribal members, communities, and nations [55].

Given that this article addresses primarily external researchers who seek to work with tribes, we did not include decolonizing methodologies as part of the initial approaches, but seek to discuss it here given its overall importance for how external researchers approach tribes. Decolonizing methodologies, or indigenous research methodologies, emphasize an orientation to research that reconfigures how researchers relate to indigenous communities. For research to truly be beneficial, the research design itself has to emerge from indigenous conceptions, cosmologies, and frameworks for empirical inquiry. Decolonizing methodologies calls for indigenous peoples to develop their own research institutions for gaining and protecting knowledge that supports their sovereignty and the well-being of their community members [56]. While this literature focuses on the development of indigenous research, it has an important implication for researchers from universities and other institutions outside of tribes. External researchers, through their research, need to support the development of both tribal capacities to produce their own research, on tribes' own terms, and also ensure that their own research institutions are organized as best as possible to match up with tribal institutions, such as the development of indigenous research institutions within U.S. colleges, universities, and research institutions [55]. Indigenous research methodologies seeks to empower indigenous persons to take on the role of researcher in ways that respect the cultural protocols, use the assumptions of empirical validity, and advance political agendas of their communities and indigenous nations; they seek to pose direct, critical questions to non-indigenous researchers about what it would really mean to decolonize the relationship between powerful universities, scientific organizations, and agencies that often sponsor research with tribes.

4.2. Collaborative Approaches to Natural Resource Management

Not only is a more collaborative approach to research with tribes good ethical and legal practice, it can also be a more effective way to make natural resource management decisions, including those about water. Research on collaborative, or participatory, research processes involving scientists, stakeholders, and policy or decision makers has shown that participants tend to trust scientific information more [57] and are more likely to use it to inform decisions when they collaborated in the process of developing the knowledge [58,59]. However, we know from research like Beierle [60], Nadasdy [61], Rowe and Frewer [62], Stern and Fireberg [63], and Ford et al. [64], among others, that not all collaborative processes are equal. The ways in which scientists engage with other parties—the processes they use—matters to the outcomes. We identified five approaches to engaging tribes on water and other natural resource management issues, including (1) Tribal Participatory Research (TPR); (2) Boundary Work; (3) Adaptive Water Governance; (4) Community Based Adaptation (CBA); and (5) Indigenous Ecological Knowledge (IEK) Engagement Structures (Table 1). Most of the approaches meet Lomawaima's [43] four simple rules of tribal research to various degrees.

Particular attention should be paid to the process of collaborating on resource management research and policy decisions with and among indigenous communities. Fisher and Ball propose an approach they call tribal participatory research (TPR), which is based on participatory action research (PAR), which emphasizes inclusion of community members throughout the research process [65]. TPR is a synthesized model created to address the historical legacy of exploitation of indigenous communities by outside researchers. They note that PAR was created based on the principle of using research to have a liberating effect and increase self-determination [66], key components that have been missing from much research involving indigenous communities in the past. TPR acknowledges the need to incorporate community-specific cultural factors, acknowledgement of historical trauma, involvement of community, and protection of tribal interests. The TPR mechanisms focus on tribal oversight, the use of a facilitator, training and employing of community members as project staff, and the use of culturally specific assessment and intervention methods. Tribal oversight includes a tribal resolution, tribal advisory committee, and tribal research protocols. Although TPR was developed for social sciences, the mechanisms outlined by TPR is significant for engaging tribes on water resources topics.

Boundary work seeks to negotiate boundaries between science and other knowledge forms in order to both make sense of different knowledges and generate new knowledge. Boundary work focuses on creating politically- and culturally-sensitive processes even when the goals of collaboration are explicitly about knowledge exchange. Robinson and Wallington examined the integration of indigenous knowledge into co-management systems in Australia and found that managers needed to take into account the institutional landscapes as well as current socio-ecological landscapes, because older management or legal structures can inhibit the development of new co-management structures. Boundaries can also create barriers to collaboration [67]. Robinson and Wallington [67] identified three factors in effective boundary work: meaningful participation in setting goals and co-producing knowledge; governance to ensure that boundary work is accountable; and co-production of boundary objects (e.g., maps, or interpretive frameworks that all participants agree to as trustworthy and respectful of their differences).

Adaptive water governance requires governance to adapt to uncertainty and change through ongoing learning, institutional integration, collaboration, and co-management partnerships [68]. It is structured as polycentricinstitutional arrangements that nest decision-making and integration of roles and activities between state and non-state entities and is useful where significant knowledge gaps exist. Adaptive water governance manages uncertainty through water entitlements that solidify indigenous claims and basin planning that incorporates indigenous knowledge. Water can be re-allocated in response to changing conditions and values and basin planning regularly reviews and updates targets. Water entitlements acknowledge indigenous claims and provide infrastructure and institutional capacity. Water planning goals, structures, and processes are based on indigenous knowledge and perspectives. Bark et al. [68] found that indigenous communities with existing land holdings were better positioned to influence water management and policy than those who lacked holdings.

Community based adaptation (CBA) is a community-led process, based on communities' priorities, needs, knowledge, and capacity to empower people to plan for and cope with the impacts of climate change. Ford et al. unpack some of the challenges and potential issues with even the most well-intentioned participatory research work focusing on CBA [64]. Based on their experiences with CBA approach to climate change impacts, they note that researchers cannot assume that research has a positive role to play in a community, just because it uses participatory processes; sometimes-new research is not the answer to a community's needs. Further, they stress the need to manage expectations—those of the community and those of the researchers—and be realistic about what research can actually achieve in a community. Finally, they caution researchers interested in the CBA approach to invest time in planning and coordinating their work well in advance, because to be truly effective a participatory approach requires significant time, effort, and resources.

Hill et al. analyzed 21 case studies of environmental management decisions in Australia that included indigenous communities to develop a typology of engagement types based on how they integrate indigenous environmental traditional knowledge (TK) with Western science, how TK is managed throughout the process, and how TK is integrated into management strategies [69]. They identify four general types of engagement: Indigenous Governed collaborations (IG); indigenous-driven Co-Governance (ICoG); Agency-driven Co-Governance (ACoG), and agency governance (AG). Indigenous-governed collaborations (IG) and indigenous-driven co-governance (ICoG) provide better prospects for integration of TK because indigenous peoples retain control over that knowledge. IGs stem from indigenous initiatives such as a confederation of indigenous nations who are focused on environmental issues, actions, and policies. IGs focus on advancing distinct indigenous society and cultures with inclusive participation of indigenous people with indigenous people retaining significant power. On the other hand, ICoG are created in response to government initiatives but empower indigenous interests and authority while recognizing the need to improve indigenous and non-indigenous people's capacity to operate and with significant power sharing with non-indigenous interests.

Cronin and Ostergren analyzed three tribal collaborative water management structures including two tribes in the Northwest and one tribe in the Southwest [70]. In this comparative study involving the analysis of 31 tribal and non-tribal interviews, Cronin and Ostergren [70] identified six factors that influence tribal engagement in water management discussions: (1) tribal cultural connection to aquatic resources; (2) political clout and legal standing of tribes; (3) relationships between tribal and non-tribal communities and relevant agencies; (4) recognition of the benefits of collaboration; (5) consistency and vision of tribal leaderships; and (6) the availability of resources to tribes. Cronin and Ostergren [70] stated that the presence or absence of any factor may not determine tribal engagement. Their recommendation of persistently seeking tribal input aligned with Lomawaima's [43] second rule, however, asking about ethics, tribal protocols, and giving back is not emphasized, instead they emphasize that tribal and non-tribal partners should seek to work together, share stories, and find whatever possible common ground there is. Notably, Cronin and Ostergren [70] suggest that non-tribal partners engage tribes early on such as in the planning phase and plan to make a long-term commitment.

Table 1. Selected five approaches of engaging tribal members, communities, and nations in water management discussions.

Tribal Participatory Research [65]	Boundary Work [67]	Adaptive Water Governance [68]	Community Based Adaptation (CBA) [64]	Indigenous Ecological Knowledge Engagement Structures ([69])
Application of participatory action research to a new context for tribal communities	Using scientific and indigenous knowledge to co-manage resources	A hybid planning model that combines scientific, institutional, and social processes	Community-led process based on communities' priorities, needs, knowledge, and capacities, which should empower people to plan for and cope with challenges	Indigenous Governed collaborations (IG), Indigenous-driven Co-Governance (ICoG), Agency-driven Co-Governance (ACoG), and Agency Governance (AG)

4.3. *Braiding Traditional Knowledge with Water Management*

When new water management structures are designed, they should explicitly allow local people to use and refine their own knowledge system in ways that maintain the integrity of the knowledge practices, according to Hill et al. [69]. This kind of flexibility requires a different type of approach to engagement and integration of new knowledge; it requires the kind of collegial engagement between scientists, decision-makers, and other participants, which, as described by Biggs [71] and consistent with Smith [56] and Whyte [55], scientists actively support the production of local or indigenous knowledge systems, not just the integration of local knowledge into Western science frameworks.

The challenge of how to braid Western and indigenous sciences in ways that are beneficial to both knowledge systems has been addressed in several places by Huntington [18,72,73]. Huntington urges Western scientists to carefully consider how they use the term traditional knowledge, to define it clearly and carefully to avoid misrepresenting the knowledge, and to recognize that there may be distinctions in types of knowledge within the community. Robinson and Wallington [67] raise a similar point and note the importance of establishing a relationship with the community that will facilitate the outsiders' understanding and use of traditional knowledge. Latiluppe claims that for "TK to advance the priorities and goals of Indigenous research partners and to be of benefit to Indigenous peoples, it is paramount that researchers consider their positionality and anticipate the outcomes of particular approaches within situated contexts" [74]. Whyte claims that "environmental scientists and policy professionals, indigenous and non-indigenous, should not be in the business of creating definitions of TK. Instead, they should focus more on creating long term processes that allow the different implications of approaches to knowledge in relation to stewardship goals to be responsibly thought through" [53].

A group of indigenous scholars with significant experience working on climate change and other resource management and planning issues have developed a set of guidelines for the use and integration of traditional knowledge into management planning efforts [16]. The purpose of these guidelines is to increase understanding for the protection and role of traditional knowledge in climate initiatives for federal and intergovernmental agencies. The eight guidelines outline principles of engagement and protocols for engaging tribes when it relates to traditional knowledge and understanding the risks tribes face when they decide to engage. As was mentioned earlier, the principles of engagement are to "cause no harm" and "free, prior and informed consent" indicating that there is fairness, early engagement, transparency, and the right to engage or disengage at any time. The eight guidelines are (1) understand traditional knowledge; (2) know that tribes have a right not to participate; (3) communicate risks to tribes and help tribes to understand risks; (4) establish an institutional interface; (5) train federal agency on traditional knowledge, its protection, and related policies and protocols; (6) establish specific directions to uphold TK protections; (7) recognize multiple knowledge systems; and (8) develop guidelines for review of grant proposals. Although these guidelines are focused on traditional knowledge in climate initiatives, the principles outlined are very applicable to environmental research more broadly.

5. Case Studies

Five case studies involving engaging Southwestern tribes on water management and water topics will be discussed. The manner in which tribes were engaged, the effectiveness of engagement, and the desires and concerns resulting from the engagement of the tribe will be discussed. The five cases studies are the Hopi Drought Study, the Pyramid Lake Paiute Tribe (PLPT) climate change project, the Water Resources Research Center (WRCC) Annual Conference focusing on "Indigenous Perspectives on the Sustainable use of Water", the WRCC's engagement of tribes to create a roadmap for considering environmental water demands, and the Colorado River Basin Tribal study and the effort to engage tribes. The first four are cases led by University of Arizona researchers. The varying levels of approval by the tribe and university research institutions are summarized in Table 2.

Table 2. Tribal and research approvals received by case study.

Case	Purpose	Tribal Council	Tribal IRB	University IRB	Community/Grassroots Support/Resolutions
(1) Hopi Drought Study	Drought planning	✔	✔	✔	✔
(2) PLPT Climate Change Project	Climate change and water planning	✔	✔	✔	✔
(3) Indigenous Water Conference	Water management				✔
(4) Environmental Demands Roadmap	Water management planning & policy			✔	
(5) Colorado River Basin Tribal Study	Water management planning & policy	✔			

5.1. Hopi Tribe Drought Study

The Hopi Department of Natural Resources (HDNR) has been collaborating with researchers from the University of Arizona (UA) to develop a drought monitoring framework to help HDNR address the drought impacts on farming, ranching, and cultural traditions occurring for at least 15 years. The framework was developed using a collaborative research project that included rapid assessment, organizational ethnography, and participant observation, as well as interviews and multiple discussions with Hopi citizens and employees for approximately 5 years. The goal of the project was to ensure that Hopi concerns were at the forefront regarding drought, existing monitoring and knowledge practices, and capacity to respond to drought impacts. By relying on local knowledge and skills, the drought monitoring framework was designed to harness local data in ways that support local decisions, rather than relying entirely on instrumental data from external sources, which is sparse across Hopi lands. For example, the team noted that the current drought plan relied on data not readily accessible to HDNR staff, making it difficult to declare (or to *un*declare) drought. However, HDNR staff were already collecting environmental status information through several programs, including water resources and range management that shed ample light on drought conditions in the region. The shift to locally controlled data, the team hopes, will place more control in the hands of local decision makers and community members who are most affected by drought impacts [75].

5.2. Pyramid Lake Paiute Tribe Climate Change Project

Researchers from The University of Arizona, University of Nevada Las Vegas, and the U.S. Geological Survey have been collaborating with the Pyramid Lake Paiute Tribe in Nevada to identify water-specific vulnerabilities to climate change [35,76]. When interviewed about their experiences working together (by one of the authors, Meadow), both the researchers and the PLPT staff discussed important practices when researchers collaborate with tribes, particularly on sensitive issues involving water and water rights. For example, although PLPT does not have its own research review board, the lead investigator (author Chief) ensured that she had the support and consent of the tribal council by formally requesting their cooperation in the project and received a letter of support from the (then) tribal chairperson that documented permission to undertake the research and the tribe's commitment to collaboration. Throughout the project, the research team checked with PLPT staff to ensure that they were following community protocols regarding meetings, interviews, or other forms of data-gathering. The researchers presented before the PLPT tribal council on an annual basis, to receive approval in research changes, and to present final reports. They worked with PLPT staff to organize community workshops to ensure that local protocols were followed. A high degree of trust developed between the researchers and PLPT staff, which was demonstrated when the potentially sensitive issue of protected cultural knowledge came up in the course of the research. Rather than become a hurdle, both groups were able to discuss what to do with cultural knowledge, should it arise in interviews or other discussions, and come to an understanding both were comfortable with.

5.3. Indigenous Perspectives on Sustainable Water Practices Conference

The University of Arizona Water Resources Research Center (WRCC) hosts an annual conference on various water management topics each year. Due to the recent highly contentious proposed Little Colorado Water Settlement, a significant number of Navajo and Hopi grassroots members advocated to have a voice at the 2014 conference on "Closing the Gap between Water Supply and Demand." To highlight the voice of the indigenous communities, the 2015 WRCC conference was focused on "Indigenous Perspective on Sustainable Water Practices" [77]. A tribal advisory committee (TAC) was formed to represent the indigenous perspectives across Arizona and provide guidance and advice on the conference agenda and speakers. TAC met monthly to plan steps. In this planning process, a survey was developed and distributed across tribes to determine conference title, topics, speakers, and provide input for the indigenous water conference. The Gila River Indian Community (GRIC) hosted the conference and over 300 individuals from 49 municipalities and 13 Native communities throughout Arizona attended the conference which was a record number of attendees for WRCC annual conferences. The GRIC Governor Stephen Lewis provided welcome remarks highlighting the celebration of the 10th Anniversary of the Gila River Water Settlement and a challenge for tribal water security into the future and an investing in the youth. John Echohawk (Pawnee), founder of the Native American Rights Fund, provided the opening remarks encouraging an active dialogue between grassroots tribal members and tribal leaders as well as non-tribal environmental managers. The conference was a convergence of traditional grassroots perspectives and environmental managers resulting in mutual learning and respect. Arizona State Senator Carlyle Begay (Diné) noted that the event represented "*... a very much needed conference, generating a lot of great discussion, a lot of great insight, and most importantly great ideas in moving our communities forward in discussions about the future of our water resources*".

5.4. *Arizona's Roadmap for Considering Water for Arizona's Natural Areas*

The University of Arizona Water Resources Research Center (WRCC) created a roadmap that outlines considerations for water for Arizona's natural areas in Arizona water management and planning decisions [78]. Often times the water demands needed for the environment, such riparian water demands and aquatic ecosystems, are not considered in state management plans. The objective was to promote a discussion on ways that stakeholders can address environmental water demands under the constraints of limited water supplies and existing water rights. The WRCC engaged a diverse set of participants including academic, business, environmental farming, mining, municipal power, ranching, and tribal perspectives through surveys, focus groups, workshops, and presentations.

Considering that tribes have a deep connection to the natural environment where water for the environment is respected, this topic was expected to be of great interest for tribal communities and nations. Among the various participants, 40 tribal persons were engaged, which was about average compared to the other non-tribal participants. The tribal participants were recruited as voluntary participants. With 22 tribes in the state of Arizona and a land base of nearly a third of the state, the recruitment of tribal participants was on an individual basis through tribal organizations, consortiums, and meetings where tribal environmental managers typically attended. Grassroots tribal environmentalists tended to be active participants seeking to be involved in this discussion. There was no report back to specifically to tribes on the results of the study, however, the results, have positive implications supporting the deep connection of tribes to the natural environment.

5.5. *Colorado River Basin Watershed Tribal Study*

The 2012 Colorado River Water Supply and Demand Study was a $5.5 million three-year study by the Bureau of Reclamation and the Colorado Basin States to develop a comprehensive plan to address risks posed by imbalances between Colorado River water supply and water needs in the basin [79]. Tribes voiced the desire for the Study to engage tribes individually on a tribe-by-tribe basis. On 18 September 2013, the Department of Interior and the BOR entered into a collaborative agreement to focus on tribal concerns and needs by working with the Ten Tribes Partnership including the Chemehuevi Indian Tribe, Cocopah Indian Community, Colorado River Indian Tribes, Fort Mojave Indian Tribe, Jicarilla Apache Nation, Navajo Nation, Quechan Indian Tribe, Southern Ute Indian Tribe, Ute Mountain Ute Indian Tribe, and Ute Tribe of the Uintah and Ouray Reservation. The Ten Tribes Partnership will focus the study on tribal water rights. BOR expects to release this report in December 2016.

6. Discussion

In considering the engagement of the five cases, we built upon the four "simple rules" of tribal research which are to ask about ethics, do more listening, follow tribal research protocols, and give back to the community. Specific engagement components were identified under each of the four simple rules. Asking about ethics means having cultural sensitivity to understand cultural protocols and to rethink one's role as a researcher in relation to the institutional context that tribes operate in that may produce risks that are not typical for researchers to consider. Doing more listening includes listening to the tribal leaders and community members to understand what their research needs are, if research is needed and useful, and allowing the community to develop and drive research goals. Following tribal protocols includes having approval by the tribe to conduct research or having tribal oversight such as a tribal advisory board to guide the project goals. Giving back means having on-going and regular communication with the tribe to update them on efforts. Giving back also means that the tribe owns the data and knowledge and data co-produced or acquired through the project should be transferred back to the tribe. Finally, giving back means reporting back to the tribe on project results (Table 3). From the literature and examining the five case studies, successful engagement with tribes involved engagement of tribal leaders and members at different levels, tribal oversight, on-going dialogue, transparency of data, and reporting back.

Table 3. Components to gain trust and build partnerships with tribes by case study [44].

Engagement Components [44]	Ask about Ethics	Do More Listening	Follow Tribal Protocols	Give-Back		
	Cultural Sensitivity	Community Driven Goals	Tribal Oversight	Ongoing Communication	Tribes Own Data	Reporting Back
(1) Hopi Drought Study	✔	✔	✔	✔	✔	✔
(2) PLPT Climate Change Project	✔	✔	✔	✔	✔	✔
(3) Indigenous Water Conference	✔	✔	✔	✔		✔
(4) Environmental Demands Roadmap	✔					
(5) Colorado River Basin Tribal Study	✔		✔		✔	

All of the cases represent the use of different elements for collaborative research. Both the Hopi drought project and the Pyramid Lake Paiute Project collaborative water management projects are university led projects that engage tribal partners in the research objectives. Both projects required the university human subjects research review approvals and tribal research approvals (Table 2). The Hopi drought project explicitly included tribal resource managers in data collection and analysis. These projects, as many of the others, sought ways to make the work relevant to improving tribes' capacities as sovereigns. Moreover, in many of the cases, researchers were both respectful to tribal governments and also paid attention to the dynamics of communities within tribes that are not associated with the government, such as the grassroots communities.

The Pyramid Lake Paiute Project engaged with a broad spectrum of tribal members and employees. The tribal chairperson, who has a background in water resources management, invited and supported the project. The chairperson participated in the climate change perception survey, individual interviews, and focus groups. The project team produced written and oral annual reports to the tribal council who guided and approved research activities. The natural resources oversight committee reviewed and approved public abstracts, presentations, and manuscript publications. The tribal environmental managers were engaged in focus groups and water management planning workshops on a regular basis. A tribal liaison from the tribal environmental was assigned to the project to receive weekly updates, participate in project meetings, and assist in coordinating meetings, workshops, and presentations with the tribe. In addition to focus groups held with the environmental managers, surveys at household levels were administered by a tribal college student, and a community workshop on water management was conducted. Finally, the project involved Native American researchers including faculty and students. This project was successful in engaging the tribe because of the involvement of tribal leadership and process of receiving tribal government research approvals through a tribal protocol. In addition, consistent communication through written, verbal (in person and via teleconference), social media, website, and newspaper articles contributed to on-going engagement. At a time when there were few tribes engaged in climate change research, PLPT became a leader in tribal climate research. Outputs from this project were fruitful including a publication in the journal of *Climatic Change*, several community videos, and two streams of funding.

The HDNR collaboration also started with an invitation to the researchers from the (now retired) director of the Department of Natural Resources to provide guidance on developing a more robust drought-monitoring plan than was in place at the time. The researchers and local resource managers collectively defined and redefined the research scope to address local concerns and decision frameworks [75]. University researchers (all who were not Native American) conducted early interviews, focus groups, and participant observations involving resource management staff. A Hopi researcher, employed by the university, conducted interviews with community members, that at times touched on traditional knowledge and practices [75].

The indigenous conference on sustainable water management was successful in engaging tribes because it provided a venue in which to discuss indigenous perspectives at a statewide conference. The conference's goals were to inform non-tribal attendees of indigenous perspectives and to engage tribal attendees. A tribal advisory committee oversaw the organizing of the conference. The advisory committed administered a post-conference survey to collect feedback on the content and structure of the conference. The conference had a record number of attendees in comparison to past versions of the conference that did not explicitly address tribal water issues.

The environmental demands project resulted in a modest number of tribal members who were engaged in discussing water management policies that would consider water for the environment. The project received university IRB approval. With a state-wide scope of engaging diverse participants, there was no requirement for a specific tribal IRB approval. A request for tribal involvement was presented at an intertribal consortium supported by the tribal water working group coordinator, however, the request was met by skepticism and hesitancy because the word "research" was mentioned in the introduction of the Water Resources "Research" Center. There was no involvement of the tribal water working group or tribes represented at the intertribal consortium. Recruitment of tribal members was broad and two Native American representatives served on the advisory steering committee. A few tribal members attended each of the many workshops and focus groups. Reporting was conducted as a whole to the entire group of participants, however, there was no specific reporting conducted for specific tribes and tribal community groups. Nonetheless, tribal grassroots members were actively involved and willing to provide feedback and contributed to understanding the grassroots perspectives on environmental demands from a tribal perspective.

The Colorado River Basin Study engaged tribes that were directly given responsibilities to provide input for tribal water rights in the Colorado River Basin. The Colorado River Basin Study did not engage tribes at the community level where the most active and vocal citizens often are found. It is unclear what types of publications directed towards the tribal citizen were created to communicate results in layman terms. Use of tribal linguists and facilitators to engage and inform tribal members at the community level would prove to be useful and productive.

7. Conclusions

Because of the deep connection tribes have to the natural environment and tribal specific challenges in water management, the manner of engaging tribal participants, from individuals to communities to nations, is important to the success of the project, goals, and dialogue. This paper synthesized approaches to engaging tribal participants from small-scale community level engagement to basin level multi-tribal engagement for various goals of tribal water management, state water management dialogue, receiving input on state water policy for the environment, and basin wide water policy development. The type of engagement is goal- and tribe- specific and there are many levels of engagement and approvals and support needed. Successful engagement with tribes involves comprehensive engagement of diverse tribal audiences, tribal oversight, on-going dialogue transparency of data, and reporting back. As climate change, along with existing issues surrounding drought, mining, and uncertain water rights, stresses tribal water resources in the southwestern U.S., engaging tribes in water management discussions becomes more critical. This paper has provided an overview of many of the social, cultural, political, and historical considerations that are key to successful collaborations between researchers, resource managers, policy makers, and tribes.

Acknowledgments: The authors would like to acknowledge the funding sources of four of the case studies. The Hopi Drought Study acknowledges support by the National Oceanic and Atmospheric Administration (NOAA) Climate Program Office Sectoral Applications Research Program (SARP) Grant (#NA10OAR4310183) and the views expressed in this publication represent those of the authors and do not necessarily reflect the views or policies of NOAA. The Pyramid Lake Paiute Tribe climate change project acknowledges support by the National Science Foundation (NSF) Nevada Experimental Program to Stimulate Competitive Research (EPSCoR) seed grant (#EPS0814372) and the United States Geological Survey Grant/Cooperative Agreement (#G12AC20506). The contents of this publication are solely the responsibility of the authors and do not necessarily represent the official views of the USGS. The Water Resources Research Center (WRCC) Indigenous Perspectives on the Sustainable Water Practices Conference acknowledges support from the Gila River Indian Community. The WRCC Arizona's Roadmap for Considering Water for Arizona's Natural Areas acknowledges support from from the Nina Mason Pulliam Charitable Trust.

Author Contributions: Karletta Chief conceived and led the writing of the article. All authors contributed to writing the outline and manuscript.

Conflicts of Interest: The authors declare no conflict of interest.

Abbreviations

The following abbreviations are used in this manuscript:

ACoG	Agency-driven Co-Governance
AG	Agency Governance
BIA	Bureau of Indian Affairs
CBA	Community Based Adaptation
CTKW	Climate and Traditional Knowledges Workgroup
GRIC	The Gila River Indian Community
HDNR	Hopi Department of Natural Resources
IEK	Indigenous Ecological Knowledge
ICoG	Indigenous-driven Co-Governance
IG	Indigenous Governed collaborations
IRB	Institutional Review Board
PAR	Participatory Action Research
PIA	Practicable Irrigable Acreage
PLPT	Pyramid Lake Paiute Tribe
TAC	Tribal Advisory Committee
TK	Traditional Knowledge(s)
TPR	Tribal Participatory Research
UA	University of Arizona
WRCC	Water Resources Research Center

References

1. Whyte, K.P. Indigenous women, climate change impacts and collective action. *Hypatia J. Fem. Philos.* **2014**, *29*, 599–616. [CrossRef]
2. McGregor, D. Honouring our relations: An Anishnaabe perspective on environmental justice. In *Speaking for Ourselves: Environmental Justice in Canada*; Agyeman, J., Cole, P., Haluza-Delay, R., Eds.; University of British Columbia Press: Vancouver, BC, Canada, 2009; pp. 27–41.
3. Black, M.S. Indian entities recognized and eligible to receive services from the Bureau of Indian Affairs. *Fed. Regist.* **2012**, *44*, 47868–47873.
4. Norris, T.; Vines, P.L.; Hoeffel, E.M. The American Indian and Native Population: 2010. 2010 Census Briefs. 2012. Available online: http://www.census.gov/prod/cen2010/briefs/c2010br-10.pdf (accessed on 4 July 2016).

5. Wilkinson, C. *Crossing the Next Meridian: Land, Water, and the Future of the West*; Island Press: Washington, DC, USA, 1992.
6. Tarlock, A.D. Prior appropriation: Rule, principle, or rhetoric. *NDL Rev.* **2001**, *76*, 881–910.
7. Winters v. United States. 207 U.S. 564 (1908). Available online: https://supreme.justia.com/cases/federal/us/207/564/ (accessed on 16 August 2016).
8. Rancier, R. Assessing Tribal Water Rights Settlements as a Means for Resolving Instream Flow Claims: A Comparative Case Approach. Master's Thesis, Oregon State University, Corvallis, OR, USA, 2012.
9. Dustan, A. With anything manmade there is going to be danger: The cultural context of Navajo opinions regarding snowmaking on the San Francisco Peaks. *Indig. Policy J.* **2010**, *21*, 2.
10. Beamer, P.; Chief, K.; Borrero, N.; Rivera, B. Water is Our Life: How a Mining Disaster Affected the Navajo Nation. Truthout News. 2016. Available online: http://www.truth-out.org/news/item/36049-water-is-our-life-how-a-mining-disaster-affected-the-navajo-nation (accessed on 4 July 2016).
11. McGregor, D. Traditional knowledge: Considerations for protecting water in Ontario. *Int. Indig. Policy J.* **2012**, *3*, 11.
12. Arquette, M.; Cole, M.; Cook, K.; LaFrance, B.; Peters, M.; Ransom, J.; Sargent, E.; Smoke, V.; Stairs, A. Holistic risk-based environmental decision-making: A Native perspective. *Environ. Health Perspect.* **2002**, *110*, 259–264. [CrossRef] [PubMed]
13. Foushee, L.; Gurneau, R. (Eds.) *Sacred Water: Water for Life*; North American Water Office: Lake Elmo, MN, USA, 2010.
14. McGregor, D. Linking traditional ecological knowledge and western science: Aboriginal perspectives. In Proceedings of the 2000 State of the Lakes Ecosystem Conference. *Can. J. Nativ. Stud.* **2008**, *28*, 139–158.
15. Hand, J.P. Co-operating to protect the shining big sea water and its siblings: Consultation with Native Peoples in protecting the Great Lakes. In *Tribes, Land, and the Environment*; Krakoff, S.A., Rosser, E., Eds.; Ashgate Publishing Company: Burlington, VT, USA, 2012; pp. 151–170.
16. Climate and Traditional Knowledges Workgroup (CTKW). Guidelines for Considering Traditional Knowledges in Climate Change Initiatives. Available online: http://climatetkw.wordpress.com (accessed on 4 July 2016).
17. Redsteer, M.H.; Kelley, K.B.; Francis, H.; Block, D. Increasing vulnerability of the Navajo people to drought and climate change in the southwestern United States: Accounts from tribal elders. In *Special Report on Indigenous People, Marginalized Populations and Climate Change*; Nakashima, D., Rubis, J., Krupnik, I., Eds.; Cambridge University Press: New York, NY, USA, 2014.
18. Huntington, H.P. Using traditional ecological knowledge in science: Methods and applications. *Ecol. Appl.* **2000**, *10*, 1270–1274. [CrossRef]
19. Cherokee Nation v. Georgia. 30 U.S. 1. Available online: https://www.law.cornell.edu/supremecourt/text/30/1 (accessed on 15 August 2016).
20. United States Court of Appeals for the Federal Circuit. Available online: https://www.narf.org/nill/bulletins/cta/documents/navajo.pdf (accessed on 15 August 2016).
21. United States v. Mitchell, 445 U.S. 535 (1980). Available online: https://supreme.justia.com/cases/federal/us/445/535/case.html (accessed on 15 August 2016).
22. Mitchell v. United States, 664 F.2d 265 Ct.Cl. 1981. Available online: https://casetext.com/case/mitchell-v-united-states-21 (accessed on 15 August 2016).
23. In the United States District Court for the District of Columbia. Available online: http://www.indiantrust.com/docs/IIM_Joint_Motion_for_Approval.pdf (accessed on 15 August 2016).
24. Indian Reorganization Act. Available online: http://heinonline.org/HOL/LandingPage?handle=hein.amindian/indireac0003&div=1 (accessed on 15 August 2016).
25. General Allotment Act, Act of 8 February 1887 (24 Stat. 388, ch. 119, 25 USCA 331). Available online: https://www.iltf.org/resources/land-tenure-history/historical-allotment-legislation/general-allotment-act (accessed on 15 August 2016).
26. Singletary, L.; Emm, S.; Hill, G. An assessment of agriculture and natural resource extension program needs on American Indian reservations in Idaho, Nevada, Oregon, and Washington. Available online: http://www.joe.org/joe/2011april/a2.php (accessed on 8 August 2016).
27. Shoemaker, J.A. Like snow in the spring time: Allotment, fractionation, and the Indian land tenure problem. *Wis. Law Rev.* **2003**, *2003*, 10–61.

28. Arizona v. California. 373 U.S. 546 (1963). Available online: https://supreme.justia.com/cases/federal/us/373/546/ (accessed on 15 August 2016).
29. 43 U.S. Code § 666. Suits for Adjudication of Water Rights. Available online: https://www.law.cornell.edu/uscode/text/43/666 (accessed on 15 August 2016).
30. Colby, B.G.; Thorson, J.E.; Britton, S. *Negotiating Tribal Water Rights Fulfilling Promises in the Arid West*; University of Arizona Press: Tucson, AZ, USA, 2005; p. 190.
31. Arizona v. California, 460 U.S. 605 (1863). Available online: https://supreme.justia.com/cases/federal/us/460/605/case.html (accessed on 15 August 2016).
32. Franks, M.C. The uses of the practicably irrigabe acreage standard in the quantification of reserved water rights. *Nat. Resour. J.* **1991**, *31*, 449–585.
33. Cosens, B. The measure of Indian water rights: The Arizona homeland standard, Gila River Adjudication. *Nat. Resour. J.* **2002**, *42*, 835–872.
34. D'Estrée, T.P.; Colby, B.G. *Braving the Currents: Evaluating Environmental Conflict Resolution in the River Basins of the American West*; Kluwer Academic Publishers: Boston, MA, USA, 2004.
35. Gautam, M.; Chief, K.; Smith, W.J., Jr. Climate change in arid lands and Native American socioeconomic vulnerability: The case of the Pyramid Lake Paiute Tribe. *Clim. Chang.* **2013**, *120*, 585–599. [CrossRef]
36. Kinney, T. Chasing the wind: Wyoming Supreme Court decision in Big Horn III denies beneficial use for instream flow protection, but empowers state to administer Federal Indian reserved water right awarded to the Wind River Tribes. *Nat. Resour. J.* **2004**, *33*, 841–871.
37. Yazzie, M. Unlimited limitations: The Navajos' Winters rights deemed worthless in the 2012 Navajo–Hopi Little Colorado River Settlement. *Wicazo Sa Rev.* **2013**, *28*, 26–37. [CrossRef]
38. Safe Water and Waste Disposal Facilities Fact Sheet. Available online: https://www.ihs.gov/newsroom/factsheets/safewater/ (accessed on 27 July 2016).
39. Ravisankar, V.A.; Seaman, R.; Mirchandani, S.; Arnold, R.G.; Ela, W.P. Solar-driven membrane distillation demonstration in Leupp, Arizona. *Environ. Health* **2016**, *31*, 79–83. [CrossRef] [PubMed]
40. Chief, K.; Artiola, J.F.; Beamer, P.; Wilkinson, S.; Maier, R.M.; Rock, C.; Sanchez, C. *Understanding the Gold King Mine Spill. Fact Sheet*; University of Arizona: Tucson, AZ, USA, 2015.
41. Cozzetto, K.; Chief, K.; Dittmer, K.; Brubaker, M.; Gough, R.; Souza, K.; Ettawageshik, F.; Wotkyns, S.; Opitz-Stapleton, S.; Duren, S.; et al. Climate change impacts on the water resources of American Indians and Alaska Natives in the U.S. *Clim. Chang.* **2013**, *2013*. [CrossRef]
42. Brubaker, M.; Berner, J.; Bell, J.; Warren, J.; Rolin, A. Climate Change in Point Hope Alaska: Strategies for Community Health. Available online: http://www.cidrap.umn.edu/sites/default/files/public/php/26952/Climate%20Change%20HIA%20Report_Point%20Hope_0.pdf (accessed on 4 July 2016).
43. Lomawaima, K.T. Tribal sovereigns: Reframing research in American Indian education. *Harv. Educ. Rev.* **2000**, *70*, 1–23. [CrossRef]
44. Davis, S.M.; Reid, R. Practicing participatory research in American Indian communities. *Am. J. Clin. Nutr.* **1999**, *69*, 755S–759S. [PubMed]
45. Sahota, P.C. Research Regulation in American Indian/Alaska Native communities: A Guide to Reviewing Research Studies. Available online: https://depts.washington.edu/ccph/pdf_files/.pdf (accessed on 27 July 2016).
46. Brugge, D.; Missaghian, M. Protecting the Navajo people through tribal regulation of research. *Sci. Eng. Ethics* **2006**, *12*, 491–507. [CrossRef] [PubMed]
47. University of Arizona Native Peoples Technical Assistance Office. *Handbook for Research in Indian Country*; University of Arizona Press: Tucson, AZ, USA, 2016.
48. Code of Federal Regulations, Title 45, Public Welfare, Department of Health and Human Services, Part 46, Protection of Human Subjects. 2009. Available online: http://www.hhs.gov/ohrp/regulations-and-policy/regulations/45-cfr-46/ (accessed on 27 July 2016).
49. The Ethics of Traditional Knowledge Exchange in Climate Change Initiatives. Available online: http://earthzine.org/2015/07/31/the-ethics-of-traditional-knowledge-exchange-in-climate-change-initiatives/ (accessed on 4 July 2016).
50. Whyte, K.P.; Brewer, J.P., II; Johnson, J.T. Weaving Indigenous science, protocols and sustainability science. *Sustain. Sci.* **2015**, *11*, 25–32. [CrossRef]

51. Burkhart, B.Y. What coyote and Thales can teach us: An outline of American Indian epistemology. In *American Indian Thought: Philosophical Essays*; Waters, A., Ed.; Wiley-Blackwell: San Francisco, CA, USA, 2004.
52. Kimmerer, R.W. *Braiding Sweetgrass: Indigenous Wisdom, Scientific Knowledge and the Teachings of Plants*; Milkweed Editions: Minneapolis, MN, USA, 2015.
53. Whyte, K.P. On the role of traditional ecological knowledge as a collaborative concept: A philosophical study. *Ecol. Process.* **2013**, *2*, 1–12. [CrossRef]
54. Hardison, P.; Williams, T. Culture, law, risk and governance: Contexts of traditional knowledge in climate change adaptation. *Clim. Chang.* **2013**, *2013*. [CrossRef]
55. Whyte, K.P. What Do Indigenous Knowledges Do for Indigenous Peoples? In *Keepers of the Green World: Traditional Ecological Knowledge and Sustainability*; Nelson, M.K., Shilling, D., Eds.; Cambridge University Press: Cambridge, UK, 2015.
56. Smith, L.T. *Decolonizing Methodologies: Research and Indigenous Peoples*; Zed Books: London, UK, 1999.
57. Jasanoff, S.; Wynne, B. Science and decisionmaking. In *Human Choice and Climate Change*; Rayner, S., Malone, E.L., Eds.; Battelle Press: Columbus, OH, USA, 1998.
58. Dilling, L.; Lemos, M.C. Creating usable science: Opportunities and constraints for climate knowledge use and their implications for science policy. *Glob. Environ. Chang.* **2011**, *21*, 680–689. [CrossRef]
59. Lemos, M.C.; Morehouse, B.J. The co-production of science and policy in integrated climate assessments. *Glob. Environ. Chang.* **2005**, *15*, 57–68. [CrossRef]
60. Beierle, T.C. The quality of stakeholder-based decisions. *Risk Anal.* **2002**, *22*, 739–749. [CrossRef] [PubMed]
61. Nadasdy, P. The anti-politics of TEK: The institutionalization of co-management discourse and practice. *Anthropologica* **2005**, *47*, 215–231.
62. Rowe, G.; Frewer, L.J. A typology of public engagement mechanisms. *Sci. Technol. Hum. Values* **2005**, *30*, 251–290. [CrossRef]
63. Stern, P.C.; Fireberg, H.V. *Understanding Risk: Informing Decisions in a Democratic Society*; National Academies Press: Washington, DC, USA, 1996.
64. Ford, J.D.; Stephenson, E.; Willox, A.C.; Edge, V.; Farahbakhsh, K.; Furgal, C.; Harper, S.; Chatwood, S.; Mauro, I.; Pearce, T.; et al. Community-based adaptation research in the Canadian Arctic. *Wiley Interdiscip. Rev. Clim. Chang.* **2015**, *7*. [CrossRef]
65. Fisher, P.A.; Ball, T.J. Tribal participatory research: Mechanisms of a collaborative model. *Am. J. Community Psychol.* **2003**, *32*, 207–216. [CrossRef] [PubMed]
66. Lewin, K. Action research and minority problems. *J. Soc. Issues* **1946**, *46*, 34–46. [CrossRef]
67. Robinson, C.J.; Wallington, T.J. Boundary work: Engaging knowledge systems in comanagement of feral animals on Indigenous lands. *Ecol. Soc.* **2012**, *17*. [CrossRef]
68. Bark, R.H.; Garrick, D.E.; Robinson, C.J.; Jackson, S. Adaptive basin governance and the prospects for meeting indigenous water claims. *Environ. Sci. Policy* **2012**, *19*, 169–177. [CrossRef]
69. Hill, R.; Grant, C.; George, M.; Robinson, C.J.; Jackson, S.; Abel, N. A typology of indigenous engagement in Australian environmental management: Implications for knowledge integration and social-ecological system sustainability. *Ecol. Soc.* **2012**, *17*. [CrossRef]
70. Cronin, A.E.; Ostergren, D.M. Democracy, participation, and Native American tribes in collaborative watershed management. *Soc. Nat. Resour.* **2007**, *20*, 527–542. [CrossRef]
71. Biggs, S.D. *Resource-Poor Farmer Participation in Research: A Synthesis of Experiences from Nine National Agricultural Research Systems*; International Service for National Agricultural Research: The Hague, The Netherlands, 1989; Volume 3, p. 37.
72. Huntington, H.P. We dance around in a ring and suppose: Academic engagement with traditional knowledge. *Arct. Anthropol.* **2005**, *42*, 29–32. [CrossRef]
73. Huntington, H.P.; Gearheard, S.; Mahoney, A.R.; Salomon, A.K. Integrating traditional and scientific knowledge through collaborative natural science field research: Identifying elements for success. *Arctic* **2011**, *64*, 437–445. [CrossRef]
74. Latulippe, N. Situating the work: A typology of traditional knowledge literature. *Altern. Int. J. Indig. Peoples* **2015**, *11*, 118–131.
75. Ferguson, D.B.; Masayesva, A.; Meadow, A.M.; Crimmins, M.A. Rain gauges to range conditions: Developing a local drought information system using locally relevant observations. *Weather Clim. Soc.* **2016**, in press. [CrossRef]

76. Smith, W.J., Jr.; Liu, Z.; Safi, A.S.; Chief, K. Climate change perception, observation and policy support in rural Nevada: A comparative analysis of Native Americans, non-native ranchers and farmers and mainstream America. *Environ. Sci. Policy* **2014**, *42*, 101–122. [CrossRef]
77. Megdal, S.M.; Chief, K.; McLain, J.E. Collaborative Conference Planning and Continuing the Dialogue. Available online: https://wrrc.arizona.edu/Collaborative_Conference_Planning (accessed on 27 July 2016).
78. Mott Lacroix, K.; Xiu, B.; Megdal, S.B. Roadmap for Considering Water for Arizona's Natural Areas. Available online: https://wrrc.arizona.edu/sites/wrrc.arizona.edu/files/RAPIDS/PDF/Roadmap_Complete_2014.pdf (accessed on 27 July 2016).
79. Bureau of Reclamation. Colorado River Basin Water Supply and Demand Study. Available online: http://www.usbr.gov/lc/region/programs/crbstudy/finalreport/index.html (accessed on 27 July 2016).

water

MDPI

Article

Public Participation in Water Planning in the Ebro River Basin (Spain) and Tucson Basin (U.S., Arizona): Impact on Water Policy and Adaptive Capacity Building

Alba Ballester [1,*] and Kelly E. Mott Lacroix [2]

1 Institut de Govern i Polítiques Públiques, Universitat Autònoma de Barcelona, Campus UAB-Edifici MRA, Bellaterra 08193, Spain

2 Water Resources Research Center, University of Arizona, 350 N Campbell Ave, Tucson, AZ 85719, USA; klacroix@email.arizona.edu

* Correspondence: alba.ballester@gmail.com or Alba.Ballester@uab.cat; Tel.: +34-628-443-001

Academic Editors: Eylon Shamir, Sharon B. Megdal and Susanna Eden

Received: 21 February 2016; Accepted: 20 June 2016; Published: 29 June 2016

Abstract: The benefits of public participation in water management are recognized by governments, scholars, and stakeholders. These benefits, however, do not result from all engagement endeavors. This leads to the question: What are the determinants for effective public participation? Given a list of criteria for achieving the transformational capacity of participation, we analyze the benefits (including the influence on public policies) gained through public participation and the determinant factors for obtaining these benefits in the Ebro River Basin in Spain and in the Tucson Basin in Arizona (U.S.). Furthermore, and considering that droughts and floods are major water management challenges in both case studies, we focus on the potential of participation to build adaptive capacity. Our analysis of these case studies concludes that influence on public policies is determined more by the context of the participatory process, i.e., legal framework, political leadership, and social awareness, whereas influence on adaptive capacity building depends more on the characteristics of the participatory process, particularly the existence of active on-site consultation and deliberation.

Keywords: water management; public participation; stakeholder engagement; adaptive capacity

1. Introduction

Water is a collective heritage with different functions, uses, and values. Water is also finite, vulnerable to contamination and the effects of climate variability, which makes its future availability uncertain. Governments and scholars alike note that decision-making for water should be through a collaborative and inclusive process, where all the voices are represented [1–3]. This recognition is reflected in requirements for public participation in different treaties and laws, particularly in natural resource management (i.e., [4–7]). Public participation is defined here as the process through which citizens seek to influence public issues [8]. The opportunities for input can be embedded within or outside of institutions, and occur with different degrees of public influence on a process. Most categorizations of public participation range from degrees of non-participation (manipulation), apparent participation (information, consultation, active involvement), to citizen power (management, decision) [9,10].

Over the past two decades, many participatory processes for water resource planning have been developed, and research about the implementation of this participation and its benefits has grown considerably (Table 1).

Table 1. Benefits of public participation.

Benefit	Citations
Building social capital, trust, and mutual respect	[11–13]
Anticipating conflict and overcoming historical tensions	[14]
Improving capacity to achieve collaborative agreements	[12,15]
Increasing the autonomy of individuals and communities	[16]
Permitting access to information	[17,18]
Fostering social learning and understanding of issues	[19–25]
Improving the efficiency of plan implementation	[26–28]
Increasing the legitimacy and quality of decisions	[29]
Increasing community cohesion and collective identity	[24,25]

These benefits are not, however, inherent to all participatory experiences. Frequently, public participation does not influence final decisions, improve knowledge or relationships that could impact the quality of decisions, or achieve the objectives of the participation process [20,30]. When these benefits do not manifest, what is the cause? Or conversely, what elements must a process have to reap the well-noted benefits of public participation? Many authors have highlighted the importance of procedural features of the participatory processes while others have focused on the importance of the context in which participation occurs (e.g., [3,31–33]). As a result of both lines of thought, Blanco and Ballester [34] identified five conditions for improving participation's transformational capacity: (1) political leadership of the participatory process and the affected resulting policies [35]; (2) a holistic approach to collective problems [36]; (3) a highly visible engagement process [37]; (4) significant citizen impact on public policies through the participatory process [9,38]; and (5) a real cultural change of politicians, bureaucrats, and citizens [39,40].

It is notable that the potential benefits of public participation coincide with the determinants of adaptive capacity. Adaptive capacity is the ability of a system to adjust, modify or change its features or actions to moderate potential damages, take advantage of opportunities, or cope with the consequences of shock or stress [41,42]. Despite a lack of agreement on the characteristics and determinants of adaptive capacity at different levels, general determinants of adaptive capacity can be found in the literature related to water issues (e.g., [43–45]). The Intergovernmental Panel on Climate Change groups these characteristics into three categories: economic development, technology, and social factors. Focusing on the social dimension of adaptive capacity, we identify three core aspects related to human, social, and political capital: knowledge, networks, and governance respectively (Table 2). Although different studies have linked public participation with adaptive capacity, and consider participation as a key factor in it [46–48], the relationship between how public participation contributes to adaptive-capacity building remains relatively unexplored empirically.

To further understanding of the potential that public participation has to increase adaptive-capacity in water management, we evaluated two case studies, Ebro River Basin in Spain and the Tucson Basin in the U.S. state of Arizona, based on the framework identified in Blanco and Ballester [34] for determining the transformational capacity of a public process. For each case study we explored the connections between participation and adaptive capacity by examining the context, process, and outcomes of public participation in light of the following questions: (1) To what extent does the legal, institutional, political, and social context of participation and the participatory process influence the results of participation; and (2) What impact does the participatory processes have on public policies and adaptive capacity building?

Table 2. Relationship between social factors of adaptive capacity and benefits of public participation.

Social Factors of Adaptive Capacity	Benefits of Public Participation
• Knowledge (Human Capital); • Improves ability to perceive and understand responses to socio-ecological systems; • React to those changes in a timely appropriate way [49]; • Permit well-informed decisions on adaptation [50].	• Permitting access to information and social learning; • Increasing public capacity and understanding of issues; • Co-creation of knowledge.
• Networks (Social capital); • Permits quicker and effective response against change [45,51].	• Building social capital, trust, mutual respect; • Increasing the autonomy of individuals and communities; • Increasing community cohesion and collective identity.
• Governance (Political capital); • Access to a greater diversity of knowledge Increased transparency, accountability, equity, and trust in water governance systems. [52–55]	• Anticipating conflict and overcoming historical tensions Improving capacity to achieve collaborative agreements; • Increasing the legitimacy and quality of decisions; • Improving efficiency of plan implementation.

2. Materials and Methods

2.1. Case Studies

This study is a part of the Sustainable Water ActioN (SWAN) project, whose primary objective is the exchange of knowledge among researchers in the United States and European Union on water management issues. This partnership provided the authors with the opportunity to jointly analyze the stakeholder engagement in water planning in the Ebro River Basin and the Tucson Basin. This assessment is focused on participation in the First River Basin Management Plan (RBMP) of the Ebro Basin after the approval of the Water Framework Directive (WFD), and the Third Management Plan of the Tucson Active Management Area (TAMA) (Figure 1).

The Ebro River is the largest in Spain in terms of volume of water discharged, with an annual average flow of 14,623 hm^3, and a total water demand of 8184 hm^3. Despite this apparent abundance, the continued growth of demands and Ebro's marked drought periods have caused regular periods of scarcity. Late 20th century land-use regulations have not been expediently implemented, i.e., delimitation of flood-prone areas was required in the 1985 "Ley de Aguas", but floodplains were not mapped until very recently, after 2007s European Floods Directive. Together with hydrological variability, this has caused high flood risk along large areas of the Ebro River's main stem, which affects both agricultural uses and urban sites.

Water use is divided into four sectors: 98% for irrigation (mainly) and farming; 2% for urban supply; >1% for livestock, and >1% for industrial uses [56]. The total population of the Ebro Basin in 2013 was 3,226,921 inhabitants, which are unevenly distributed across an area of 85.362 km^2: The population is concentrated in 5 large cities, representing 50% of the total population.

The Ebro River is considered to be a "transfer" basin, meaning that water is removed from it to other basins. At this time just 1% of annual average flow is transferred out of the basin, however, various inter-basin transfers, constituting up to 10% of annual average flow, have been recently proposed.

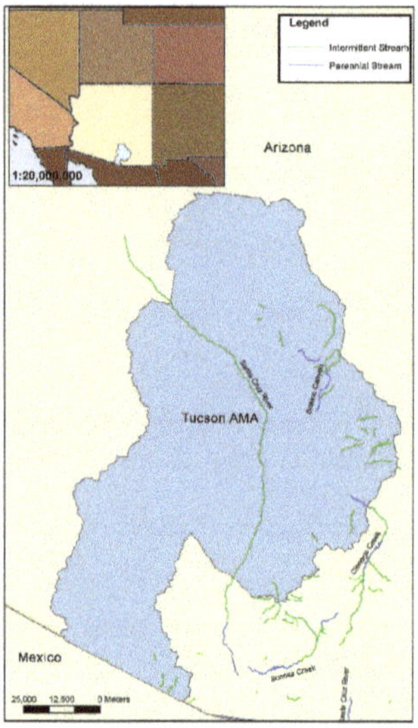

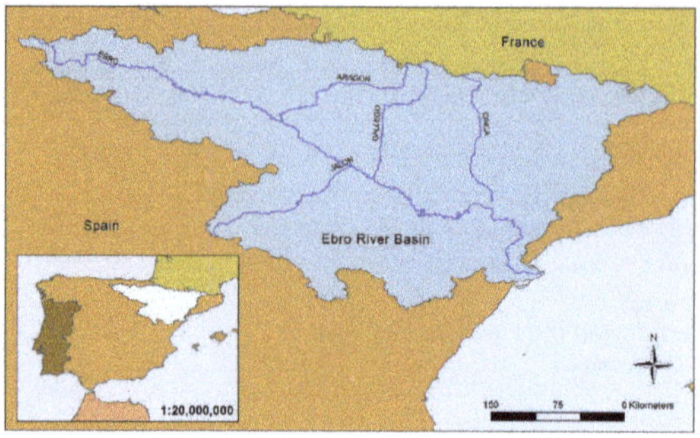

	Ebro	Tucson
Location	Spain	Arizona, US
Area (km²)	85.362	4402.98
Population (hab)	3.226.921	1.000.934
Annual average flow (hm³)	14.623	Intermittent River
Main social conflict	Water quantity, water transfers	Water quantity, water quality
Main water powers	Country	State
Condition	Transfer basin	Recipient basin

Figure 1. Case study areas and key aspects.

Tucson Basin's primary river, the Santa Cruz, is an intermittent stream. Although sections of the Santa Cruz River have always been intermittent, historically there were perennial reaches where groundwater always met the surface [57]. The current drought, while not uncommon historically, has exacerbated water supplies [58] and growth in the urban areas has contributed to the risk of scarcity [59].

The population of the Tucson Basin in 2013 was approximately 1,000,000. Of this population, 53% live in the City of Tucson. Aside from Tucson, there are four other cities with populations greater than 20,000. While population is one-third the one in the Ebro Basin, total water demand is as much as 20 times smaller than in the Ebro Basin with 421 hm^3 in 2013. This demand is distributed as 48% for urban supply; 32% for farming and ranching; 14% for industry; and 5% for Native American tribes. These demands are met by groundwater (50%), imported water from the Colorado River via Central Arizona Project (CAP) (38%), and reclaimed water (14%). It is notable that in 1985 the water supply of the basin was 99% groundwater [60].

Despite their different scales, water uses, and population, the Ebro River and Tucson Basin have and will continue to face similar water management challenges, which means that building adaptive capacity is and will be an issue of great importance. Besides this, participatory processes under diverse legislation, social and cultural patterns, have occurred in each of the basins to try to cope with the aforementioned challenges. A comparative analysis of the cases contributes to an understanding of the factors that may encourage adaptive-capacity building within a participatory process.

2.2. Methods

The methods are designed to answer the aforementioned questions: (1) To what extent does the context of participation and the participatory process influence the results of participation; and (2) What impact does the participatory processes have on public policies and adaptive capacity building? To answer these questions we examined (1) the legal, political, and social context of participation; (2) the nuts and bolts of the participatory process; and (3) outcomes of participation. A detailed description of the context and process are provided in Table 3.

Table 3. Context and process aspects and study questions.

Variable	Aspects	Questions
Context	Legal framework	Is there legislation that requires public participation in water management? How detailed are the requirements?
	Political leadership	Does the convener provide the resources necessary for effective participation? Is the participatory process central to water planning decisions?
	Social awareness	Is there any conflict associated with the participatory process? Are there any social movements specifically devoted to the reason of the participatory process?
Process	Objectives	What are the objectives of the participatory process? Were these objectives clearly outlined for the participants?
	Participants	How many participants were there? Did the participants represent diverse water interests and perspectives?
	Meetings	How many meetings have been organized? Was there sufficient time for deliberation?
	Consultation	Was participation in person or remote?
	Methods	What methods of outreach and engagement were used? (e.g., written, conferences, round table, deliberation)
	Visibility	Was the process widely visible for general public?

The outcome of participation's impact on public policy was straightforward: Did the participation change the policy? For adaptive capacity, however, it was necessary to take a more detailed approach. For the adaptive capacity outcomes, we propose a list of indicators based on the adaptive management literature. To build these indicators, we started with the social dimensions of adaptive capacity (knowledge, networks, and governance) and then developed indicators based on the elements of a process associated with increasing or improving knowledge, networks, and governance [48–55]. For each indicator we have defined a list of key research questions to be answered (Table 4).

Methods and data used for this study include: (1) analysis of documents related to the two participatory processes: minutes, planning documents, and previous research related to the case studies [8,61,62]; and (2) qualitative data obtained through 18 interviews with key stakeholders who actively participated in either the Ebro or Tucson basin participatory processes (Table 5). Of the interviews, 14 are from participants in the Ebro Basin and 4 are from the Tucson Basin. While the number of interviewees from the Tucson Basin is small, we were able to rely on a previous study [62] that interviewed a broad set of stakeholders about the planning process for the Third Management Plan to supplement our information. For the Ebro process, the author was a facilitator for recreational users at the basin scale, and participated in other plan-related meetings, while for the Tucson process the author was not involved in any part of the process.

Table 4. Outcomes: Indicators and study questions.

Variable	Aspect	Indicators		Questions
Outcomes	Impact on public policies	Existence of minor or major revisions or changes in the final plan: (1) Minor: The incorporation of measures that usually are widely agreed upon and do not require a significant investment in time or money. (2) Major: The incorporation of substantive changes in the approach to water management, or the incorporation of relevant and sensitive measures in water planning (i.e., environmental flows, revision of agricultural water concessions, link to urban development, etc.).		Did public participation change the final plan or policy?
	Impact on adaptive capacity building	Learning	Existence of learning, both in water authorities and the public	Was the participation a learning process?
		Knowledge	Generate new knowledge	Did it generate any new knowledge?
		Collective vision	Reach collective vision, the transformation of private interests to a collective public vision	Did the participants change their perspective on key issues? Did the process permit to transform private interests into a public position that incorporates a collective perspective?
		Willingness	Gain willingness to be involved in public issues	Did participation impact on the public's willingness to be involved in water issues and collaborate with other groups?
		Networks	• Improve networks between stakeholders • Existence of lasting groups emerging from the process	• What impact did the process have on communication among water interest groups? • Where there any lasting groups, associations or networks that emerged from the process?
		Trust	Increased trust among the involved stakeholders and practitioners	What impact did the process have on trust among water community?
		Continuity	Continuity of the process in subsequent planning cycles	Are new planning efforts improving public participation?

Table 5. Interviewees profile in Ebro and Tucson basins.

Stakeholder Type	Ebro	Tucson
Recreation	3	
Environment	3	
State government	1	2
Agriculture	2	1
Facilitator	2	
Municipal government	2	1
Power generation	1	

3. Results

3.1. Context

3.1.1. Legal and Institutional Framework

The legal and institutional frameworks that establish water management process and public involvement in Ebro and Tucson basins are distinct. Though some aspects of the frameworks are not directly applicable to the case studies, we examined the complete framework to gain insight into the overall context in each basin and its potential to influence participatory outcomes in each basin. One difficulty in contrasting the basins is that the management scales do not coincide. Theoretically, the comparison between the two should be at the country level, i.e., Spain vs. U.S. In practice, however, the regulatory framework of water management is first established at European and U.S. levels. For this reason, our analysis begins at the geographic scope of the United States and the European Union (Figure 2).

The European Union adopts regulations, directives, and decisions that apply to all EU Member States, which prevail over the rules of individual Member States. Since 2000, the main legislation applicable to water is the WFD (2000/60/EC), which is premised on ecosystems protection, sustainable use of water, and public participation. In accordance with these principles, the WFD establishes one main objective for water policy: The good ecological status (ecological, quantitative and chemical) of water bodies by 2015. To achieve this goal there are a suite of tools: Water planning, programs of measures, and monitoring programs [16].

Spain is politically divided into seventeen regional governments (autonomous communities) with large political, economic, and administrative capabilities. Spanish water management is decentralized, and authority is distributed between national, autonomous communities, and local administrations. Water planning is organized through 18 river basins, nine of them are intercommunity basins under national authority; seven are intracommunity basins, under autonomous competences; and two special cases correspond to the independent cities of Ceuta and Melilla, which are located in Northern Africa [63]. Each basin has its own water authority, called a River Basin Organization (RBO) in the basins that belong to the nation, and Water Agencies in the basins that belong to an Autonomous Community. Each basin authority is responsible for developing a River Basin Management Plan, which is reviewed by the National Water Council, an advisory body ascribed to the Ministry of Environment, and approved by Council of Ministries.

The Ebro River Basin is an intercommunity basin under national control. It has a governing board (Water Council) formed by representatives of national and RBO administrations, all regional administrations from within the basin, and some supra-local administrations. It also includes representatives of water users, and more recently, other water stakeholders that are invited by the RBO. Traditionally, representation of water users has been proportional to the amount of water allocated [63]. As noted before, agriculture uses over 90% of the water concessions in the region, therefore agricultural users traditionally have had more influence than others on the governing board and therefore in the decision-making process. The approval of WFD has required a more open participation that goes beyond the traditional users to include all stakeholders and the general public.

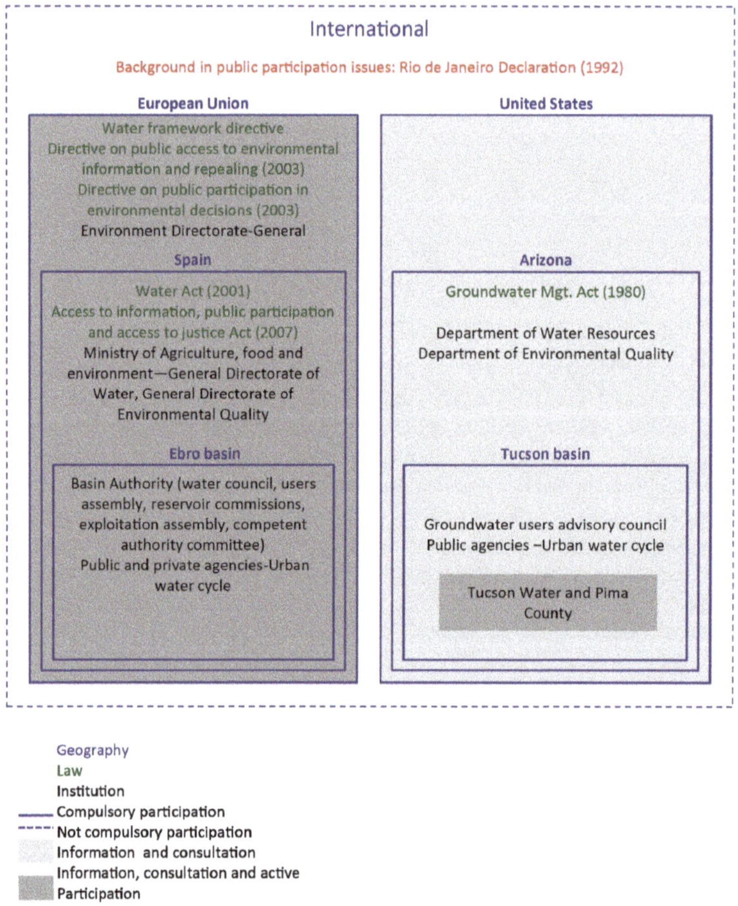

Figure 2. Legal and institutional frameworks for water management and public participation in Ebro and Tucson basins.

The legal framework for public participation in environmental matters in Europe is regulated by the Aarhus Convention (1998) and by the WFD. The particular obligations of public participation in water policy are regulated through article 14 of WFD: "Member States shall encourage the active involvement of all interested parties in the implementation of this Directive, in particular in the production, review and updating of the river basin management plans...". Within this context information supply and consultation are ensured and active involvement is encouraged [64].

In the U.S., water quantity is regulated on a state-by-state basis, and many states do not have laws regulating water quantity. Water quality, on the other hand, is regulated through the Clean Water Act, which is either administered by the Environmental Protection Agency (EPA) or through a statewide water quality agency for whom the EPA has delegated authority. Water management in the Tucson Basin is at the state and local levels via the Arizona Department of Water Resources (ADWR), county, and local jurisdictions. Regional water management is through the Tucson Active Management Area (AMA), which is part of ADWR. The regulatory framework is at the state level and results in

(1) strict regulation of groundwater pumping; (2) requirements for renewable water supplies; and (3) achievement of "safe-yield" of the aquifer by 2025 [65].

Arizona is often cited for its forward thinking regulation of groundwater in the populous portions of the state through the 1980 Groundwater Management Act (GMA) [62,66]. Public participation in the GMA (ARS 45–420) management plan process consists of public hearings to review all draft management plans prior to approval. There are no other provisions for public participation. The GMA does, however, establish a groundwater user's advisory council (GUAC) in each basin (referred to as active management areas (AMAs)) that consists of five members appointed by the governor of Arizona "on the basis of their knowledge of, interest in and experience with problems relating to the development, use and conservation of water". The role of the GUAC is to: "Advise the area director for the active management area, make recommendations on groundwater management programs and policies for the active management area and comment to the area director and to the director on draft management plans for the active management area before they are promulgated by the director". In addition, "Provide comment to the Arizona water banking authority with regard to draft plans for additional storage facilities and draft plans of operation in accordance with sections 45–2453 and 45–2456". Finally, although more generally applicable, citizens in Arizona have the ability to refer issues to the ballot, through collecting signatures equal to 5% of the registered voters in the jurisdiction where the election will be held. Draft plans for the Arizona water banking authority are provided for public comment each year and citizen ballot initiatives are common. Most notably is a 1990s citizen initiative that restricts how water from the CAP canal is delivered to Tucson residents.

Examining the two cases side by side (Table 6), there is more specific legal framework on public participation in water policy in European Union, which is especially true when compared to the federal level in the United States. For example, the Groundwater Management Act does not mention active involvement, whereas active involvement is included in the EU Water Framework Directive.

Table 6. Comparison of case studies legal frameworks.

	Main Laws	Relevance of Participation	Participation Degree	Degree of Obligation	Public Involved	Participation Timing
Ebro Basin	European Water Framework Directive	Considered a Principle of Water Management	Information and Consultation	Must be ensured	General public & stakeholders	Whole planning cycle
			Active involvement	Must be encouraged	Stakeholders	
Tucson Basin	Groundwater Management Act	No specification	Information and Consultation	Must be ensured	General public & stakeholders	Draft management plans

3.1.2. Political Leadership

The Ebro RBO is responsible for implementing the public participation process during the elaboration, review, and implementation of RBMP. Thus, public participation during the first planning cycle after WFD was funded through the basin organization (national funding), and promoted through the Hydrological Planning Office, which is in charge of the elaboration of the RBMP. The Hydrological Planning Office did not have any experience with a public participation process, were not devoted to the process, and participation in plan development was perceived as more of an obligation than a priority [67] Nevertheless, according to an interview with the process managers, a significant amount of money (~1 M€), was allocated for the participatory process.

In the Tucson AMA the Arizona Department of Water Resources, is responsible for the creation, review, and implementation of the AMA management plans. At the time of the Third Management Plan, the Tucson AMA had a local field staff, who were responsible for the planning process, however, none of the staff were dedicated only to managing the public engagement process and the members of the groundwater users advisory committee (GUAC) were not compensated for their time.

While explicit funding for the participatory process, as indicated in the Ebro River case, could be a reflection of a greater political leadership, our research did not indicate that political leadership was significant in either case.

3.1.3. Social Awareness

Historically, the Ebro Basin has a significant tradition of social movements fighting for an improved water management in the basin [68]. Huge public demonstrations, mostly related to building of dams and water transfers, have been organized in the last 20 years resulting in high media visibility and impact on water policies [69–71]. As a consequence, social awareness and an involvement of a significant number of people on water issues has occurred, creating a network of over one hundred of green social organizations along the basin. On the other hand, collaboration between environmental organizations and water authority has been almost non-existent, while the importance of irrigation uses in the basin has resulted in a close relationship between the sector and the water authority [67,72]. Implementation of the Water Framework Directive during the first planning cycle did, however, cause the RBO to try to open new paths for collaboration with all the stakeholders. Based on responses from interviewees, and the fact that no participatory activities at a basin or sub-basin level have been implemented after the first RBMP planning cycle, it seems that this path has been lost in the second planning cycle. Social movements initiated in the first planning cycle have, however, continued.

In Tucson, there is an active and involved water community that includes citizen advocates and water professionals from the private and public sectors. The region has a history of grassroots action on water issues [73,74]. In the 1990s, issues with the delivery of CAP water that caused brown water to flow from some resident's taps resulted in a citizen's initiative that restricts how CAP water is delivered to this day [75]. Most conflict in the basin has been centered around decisions to deliver CAP water and the use of reclaimed water, not the AMA management plans per se [76]. The community has a long-standing and strong water conservation ethic [76], and is focused primarily on collaboration with water managers. However, according to interviews, at the time of the Third Management Plan the environmental community was not involved in planning.

3.2. *The Participatory Process*

Among the different degrees of public participation mentioned previously (e.g., [9,10]), we will focus our analysis on the apparent degree of participation using the classifications proposed by Organization for the Economic Co-operation and Development (OECD) and the WFD in its guidance document on public participation [64], from which we identify the following degrees of participation: (1) Information: One-way relationship, where government produces and delivers information for use by the public; (2) Consultation: Two-way relationship, where government defines the issues for involvement, sets the questions, and manages the process; the public contributes with their views and opinions; (3) Active involvement, a partnership among government and the public, where the public is actively engaged in defining the process and content of policy-making, setting the agenda, and proposing policy options; the responsibility for the final decision rests with government. Taking into account this framework we review: the clarity of participation objectives; number and diversity of participants; number and duration of meetings; type of involvement and participatory methods used; and the visibility of the process.

3.2.1. Ebro Basin

The general aim of the Ebro's participatory process in the first water planning cycle after the approval of the WFD was to inform the general public about the planning process and its contents, and to collect possible metrics to be included in a potential program of measures. There was no commitment to include suggested measures in the final Ebro River management plan.

The quantity, quality, and access to information were significantly improved from previous planning efforts and provided: hydrographic information in digital and printed format, and a specific section of participatory process and related documents was created in the website of the RBO, including technical reports, meeting minutes and legal information. However, the reports prepared for the regional meetings were excessively technical, long, and were sent with insufficient time for review prior to the meetings. Interestingly, the index of transparency in water management in the Ebro Basin was considered to be up to 80% in 2011, but has decreased to 60% in 2015, when significant gaps on financial and economical information and information regarding relationships among the RBO and the citizens were identified [72]. These percentages coincide with moments of high participation (year 2011) and low participation (year 2015) in the planning cycle.

With regards to the consultation process, there were three different typologies of involvement: (1) formal periods for written comments (6 months for each consultation, 3 in total: timetable and work program for the production of the plan, overview of the significant water management issues in the river basin, draft of the river basin management plan, as is established in WFD); (2) institutional participation through Ebro's RBO Water Council (1 annual meeting, with 40 representatives of stakeholders in the basin) for making decisions about water planning; and (3) non-institutional or informal participation, through the implementation of an on-site participatory process. This process was organized between 2006 and 2008 (during 19 months) at the basin (17 meetings) and 27 sub-basins (107 meetings) levels. These meetings were organized with different sectors separately: social sector (environmental orgs, neighborhood orgs, educational orgs, recreational users); economic sector (irrigators, industry, etc.); administration sector (local councils, regional governments). The total number of participants was 1483, but only 172 participated in more than one meeting. Each meeting lasted for two hours, including a presentation by the basin authority and time afterwards for participant comments and proposals. Taking into account the little time for discussion and the absence of intersectorial debates, no opportunity for deliberation or opinion exchange was promoted, nor any actions for general public involvement. However, a diversity of stakeholders and water interests were represented in the meetings. Moreover, at basin scale, two bottom-up participatory processes were implemented, one with environmental sector, another with recreational users [64]. Both were proposed to the basin authority by the environmental community in one hand, and the recreational users on the other. These processes were supported and funded by the RBO and managed by independent consultants. They had a deliberative approach and can be identified as active involvement. These two participatory processes were considered best practices by interviewees.

The visibility of the planning and participatory processes have improved significantly as compared to previous planning processes, which were elaborated from within the confines of the governing board of the RBO. To date, however, the current process has only reached expert stakeholders, as the general public was not involved in the process and no active dissemination was promoted for them.

3.2.2. Tucson AMA

In the Tucson AMA, management plans provide a regulatory context and framework, or the rules of engagement, but they are not operational plans for the regulated water users in the region. Within this framework water companies and other regulated water users determine how they will meet these rules. The objective of public participation in the Third Management Plan was therefore to gain approval by the regulated community of the proposed framework and this objective was clearly outlined within the plan. It is notable, however, that Arizona statute (§ 45–421(1)) requires that the GUAC comment, but not approve the plan before it is promulgated. Therefore, there are no requirements for approval by the citizen body (the GUAC) or the public prior to adoption of the plan.

The creation of the Third Management Plan was between informational and consultational on the spectrum of participation. The plan was developed in three stages beginning in 1994 and culminated with adoption of the plan in 1999. Participatory methods used were meetings, public hearings, and written comments. The first stage of development was to collect and internally analyze data. These data were released for public review and comment through a "State of the AMA" report [77]. Issues identified in the report and raised by the community were addressed by the Department through a series of white papers that identified recommended alternatives. These alternatives were then formulated into program concepts and then the chapters within the plan. Public input through particularly the second and third stage of this process was via the GUAC and eight formal technical advisory committees in different subject areas relevant to the plan. The technical committees met on a regular basis for almost two years during the development of the plan. In total, there were 88 people involved representing a broad array of water management interests. In the opinion of some interviewees, however, the involvement of the advisory committees was mostly after management decisions have been made based on the agency research associated with the background technical report.

The GUAC rejected the initial plan presented by the Department because it did not bring the Tucson AMA to safe-yield of the aquifer. Once the plan was released, public hearings where ADWR described and answered public questions on the plan were conducted to give an opportunity for public oral and written comments. The Department evaluated these comments and then issued a summary of the hearings and findings prior to the first-order for adoption for the plan. The final management plan was adopted in 1999. The majority of participation in the plan was in person.

Despite the assertion within the plan that "The Department [ADWR] philosophy is to maximize public input on the content of the management plans to ensure that the public's concerns and ideas are adequately incorporated", our interviews confirmed those of a previous review of AMA management plans by Megdal et al. [62] who found that overall the public process in this planning effort was insufficient. Trust between ADWR and the regulated community was of particular concern in our findings and that of Megdal et al. [62]. In their report Megdal et al. [62] suggest that the public participation process could be improved by "empowering (ADWR) planning staff so it is clear to stakeholders that they are influencing the decision-making process". It was also noted that ADWR's ability to engage with stakeholders is hampered by a lack of expertise and experience with the management plan process by current staff as a result of staff turnover. Like the Ebro River Basin process, the Third Management Plan was not visible to the general public, but did incorporate expert stakeholders that are impacted by the Plan, namely the regulated water community.

3.2.3. Comparative Analysis

With regards to the participation process, the cases appear very similar (Table 7), which is especially notable given the varied scales of the two case-study basins. A more in-depth review would undoubtedly show more differences, i.e., information to the public was further developed and systematically organized in the Ebro Basin (probably as a result of the WFD obligations), and the structure and participative procedures of the regional meetings and active involvement.

Table 7. Comparison of case studies' participatory processes.

	Ebro Basin	Tucson Basin
Clarity of Objectives	Yes	Yes
Participants	1483/high diversity	88/high diversity
Meetings	124	Exact number not available, frequent meetings over 2-years
Participation	Remote and in person	Remote and in person
Methods	Information, consultation and active involvement (written comments, institutional and non institutional participation)	Information and consultation (written comments, institutional and non institutional participation)
Visibility	Low	Low

3.3. *Outcomes*

The outcomes of the participatory processes are examined against the impacts on public policy and elements of adaptive capacity building discussed previously in the methods. The impact on public policies is measured on the ability for the public to influence the final plan and adaptive-capacity building is measured based on seven indicators: learning, knowledge, collective vision, willingness, networks, trust, and continuity.

3.3.1. Ebro Basin

Regarding the impact on public policies, only a few measures proposed during the participation process were included in the plan, and no changes were made to the strategic vision of the plan. There is a general perception that stakeholder impact on policy does not depend on the quality of the process but rather on the limitations of management and political will. According to interviewees, the boundaries of negotiation and a mechanism for prioritizing management measures suggested during the participatory process need to be defined at the beginning of the process. Without these the process is considered to be inefficient. Nevertheless, the more optimistic interviewees noted that despite minimal impact on the final plan, the consultation process alone forced the agency to respond and justify with more details the contents of the plan, and in doing so improved the final contents of the plan.

In contrast to the influence of stakeholders on the final plan, interviewees identified considerable impact on adaptive capacity by the process. Engagement was seen as a learning process in both technical (i.e., planning cycle, the main issues in the Ebro Basin) and social aspects (i.e., stakeholder opinions, management of disagreements, how to participate), which also generated knowledge about the issues and new proposals for management. However, there was no transformation of private interests into a public position that incorporated a collective perspective, as evidenced by the fact that there was neither the time nor the space for a common discussion or deliberation among stakeholders, with the exception of the aforementioned participatory processes at basin scale promoted by the RBO with the environmental and recreational sectors.

The process has also increased stakeholder willingness to be involved in water issues and their collaboration with other groups. However, the fact that no participatory process has been promoted in the subsequent water planning cycles has caused frustration and a decrease in public involvement in water issues in the basin. Even so, two lasting groups emerged from the participatory process at the basin scale: Cuenca Azul (network of the environmental sector), and AURA (association of recreation users). The first is still active, while the second exists but is not currently active. Furthermore, the process improved coordination within the existing groups. At sub-basin scale, the participatory process was not conducive to the emergence of lasting groups or the promotion of coordination among stakeholders, as no deliberation and no multi-sectoral meetings were held, and most of the time, participants only attended one meeting, as there were no other opportunities to meet. Along the same lines, trust was built among members of the same sector but not across different stakeholder groups. According to the interviewees responses, the absence of deliberation and the insufficient time and opportunities to meet have been identified as one of the weakness of the process, while the creation of close relationships among stakeholders and water agency is considered one of its positive results.

As previously noted, most of the impacts on adaptive capacity building are attributed to informal and on-site participation processes. In the following planning cycle participation was relegated to the formal periods for written comments and participation in the Water Council. As one of the key elements of adaptive capacity building is the continuity in time, presumably, the gaps in opportunities for participation in the forthcoming plans will involve that adaptive capacity will not be maintained in the future at current levels.

3.3.2. Tucson AMA

As described by one interviewee, the Third Management Plan was the result of a "decide, present, defend" approach to public participation. In contrast, more recent water planning efforts in the region, led by the City of Tucson and Tucson Water, have been more of a bottom-up process and incorporated public, or at least key stakeholder information from the outset. While ADWR made an effort to respond to comments and suggestions collected during the process, this participation does not appear to have had a significant impact on the final plan per se. Subsequent litigation over plan requirements for municipal water use reductions (gallons per capita per day (GPCD) program) has, however, made changes the Tucson AMA Third Management Plan. The dispute over the GPCD program did not resolve until 2004 and it was not until 2008 when the plan was changed as a result of the litigation.

It is unclear through the interviews if the process of creating the Third Management Plan was a learning process overall. From the standpoint of collection and dissemination of data to the working groups it was a learning process, however, whether or not ADWR learned from the stakeholders and/or the stakeholders learned from the Department is unclear. Similar to the Ebro River Basin experience the process had an impact on communication among water groups as it resulted in the formation of a new community-led group, the Safe-Yield Task Force. Different from the Ebro Basin experience, however, the community-led group is broad based, not just among environmental or recreational interests. The Safe-Yield Task Force was a direct result of stakeholders concern that the working group and GUAC models of participation were not sufficient to address the local concerns about maintaining (and improving) aquifer water levels. The overall process appears to be adequate at incorporating the thoughts of the regulated community, it did not, however, include the general public in any meaningful way, as it is not necessarily intended to, and certainly not required to be, a public process. Interviewees noted that during the Third Management Plan time trust between ADWR and the public was relatively low. This trust has increased overtime, as perhaps indicated by the disbanding of the Safe-Yield task force in the mid-2000s. This group did, however, come together again after a period of almost 10 years as ADWR began the Fourth Management Plan process. It is unclear if the process impacted stakeholder's willingness to be involved in water issues or changed participant perspectives on key issues, but given the robust engagement seen in Tucson today it does not appear, at the very least, to have hindered involvement. Finally, there is no evidence that the final plan incorporated collective perspectives gathered from interactions among the different stakeholders and the Department.

Current planning efforts under the Fourth Management Plan, which should have been released almost five years ago, has taken more of a bottom-up approach, however, the plan still lacks a mechanism for input beyond the GUAC. As noted above, the safe-yield taskforce, created out of frustration with the Third Management Plan public input process, had stopped meeting for a period of almost 10 years, and then re-started in late 2008 in response again to frustrations with the timeliness of the Fourth management plan and concerns over ADWR capacity. One interviewee noted frustration with the participation process in the Fourth management plan because it has been under development for so long and has not been widely discussed. This person noted, however, that while opportunities for participation and input have been slight, ADWR has been straightforward about their intentions to not make significant changes from the Third Management Plan and to not have public participation outside of GUAC meetings. It should also be recognized that during the Third Management Plan ADWR had a staff of over 10 people, physically located in Tucson, dedicated to creating the plan (although not necessarily dedicated to public engagement). For the Fourth management plan there is no one person dedicated to the Tucson management plan alone and all staff are in Phoenix. In other words, operational (person) capital has been severely weakened, as has leadership capital.

3.3.3. Comparative Analysis

The impacts of public participation on public policies and indicators of adaptive capacity are similar across the case studies (Table 8). In both of them only minor revisions to the plans occurred as a result of public input during the process. In the case of the Tucson AMA, major revisions to the plan were made after the promulgation of the plan as a result of a lawsuit, which suggests a failure of the participation process to address stakeholder concerns. In the case of the Ebro, more than 3000 proposals were incorporated in a "potential measures" document, but just a few of them were finally contemplated in the resultant Basin Management Plan.

With regards to adaptive capacity, both processes increased knowledge of the physical system and stakeholder concerns, encouraged willingness to be involved, and contributed to the creation of lasting groups which have served to increase the mechanisms for engagement in water management.

Table 8. Impacts of public participation on public policies and adaptive capacity.

Impact	Ebro		Tucson	
	Yes	No	Yes	No
	Public policies			
Minor revisions	X		X	
Major revisions		X		X
	Adaptive capacity			
Learning	X		Unclear	
New knowledge	X		X	
Collective vision		X		X
Willingness to be involved	X		X	
Lasting groups	X		X	
Trust	X			X
Continuity		X		X

In interviews for the Tucson AMA some participants felt that learning occurred, while others felt strongly that it did not: It is, therefore, unclear if learning actually occurred. This result is intriguing given the length of engagement (over two years) that would seem to be sufficient for learning to occur. Interviewees for the Ebro, on the other hand, did indicate that learning had occurred and that trust had increased. The feeling that trust was not increased in the Tucson process could be due to the low levels of antecedent trust indicated by interviews or because the process did not appear to successfully incorporate stakeholder perspectives. Finally, neither of the case study processes achieved a collective vision through the process. This indicator is frequently linked to deliberation [78]. Thus, it is probable that by improving deliberation, a common public perspective will be an outcome of the participatory processes.

4. Discussion

The aforementioned conditions established by Blanco and Ballester [36] (political leadership, holistic approach, highly visible process, significant citizen impact on policies or a real cultural change) for achieving the "transformational capacity" of participation processes were largely absent in the case studies. According to the authors, without these conditions the benefits of participation should not be significant. However, some changes to the basin management plans and social adaptive capacity have been identified in our research as a result of both participatory processes.

Presumably, higher impacts would have been achieved if the conditions of "transformational capacity" were present, but it appears that they are not a necessary condition for change per se, but rather for achieving a higher degree of transformation.

4.1. Role of the Context in Influencing Participatory Process and Its Outcomes

Despite the more developed participatory legal framework for water management in Europe, the impact of the participatory process is similar in both case studies (Table 6). This may indicate that a higher democratic culture in Tucson case is compensating for its less developed participatory legal framework. On the other hand, the mere existence of a participatory legal framework guarantees the development of a participatory process, and therefore the beginning of a more or less productive interaction between the public and regulators, which otherwise would not have taken place in a context of less a democratic culture, such as in the Ebro Basin.

To date there have been no specific guidelines on implementing a participatory process in either case. While this provides flexibility in approach, it also enables a broad interpretation of the regulatory framework and disparate implementation under the same regulatory context. Less ambiguous legislation or rules on the features of the participatory processes could improve their design, development, and impacts [79]. For example, in the case studies stakeholder feedback was primarily carried out through written comments because of the legal obligation to respond them. This did not happen with the other types of consultation, such as the informal participation in the Ebro Basin.

Regarding the political leadership and the impact of participation on public policies, interviewees of both case studies have identified a direct relation between these two elements, which is consistent with other studies [37]. In fact, we would argue that the absence of political leadership is likely one of the most important reasons for the minimal impact of participatory processes in the Ebro and Tucson basins on the final basin management plans.

The Tucson and Ebro experiences, however, demonstrate how a lack of political leadership can be a starting point for the emergence of effective and grass-roots participation that increases stakeholder capacity to influence public policy. This is so for social movements regarding water management in the Ebro River Basin and their successes [79], but also the rise of social capital around water management in Tucson Basin through groups like the Safe-Yield Task Force. Both demonstrate the value of connections created as a result of lack of political leadership and an unsatisfactory public participation processes.

Finally, high levels of social awareness in both case studies have significantly influenced the outcomes of the participatory process. In the Ebro Basin in particular, social awareness led to additional capacity to exert political pressure and have an impact on mass media, therefore increasing visibility of the process. This increased awareness and visibility resulted in a broader set of informed stakeholders and increased technical capacity to participate, which enabled a deeper understanding of the issues and the creation of new knowledge.

Social awareness can also influence a participatory process' design and development (i.e., the creation of two ad-hoc participatory processes demanded by environmental and recreation users in the Ebro Basin). It can also guarantee of continuity of public involvement in water policies. Although the Safe-Yield Task Force in Tucson did not meet for almost a decade, the foundation laid in the late 1990s allowed it to reform when needed for the current planning effort. This group and the environmental network of the Cuenca Azul also prove how social awareness encourages the continuity required by adaptive capacity building.

Beyond the influence of social awareness on participation processes outcomes, it is interesting to observe that social awareness can manifest in different ways: a "reactive way" in the Ebro Basin (i.e., public demonstrations and demands), and a "collaborative way" in the Tucson Basin (i.e., inclusive approach through dialogue with stakeholders). While we do not have empirical data here to analyze the effectiveness of either approach, others have found that collaborative methods usually have better long-term results [80].

4.2. Influence of the Participatory Process' Features in Its Outcomes

There is a significant body of literature outlining ideal features of a participatory process (e.g., [1,81,82]). Of all the factors identified, two appear to be most salient to the results in our case studies: the existence of an active and on-site consultation process and that of deliberation. Even though neither of the participation processes had a meaningful impact on public policies, both had a significant influence on adaptive capacity. While adaptive capacity can also be achieved through other consultation means (e.g., online consultation, written comments) it is not to the degree and/or depth encouraged by the mentioned factors in our case studies (e.g., deeper social learning, additional knowledge, more and stronger networks, and more intense exchange of opinions).

Another question emerging from our case-study comparison is related to the continuity of the processes, which is a key component in maintaining adaptive capacity and encouraging social learning. In both cases there has been minimal effort to continue engagement efforts after the participation processes presented in our study and participation mechanisms are currently relegated to formal consultations through written comments and institutional participation. While in the case of the Tucson Basin this is certainly driven in part by the funding cuts made to ADWR in the late 2000s, in both basins the economic crisis is not the sole determining factor.

The lack of continuity has been justified in the Ebro case because of the high human and financial expense and the significant time needed for organization of the participation process. New social innovation mechanisms and resources, such as online applications and networks that permit remote participation by stakeholders, can be implemented to reduce these costs and facilitate participation [83,84]. These methods should not be seen as a substitute for in-person consultation and deliberation, however, a combination of approaches is increasingly necessary to ensure a more effective and wider participation. While this is true for large basins like Ebro where the costs of travel and organizing events across the geography is resource intensive, it also applies to smaller basins like Tucson where there are no longer regional offices and the conveners, in this case ADWR, must travel to hold in-person meetings.

5. Conclusions

Stakeholder engagement is essential for effective and adaptive water management; however, the determinants of effective public participation are not well understood. In the case studies outlined here, we found that the influence of a participatory process on public water policies is mainly determined by the context of the process, i.e., legal framework, political leadership, and social awareness, whereas ability of a process to build adaptive capacity depends upon how the engagement process is implemented, in particular the existence of active on-site consultation and deliberation.

Acknowledgments: Funding for this research was provided by the University of Seville in the context of SWAN—Sustainable Water Action Project (FP 7 Grant Agreement: INCO-20011-7.6) (Alba Ballester) and the Arizona Technology and Research Initiative Fund (Kelly Mott Lacroix).

Author Contributions: Alba Ballester and Kelly Mott Lacroix conceived and designed the approach to this article, analysis of documents and interviews from their respective case studies, and co-wrote the paper.

Conflicts of Interest: The authors declare no conflict of interest.

References

1. The Organisation for Economic Co-operation and Development (OECD). *Stakeholder Engagement for Inclusive Water Governance, OECD Studies on Water*; OECD Publishing: Paris, France, 2015.
2. Lennox, J.; Proctor, W.; Russell, S. Structuring stakeholder participation in New Zealand's water resource governance. *Ecol. Econ.* **2011**, *70*, 1381–1394. [CrossRef]
3. Antunes, P.; Kallis, G.; Videira, N.; Santos, R. Participation and evaluation for sustainable river basin governance. *Ecol. Econ.* **2009**, *68*, 931–939. [CrossRef]
4. Río Declaration. *Río Declaration on Environment and Development*; ONU Doc. A/CONF. 151/5/Rev.1; ONU: Río de Janeiro, Brasil, 1992.
5. United Nations Economic Commission for Europe (UNECE). *Aarhus Convention on Access to Information, Public Participation in Decision-making and Access to Justice in Environmental Matters*; UNECE: Aarhus, Denmark, 1998.
6. Directive 2000/60/EC of the European Parliament and of the Council of 23 October 2000 establishing a framework for Community action in the field of water policy.
7. Directive 2001/42/EC of the European Parliament and of the Council of 27 June 2001 on the assessment of the effects of certain plans and programmes on the environment.

8. Ballester, A.; Subirats, J. Nuevos enfoques institucionales en la gestión del agua. Directiva Marco del Agua: Participación pública y gestión de conflictos. In *El Agua: Perspectiva Ecosistémica y Gestión Integrada de Aguas;* Fundación Nueva Cultura del Agua: Zaragoza, Spain, 2015; Available online: http://fnca.eu/biblioteca-del-agua/directorio/file/2767?search=1 (accessed on 21 June 2016).
9. Arnstein, S.R. A Ladder of Citizen Participation. *J. Am. Inst. Plan.* **1969**, *35*, 216–224. [CrossRef]
10. International Association for Public Participation. Spectrum of Public Participation. 2007. Available online: http://c.ymcdn.com/sites/www.iap2.org/resource/resmgr/imported/IAP2%20Spectrum_vertical.pdf (accessed on 19 February 2016).
11. Putnam, R.D. *Bowling Alone: Americas Declining Social Capital;* Aula: Budapest, Hungary, 2006; pp. 207–219.
12. Reed, M.S. Stakeholder participation for environmental management: A literature review. *Biol. Conserv.* **2008**, *141*, 2417–2431. [CrossRef]
13. Bal, M.; Bryde, D.; Fearon, D.; Ochieng, E. Stakeholder Engagement: Achieving Sustainability in the Construction Sector. *Sustainability* **2013**, *5*, 695–710. [CrossRef]
14. Garin, P.; Rinaudo, J.D.; Ruhlmann, J. Linking expert evaluations with public consultation to design water policy at the watershed level. *Water Sci. Technol.* **2002**, *46*, 263–271. [PubMed]
15. Barraqué, B.; Le Bourthis, J.P.; Maurel, P.; Raymond, R. *Public Participation in the Dordogne River Basin;* Case study report produced under work package 5 of the HarmoniCOP project. LATTS-CNRS; Sorbonne University & CEMAGREF: Paris, France, 2004.
16. Christenson, J.A.; Robinson, J.W., Jr. In search of community development. In *Community Development in America;* Iowa State University Press: Ames, IA, USA, 1980; pp. 3–17.
17. Chase, L.C.; Decker, D.J.; Lauber, T.B. Public participation in wildlife management: What do stakeholders want? *Soc. Nat. Res.* **2004**, *17*, 629–639. [CrossRef]
18. Ballester, A.; Pares, M. Democracia deliberativa y política de agua. In *Experiencias de Participación en el Contexto de la Directiva Marco del Agua en España;* Congreso de Gestión y Planificación de Agua: Lisboa, Portugal, 2013.
19. Garmendia, E.; Stagl, S. Public participation for sustainability and social learning: Concepts and lessons from three case studies in Europe. *Ecol. Econ.* **2010**, *69*, 1712–1722. [CrossRef]
20. Tippett, J.; Searle, B.; Pahl-Wostl, C.; Rees, Y. Social learning in public participation in river basin management-early findings from HarmoniCOP European case studies. *Environ. Sci. Policy* **2005**, *8*, 287–299. [CrossRef]
21. Mosert, E.; Craps, M.; Pahl-Wostl, C. Social learning: The key to integrated water resources management? *Water Int.* **2008**, *33*, 293–304. [CrossRef]
22. Monroe, M.; Plate, R.; Oxarart, A. Intermediate collaborative adaptive management strategies build stakeholder capacity. *Ecol. Soc.* **2013**, *18*, 24. [CrossRef]
23. Jolibert, C.; Wesselink, A. Research impacts and impact on research in biodiversity conservation: The influence of stakeholder engagement. *Environ. Sci. Policy* **2012**, *22*, 100–111. [CrossRef]
24. Innes, J.E.; Booher, D.E. Reframing public participation: Strategies for the 21st century. *Plan. Theory Pract.* **2004**, *5*, 419–436. [CrossRef]
25. Muro, M.; Jeffrey, P. Time to Talk? How the Structure of Dialog Processes Shapes Stakeholder Learning in Participatory Water Resources Management. *Ecol. Soc.* **2012**, *17*. [CrossRef]
26. Pelling, M. Learning from others: Scope and challenges for participatory disaster risk assessment. *Disasters* **2007**, *31*, 373–385. [CrossRef] [PubMed]
27. Bankoff, G.; Frerks, G.; Hilhorst, D. *Mapping Vulnerability: Disasters, Development and People;* Earthscan: London, UK, 2004.
28. Walker, G.B.; Senecah, S.L.; Daniels, S.E. From the forest to the river: Citizens' views of stakeholder engagement. *Hum. Ecol. Rev.* **2006**, *13*, 193.
29. Burton, P. Conceptual, Theoretical and Practical Issues in Measuring the Benefits of Public Participation. *Evaluation* **2009**, *15*, 263–284. [CrossRef]
30. Newig, J.; Günther, D.; Pahl-Wostl, C. Synapses in the network: Learning in governance networks in the context of environmental management. *Ecol. Soc.* **2010**, *15*, 24.
31. Mott Lacroix, K.E.; Xiu, B.C.; Megdal, S.B. Building Common Ground for Environmental Flows using Traditional Techniques and Novel Engagement Approaches. *Environ. Manag.* **2016**, *57*, 912–928. [CrossRef] [PubMed]

32. Hermans, F.L.P.; Haarmann, W.M.F.; Dagevos, J.F.L.M.M. Evaluation of stakeholder participation in monitoring regional sustainable development. *Reg. Environ. Chang.* **2011**, *11*, 805–815. [CrossRef]
33. Booth, A.; Halseth, G. Why the public thinks natural resources public participation processes fail: A case study of British Columbia communities. *Land Use Policy* **2011**, *28*, 898–906. [CrossRef]
34. Blanco, I.; Ballester, M. Participar para transformar? La experiencia de los Presupuestos Participativos en la provincia de Barcelona. *Gest. Anál. Políticas Públicas* **2011**, *5*, 117–144.
35. Haus, M.; Heinel, H.Y.; Stewart, M. Introduction. In *Urban Gover Nance and Democracy: Leadership and Community Involvement*; Londres: Routledge, UK, 2005.
36. Wagenaar, H. Governance, complexity and democratic participation: How citizens and public officials harness the complexities of neighbourhood decline. *Am. Rev. Public Adm.* **2007**, *37*, 17–50. [CrossRef]
37. Font, J.; Blanco, I. *Polis, la Ciudad Participativa*; Diputació de Barcelona: Barcelona, Spain, 2006.
38. Fung, A. Varieties of Participation in Complex Governance. *Public Adm. Rev.* **2006**, *66*, 66–75. [CrossRef]
39. Brugué, Q. Modernizar la administración desde la izquierda: Burocracia, nueva gestión pública y administración deliberativa. *Rev. CLAD Reforma Democr.* **2004**, *29*, 1–16.
40. Subirats, J. *Elementos de Nueva Política*; Centro de Cultura Contemporánea de Barcelona: Barcelona, Spain, 2003.
41. Brooks, N.; Adger, W.N.; Kelly, P.M. The determinants of vulnerability and adaptive capacity at the national level and the implications for adaptation. *Glob. Environ. Chang.* **2005**, *15*, 151–163. [CrossRef]
42. Jones, L.; Ludi, E.; Levine, S. *Towards a Characterization of Adaptive Capacity: A Framework for Analysing Adaptive Capacity at Local Level*. Overseas Development Institute. 2010. Available online: https://goo.gl/CRvNsd (accessed on 21 June 2016).
43. Eakin, H.; Lemos, M.C. Adaptation and the state: Latin America and the challenge of capacity-building under globalization. *Glob. Environ. Chang.* **2006**, *16*, 7–18. [CrossRef]
44. Hobson, K.; Niemeyer, S. Public responses to climate change: The role of deliberation in building capacity for adaptive action. *Glob. Environ. Chang.* **2011**, *21*, 957–971. [CrossRef]
45. Lockwood, M.; Raymond, C.; Oczkowski, E.; Morrison, M. Measuring the dimensions of adaptive capacity: A psychometric approach. *Ecol. Soc.* **2015**, *20*, 37. [CrossRef]
46. Robinson, L.; Berkes, F. Multi-level participation for building adaptive capacity: Formal agency-community interactions in northern Kenya. *Glob. Environ. Chang.* **2011**, *21*, 1185–1194. [CrossRef]
47. Emerson, K.; Gerlak, A. Adaptation in collaborative governance regimes. *Environ. Manag.* **2014**, *54*, 768–781. [CrossRef] [PubMed]
48. Pahl-Wostl, C.; Craps, M.; Dewulf, A.; Mostert, E.; Tabara, D.; Taillieu, T. Social learning and water resources management. *Ecol. Soc.* **2007**, *12*, 5.
49. Olsson, P.; Folke, C.; Berkes, F. Adaptive co-management for building resilience in social-ecological systems. *Environ. Manag.* **2004**, *34*, 75–90. [CrossRef] [PubMed]
50. Hagmann, J.; Chuma, E. Enhancing the Adaptive Capacity of the Resource Users in Natural Resource Management. *Agric. Syst.* **2002**, *73*, 23–39. [CrossRef]
51. Adger, W.N. Social Aspects of Adaptive Capacity. In *Climate Change, Adaptive Capacity and Development*; Smith, J.B., Klein, R.J., Eds.; Imperial College Press: London, UK, 2003; pp. 29–50.
52. Gupta, J.; Termeer, C.; Klostermann, J.; Meijerink, S.; Van den Brink, M.; Jong, P.; Nooteboom, S.; Bergsma, E. The adaptive capacity wheel: A method to assess the inherent characteristics of institutions to enable the adaptive capacity of society. *Environ. Sci. Policy* **2010**, *13*, 459–471. [CrossRef]
53. Pahl-Wostl, C.; Newig, J.; Ridder, D. Linking public participation to adaptive management. In *Groundwater Science & Policy: An International Overview*; Quevauviller, P., Ed.; Springer: Berlin, Germany, 2008; pp. 150–173.
54. Tompkins, E.L. Successful adaptation to climate change across scales. *Glob. Environ. Chang.* **2005**, *15*, 77–86.
55. Engle, N.; Lemos, M.C. Unpacking governance: Building adaptive capacity to climate change of river basins in Brazil. *Glob. Environ. Chang.* **2010**, *20*, 4–13. [CrossRef]
56. Confederación Hidrográfica del Ebro. *Plan Hidrológico de la Cuenca del Ebro*; Ministerio de Medio Ambiente: Madrid, Spain, 2014.
57. Webb, R.H.; Betancourt, J.L.; Johnson, R.R.; Turner, R.M. *Requiem for the Santa Cruz: An Environmental History of an Arizona River*; University of Arizona Press: Tucson, AZ, USA, 2014.
58. Overpeck, J.; Udall, B. Dry Times Ahead. *Sci. Signal.* **2010**, *328*, 1641. [CrossRef] [PubMed]

59. Marshall, R.M.; Robles, M.D.; Majka, D.R.; Haney, J.A. Sustainable Water Management in the Southwestern United States: Reality or Rhetoric? *PLoS ONE* **2010**, *5*. [CrossRef] [PubMed]
60. Arizona Department of Water Resources. *Personal Communication with Michelle Moreno*; Public Information Officer: Phoenix, AZ, USA, 2015.
61. Ballester, A.; La Calle, A.; Subirats, J.; Espluga, J. Participación pública en la demarcación hidrográfica del Ebro. In Proceedings of the VI Congreso Ibérico de Gestión y Planificación del Agua, Vitoria, Spain, 4–7 December 2008.
62. Megdal, S.; Zachary Smith, A.; Lien, A. *Evolution and Evaluation of the Active Management Area Management Plans*; University of Arizona Water Resources Research Center: Tucson, AZ, USA, 2008.
63. Rabelo, D.; Espluga, J.; Edmilson, C.; Brugué, Q. Citizenship participation in water management plans in the Doce River Basin, Brazil and Catalonia, Spain. *Water Policy* **2014**, *16*, 205–221. [CrossRef]
64. European Commission. *Guidance on Public Participation in Relation to the Water Framework Directive*; Office for Official Publications of the European Communities: Luxembourg, Luxembourg, 2003.
65. Arizona Department of Water Resources (ADWR). *Tucson Active Management Area—Third Management Plan*; ADWR: Phoenix, AZ, USA, 1999.
66. Jacobs, K.L.; James, M.H. Managing for Sustainability in an Arid Climate: Lessons Learned from 20 Years of Groundwater Management in Arizona, USA. *Hydrogeol. J.* **2004**, *12*, 52–65. [CrossRef]
67. Ballester, A. Análisis de Las Entrevistas Realizadas en Relación Con el Proceso de Participación Asociado a la Elaboración del Plan Hidrológico de la Demarcación Hidrográfica del Ebro. PART-DMA Project Unpublished Report. 2012. Available online: http://blogs.uab.cat/partdma/documentacion/ (accessed on 20 February 2016).
68. Ballester, A.; Ferrer, G. Conflictividad en la Gestión Hídrica. Guía Nueva Cultura del Agua: Una Guía de Referencia y Consulta Sobre la Gestión y Conservación de Los Recursos Hídricos en España, Fundación Nueva Cultura del Agua. 2012. Available online: http://www.fnca.eu/guia-nueva-cultura-del-agua/ (accessed on 20 February 2016).
69. Parés, M. River basin management planning with participation in Europe: From contested hydro-politics to governance-beyond-the-state. *Eur. Plan. Stud.* **2011**, *19*, 457–478. [CrossRef]
70. Arrojo, P.; Casajús, L.; Copitzy, A. *La Rebelión de la Montaña: Los Conflictos del Agua en Aragón*; Bakeaz Centro Documentación Estudios Para La Paz: Bilbao, Spain, 2010.
71. Casademunt, A.; Alfama, E.; Coll, G.; Cruz, H.; Martí, M. *Per Una Nova Cultura del Territorio*; Icaria, Antrazyt: Barcelona, Spain, 2007.
72. Transparency International España. Índice de Transparencia en la Gestión del Agua (INTRAG). 2015. Available online: http://transparencia.org.es/intrag-2015/ (accessed on 19 February 2016).
73. Logan, M.F. *The Lessening Stream: An Environmental History of the Santa Cruz River*; University of Arizona Press: Tucson, AZ, USA, 2002.
74. Tarlock, A.D.; Van de Wetering, S.B. Water and western growth. *Am. Plan. Assoc. Plan. Environ. Law* **2007**, *59*, 3–13. [CrossRef]
75. Megdal, S.B.; Forrest, A. How a Drought-Resilient Water Delivery System Rose Out of the Desert: The Case of Tucson Water. *J. Am. Water Works Assoc.* **2015**, *107*, 46–52. [CrossRef]
76. Akhter, M.; Ormerod, K.J.; Scott, C.A. Lost in translation: Resilience, social agency, and water planning in Tucson, Arizona. *Crit. Plan.* **2010**, *17*, 47–65.
77. Arizona Department of Water Resources. *State of the AMA: Tucson Active Management Area*; Arizona Department of Water Resources: Phoenix, AZ, USA, 1996.
78. Gutmann, A.Y.; Thompson, D. *Why Deliberative Democracy*; Princeton University Press: Princeton, NY, USA, 2004.
79. De Stefano, L. Facing the Water Framework Directive challenges: A baseline of stakeholder participation in the European Union. *J. Environ. Manag.* **2010**, *91*, 1332–1340. [CrossRef] [PubMed]
80. Ridder, D.; Mostert, E.; Cernesson, F.; HarmonyCop Team. *Learning Together to Manage Together: Improving Participation in Water Management*; University of Osnabrück: Osnabrück, Germany, 2005.
81. Mott Lacroix, K.E.; Megdal, S.B. Explore, Synthesize, and Repeat: Unravelling Complex Water Management Issues through the Stakeholder Engagement Wheel. *Water* **2016**, *8*. [CrossRef]
82. Jeffery, N. *Stakeholder Engagement: A Road Map to Meaningful Engagement*; Doughty Centre, Cranfield University School of Management: Cranfield, UK, 2009.

83. Sarzynski, A. Public participation, civic capacity, and climate change adaptation in cities. *Urban Clim.* **2015**, *14*, 52–67. [CrossRef]
84. Hernandez-Mora, N.; Cabello, V.; De Stefano, L.; Del Moral, L. Networked water citizen organizations in Spain: Potential for transformation of existing power structures in water management. *Water Altern.* **2015**, *8*, 99–124.

water

MDPI

Article

Staying the Course: Collaborative Modeling to Support Adaptive and Resilient Water Resource Governance in the Inland Northwest

Allyson Beall King * and Melanie Thornton

School of the Environment, Washington State University, Pullman, WA 99164-2812, USA; melanie.thornton@wsu.edu

* Correspondence: abeall@wsu.edu; Tel.: +1-509-335-4037

Academic Editors: Sharon B. Megdal, Susanna Eden and Eylon Shamir

Received: 22 February 2016; Accepted: 18 May 2016; Published: 31 May 2016

Abstract: Water resource governance, much like the systems it endeavors to manage, must be resilient and adaptive. Effective, resilient and adaptive water resource governance requires continuing stakeholder engagement to address the complex nature of human and natural systems. Engagement is an adaptive and iterative process of education and empowerment, building relationships and trust, and facilitating collaboration. Collaborative modeling is a methodology that integrates diverse stakeholder perspectives, fosters discussions, and creates space for problem identification and consensus-based strategies and solutions to current water resource challenges. We define collaborative modeling broadly, such that it includes a wide range of systems thinking exercises, as well as dynamic models. By focusing on the relationships and interconnections in the system, collaborative modeling facilitates clarification of mental models and the communication of science. We will describe our work in two interstate basins and how it has evolved over time as these basins strive to develop collaborative governance, and find solutions for their water resource challenges.

Keywords: collaborative modeling; integrated water resource management; systems thinking; stakeholder engagement; public participation

1. Introduction

Water resource governance, much like the systems it endeavors to manage must be adaptive and resilient. *Resilient* refers to the capacity of the socio-ecological system to absorb disturbance and reorganize while undergoing change so as to still retain essentially the same function, structure, identity, and feedbacks [1]. We use the term *adaptive* to describe an approach that adjusts in response to emerging knowledge and has the capacity to support system resilience [1,2]. The foundations for adaptive and resilient *management* are rooted in the social, legal and scientific recognition that it is crucial to integrate the values and worldviews of diverse stakeholders into decision processes for resource management that utilizes best available science [3–5]. These decision processes must be iterative, ongoing, and capable of dealing with scientific uncertainty within a constantly changing social and biophysical system.

We are in the Anthropocene [6]; human impacts are now the dominant driver of global change creating ever more complex environmental conditions that impact both ecosystem and human health. In the 1960s and 1970s, the United States developed policies and laws, such as the National Environmental Policy Act (NEPA), the Clean Air Act (CAA), the Clean Water Act (CWA) and the Endangered Species Act (ESA), which endeavor to assess risk and protect the environment while, at the same time, consider the economic and social needs of the nation's human population [7–10]. Indeed, the NEPA purpose statement declares that it is "a national policy which will encourage productive and enjoyable harmony between man and his environment" [7]. This foundation of policy and law has helped mitigate problems, such as point source pollution; however, managing complex and cross jurisdiction environmental challenges related to climate change impacts, non-point source pollution and other ecosystem service tragedies of the commons [11] are still, if not more problematic today.

Policies, such as NEPA, and laws, such as the CWA, CAA, and ESA, were passed in response to anthropogenic environmental degradation and in recognition of "the critical importance of restoring and maintaining environmental quality" [7]. Years of case law have defined the roles and responsibilities of national and state agencies and the manner in which these agencies integrate science into management. The scientific approaches that have been adopted require social involvement in the problem-identification and solution-producing process such that decision development is transparent and respectful of social needs and diverse world-views [12–15]. The society-science nexus is so completely intertwined that "neither science nor democracy can enhance human welfare in the absence of the other" [16] (p. xiii). With respect to science, the field of ecosystem ecology contributes to an understanding that humans are integral parts of all ecosystems, and that ecosystems and our knowledge of them change over time. Natural resource management has adopted adaptive management as a theoretical framework to manage systems for which scientific knowledge is provisional. It tells us that we should learn by monitoring the results of scientifically designed management plans and then adjust accordingly [17]. It builds on key premises which include: (1) significant connections need to be determined; (2) structural features are more important to measure than numbers; (3) changes in one variable can have unexpected impacts; (4) monitoring of one variable can seem to indicate no change when drastic change is imminent [2,12,18]. To adaptively manage a system, we must acknowledge uncertainty, surprise, and the dynamic variability and intricate feedbacks that are inherent to both the system and to its dynamics [19]. Integrating the concept of resilience into this framework encourages us to manage for function and structure such that the ecosystem services upon which we are reliant are available for both current and future generations [20]. Finally, adaptive and resilient management of bio-physical systems occurs within dynamic social systems that are governed by historical context, diverse values and worldviews.

Water resource management in the United States has evolved from what was initially concerned only with the delivery of sufficient quantity and quality. This traditional approach has been largely supplanted by Integrated Water Resource Management (IWRM), that strives to integrate quantity, quality, surface and groundwater dynamics, as well as ecosystem services into an adaptive, resilient governance framework that is capable of spanning jurisdictional boundaries [21]. Hydrologic systems have multiple actors and entities involved in management decisions and legacy issues such as appropriated water rights and instream infrastructures. Making decisions within the complexity of IWRM requires that people involved in water resource governance span boundaries that may be institutional, cultural, spatial and or temporal, and it requires viewing the social-ecological-hydrological system as dynamic rather than linear [22]. Furthermore, IWRM necessitates bridging the gaps between water resource science and multi-jurisdictional decision making [23,24].

Models and modeling processes have long been recognized as useful tools for integrating social, economic and environmental information to facilitate complex environmental management issues and decisions [12,25–31]. We describe "a model [as] a substitute for a real system" [32]. The model may be used for communication or for learning about how the system behaves over time. The model could be as simple as a picture, or a diagram. A model could be a physical representation of natural processes such as wind tunnel or stream table. A model could portray relationships through a causal loop diagram (CLD) that describes causation. A mathematical computer model can be used to simulate behavior of multiple parameters over time. These representations of reality may be used to communicate the perceptions of an individual or group of individuals, or they may be used to calculate physical processes, such as the flow of water, to help us understand how water moves across and through a landscape. The model building process begins with the creation of mental models. Mental models are representations of how an individual understands the surrounding world based on their own experiences [33–35]. These can be "deeply ingrained assumptions, generalizations, or even pictures or images [that] influence how we understand the world around us, and how we take action" [34]. We communicate mental models with language and images. Communicating and analyzing our understanding of system complexity may also require the use of systems thinking tools such as causal loop diagrams that describe relationships and causation [34,36]. While mental models may serve us well as we conduct our daily lives, and systems thinking may help us see the complex relationships around us, calculating the relationships between multiple parameters and the feedbacks between these parameters simultaneously and over time requires the use of computers [27,32,36,37]. Computer models that integrate social, economic and environmental information are effective for synthesizing scientific research and decision making [25,30,37–40]. Building models in collaborative social processes that include hydrologic experts and local stakeholders enables transparency and salience, and helps bridge the gap between science and water resource management [12,31,37,39,41–49]. Best practices for creating such models include involving stakeholders early and often in the model building process [37,41,43].

Collaborative modeling (CM) is a methodology that integrates diverse stakeholder perspectives, fosters discussions, and creates space for problem identification and the production of consensus-based, scientifically sound management strategies. Balancing the need for dynamic solutions while maintaining respect for values creates space for exploration and innovation that is served by the tension between *what was or is* and *what could be*. As stakeholder groups negotiate the 'problem to solution' space, CM facilitators seek to transition people away from linear thinking and conflicts that are often the result of entrenched value systems or worldviews [42–44,48]. While conflict can progress in an unproductive manner, and create unfavorable outcomes it is important to note that conflict, in of itself is not necessarily bad. Conflict has the potential to drive needed change and without the potential for conflict people may not be motivated to adapt their system in advance of crisis. This paper will describe the use of CM as a methodology that provides the opportunity for people to visualize possible futures and reflect on which possibilities are favorable and which are undesirable in *advance* of potential conflict. The gap between favorable and undesirable futures creates tension that provides impetus for change, innovation, and creativity. We will refer to this type of tension as creative tension [34]. It is through the process of engaging in this state of "creative tension" that resilient and adaptive solutions arise [34].

In this paper, we describe experiences from two collaborative modeling processes that integrate stakeholder engagement, systems thinking and system dynamics, the disciplines of water and natural resource modeling, and resource planning methodologies. For the purposes of this research, we define stakeholders broadly. Participants in our CM processes include water resource professionals involved in decision-making, local and state agency personnel, purveyors, consultants, NGOs, legal experts, elected officials, academic researchers and self-selected members of the general public who are interested in water resource management. These CM processes have occurred within the inland Northwest, in the Palouse basin and the Spokane basin. Both basins are bi-state and straddle the Idaho and Washington border. We will describe our work in these basins and how it has evolved over time as stakeholders strive to develop collaborative governance for adaptive and resilient water resource management.

2. Palouse and Spokane Basins: The People and the Basins

We are continuing to engage stakeholders in two interstate basins in the inland Northwest, the Palouse and Spokane basin. Both basins are designated as sole source aquifer systems [50–52]. The Palouse basin and Spokane River-Spokane Valley Rathdrum Prairie Aquifer basin have from many perspectives an abundance of water. Despite the apparent abundance of freshwater, system leaders in the region understand the importance of institutionalizing resilient and adaptive management in advance of a potential water management crisis. They seek to institutionalize collaboration, and to utilize the creative tension that the potential for conflict creates, as well as facilitate the communication of science and education and outreach through annual water summits or forums. In the Palouse and Spokane basins, system leaders invited collaborative modelers to work with diverse regional stakeholders to help people move out of entrenched mental models through systems thinking exercises, the development of computer models, and involvement in annual summits and forums. It is hoped that this early and iterative engagement with collaborative modeling will serve to prevent unproductive conflict over water resources in the future. System leaders sought the involvement of collaborative modelers at Washington State University to assist basin stakeholders as they plan for the unthinkable. It was in this context that a facilitated collaborative modeling process was implemented beginning in 2008 in the Palouse and in 2010 in the Spokane basin.

In the Palouse basin, there has been growing concern over water supply and declining well levels, which is compounded by the scientific uncertainty related to total supply and potential for recharge. In the late 1980s, stakeholders from the Palouse Basin Aquifer Committee, including major groundwater pumpers, developed a groundwater management plan, that when combined with engineering and technology has effectively kept demand steady at 1990s levels, in spite of an average population growth rate of one percent [53]. Rates of decline in wells have decreased since the late 1980s, however scientific explanation for this behavior is still primarily in the form of hypotheses [50,53,54]. Research to explain well behavior, aquifer storage, flow and potential for recharge is ongoing, yet it is limited by complex hydrogeology, the number of monitoring wells, and limited fire suppression storage that restricts the length of pump tests. Although there is significant concern about declining well levels at current rates of growth, the water supply will likely last decades. In the near term, conflict in this basin will likely not arise from a limited supply but rather from differing social values for the resource, development, and conservation. The potential for conflict as well as the need for sustainable long-term thinking led to our involvement [53]. We began our work in the Palouse in 2008 with a collaborative model building process that resulted in a web based simulation model for education and outreach [53]; it has progressed with the development of systems thinking exercises at the annual water summit. To contrast, the Palouse basin has a fossil water supply that will slowly be compromised by development while the Spokane basin has a highly dynamic resource in what has historically been a snow-dominated basin [55].

Our work began in the Spokane basin in 2010 with system thinking and mental model assessment exercises; we then added a simulation model-building component in 2013. The process was first initiated during our National Science Foundation Water, Sustainability and Climate planning grant in 2010, and then expanded as a part of Watershed Integrated System Dynamics Modeling (WISDM), a Water, Sustainability and Climate grant funded by National Institute for Food and Agriculture. The goal of our contribution to WISDM was to develop a collaborative modeling methodology to better understand how and when institutions change. In 2009, the Spokane basin finished a 10 year long, contentious total maximum daily load (TMDL) process for phosphorous and dissolved oxygen levels in the Spokane River. People were tired of hearing about phosphorous, tired of conflict and still feeling bruised from compromise and the process in general. Although there was a sense of accomplishment, relationships were fragile. We were invited to help facilitate continued collaboration that would help maintain and advance cooperation until the next big issue needed to be addressed [56]. In addition, hydrologic experts are concerned about the lay perception that the "basin has abundant water" and that "talk of water conservation is bad for business" [56,57], even though there is evidence that the basin is vulnerable to climate change [55,58].

2.1. The Palouse Basin: Issues and Hydrology

The Palouse basin aquifer system supports approximately 50,000 residents and includes the cities of Moscow and Pullman, and two research universities, University of Idaho (UI) and Washington State University (WSU). The two cities and two universities, collectively known as the four "big pumpers" provide the bulk of supply for the basin. The basin has a strong agricultural base, however it is primarily dry land farming with very limited irrigated agriculture. Baseline indoor use across the basin is about 90 gallons per person per day, while total use averaged over the year is about 120 gallons per person per day. Weather has a significant impact on summer pumping in the inland Northwest, where there is little summer rainfall. The aquifer system is comprised of two heterogeneous basalt aquifers named for their respective geologic formations. The upper aquifer called the Wanapum has limited recharge; the deeper, much larger Grande Ronde aquifer is confined, contains fossil water and has little evidence of recharge on time scales relevant to resource management [50,53,54]. The Wanapum, primarily used by the City of Moscow, exhibited sharply declining well levels through the 1960s at which time Moscow added additional wells into the Grande Ronde. Wanapum wells now sustainably support about 6000 people [53]. Concern over declining well levels also led Moscow to add a reuse plant for irrigating landscape at the UI in 1976. Basin wide water demand on the Grande Ronde has increased seven fold since the 1930s, and although decline in water levels do not follow demand in a linear manner, there is nevertheless concern over unsustainable use.

2.2. The Spokane Basin: Issues and Hydrology

To the north of the Palouse basin in a very different geologic system is the Spokane River basin and the Spokane Valley Rathdrum Prairie (SVRP) aquifer. The Spokane River Basin covers roughly 2400 square miles. While the headwaters lie in the Northern Rockies, the Spokane River begins at the outlet of Coeur d'Alene Lake in Idaho and flows westward, crossing the border into Washington 10 miles downstream then a flowing an additional 110 miles to its confluence with the Columbia River. The river is used for fishing, swimming, boating and hydropower. The river also provides aquatic, riparian and upland habitat for an assortment of species. The Spokane River interacts with the Spokane Valley Rathdrum Prairie (SVRP) aquifer, a sole source aquifer for over 600,000 people that is utilized for domestic, irrigation, and industrial water [52]. The SVRP aquifer is an unconfined aquifer composed of unconsolidated coarse-grained sand, gravel, cobbles and boulders deposited by the series of catastrophic glacial floods with hydraulic conductivity estimates ranging from 100 to 6200 ft./day, with most values larger than 1000 ft./day [59]. The SVRP and the Spokane River interact at rates upwards of millions of gallons per day, with losing and gaining reaches throughout the basin floor [52,60]. This river basin has experienced low flow conditions on the Washington side of the border that has resulted in a Low Flow Rule [61], phosphorous loading from sources in both Washington and Idaho resulting in a Total Maximum Daily Load (TMDL) rule for phosphorous and dissolved oxygen on the Spokane River [62], metals contamination from mining, polychlorobiphenyl (PCB) contamination, and multiple jurisdictions involved in management. While these issues may appear extensive, there has been a significant improvement in water quality since the 1950s, and local stakeholders describe the basin as providing them with a high quality of life. Aside from the arduous phosphorous-dissolved oxygen TMDL process that culminated in 2010, there is relatively low conflict over water due in part to the abundant supply.

After a description of collaborative modeling we will describe our collaborative modeling experience in each basin that includes exercises for systems thinking and mental model assessment, and the model building and simulation processes.

3. The Breadth and Depth of Collaborative Modeling

Collaborative modeling is a methodology that helps us embrace adaptive theory while striving to holistically address the complexity of details concerning system behavior including feedback and time delays [25,30,32,42,63]. Collaborative modeling has foundations in group model building [35,64,65] that utilizes systems thinking, model building and computer simulation. The process of building a collaborative model helps water resource stakeholders clarify their own mental models and gain a better understanding of the system as they are facilitated through the process of integrating technical, scientific information with local knowledge and expertise in a computer simulation model [37,39–49]. The CM process develops one or more types of systems models ranging from qualitative, descriptive visual maps or causal loop diagrams that seek to identify the archetype of behavior, to quantitative, spatial and or temporal modeling. CM processes can be used to mitigate conflict through gaming and roll playing, even having stakeholders represent the opposition in a simulation exercise [66]. Many practitioners have a preferred modeling platform for building models in methods for which they were trained or that work well with their own mental models of how the world works. The choice of models may be related to the problems people are trying to define and or solve, or by the needs and preferences of stakeholders. Costanza *et al.* describe the purpose of modeling building processes as ranging from scoping, to research, to management [40]. Specifically with respect to water resource management we see the purpose of this iterative process ranging from research and basic science, to education and outreach, and to planning and management of complex hydrologic systems. The development process is not linear but rather context and stakeholder specific.

We approach collaborative modeling from the perspective of the eight steps of model formation adapted from Beall and Ford, as depicted in Figure 1 [25]. Early stages of iterative research, problem development and or education and outreach activities may focus on problem familiarization using systems thinking exercises (Figure 1a). The development of simulation models focuses on extending the process to include policy and sensitivity analysis (Figure 1b). In both contexts iteration over time is necessary for the development of robust models that reflect system complexities. These exercises offer the opportunity for collaborators to explore the structure and behavior of a system and create space for reflection, exploration and discussion. Finally, the process of collaborative modeling creates opportunity to explore, challenge, iterate and advance mental models.

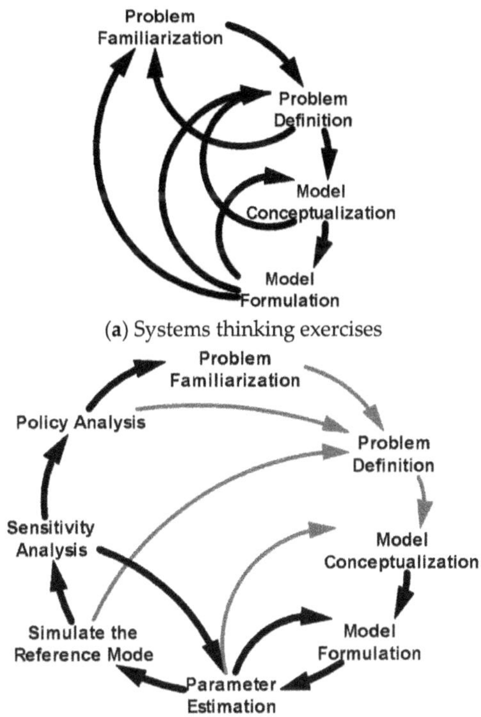

(**a**) Systems thinking exercises

(**b**) Systems models for policy analysis

Figure 1. The eight steps of model building: (**a**) Iterative system thinking exercises for identifying and clarifying problems; (**b**) building simulation models expands the iterative process to include sensitivity analysis and policy analysis. Source: Beall and Ford [25].

3.1. Iteration over Time and Collaboration

We argue, as does Senge and others, that advancing mental models require iterative *investment in systems thinking* [32,67,68]. Systems thinking enables visualization of complexities, structure and feedbacks that cause behavior, and in so doing creates a more complete understanding of a system. While this mode of thinking is intuitive for some, for other individuals system thinking may require intentional and thoughtful practice, and a personal decision to value and invest in a more complete understanding of the system. Investing in systems thinking supports system resilience and the ability for self-sustaining adaptation over time because it helps us organize and evaluate uncertainty, surprise, dynamic variability, feedbacks and time delays. With respect to water resource management and who would benefit from such an investment, we consider the question: who are the decision makers in water resource management? Water use decisions are made at multiple spatial scales, ranging from large-scale hydropower and agricultural management, to individual indoor and outdoor residential water use. Therefore, we argue that all water users should be considered decision makers. Creating a suite of models relevant to the scope and scale at which decision makers operate can help bridge the gap between policy makers, resource managers, and end users. For example, collaboratively built education and outreach tools that simulate decisions made at the individual or household level highlight the impacts of individual water use behaviors on the system at large. Hydrologic models that simulate large-scale decisions may be used to illustrate the potential impacts of policies on individual users at a variety of scales. Collaborative modeling integrates multiple scales of water users into the decision development space, which improves transparency and the likelihood of politically acceptable water management outcomes [25,37,39–48].

Perceived problems and acceptable outcomes change over time, based on environmental factors including weather and climate, and social factors including cultural values and economics. Because concerns and management objectives change over time, long-term *investment in systems thinking* is an especially critical component of CM. Revisiting places, people and projects over time creates the opportunity for continuing discussion, education and outreach. A long-term process also facilitates model revision and iteration as new data and new knowledge develop. The act of revisiting may involve moving from systems thinking exercises for the stage of problem familiarization or definition (Figure 1a) to the stage of developing dynamic quantitative models (Figure 1b). Quantitative models are then iteratively expanded and improved. In some cases, the revisiting process helps participants iterate through a new problem familiarization loop. Invitations to return often come from *system leaders* in the communities who seek to create adaptive and resilient communities. These system leaders may work for industry, NGOs, city, county or state agencies. These individuals are invested in managing their systems over the long term, which often necessitates navigating a political environment characterized by regularly changing elected officials who may not have a sophisticated understanding of what 'long-term' water resource planning means [57]. As collaborative modelers, we emphasize the importance of sharing this long-term perspective, and share in the time commitment it takes for institutional, organizational and ecological change. Such change may be gradual or it may happen rapidly when the system reaches a tipping point or when a collective moment of realization and insight delivers inspiration.

The collaborative modeling process in the Palouse and Spokane basins includes an ongoing, iterative and flexible approach to stakeholder engagement and model building [25,69]. Engagement activities include systems thinking exercises, and model development that ranges from problem scoping, basic research, education and outreach, to planning and management. It is important to note that this was not a linear process, and our iterative approach varied in each basin, based on relevant stakeholder issues and needs. Table 1 lists our engagement activities and the years in which activities were performed in each basin.

Table 1. Engagement activities over time in the Palouse and Spokane basins (2008 to 2016).

Palouse Basin	08	09	10	11	12	13	14	15	16
Systems Thinking									
Meetings									
CLD									
Summit/clickers									
Modeling and Simulation									
Conceptualization									
Formulation									
Parameter Estimation									
Model Simulation									
Scenario Analysis									
Spokane River Basin	**08**	**09**	**10**	**11**	**12**	**13**	**14**	**15**	**16**
Systems Thinking									
Meetings									
CLD									
Forum/clickers									
Modeling and Simulation									
Conceptualization									
Formulation									
Parameter Estimation									
Model Simulation									
Scenario Analysis									

3.2. Systems Thinking: Using Causal Loop Diagrams for Problem Identification

Early stages of the CM process include systems thinking exercises to begin to clarify issues and to define the problem (Figure 1a). In the Palouse basin, the first several group meetings were dedicated to causal loop diagraming (CLD), while in the Spokane we utilized a group breakout session at the regional water conference to begin CLD. Participants were asked to work together to draw causal loop diagrams to conceptualize the subsystems of their regional human-natural system. They were asked to identify stocks, such as water in the aquifers or population living in the basin, and then they were asked to draw the stocks and describe connections between the stocks. We utilize CLD for multiple purposes for our systems thinking sessions. First, we use them to gain an understanding of the problems and issues as viewed by the participants. This is important so that we better understand the basin and can be equipped to build model simulations of important and relevant issues. CLD sessions are beneficial to participants because they help people view the basin broadly as an integrated socio-bio-physical system. Finally, the process facilitates discussion about mental models and promotes reflection about personal socio-scientific knowledge. In addition to CLD, participants were also asked to draw a timeline of the basin to facilitate discussion of how the basin had changed over time. In short, systems thinking exercises aim to help broaden participant's perspectives.

3.3. Systems Thinking: Clickers and Mental Models

We also utilize systems thinking exercises with broader audiences in each basin during annual regional water conferences that bring together local, state and federal agencies, elected officials and the public to share relevant water resource science and to discuss pertinent water resource challenges. These conferences are hosted by the Palouse Basin Water Summit and the Spokane River Forum, non-profit organizations founded at the request of basin stakeholders for the purpose of education and outreach, dissemination of science, and for talking about "all things related to water" [70,71]. Before each conference, we work with system leaders to develop multiple-choice questions about science, values and preferences, and management challenges. We are careful to use the iterative question development process as a transparent, trust building exercise. Questions are elicited from individuals and shared with a large group in advance of the conferences without identifying who posed the question. The group scrutinizes questions for clarity and to reduce the likelihood of controversial wording. During the conference, attendees are supplied with individual audience response technology, that we generically call clickers, to anonymously answer the questions [72]. The multiple-choice questions are imbedded in a power point presentation on a computer that receives information from each clicker. When a question is opened in the response system, participants are free to discuss the questions with one another before responding with their clickers. After the responses have been received, the cumulative responses are revealed instantly to the group. Depending on the type of question and level of discussion among participants, the session facilitator may leave the question open for any length of time. Questions related to science help participants learn about the resource when the correct answer is revealed. The number of correct answers helps facilitators, as well as participants, assesses the level of general knowledge. Questions related to values help the groups understand potentially diverse attitudes and preferences. Using clickers with multiple types of questions provides the opportunity to learn about individual and group mental models. The response system provides the opportunity to track how an individual operating the clicker that day answered all of the questions. Anonymity is preserved by random distribution of clickers as well as the size of the groups with whom we work. Over time and multiple conferences, we revisit specific questions of interest, revise questions to improve the educational experience, and design new questions as the basins adapt and or are faced with new challenges. These sessions provide the opportunity for people to learn about the resource, the values of others and they create space for reflection and discussion.

In the Palouse basin, system leaders hold an annual Water Summit for the local community as an opportunity to "continue the conversation about all things related to water in the basin" [70]. The gathering of approximately 200 people includes water resource managers from local and state agencies, elected officials and the public. Each year since 2013, we have been invited to facilitate clicker sessions. The invitation came after stakeholders in Spokane basin relayed their optimism about their own 2011 clicker experience to system leaders in the Palouse. We use clickers as an educational tool that assesses the level of knowledge about the aquifer system and well behavior. In addition, we use this opportunity to help this group of people better understand one another's mental models and preferences for resource management. These sessions are preceded by a state of the aquifer talk that includes well levels and rates of decline, changes in demand for the previous year, and weather conditions that impact demand. Clicker questions designed by stakeholders and answered anonymously include education, perceptions, values and policy preferences (Table 2).

In the Spokane River Basin, system leaders convene the Spokane River Forum every 12–18 months for water resource professionals to share information about their regional water resources. In fall 2010, we were invited to devise an interactive session at the May 2011 Spokane River Forum on the potential development of a PCB TMDL for the Spokane River. We spent several months in individual meetings with key stakeholders exploring perceptions, and formulating and screening a set of questions that we would present in May. We began by asking stakeholders what they would like to ask each other. Clicker questions designed by stakeholders and answered anonymously included preferences on a PCB TMDL. At subsequent Forums in 2012 and 2014, we have used this process to gauge the level of interest for collaborative bi-state water resource management. Similar to the session in the Palouse, at the 2012 Forum we also use this method to assess the level of the participants' hydrologic knowledge. At the conclusion of the 2012 Forum, we included questions that evaluated how well participants understood river-aquifer dynamics. Organizers were concerned that the audience did not perform well on the river-aquifer basics questions and perhaps left the conference with misinformation [56]. In response to this concern, at the 2014 event organizers began the Forum with "Hydrology 101", a plenary session on hydrology and hydrogeology basics presented with a panel of local hydrologic experts [56]. After a short explanation of correct answers attendees were guided to breakout sessions that would help explain complexities of basin hydrology. Table 3 provides a partial list of questions that were asked at the Forums between 2011 and 2014.

Table 2. Partial list of clicker questions presented to participants at the Palouse Basin Water Summit between 2013 and 2015 (for additional questions please contact authors). Multiple-choice answer alternatives to six of these questions and the participant's responses are listed in Table 4.

Clicker Question Category	Palouse Basin—Clicker Questions
Education	• Pullman, WSU and UI utilize the Grande Ronde. Moscow uses both the Grande Ronde and the Wanapum. How much of Moscow's water supply comes from the Wanapum? • In the Palouse Basin what is the average per capita use per day averaged over the year?
Perceptions	• What do you think is the primary reason that pumping rates have not followed population growth? • Based on our current rates of use, how long do you think we can continue to use the Grande Ronde? • Rates of decline changing from 1.5 > 0.9 > 0.6 ft./year ... What does this mean to you? • If the population growth rate increases to 2% across the basin will we be able to maintain current pumping rates?
Values	• At what point would you consider water resources in the basin to be in a critical situation? • Is it important to take a regional, collaborative approach among governments, agencies and stakeholders to address water resource issues? • As an individual water consumer on the Palouse what are you going to do in the coming year?
Preferences and Priorities	• Where do you see water resource management in the Palouse Basin in 20 years? • Is the development of a surface water reservoir a priority? • Is the development of a reclaimed water plant in Pullman a priority? • Would you prefer to have your water bill state volume of use based on gallons per month or in cubic feet?

Table 3. Partial list of clicker questions presented to participants at the Spokane River Forum between 2011 and 2014 (for additional questions please contact authors). Multiple-choice answer alternatives to six of these questions and the participant's responses are listed in Table 5.

Clicker Question Category	Spokane Basin—Clicker Questions
Education	• If flows at Post Falls Dam were increased during summer low-flow periods, what would the result be? • The aquifer interacts with the Spokane River. In some places the river loses water to the aquifer, in other places the river gains water from the aquifer. How many gaining and losing reaches of the river are there?
Perceptions	• If WA Dept. of Ecology started developing a PCB TMDL today, how long do you think it would take to complete? • How much do you think a PCB TMDL would cost? • Considering an estimated population growth rate of Washington (~3.88%) and Idaho (~2.3%), do purveyors currently own enough water rights to meet user needs for the next thirty years? • What do you think is the primary purpose of an instream flow rule?

Table 3. *Cont.*

Clicker Question Category	Spokane Basin—Clicker Questions
Values	• When do you predict there will be a surface water availability crisis in the Spokane River Basin? • When do you predict there will be a ground water availability crisis in the Spokane River Basin? • Are you interested in learning more about bi-state water management agreements used in the US? • What is your comfort level with a regional board/task force deciding where PCB removal and reduction funding should be targeted?
Preferences and Priorities	• Where do you see management of water resources in the Spokane River Basin in 20 years? • Should establishing a plan for basin-wide water resources management be a priority? • Do you consider PCBs unimportant, a WA issue, a ID issue or a bi-state issue? • If a water supply crisis is imminent, which of these (1) infrastructure measures and (2) behavior measures should we take to address it?

Table 4. Partial list of responses to clicker questions presented to participants at the Palouse Basin Water Summit (for additional questions please contact authors).

What do you think is the primary reason that pumping rates have not followed population growth?	**2013** $n = 100$	**2014** $n = 100$	**2015** $n = 100$
Changes in human behavior	5%	15%	3%
Changes in engineering and infrastructure	20%	20%	9%
A mix of the above answers	75%	65%	86%
I don't know	0%	0%	3%
Where do you see water resource management in the Palouse Basin in 20 years?	**2013** $n = 100$	**2014** $n = 100$	**2015** $n = 100$
Increased "collaboration and cooperation"	42%	52%	56%
Formal bi-state agreement"	24%	30%	23%
"Business as usual"	18%	9%	4%
"Increased conflict"	15%	10%	16%
At what point would you consider water resources in the basin to be in a critical situation?	**2013** $n = 100$	**2014** $n = 100$	**2015** $n = 100$
The basin is already in crisis	23%	36%	47%
Any increase in pumping	24%	29%	19%
Year 2000 pumping rates	27%	16%	13%
Above the 1% rate set by the Groundwater Plan	20%	16%	21%
Increased pumping is not a crisis	6%	4%	0%
Is it important to take a regional, collaborative approach among governments, agencies and stakeholders to address water resource issues?	**2013** $n = 100$	**2014** $n = 100$	**2015** $n = 100$
Strongly Agree	N/A	N/A	93%
Agree	N/A	N/A	7%
Neutral	N/A	N/A	0%
Disagree	N/A	N/A	0%
Strongly Disagree	N/A	N/A	0%

Table 4. *Cont.*

Is the development of a reclaimed water plant in Pullman a priority?	2013 n = 100	2014 n = 100	2015 n = 100
The completion of this plant is a high priority	87%	91%	94%
This plant is a waste of money	3%	4%	6%
I have no opinion	10%	4%	0%
As an individual water consumer on the Palouse what are you going to do in the coming year?	**2013 n = 100**	**2014 n = 100**	**2015 n = 100**
Reducing my indoor water use by changing out fixtures	N/A	N/A	12%
Change my personal water behavior	N/A	N/A	23%
Talk to friends and neighbors about water	N/A	N/A	12%
Make my home/business landscaping an example of low water use	N/A	N/A	27%
Get involved in education and outreach	N/A	N/A	26%

Table 5. Partial list of responses to clicker questions presented to participants at the Spokane River Forum in 2011, 2013 and 2014 (for additional questions please contact authors).

What is your comfort level with a regional board/task force deciding where PCB removal and reduction funding should be targeted?	2011 n = 100	2013 n = 100	2014 n = 100
Very comfortable	25%	N/A	N/A
Somewhat comfortable	51%	N/A	N/A
Not comfortable	21%	N/A	N/A
I'm not sure	3%	N/A	N/A
Where do you see management of water resources in the Spokane River Basin in 20 years?	**2011 n = 100**	**2013 n = 100**	**2014 n = 100**
Managed as a basin through regulation	N/A	N/A	31%
Managed as a basin through collaboration	N/A	44%	46%
Increased conflict	N/A	40%	12%
Business as usual	N/A	16%	11%
When do you predict there will be a groundwater availability crisis?	**2011 n = 100**	**2013 n = 100**	**2014 n = 100**
There is currently a water crisis	N/A	28%	N/A
Within 5 years	N/A	9%	N/A
Within 10 years	N/A	7%	N/A
Within 20 years	N/A	16%	N/A
Within 50 years	N/A	19%	N/A
No reason to think there will be a water supply crisis in the future	N/A	21%	N/A
Should establishing a plan for basin-wide water resources management be a priority?	**2011 n = 100**	**2013 n = 100**	**2014 n = 100**
Yes		90%	
No		8%	
Not sure / have not thought about it		2%	
The aquifer interacts with the Spokane River. In some places the river loses water to the aquifer, in other places the river gains water from the aquifer. How many gaining and losing reaches of the river are there?	**2011 n = 100**	**2013 n = 100**	**2014 n = 100**
The statement is not true. The river does not interact with the aquifer.	N/A	N/A	2%
1–3	N/A	N/A	16%
More than 3	N/A	N/A	82%
Which do you value most?	**2011 n = 100**	**2013 n = 100**	**2014 n = 100**
River water quality above all	N/A	9%	3%
River flows above all	N/A	4%	1%
Groundwater quality is the greatest priority	N/A	27%	18%
Groundwater quality and river quality are equal priorities	N/A	N/A	23%
River flows, aquifer quality and river quality should be balanced equally	N/A	61%	55%

3.4. Building Simulations

In conjunction with the initial and ongoing stakeholder engagement activities, stakeholders were involved in the model building and model simulation phases. In the Palouse Basin, Stella [73] systems modeling software was used for simulation. In the Spokane basin, OASIS [74] linear programming model was used for simulation. The development of simulation models to explore future changes to water resources enables stakeholders to share insights into potential solutions to better understand and or manage their local water resource system. Iterations over time are necessary to develop robust models that incorporate the complexities of water resource systems. Iteration provides opportunity to clearly communicate assumptions and uncertainties and to test sensitivity and policies (Figure 1b).

In 2008, collaborators began a model building exercise in the Palouse Basin resulting in the 2010 version of the Palouse Basin Water Resource Visioning Tool that integrated current scientific understanding of the aquifer system and a suite of options for simulating growth, conservation and supply augmentation [53]. The model was built with system dynamics theory and Stella software. We chose to use Stella, because of its module and storytelling features. Modules were used to separate the structures for each aquifer, demand and potential new supplies so as to more easily display these structures to stakeholders. The story telling feature was integrated into the online version of the tool to allow users to click through model structures that, when supported with pop up text boxes, explain how supply is determined. The online version utilizes NetSim software and is housed on a publically available Forio server [75]. During the two year model development process, we met every month to six weeks with four to eight representatives from the Palouse Basin Aquifer Committee (PBAC) and two to six members of the interested public. Between meetings the lead modeler met individually with hydrogeologic experts from University of Idaho, Washington State University and Washington Department of Ecology to develop consensus on model structure, assumptions and uncertainty. The latest iteration of the model includes hypothetical groundwater dynamics (data is limited), historic pumping data, projected future demand under user varied growth and conservation scenarios, individual water use calculator, the opportunity to add three different reservoir sizes to augment supply, and information on basin hydrogeology and model assumptions [75].

In the Spokane basin, the model building process began in 2013. The trust we built in the 2008–2010 with Inland Northwest hydrogeologic experts during the Palouse project was instrumental in the acceptance of this process. Unlike the Palouse, the Spokane system has abundant data, an extensive network of monitoring wells, and a USGS MODFLOW model. We approached stakeholders with the idea of using this information to create a dynamically integrated ground and surface water model that could be used as a collaborative planning tool in, for what would initially be, a mock planning process. The model has been built using OASIS HydroLogics software that uses a fully configurable linear programming solver to simulate water routing and optimize system operations on a daily time step [74]. Individual and small group meetings took place monthly over the course of two years between researchers and experts from Washington State University, Washington Department of Ecology, Idaho Washington Aquifer Collaborative, and Spokane and Kootenai Counties. The collaboratively built model includes surface and groundwater dynamics, and a suite of stakeholder relevant issues that include growth, conservation, and climate change. The model simulates a range of potential impacts to the hydrologic system, including changes to lake levels during the summer or violation of the minimum instream flow rule. Through this process, participants have the ability to suggest simulation of alternative strategies or scenarios. The capacity to quickly modify parameters for new scenarios provides stakeholders with immediate feedback about their perspectives, ideas and concerns [76].

4. Collaborative Modeling in Action

4.1. Regional Water Conferences: Using Clicker Surveys to Understand Group Mental Models

Regional water conferences provided an opportunity to anonymously survey an audience to better understand participant's knowledge of basic water science, values, preferences and priorities on water management challenges. The collaborative development of these questions and the use of clickers at the conference is a useful exercise for exploring mental models of both individuals and groups of people interested in their local water resources. Over time, we have had the opportunity to iterate useful, illuminating questions and to remove questions that offer little insight. Each water conference used the clickers and developed questions that were relevant to current water resource challenges in each basin. Tables 4 and 5 provide results from a limited number of clicker questions to highlight insights from these sessions.

4.1.1. Palouse Water Summit

In the Palouse basin, conference participants answered questions on hydrology basics, preferences on infrastructure development and local water management in 2013, 2014 and 2015. Answers to the questions on basic water science indicate that individuals understand that growth and increase in water demand are not linear, and that water levels are dependent on multiple internal (aquifer properties) and external (pumping) factors. Participants indicate that there is significant interest in the City of Pullman and WSU moving forward with a proposed reclamation plant that will supply irrigation water for landscaping on campus (Table 4). Participants were also asked where they viewed water resource management in 20 years. Summit attendees increasingly have the opinion that the basin is already in crisis, however they also value bi-state cooperation (Table 4). In addition, participants have consistently preferred increased collaboration and cooperation, rather than a formal bi-state agreement or increased conflict (Table 4). Participants are indicating a willingness to proactively communicate information about the resource to others in the community and to make a personal effort to conserve (Table 4).

These sessions spark discussion among participants that often continue post Summit [76]. Individuals are able to grasp how well others understand a system and or feel about its future, which helps individuals evaluate how and why people are behaving and making choices. Based on individual stakeholder meetings and collaboratively developing the Summit questions, it appears that there are at least two categories of mental models that stakeholders view as a potential source of conflict. Some individuals believe "we have lots of water, so why worry about it", and "when there is a problem, they [future generations] can pay for it." Others believe "we need to conserve more water *now*", and "the big four pumpers are using water unsustainably, they need to change use immediately." The "we have plenty of water" *versus* "conserve more now" mental models often become part of the "pro development" *versus* "limit development" conversation. While pro development contingents are not promoting high water demand industry, there is a significant contingent of people who indicate that conservation should be a high priority. Many of the pro conservation people also value limited development.

While collaboration across state lines is supported, the potential for conflict over values, growth and water for future generations is resulting in creative tension. With no immediately imminent supply crisis, the tension over limited future supply and the sense that the basin is already in crisis is driving not only conservation of water resources, but also the desire for proactive exploration into alternative infrastructure developments to support a sustainable future. The Palouse Basin Aquifer Committee has utilized state funds to support a study to explore multiple types of augmentation including aquifer storage and recovery, reservoirs and reuse with a potential target for implementation in the 2020s [50,77].

4.1.2. Spokane River Forum

In the Spokane Basin, conference participants used clickers at the 2011, 2013 and 2014 Forums to answer questions related to education, perceptions, values and priorities. In 2011, the focus on the clicker questions was related to potential future collaboration on a PCB TMDL and the potential for collaborative basin wide management (Table 5). As we approached the end of the session in 2011, people were beginning to look around at one another with quizzical expressions. Based on data that included answers to questions such as level of comfort for a regional board deciding where PCB removal and reduction should be targeted, and post session discussion, we discovered that people were very surprised at the degree of consensus in the room. To continue to support and promote collaboration across state lines, at subsequent Forums questions were asked about the interest for collaborative bi-state water resource management. Responses related to collaborative bi-state water resources management indicate the level of interest is high (Table 5). The decision to not continue a question related to crisis was driven by the organizer's value for emphasizing collaboration over crisis.

At the end of the 2013 Forum, conference participants answered questions on basic water science. We found that there was a significant level of misunderstanding about the aquifer and how it interacted with the river, even though there were two days of presentations on aquifer science, hydrologic research and policy questions based on hydrology. As a result, conference organizers changed strategy at the next Forum held in 2014, and held a plenary session on "hydrology 101" early on the first day. While the number of participants indicating that there were at least 3 gaining and losing reaches in the river was high this type of question provided the opportunity for the panel to reinforce river-aquifer interaction. Additionally at both the 2013 and 2014 Forums, conference participants answered questions on water availability, their values on water, and collaboration (Table 5).

Questions are tailored for the specific topic of the Forum therefore questions are not consistent across time. The most prominent mental model that has surfaced at the Forums is the desire for basin wide collaborative management. The final panel session at the 2014 Forum included state agency leads, tribal representatives and a local environmental attorney. The discussion highlighted the recognition that this basin could collaborate on difficult issues and that this value should be integrated into the training of next generation of managers. As collaborative modelers we recognize that this desire to collaborate is the result of multiple factors however using clicker sessions to intentionally recognize the number of people who value collaboration helps build confidence in the possibilities that collaboration brings.

4.2. Building Computer Simulations: Dialogue and Insights

The collaborative modeling process has provided the opportunity for model building and facilitated group problem solving to take place simultaneously [53,76]. In these individual and group meetings, stakeholders have continuously been encouraged to communicate their own values, interests and knowledge of the system, and willingly have been active participants in discussion. These lively discussions have been important to incorporating stakeholder's interests into the model.

4.2.1. Development of the Palouse Basin Water Resource Visioning Tool

During the two-year CM process our discussions often included issues associated with scientific uncertainty and needed research. One manager asked how we "could create a model without more information." This experienced manager had a mental model of how the system would respond to pumping over the long term; however, he conveyed that communicating that experience was challenging. The model and CM process created an opportunity for managers to communicate their mental models to one another and to communicate to model users the level of scientific uncertainty under which managers made decisions. Political sensitivity and the manner in which the model portrayed information were of concern. For example, well levels were described graphically as feet above mean sea level, yet depending upon the scale, there was concern by several stakeholders that the model may indicate that the "basin was running out of water". Many of the graphical interfaces were iterated multiple times in response to this concern. The potential impacts of population growth on future demand were also of political concern. The model utilizes per capita use and has the option for users to choose population growth rates. During a presentation to the entirety of PBAC, one stakeholder stated that we "could not talk about development", again indicating the concern about pro growth *versus* the no growth mental models in the basin. Stakeholders have stated that lack of imminent supply crisis, and the concern about scientific uncertainty and sensitive political relationships has prevented the use of this tool for planning. With this in mind, the interface was designed primarily as an education and outreach tool, however an interface suitable for technicalities related to planning could be added at any time [53,75]. The tool has been used extensively as an educational tool for university students, who comprise almost half of the population of the basin. Since its launch in May 2010, at least 3500 WSU students have utilized the model for aquifer education and as a water footprint calculator.

Additional well data, interest in potential new supply, and requests for revision of the model interface are leading to a major revision of the tool that is taking place in 2016–2017. Familiarity with the modeling platform and the long-term relationships we have built in the basin is expediting the CM process and should expand the model scope and usefulness.

4.2.2. Development of the Spokane River Basin OASIS

The ongoing collaborative modeling process has been iterative and is deepening understanding of the complex dynamic interaction of the surface and groundwater systems in the region in the context of climate change and increasing demand [76]. We have often returned to discussions that focus on identifying and defining the regional water resource problem, a valuable step in the systems thinking to policy development process. There has been lengthy dialogue about the complexity of interactions between surface and groundwater at different reaches along the Spokane River. At times, stakeholder perspectives and knowledge about ground-surface water interaction have been conflicting and or dissimilar. For example, most stakeholders generally agree that groundwater pumping impacts streamflow, but during one discussion, a stakeholder stated "groundwater pumping does not impact flows in the river." This particular circumstance was met as an opportunity to build consensus. We asked stakeholders to teach us about their system to be better reflected it in the model. Stakeholders clarified their scientific understanding and related assumptions about the integrated surface and groundwater system. Although some have said that pumping did not impact streamflow, in general, we found the issue had more to do with current pumping rates than the issue itself. This non-confrontational manner of facilitating discussion diverted a potentially divisive moment. We have observed through these types of facilitated discussions, that stakeholders realize there is significant scientific consensus and that disagreements may often be the result of semantics. We observe that this consensus is helping to build trust among bi-state stakeholders.

5. Discussion

5.1. Benefits of Collaborative Modeling before Conflict or as a Pre-Planning Tool

Water resource managers, in both the Spokane and the Palouse, have stated that they know that tough decisions will have to be made in the future. They have stated that providing the opportunity for discussion, in advance of the need for a specific, potentially legally binding decision, opens up opportunity for critical and creative thinking. They have repeatedly made requests for us "to help keep the conversation going" because as systems leaders they understand the value of conversation for education, outreach and for shifting entrenched mental models. As modelers outside of the decision-making framework, we are able to facilitate exploration that may include options that are not possible in current regulatory regimes. For example, in the Spokane basin we are able to explore what if the system was not constrained by the current FERC license. While we are not advocating that the licensing process was problematic, we do want to offer a full suite of potential management options for exploration as we help basin stakeholders plan for the unthinkable, such as an extended drought. In the Palouse basin, we included the option for reservoirs in Idaho to augment supply in the entire basin. Delivering water across state lines is not possible under the current regulatory framework. Stakeholders are enthusiastic about having these options available, however they may be reluctant to bring them forward while representing their respective entities. Pre-conflict, pre-decision simulation and gaming exercises provides a space where no option is "off the table", or where no particular set of values takes precedent, and in doing so facilitates the creative tension between *what was or is* and *what could be.*

The development of clicker sessions that pose questions from "stakeholders to stakeholders" (through the modeling team) illuminates mental models and assumptions of individuals, as well as their perception of what others may value. Question development with basin stakeholders offers individuals the opportunity to think about the greater system. For example, at the end of a question development meeting, a wastewater plant manager thanked us for the opportunity to think beyond the operation of "pumps, pipes and the regulation of effluent for water quality" to the production and impact of wastewater on the greater system. His view of the greater system included fisheries and recreation; the value that people place on household products and the risk assessment process they use to make choices about those products; the economic legacy of the area that includes impacts from agriculture, timber, mining, aluminum foundries; and finally a vision of how today's legacy will be viewed in the future. Opportunities such as this support individual *system thinking investments* and help advance mental models well beyond typical daily activities.

As preferences of groups of people are revealed through the clickers, group mental models are supported. For example, although conflict in the Palouse basin has been fairly limited, clicker results consistently indicate that the potential for conflict is high, any increase in pumping is not acceptable, and any increase in growth would necessitate large changes in supply infrastructure. The group mental model "we need to do something soon, and do it together" is supporting system leaders in the Palouse as they explore a diverse array of supply augmentation strategies [77].

Clicker sessions in the Spokane have continually reinforced the value that people place on collaboration and the value that people place on leaving a legacy of collaboration in their respective agencies and organizations. Additionally, we are finding that this desire to collaborate is altering how water professionals and purveyors view the public they serve. For example, we have been facilitating informal collaborative modeling workshops with a group of water resources professionals from both Idaho and Washington with the intent of developing future scenarios. The long term purpose of these discussions is to develop future planning scenarios relevant to managers, however focus within this group has in the short term shifted to public perception, awareness and understanding of local water supply. This apparent deviation from scenario development reflects the sensitivity that managers have for developing scenarios for what they describe as an "unsuspecting" public. They have stated that it is "difficult to justify the importance of conservation to a public who perceives the water supply as endless." Educating the public is now a central focus for this group. While some may consider that our CM scenario development process was side tracked, we view this as an opportunity to explore perceptions about public tolerance for potential policies. Additionally, this reinforces the concept that all water users are decision makers. Our long-term investment with this basin offers us this important opportunity, even though the time-table for model scenario development has been disrupted.

When we first began visiting stakeholders in the Spokane basin, we were told "Idaho did not have a supply or quality problem, it has a Washington problem." While these issues are certainly not completely resolved, it is now rare to hear about them when working with stakeholders. We are now able to include future scenarios that include growth, conservation, and climate change. We are able to include potential decisions about these issues that are not state or community specific. Finally, we are hearing stakeholders make statements including "lets think about potential solutions for supply augmentation" in the Palouse basin and "how can we get the public on board with conservation" in the Spokane basin.

5.2. Challenges of Pre-Conflict Collaborative Modeling

Participatory processes have many benefits that engage stakeholders, however it is often challenging to continuously encourage stakeholder participation, manage expectations of stakeholders and scientists and effectively communicate among stakeholders. Stakeholder participation and modeling is time consuming, and the quality of work is dependent on the quality of available data, the techniques and skills of the facilitators and modelers, and the willingness of the stakeholders to work together to find potentially mutually beneficial solutions [37,43]. Stakeholders are often volunteers, attendance may be inconsistent, and the composition of the group may change over time. It is difficult to encourage stakeholders to actively participate and maintain interest throughout the timespan of the project when there is not an active crisis or a decision to be made. Absence of mutual trust among stakeholders can hinder effective participation during the participatory process [43]. Participants may not be willing to change their mental models, and they may have personal objectives that reduce their willingness to change their opinions. However, both of the participatory process in each of these basins has allowed for ongoing communication of stakeholders and scientific information. The continued communication among stakeholders has been crucial in each basin.

Another challenge that is faced, in many if not all collaborative modeling processes is measuring success. Even in processes that have resulted in a specific decision, long-term success is not guaranteed. In the absence of a specific decision, we are left with a qualitative assessment of success that is potentially biased by those whose mental models are supported. However, even in the absence of empirical evidence, we argue in support of the ideals that resulted in NEPA and the inclusion of adaptive management principles in natural resource management. Discussion that includes social and ecological values, and integrates best available science increases the likelihood of outcomes that support adaptive and resilient social and ecological systems.

6. Conclusions

These case studies focus on the value of collaborative modeling processes prior to crisis driven decision-making. System leaders in both basins realize that conversation and education in advance of crisis builds relationships and trust. In both of the Palouse and Spokane systems, CM has helped facilitate collaboration and improved group dynamics, and it has created space for exploring the dynamic problem, potential futures and potential opportunities for improving long-term resilience. Throughout the simulation model building process in both basins, we have found that there is an increasing willingness to discuss, and model what can be divisive topics such as climate change, growth or scientific uncertainty. We are hopeful by supporting stakeholders within a pre conflict exploratory space that resilient governance will be discovered and designed.

The need for adaptive and resilient water resource management and governance is challenging resource managers and water users to think about their resources in context of highly integrated systems. Issues such as increasing demand, decreasing quality, development, storm water, runoff, as well as unanticipated feedbacks, time delays, changing weather patterns, and changing social values are characteristics that these integrated systems often share. As we learn to do adaptive management in these systems, it is easy to get caught in the minutia of the task at hand and lose sight of the integrated system. Entrenchment in a single part of the system or in mental models that only reflect the past have the potential to restrict the creative space needed for adaptation, and in so doing increases potential for conflict. It is also important to understand that we are all trying to design governance for which no blueprint exists and that policies that may work in one basin may not necessarily work in another basin. Finally, our stakeholders have often said that we must realize that we never solve all of our problems and challenges, we just move from solution to the next challenge. Throughout this process, maintaining relationships is critical.

As collaborative modelers, we seek to support our stakeholders and their communities, through opportunities for ongoing, iterative systems research, education and outreach. Supporting system thinking investment within iterative decision processes improves the capability for spanning institutional, cultural, spatial and or temporal boundaries, and simultaneously improves the potential for adaptive and resilient water resource governance within constantly changing social and biophysical systems.

Acknowledgments: Acknowledgement to USDA National Institute of Food and Agriculture, Spokane River Forum, Palouse Basin Water Summit and all of the stakeholders in the Palouse and Spokane Basins from whom we have learned so much.

Author Contributions: Author Beall King conceived of the collaborative modeling projects in the Palouse and Spokane basins and has been lead on the exercises described in this paper with co-author Thornton often serving as co-facilitator (2013–present). Author Thornton built the OASIS model described herein and is lead facilitator for the Spokane basin OASIS exercise. Beall King and Thornton collaborated on the text of this paper.

Conflicts of Interest: The authors declare no conflict of interest.

References

1. Walker, B.; Holling, C. Resilience, adaptability and transformability in social—Ecological systems. *Ecol. Soc.* **2004**, *9*, 5.
2. Holling, C.S. *Adaptive Environmental Assessment and Management*; John Wiley & Sons: Chichester, UK, 1978.
3. Cortner, H.; Moote, M. Trends and issues in land and water resources management: Setting the agenda for change. *Environ. Manag.* **1994**, *18*, 167–173. [CrossRef]
4. Pahl-Wostl, C.; Craps, M. Social learning and water resources management. *Ecol. Soc.* **2007**, *12*, 2007.
5. Kallis, G.; Videira, N.; Antunes, P. Participatory methods for water resources planning. *Environ. Plan. C Gov. Policy* **2006**, *24*, 215–234. [CrossRef]
6. Crutzen, P. Geology of mankind. *Nature* **2002**, *415*, 23. [CrossRef] [PubMed]
7. *National Environmental Policy Act (NEPA)*; 42 U.S. Code §4321; United States Federal Government: Washington, DC, USA, 1969.
8. *Clean Air Act (CAA)*; 42 U.S. Code §§7241-7671; United States Federal Government: Washington, DC, USA, 1970.
9. *Clean Water Act (CWA)*; 33 U.S. Code §§1251-1387; United States Federal Government: Washington, DC, USA, 1972.
10. *Endangered Species Act (ESA)*; 16 U.S. Code §§1531-1544; United States Federal Government: Washington, DC, USA, 1973.
11. Hardin, G. The tragedy of the commons. *Science* **1968**, *162*, 1243–1248. [CrossRef] [PubMed]
12. Beall, A.M. Participatory Environmental Modeling and System Dynamics: Integrating Natural Resource Science and Social Concerns. Ph.D. Thesis, Washington State University, Pullman, WA, USA, 2007.
13. Guttman, N. Bringing the mountain to the public: Dilemmas and contradictions in the procedures of public deliberation initiatives that aim to get "ordinary citizens" to deliberate policy issues. *Commun. Theory* **2007**, *17*, 411–438. [CrossRef]
14. Reed, M.S. Stakeholder participation for environmental management: A literature review. *Biol. Conserv.* **2008**, *141*, 2417–2431. [CrossRef]
15. O'Leary, R.; Bingham, L.B. *The Promise and Performance of Environmental Conflict Resolution*; RFF Press: New York, NY, USA, 2003.
16. Dietz, T.; van den Belt, M. Forward. In *Mediated Modeling*; Island Press: Washington, DC, USA, 2004; p. xiii.
17. Grumbine, R.E. What Is Ecosystem Management? *Conserv. Biol.* **1994**, *8*, 27–38. [CrossRef]
18. Nagle, J.; Ruhl, J. *The Law of Biodiversity and Ecosystem Management*; Foundation Press: New York, NY, USA, 2002.
19. Folke, C. Resilience: The emergence of a perspective for social–ecological systems analyses. *Glob. Environ. Chang.* **2006**, *16*, 253–257. [CrossRef]
20. Millennium Ecosystem Assessment. Available online: http://www.millenniumassessment.org/en/index.html (accessed on 15 May 2016).
21. Global Watership Partnership Technical Advisory Committee. *Integrated Water Resources Management*; TAC Background Paper; Global Water Partnership: Stockholm, Sweden, 2000.
22. Dörendahl, E.I. *Boundary Work and Water Resources: Towards Improved Management and Research Practice?*; ZEF Working Paper Series, No. 122; Zentrum für Entwicklungsforschung: Bonn, Germany, 2013.
23. Liu, Y.; Gupta, H.; Springer, E.; Wagener, T. Linking science with environmental decision making: Experiences from an integrated modeling approach to supporting sustainable water resources management. *Environ. Model.* **2008**, *23*, 846–858. [CrossRef]

24. Jacobs, K. *Connecting Science, Policy, and Decision-Making: A Handbook for Researchers and Science Agencies*; NOAA Office of Global Programs; NOAA Climate Diagnostics Center: Boulder, CO, USA, 2002.
25. Beall, A.; Ford, A. Reports from the Field: Assessing the Art and Science of Participatory Environmental Modeling. *Int. J. Inf. Syst. Soc. Chang.* **2010**, *1*, 72–89. [CrossRef]
26. Bourget, E.C.; Langsdale, S.M.; van den Belt, M. Featured Collection Introduction: Collaborative Modeling for Decision Support as a Tool to Implement IWRM. *J. Am. Water Resour. Assoc.* **2013**, *49*, 605–608. [CrossRef]
27. Meadows, D.; Robinson, J. *The Electronic Oracle*; System Dynamics Society: Albany, NY, USA, 1985.
28. Perez, E.M.; Viessman, W. *The Role of Technology in Water Resources Planning and Management*; Brown & Caldwell: West Palm Beach, FL, USA, 2009.
29. Voinov, A.; Bousquet, F. Modelling with stakeholders. *Environ. Model. Softw.* **2010**, *25*, 1268–1281. [CrossRef]
30. Winz, I.; Brierley, G.; Trowsdale, S. The use of system dynamics simulation in water resources management. *Water Resour. Manag.* **2009**, *23*, 1301–1323. [CrossRef]
31. Voinov, A.; Gaddis, E.J.B. Lessons for successful participatory watershed modeling: A perspective from modeling practitioners. *Ecol. Model.* **2008**, *216*, 197–207. [CrossRef]
32. Ford, A. *Modeling the Environment: An Introduction to System Dynamics Modeling of Environmental Systems*, 2nd ed.; Island Press: Washington, DC, USA, 2009.
33. Johnson-Laird, P. Mental models in cognitive science. *Cogn. Sci.* **1980**, *4*, 71–115. [CrossRef]
34. Senge, P. *The Fifth Discipline: The Art and Practice of the Learning Organization*; Double Day: New York, NY, USA, 1990.
35. Vennix, J.A.M. *Group Model Building: Facilitating Team Learning Using System Dynamics*; Wiley: Chichester, UK, 1996.
36. Richmond, B.; Peterson, S. *An Introduction to Systems Thinking*; High Performance Systems: Hanover, NH, USA, 2001.
37. Bourget, L. *Converging Waters: Integrating Collaborative Modeling with Participatory Processes to Make Water Resources Decisions*; Bourget, L., Ed.; Institute for Water Resources: Alexandria, VA, USA, 2011.
38. Parson, E. Integrated assessment and environmental policy making: In pursuit of usefulness. *Energy Policy* **1995**, *23*, 463–475. [CrossRef]
39. Cardwell, H.; Langsdale, S.; Stephenson, K. *The Shared Vision Planning Primer: How to Incorporate Computer Aided Dispute Resolution in Water Resources Planning*; Institute for Water Resources: Alexandria, VA, USA, 2008.
40. Costanza, R.; Ruth, M. Using dynamic modeling to scope environmental problems and build consensus. *Environ. Manag.* **1998**, *22*, 183–195. [CrossRef]
41. Cardwell, H.E.; Langsdale, S. Collaborative Modeling for Decision Support—Definitions and Next Steps. In Proceedings of the World Environmental and Water Resources Congress; Beighley, R., II, Kilgore, M., Eds.; ASCE: Reston, VA, USA, 2011; pp. 2805–2814.
42. Van den Belt, M. *Mediated Modeling: A System Dynamics Approach to Environmental Consensus Building*; Island Press: Washington, DC, USA, 2004.
43. Langsdale, S.; Beall, A.; Bourget, E.; Hagen, E.; Kudlas, S.; Palmer, R.; Tate, D.; Werick, W. Collaborative Modeling for Decision Support in Water Resources: Principles and Best Practices. *J. Am. Water Resour. Assoc.* **2013**, *49*, 629–638. [CrossRef]
44. Videira, N.; Antunes, P. Participatory modelling in environmental decision-making: The ria Formosa natural park case study. *J. Environ. Assess. Policy Manag.* **2003**, *5*, 421–447. [CrossRef]
45. Cockerill, K.; Passell, H.; Tidwell, V. Cooperative Modeling: Building bridges between science and the public. *J. Am. Water Resour. Assoc.* **2006**, *42*, 457–471. [CrossRef]
46. Creighton, J. *How to Conduct a Shared Vision Planning Process*; IWR Report 10-R-6; U.S. Army Institute for Water Resources: Alexandria, VA, USA, 2010.
47. Lund, J.; Palmer, R. Water resource system modeling for conflict resolution. *Water Resour. Update* **1997**, *3*, 70–82.
48. Langsdale, S.; Beall, A.; Carmichael, J.; Cohen, S.; Forster, C.; Neale, T. Exploring the implications of climate change on water resources through participatory modeling: Case study of the Okanagan Basin, British Columbia. *J. Water Resour. Plan. Manag.* **2009**, *135*, 373–381. [CrossRef]
49. Sheer, D.P.; Baeck, M.L.; Wright, J.R. Computer as negotiator. *J. Am. Water Work. Assoc.* **1989**, *81*, 68–73.
50. Palouse Basin Aquifer Committee (PBAC). Available online: http://www.webpages.uidaho.edu/pbac/ (accessed on 15 January 2016).
51. Leek, F. Hydrogeological Characterization of the Palouse Basin Basalt Aquifer System. Master's Thesis, Washington State University, Pullman, WA, USA, 2006.

52. Kahle, S.; Bartolino, J. *Hydrogeologic Framework and Ground-Water Budget of the Spokane Valley-Rathdrum Prairie Aquifer, Spokane County, Washington, and Bonner and Kootenai Counties, Idaho*; U.S. Geological Survey Scientific Investigations Report; USGS: Reston, VA, USA, 2007.
53. Beall, A.; Fiedler, F.; Boll, J.; Cosens, B. Sustainable water resource management and participatory system dynamics. Case study: Developing the Palouse basin participatory model. *Sustainability* **2011**, *3*, 720–742. [CrossRef]
54. Smoot, J.; Ralston, D. *Hydrogeology and a Mathematical Model of Ground-Water Flow in the Pullman-Moscow Region, Washington and Idaho*; Idaho Water Resources Research Institute: Moscow, ID, USA, 1987.
55. Luce, C.; Abatzoglou, J.; Holden, Z. The missing mountain water: Slower westerlies decrease orographic enhancement in the Pacific Northwest USA. *Science* **2013**, *342*, 1360–1364. [CrossRef] [PubMed]
56. Dunau, A.; Spokane River Forum, Spokane, WA, USA. Personal communication, 2009–2016.
57. Idaho Washington Aquifer Collaborative (IWAC). Available online: http://www.iwac.us (accessed on 16 February 2016).
58. Fu, G.; Barber, M.; Chen, S. Impacts of climate change on regional hydrological regimes in the Spokane River Watershed. *J. Hydrol. Eng.* **2007**, *12*, 452–461. [CrossRef]
59. Hsieh, P.; Barber, M.; Contor, B. *Ground-Water Flow Model for the Spokane Valley-Rathdrum Prairie Aquifer, Spokane County, Washington, and Bonner and Kootenai Counties, Idaho*; U.S. Geological Survey Scientific Investigations Report; USGS: Reston, VA, USA, 2007.
60. Molenaar, D. *The Spokane Aquifer, Washington: Its Geologic Origin and Water-Bearing and Water-Quality Characteristics*; U.S. Geological Survey Scientific Investigations Report; USGS: Reston, VA, USA, 1988.
61. Washington Department of Ecology. Water Resource Program: Rulemaking. Available online: http://www.ecy.wa.gov/programs/wr/rules/557-ov.html (accessed on 16 February 2016).
62. Washington Department of Ecology Water Quality Improvement Project Spokane River and Lake Spokane Area: Dissolved Oxygen. Available online: http://www.ecy.wa.gov/PROGRAMS/wq/tmdl/spokaneriver/dissolved_oxygen/index.html (accessed on 16 February 2016).
63. Stave, K. Using system dynamics to improve public participation in environmental decisions. *Syst. Dyn. Rev.* **2002**, *18*, 139–167. [CrossRef]
64. Vennix, J.A.M.; Andersen, D.F.; Richardson, G.P. Foreword: Group model building, art, and science. *Syst. Dyn. Rev.* **1997**, *13*, 103–106. [CrossRef]
65. Andersen, D.F.; Richardson, G.P.; Vennix, J.A.M. Group model building: Adding more science to the craft. *Syst. Dyn. Rev.* **1997**, *13*, 187–201. [CrossRef]
66. Rivera, M.W.; Sheer, D.P.; Miller, A.J. Computer-Aided Negotiations of Water Disputes: An Interdisciplinary Role-Playing Course. *J. Am. Water Resour. Assoc.* **2013**, *49*, 700–714. [CrossRef]
67. Forrester, J. System dynamics, systems thinking and soft OR. *Syst. Dyn. Rev.* **1994**, *10*, 245–256. [CrossRef]
68. Meadows, D.; Wright, D. *Thinking in Systems: A Primer*; Chelsea Green Publishing: Danvers, MA, USA, 2008.
69. Mott Lacroix, K.; Megdal, S. Explore, Synthesize, and Repeat: Unraveling Complex Water Management Issues through the Stakeholder Engagement Wheel. *Water* **2016**, *8*. [CrossRef]
70. Palouse Basin Water Summit. Available online: http://www.palousewatersummit.org/ (accessed on 12 December 2015).
71. Spokane River Forum. Available online: http://spokaneriver.net (accessed on 12 December 2015).
72. Stowell, J.; Nelson, J. Benefits of electronic audience response systems on student participation, learning, and emotion. *Teach. Psychol.* **2007**, *34*, 253–258. [CrossRef]
73. *Stella Systems Modeling Software*; ISEE Systems Inc.: Lebanon, NH, USA, 2005.
74. Hydrologics Inc. *User Manual for OASIS with OCL*; Hydrologics Inc.: Columbia, MD, USA, 2009.
75. Beall, A. The Palouse Basin Water Resource Visioning Tool. Available online: http://forio.com/simulation/ns/allysonbeall/palouse_basin_model/ (accessed on 18 February 2016).
76. Thornton, M. Washington State University: Pullman, WA, USA, 2017. Unpublished Work.
77. Kimmell, P.; Palouse Basin Aquifer Committee, Latah County, ID, USA. Personal Communication, 2013–2015.

water

MDPI

Article

Opening the Black Box: Using a Hydrological Model to Link Stakeholder Engagement with Groundwater Management

Susanna Eden [1,*], Sharon B. Megdal [1], Eylon Shamir [2], Karletta Chief [3] and Kelly Mott Lacroix [1]

1 Water Resources Research Center, College of Agriculture and Life Sciences, University of Arizona, Tucson, AZ 85719, USA; smegdal@email.arizona.edu (S.B.M.); klacroix@email.arizona.edu (K.M.L.)
2 Hydrologic Research Center, San Diego, CA 92130, USA; EShamir@HRCwater.org
3 Department of Soil, Water and Environmental Science, University of Arizona, Tucson, AZ 85719, USA; kchief@email.arizona.edu
* Correspondence: seden@email.arizona.edu; Tel.: +1-520-621-5670

Academic Editor: Ashok K. Chapagain
Received: 7 March 2016; Accepted: 10 May 2016; Published: 23 May 2016

Abstract: Stakeholder participation is a foundation of good water governance. Good groundwater governance typically involves the co-production of knowledge about the groundwater system. Models provide a vehicle for producing this knowledge, as well as a "boundary object" around which scientists and stakeholders can convene the co-production process. Through co-production, stakeholders and scientific experts can engage in exchanges that create system knowledge not otherwise achievable. The process involves one-way transfer of information, active two-way conversations, and integration of multiple kinds of knowledge into shared understanding. In the Upper Santa Cruz River basin in Arizona, USA, the University of Arizona Water Resources Research Center (WRRC) convened a project aimed at providing scientific underpinnings for groundwater planning and management. This project, entitled Groundwater, Climate, and Stakeholder Engagement, serves as a case study employing the first two stages of knowledge co-production using a hydrological model. Through an iterative process that included two-way communication, stakeholders provided critical input to hydrologic modeling analyses. Acting as a bridging organization, the WRRC facilitated a co-production process, involving location-specific and transferability workshops, which resulted in new knowledge and capacity for applying the model to novel problems.

Keywords: groundwater modeling; groundwater management; stakeholder engagement; knowledge co-production; Santa Cruz River Basin

1. Introduction

Stakeholder participation is a foundational principle of good water governance [1]. Groundwater governance has varied definitions, but in general it refers to the overarching framework, usually spanning multiple jurisdictional levels, of laws, regulations, and customs for groundwater management [2]. It is distinguished from government in emphasizing processes of influence rather than control [3]. Principles of good water governance call for replacing hierarchy with networks of relationships in which stakeholders play a major role [4,5].

Groundwater management operates within the governance framework. Lautze *et al.* argued that good water governance and good water management are separate goals that should be evaluated on different criteria [6]. The goals of water management are outcome oriented, while the goals of good governance are process oriented. The definition of groundwater governance adopted by Groundwater

Governance, Global Framework for Action to achieve the vision on Groundwater Governance, includes good management outcomes: "control, protection and socially-sustainable utilization of groundwater resources and aquifer systems for the benefit of humankind and dependent ecosystems" [7]. There is still debate, however, on whether the processes of good governance are necessarily linked with positive management outcomes. The contrary examples of India, China, Jordan and Saudi Arabia provided by Lautze *et al.* make the claim difficult to support [6]. Individual good governance practices, such as stakeholder engagement, however, have been linked with improved implementation of groundwater management actions.

Stakeholders include any individual or group with a stake in the outcome of a project or action. In general, stakeholder engagement aims to include a wide range of interests influenced by or influencing a decision [8] for reasons of effectiveness [9], as well as legitimacy [10]; that is, engaged stakeholders are more likely to view policy outcomes as legitimate and less likely to impede their implementation. Many economic, environmental, and social benefits are gained from effectively engaging stakeholders in water resource decision making. Systematic and inclusive stakeholder engagement improves the chances that the time and resources invested will yield the desired return and stakeholder issues can be handled effectively [11]. Outcomes, such as support for reforms, are more likely. Less tangible benefits also can evolve from better co-operation, including knowledge development, conflict avoidance, and social cohesion [12].

It is recognized that stakeholders may engage for the purpose of controlling or disrupting collective efforts. Interest conflicts may dominate, especially where potential outcomes are likely to produce winners and losers. In these instances, developing a shared understanding of the issue context may be insufficient to resolve differences, and the benefits of stakeholder engagement will be more difficult to realize [13].

Despite this possibility, it is generally accepted that scientific information and system understanding are associated with better management. The many endeavors producing groundwater science in support of effective management are evidence of this association, and many have successfully translated scientific information into improved management outcomes [14,15]. However, good science is likely to be ineffective in producing better management when there are communication failures between scientific knowledge producers and users [16,17]. To overcome this disconnection, it is necessary to engage science users in jointly producing knowledge appropriate to the management context [16,18,19].

For effective groundwater management, many kinds of knowledge are essential. Knowledge of the physical groundwater system, its interactions with other physical systems, the social, political, and economic context, and other case-specific relevant factors are needed to manage the resource. Relevant knowledge does not reside in a single source but emerges through an interactive process. Scientists may take a leading role in this process, but they take part in a conversation in which people having other types of expertise, e.g., practical or local knowledge, are heard [20–23]. Co-production occurs when scientists and stakeholders interact to produce knowledge that can be used in the relevant water management context [19].

Co-production in the context of scientific research involves establishing linkages between researchers and research users [23]. A survey by Phillipson *et al.* of 21 research projects engaging stakeholders found "a complex and diverse range of knowledge exchange relations" [24]. In an assessment of conditions for successful joint knowledge production, Hegger *et al.* proposed in successful co-production, relevant resources are available to support the process. These may include so called "boundary objects", such as models, that lie at the cultural boundary between scientists and stakeholders and provide a focus for exchange [18].

Engaging stakeholders in modeling a natural system such as an aquifer is a recognized means of transferring knowledge about the modeled system [25]. Scientists engage in describing and explaining by presentations and responding to questions. Stakeholders contribute knowledge regarding the system only they can provide from experience with the system in their various roles. The model also provides a neutral platform on which system knowledge is built and agreed upon by stakeholders who may disagree about desired policies or practices.

Models present special challenges to engagement of stakeholders. The model structure must be scientifically defensible. It must also be explained well and as transparent as possible. This means communicating model assumptions and data, its reliability limits and uncertainty. Otherwise, trust in the model and trust between the scientists and stakeholders are undermined. Such communication achieves its goals more effectively when done through engagement and interaction [9,26]. Stakeholders who participate in posing questions and evaluating results of simulations are in a position to assess potential future states of the system under different scenarios [27]. Revealing the workings of the model builds confidence in the model as a platform for objectively answering stakeholder relevant questions.

Another element with a positive effect on knowledge co-production is the intervention of an individual or organization that carries out the functions of facilitating communication among the various parties. Universities commonly provide these functions in stakeholder engagement [28]. Academic examples include the Regional Integrated Science Assessment Centers, National Institutes for Water Resources, and Southwest Ecological Restoration Institutes [29]. Perhaps the most longstanding example is cooperative extension, which began as an institution to link scientists and farmers for information exchange, mutual learning, and decision support [30]. There are multiple designs for achieving this functional goal, including bridging organizations. Bridging organizations have a broad scope and engage in multiple functions [31]. A bridging organization serves as an entity that links people from multiple interests in order to help them solve problems that they would not have been able to solve if acting alone [32]. It is "a conduit of ideas and innovations, a source of information, a broker of resources, a negotiator of deals, a conceptualizer of strategies, [and] a mediator of conflicts" [33]. Bridging organizations might assist collaborative initiatives by gathering and interpreting technical information or providing legal, financial, or simply moral support. These organizations also can help balance power asymmetries in discussions [29], which is particularly important in water management engagement [34]. Bridging organizations also are important because they provide an arena where groups that assemble knowledge in very different ways can form productive relationships with one another [35,36]. Finally, bridging organizations have been widely cited as promoting learning [37].

2. Approach

This paper examines stakeholder engagement with an innovative modeling framework for the purpose of co-producing knowledge relevant to groundwater management. In so doing the paper seeks to fill the need for case studies that illustrate challenges and outcomes of knowledge co-production [38]. The project, Groundwater, Climate, and Stakeholder Engagement (GCASE) was defined initially with three goals: (1) Incorporating projected future climate change patterns into the modeling framework; (2) Engaging stakeholders to develop an understanding of the relevance of the model to water management problems; and (3) Establishing the transferability of the modeling framework and stakeholder approach to other areas. The technical accomplishment of the first goal is described elsewhere [14].

The hydrologic modeling framework was constructed over a period of years prior to GCASE by the Hydrologic Research Center in collaboration the Arizona Department of Water Resources. This modeling framework was specifically constructed to represent the dominant meteorological-hydrological-hydrogeological processes in the study region with an emphasis on climate variability and projected climatic changes. The hydrologic modeling framework has been used as a tool to assess the benefit and impact of various water resources strategies. The model selection, construction verification and case studies are available in following references [14,15,39–41], and a concise description of the modeling framework is in Section 3.3. The GCASE project sought to demonstrate the usefulness to stakeholders of the existing model with its climate enhancements. Use of an existing model also facilitated a focused approach made necessary by limitations of time and resources.

Engagement of stakeholders and scientists in developing management-relevant information places the project in the realm of knowledge co-production. Meadow *et al.* (2015) listed three principles of knowledge co-production: (1) building ongoing relationships between scientists and stakeholders; (2) ensuring two-way communication between scientists and stakeholders; and (3) maintaining a focus on usable science [38]. In this context 'usable science' refers to science produced in a form that can be used by stakeholders in their management and decision making roles. The project team employed an approach that emphasized these three principles.

The literature of co-production emphasizes principles, approaches, modes, and factors that foster successful joint knowledge production, but does not identify process steps [18,21,24,38,42,43]. Although Voinov and Gaddis provide twelve lessons learned regarding the participatory modeling process [9] and Michaels describe six strategies for knowledge brokering [19], steps are left to the project designer. This lack of prescribed steps is partially explained by the iterative nature of the co-production process. As a process involving stakeholder engagement, knowledge coproduction can be visualized as a wheel, in which actions are revisited, reevaluated, revised and repeated [44]. Knowledge co-production requires an adaptive approach in which actions are adopted and adjusted based on the nature of the project as it may change over time [17,27].

The GCASE project sought to demonstrate the potential of a specific combination of stakeholder engagement approach and hydrological model. The approach used was similar to the participatory integrated assessment as described by Meadow *et al.* who mined the current literature to provide typologies of modes of engagement and approaches to collaboration [38]. The principal participation mechanism used within this approach was the workshop because of it provides venue in which to work in a collaborative mode.

In keeping with guidance from the literature, the project's design aims included advice of key stakeholders, on-going and expanding stakeholder consultation, diverse knowledge, negotiated communication, effective dialogue on the model and its use, model analyses responsive to stakeholder queries, and transferability assessments. These aims were addressed in the construction of the stakeholder process as described in this case study.

GCASE engaged a diverse group of stakeholders in a series of workshops that acquainted stakeholders with the modeling framework and elicited questions and comments to guide scenario development and analysis. Through an iterative process that emphasized two-way communication, stakeholders provided critical input on hydrologic modeling analyses that incorporated climate uncertainty and simulations of alternative pumping scenarios. In addition, stakeholders provided suggestions on model modifications that could make the GCASE methodology transferable to other locations with different ground and surface water management issues. Scientists and stakeholders shared knowledge through workshop interactions, email correspondence, and a dedicated web site. The University of Arizona Water Resources Research Center (WRRC) acted as the stakeholder convener or bridging organization to ensure that requirements for effective stakeholder engagement, such as open, consistent communication, inclusiveness, and interactivity, were met. In order to avoid the pitfalls of black-box science, workshops were designed to lay open the model and its assumptions. The WRRC monitored stakeholders' understanding of the information presented and adapted communications based on their input.

3. Case Study

3.1. Setting

The Santa Cruz Active Management Area (SCAMA) covers 1850 square kilometers of river basin along the Upper Santa Cruz River, a desert stream characterized as ephemeral, except from an effluent fed river section downstream from the Nogales International Wastewater Treatment Plant (NIWTP) (see Figure 1). The Santa Cruz River flows southward from its headwater in Arizona into Mexico, where it continues southward for a short distance before turning northward to reenter Arizona five miles east of the City of Nogales, Arizona. From there it enters the SCAMA and flows north and west into the Tucson Active Management Area.

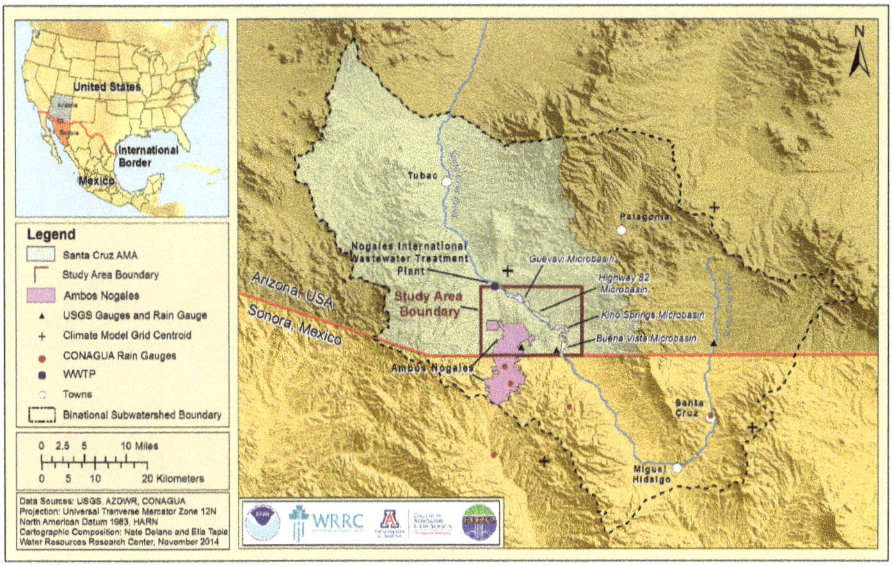

Figure 1. Map of case study area. Nogales, Arizona and Nogales, Sonora are collectively refered to as "Ambos Nogales".

The SCAMA aquifer can be roughly categorized as two systems. The first is the "Microbasin" region along the river from the International Boundary to the NIWTP. This region consists of a series of relatively shallow alluvial aquifers that are tightly dependent on the Santa Cruz River's intermittent rain driven streamflow events. The second system, which is located north of the NIWTP and extends to the northern SCAMA boundary, is a wider and much deeper aquifer. This aquifer also relies on recharge from the Santa Cruz River and it receives treated wastewater effluent from the NIWTP.

Nogales, Arizona is the largest city in the SCAMA, with a population of little more than 20,000 people. In contrast, Nogales, Sonora has a population of approximately 200,000 people. The two cities are jointly referred to as Ambos Nogales. Industrial growth south of the border has spurred population growth since the passage of the North American Free Trade Agreement in 1993, while the population of Nogales, Arizona grew much more slowly in the same period. Water supplies for Nogales, Sonora come from local aquifers in the Santa Cruz River and the Los Alisos River. Future increases in water use from the Santa Cruz watershed in Sonora could reduce flows into Arizona. Although the Nogales, Arizona water demand is expected to remain stable [45], the Arizona Department of Water Resources (ADWR) projected increase in urban water demand for the entire SCAMA by 2025.

The City of Nogales, Arizona, provides water for its population from two groundwater sources: the Santa Cruz River microbasins and the Potrero aquifer on the west side of the city. The microbasins aquifer is recharged by flows in the Santa Cruz River triggered by precipitation events. The Potrero aquifer, which is deeper, is replenished through slower recharge processes. The city's water managers prefer to use the relatively shallow microbasins aquifer because they determined it provides better quality water at lower pumping costs. Some wells in the Potrero aquifer produce water with arsenic concentrations higher than allowed by United States federal drinking water standards. Reliability of the microbasins as a water supply depends largely on climate and understanding the regional climate variability and uncertainty could improve water supply practices for the city.

The Santa Cruz River aquifer is a binationally shared resource and developments upstream in Mexico have the potential to affect supplies in Arizona. Water quality is a chief concern, but potential quantity implications exist. Of special interest is the ongoing development of the El Pilar mine concessions in the Santa Cruz River basin southeast of Nogales, Sonora. In addition, flow in the Santa Cruz River downstream from the NIWTP is dependent on discharges from the plant, which treats wastewater from both Nogales, Arizona, and Nogales, Sonora. Mexico's portion of the discharge is approximately 55 percent and could be diverted away from the NIWTP and the river if Mexico decides such action would be in its best interest. This action would likely reduce flow in the river, with a negative impact on the downstream riparian ecosystem in Arizona [46].

3.2. Management Framework

The Arizona the Groundwater Management Act of 1980 established ADWR as the agency with regulatory responsibility for water resources management. The Act created Active Management Areas (AMAs) in those groundwater basins where groundwater pumping was leading to severe water level declines. Approximately 80 percent of Arizona's population resides in an AMA. Within these AMAs, groundwater withdrawals are limited based on historic use, and conservation requirements restrict growth in groundwater use. Four AMAs were created in 1980 and a fifth, the SCAMA, separated from the Tucson AMA in 1994. For each AMA, a groundwater management goal was specified, and for three of the five AMAs that goal includes reaching "safe-yield" by 2025. Safe-yield, as defined in statute, is an attempt "to achieve and thereafter maintain a long-term balance between the annual amount of groundwater withdrawn in an active management area and the annual amount of natural and artificial recharge in the active management area" [47]. The Act mandated that periodic management plans be written that prescribe actions toward achieving the management goal for each AMA by 2025.

The SCAMA separated from the Tucson AMA in part because the SCAMA region was already in safe-yield. The SCAMA has a two-part management goal. In addition to the safe-yield goal, the SCAMA aims to prevent local water tables from experiencing long-term declines. The management goal did not specify the protection of riparian natural resources. Although many people maintain that such protection was the intent of the goal, others reject that interpretation. Developing SCAMA rules to achieve the management goal depends on definition of the term "long-term decline", which is undefined in the statute. One interpretation would prevent declines of sufficient duration to endanger riparian vegetation such as cottonwood trees. Understanding the likely effects of pumping in the microbasins could inform this discussion.

The Act also mandated rulemaking to ensure that growth within AMAs is supported by an assured supply of water for 100 years. These Assured Water Supply rules deny a developer permission to subdivide land for sale or lease without showing water availability. Unless a subdivision developer has obtained a written commitment from a designated water service provider, he/she must obtain a certificate of assured water supply from ADWR before the subdivision can be approved. Criteria for granting a certificate of assured water supply include proof of physically, legally, and continuously available supply for 100 years. In addition, the developer must show consistency with the AMA management goal and management plan. Water service providers must obtain a designation of assured water supply from ADWR in order to support growth within their service areas. The same criteria apply.

Prior to a moratorium on rulemaking in Arizona issued in 2009, and still in effect after this project's completion, there was a major effort to establish assured water supply criteria consistent with the SCAMA goal. Recommended rules were developed based on a historical analysis of the aquifer [48], but review and adoption were halted before any action could be taken. Deliberations focused in part on interpretation of that section of the management goal that would prevent long-term water level declines.

Regulating groundwater use for growing population centers remains a major water management concern for stakeholders in the SCAMA. Other concerns focus on vulnerability to climate variations and the need for water storage. In addition, the City of Nogales desires to optimize use of the microbasins.

As a statement of policy, the SCAMA endorsed consideration of "community and stakeholder input to the highest degree possible" [49]. Stakeholder involvement in the development of AMA management plans involves public consultation and opportunities for comment. Meetings at which the plans are discussed are open to the public and notifications of meetings are sent to interested stakeholders. Draft plans receive review and comment from a range of stakeholders. However, outreach to stakeholders is limited by departmental resources and engagement does not reach the level of knowledge co-production.

Groundwater Users Advisory Councils (GUACs) to the ADWR Director are groups consisting of five members, appointed by the governor on the basis of their knowledge of, interest in, and experience with the local water resource issues. A vehicle to solicit stakeholder input, the GUACs have no authority to implement recommendations, but they are kept abreast of relevant agency activities and provide input to plans and regulatory proposals developed by ADWR.

3.3. Modeling Framework

In order to have sufficient information on water supply and demand for planning and regulation of groundwater management, ADWR undertook the development of physically-based groundwater models in the AMAs. Beginning in 1997, ADWR initiated a field monitoring program in the SCAMA to provide data for conceptual model design and model calibration. In the mid-2000's ADWR developed a MODFLOW groundwater model simulating conditions south of the NIWTP to the Mexican border [50] and north of the NIWTP to the AMA northern boundary [51]. The capabilities of this physically-based model were enhanced to accept stochastic input from ensemble simulations of likely-to-occur rainfall and streamflow scenarios, developed by the Hydrologic Research Center (HRC) (Hydrologic Research Center is a public benefit nonprofit organization with 501(c) (3) status in the United States) with ADWR funding and cooperation [40,41]. The rainfall-streamflow scenarios in conjunction with the groundwater model enabled a risk-based assessment of various water withdrawal scenarios and their effect on the aquifer [15,39].

As part of the GCASE project, HRC, and the University of Arizona, Department of Atmospheric Sciences, examined regional dynamically downscaled climate models to project future climatic changes in the SCAMA. These projections were used to modify the rainfall-generating component of the modeling framework. The key factor affecting changes in rainfall patterns was projected changes in inter-annual trends in both the summer and winter seasons, when most of the rain falls in typical years [14].

The modeling framework employed in GCASE included simulation of the hydrologic system from rainfall, through streamflow to groundwater response for the microbasins portion of the SCAMA. The hydrologic model components reproduce well the working of the ADWR MODFLOW model and allowed for multiple queries to be analyzed with relatively short turn-around times. This modeling framework provided a tool to examine various groundwater management strategies under historic and projected rainfall regimes.

With knowledge of how climate and groundwater management strategies interact, decision makers can optimize their pumping to achieve groundwater management goals. They are also able to make tradeoffs with respect to factors such as system reliability, storage needs, and groundwater recharge optimization.

3.4. Stakeholder Engagement

The study team began identifying stakeholders to participate in project workshops by exploiting established connections in the study region. Relationships built over time formed the core of the outreach. The WRRC has undertaken engagement activities throughout the state of Arizona that created relationships of trust and respect with a numerous community of actors in the water arena. These relationships are cultivated consistently through personal contacts, meetings, and various forms of outreach. Extending these relationships is a strategic goal of the WRRC.

3.4.1. Stakeholder Identification and Recruitment

Known stakeholders were polled to identify additional individuals and organizations to be invited to participate in workshops. Because the goal of engagement was to develop an understanding of the relevance of the model to water management problems, the focus of the engagement was on expert stakeholders. Expert stakeholders are people who are professionally or institutionally involved with the issue and have "gained domain specific expertise through their profession" [52]. This focus eliminated the option of advertising workshops broadly. Invitations were sent to individuals and organizational representatives with a known or apparent interest in the water resources of the study region.

Efforts in stakeholder identification and recruitment aimed at broad participation among expert stakeholder interests. The goal of stakeholder recruitment was to provide a large pool of ideas and perspectives for input into the modeling process. The number of invitations sent ranged from 39 for the first workshop to 81 for the third workshop. Table 1 shows the range of stakeholder interests at the case study workshops. An average of 21 stakeholders participated in each of the three case study workshops. Approximately half of all participants attended two or more workshops.

Table 1. Categories of stakeholders present at case study workshops.

Attorney	Binational Commission	City Government
Council of governments	County government	GUAC member
Mexican water utility	Non-Governmental organization	Private consulting firm
State government	United States federal government	United States water utility
University		

Individual communications were used to activate stakeholder networks and generate interest in the study research. Invitations emphasized the interactive nature of the research and the importance of stakeholder input. Follow-up messages and telephone calls encouraged participation. To provide incentives for key stakeholders to attend and to make sure their knowledge was included, they were asked to be workshop speakers. For example, the Kickoff Meeting agenda included representatives from ADWR, the City of Nogales, a National Park, an NGO focused on protection of the Santa Cruz River, and representatives of regional and federal agencies, in addition to project scientists.

Formation of an advisory committee was a key to developing a project that encouraged co-production of relevant knowledge. Members were identified as essential links between the science and management contexts. The committee was kept small—four external members—to minimize cumbersome logistical arrangements for meetings and reviews. The Deputy Director of ADWR and the Director of Public Works for the City of Nogales were joined by representatives from the United States Geological Survey and the Salt River Project, a major water supplier in central Arizona. In addition to influencing the direction of the project and design of workshop, they provided advice on transferability of the project to other regions in Arizona. They also were consulted on development of communications with other stakeholders.

3.4.2. Stakeholder Communication

Throughout the project, stakeholders were kept informed about the project and its progress through intermittent email communication. In addition, a website was created for posting of project-relevant information, including workshop presentations and summaries. An effort was made to engage stakeholder in an on-line conversation about the project through an access-limited dialogue page.

One key project innovation was the preview of technical workshop presentations to evaluate their ability to communicate to stakeholders. Stakeholder or non-technical surrogates (staff and students) previewed workshop presentations and provided feedback on understandability and effectiveness in communicating key messages. Use of scientific jargon presented an initial challenge and engendered conversations on how to translate terms or define them in everyday language. It was also necessary to explore the purpose information—to answer the question "Why are you telling me this?" Slides were critiqued to promote simple, self-explanatory graphics and discourage complex charts and graphs that were difficult to explain. These exercises were appreciated by all the presenters whose presentations were previewed, who reported that their presentations were improved without sacrificing scientific credibility.

3.4.3. Case Study Workshops

Workshops were designed to maximize stakeholder participation. The first section of the workshops emphasized information sharing; researchers and key stakeholders presented background needed for discussing research goals and strategies. Ample time was allotted for question and answer, which delved into the workings of the modeling framework and the climate projections. At this stage the emphasis was on learning through interactive listening, with a goal of one-way information transfer. The model must be explained and understood before it can be usefully queried.

The second workshop session consisted of facilitated discussion in which stakeholders were presented with a series of open-ended questions about their groundwater management concerns, their expectations of the modeling framework, and what questions they would like the model to answer. Here the emphasis is on two-way communication in which knowledge pertinent the model's potential role in groundwater management is co-produced. GCASE analyses depended on this knowledge to structure relevant scenarios.

The workshop participants examined key aspects of the modeling framework, such as the use of the observed rainfall records and the nature of the uncertainty in projections. They were instrumental in developing a list of questions to explore in simulations based on assumptions regarding future conditions. These questions were used in the development of case study scenarios that provided a range of possible futures for groundwater use and recharge under current and future climate.

At the three location-specific (SCAMA) workshops, stakeholders raised questions that sought to explore the physical limits of the groundwater system, the range of climate impacts, and model capabilities. They were more reticent about suggesting potential water demands and allowable drawdown of the aquifers. There was little overt conflict over management goals, although opinions were divided on the interpretation of those goals as referring to maintenance of riparian vegetation. Comments and questions focused on potential uses of the model and the kinds of information it could provide. The possibility of significant change in water management in the Santa Cruz watershed in Mexico was raised, and it was admitted that this was a major uncertainty that could not be addressed in the GCASE project.

3.5. *Transferability*

An additional series of five workshops engaged stakeholders in discussions designed to investigate the transferability of the GCASE methodology to other locations and groundwater management contexts. The transferability workshops necessarily involved a broader set of stakeholders from outside the SCAMA. Some SCAMA workshop stakeholders attended a transferability workshop, while the majority of stakeholders at each of those workshops were new to the project. The new stakeholders possessed knowledge about the water management issues particular to each workshop region and were interested in the potential of the project methodology to be applied locally. The project team engaged organizations in each location to co-host the transferability workshops. The workshop co-hosted by the International Boundary and Water Commission was bilingual and involved stakeholders from Mexico. Agreeing to cohost demonstrated the interest of these organizations and provided a mechanism to reach more local stakeholders. Table 2 shows the location and co-host for each of the five transferability workshops.

Table 2. Locations and co-hosts, transferability workshops.

Location	Co-Host
Prescott, Arizona	Yavapai County Cooperative Extension
Phoenix, Arizona	Arizona Department of Water Resources
Tucson, Arizona	Pima Association of Governments
Sierra Vista, Arizona	Upper San Pedro Partnership
Rio Rico, Arizona	International Boundary and Water Commission

The five workshops employed formats similar to the SCAMA workshops. The first part of the workshop was devoted to presenting a description of the case study and its results. The case study presentation was revised following each workshop based on the participants' questions and comments. Stakeholder input was thereby used to increase the effectiveness of communicating the impacts of climate change on water resource planning, the capabilities of the modeling framework, and its potential for informing planning and management efforts. Questions of clarification were answered during the presentations, often leading to discussion of the modeling framework and climate projections.

Stakeholders were interested in the capability of the modeling framework to link climate projections with hydrological system responses. It was acknowledged that hydrologic projections based on historical data were not likely to reflect future conditions. The capability of the framework to capture the range of uncertainty was considered useful to water resource planning.

A short presentation on transferability criteria was followed by a discussion on the potential for transferability to the region. Participants engaged in assessing the potential according to the following five criteria: (1) local climate is a major factor in the state of the local water resources; (2) rainfall and streamflow are highly variable and difficult to predict; (3) future climate projections indicate increased variability and uncertainty; (4) informative datasets are available for the region; and (5) local agencies and stakeholders are engaged.

Stakeholder transferability concerns centered on whether the transferability criteria were met. In the greater Phoenix area, dependence on large regional aquifers means local climate plays a relatively minor role in the state of local groundwater resources. The lack of compatible data sets was raised by Mexican stakeholders, and a range of solutions were discussed involving additional data manipulation within the modeling framework. It was concluded that an increased level of stakeholder engagement may be needed where data are scarce.

A discussion of potential uses of the project methodology followed. Stakeholder demonstrated a grasp of the modeling framework's capabilities by suggesting feasible location-relevant applications. Many of these applications challenge the model to produce analyses different in kind from the SCAMA case study. For example, one suggestion was to use the model to investigate the impact of development on baseflow in a protected creek fed by shallow groundwater. Another suggestion identified the need to determine the impact of a new well field in an aquifer at the headwaters one of the few perennially flowing rivers in Arizona. These applications would necessitate modifications to the modeling framework, but they are well within its capabilities, assuming transferability criteria are met. It was acknowledged that transferring the methodology would require stakeholder engagement to drive integration of new questions into modifications of the modeling framework.

3.6. Outcomes

For the SCAMA, model outputs clarified groundwater management issues relating to aquifer resilience, potential for maximizing recharge, and increasing supply reliability. Results clearly showed greater uncertainty under future climate conditions that challenge reliability, increase groundwater deficits, and decrease recharge. Results also indicated, however, that a management strategy that included high pumping rates and deep pumping cessation thresholds could be employed to minimize these effects. The implications of this strategy, however, must be carefully considered in the context of SCAMA management goals.

Stakeholders demonstrated in their discussions an understanding of the results and their implications. Maximizing water supply from the microbasins would have impacts on infrastructure needs and the environment. Climate uncertainty, which exacerbates planning uncertainties, fueled the desire for better data and forecasts.

In follow-up discussions, local stakeholders agreed that access to the types of data used in the modeling framework would improve the capacity of local operators to manage groundwater withdrawals. Meetings with the SCAMA GUAC and other stakeholders were instrumental in conceptualizing a new project. Specifically designed to meet the needs of groundwater users in the SCAMA, the project will provide a web-based data service based on data analyses and visualization tools of pertinent local hydro-meteorological datasets that are updated in real time. An additional component of the web site will include analysis of the Climate Forecast System from the National Centers for Environmental Prediction to provide a seasonal forecast of rainfall. Stakeholders will take part in workshops to provide input on project development and products.

Transferability discussions produced a set of options for future work demonstrating the effectiveness of stakeholder engagement with the GCASE modeling framework. Stakeholder connections developed in the context of these discussions form a base for partnerships in linking scientific models with groundwater governance.

4. Discussion

The centrality of stakeholder engagement to the GCASE project was explicit from the beginning. Meadow *et al.* (2015) listed three principles of knowledge co-production: (1) building ongoing relationships between scientists and stakeholders; (2) ensuring two-way communication between scientists and stakeholders; and (3) maintaining a focus on usable science [38]. In this context 'usable science' refers to science produced in a form that can be used by stakeholders in their management and decision making roles. The project team employed an approach that emphasized these three principles.

Identifying stakeholders who are willing and able to contribute input requires knowledge of the context and will benefit from access to existing networks that can extend the limits of prior knowledge [53]. Stakeholder identification depends strongly on the context of the engagement process and should be sensitive to its specific topic and goals [54]. Stakeholders who are likely to participate in co-production processes, which may require repeated engagement, typically are recruited from the ranks of paid representatives of agencies and organizations with a stake in decisions and unpaid citizens with a deep interest in sustainable water management and related issues [55]. They tend to be connected through networks of interest that can be tapped through involvement of key known individuals.

It was a challenge to capture the attention of busy stakeholders for engagement in the GCASE project. It helped that the WRRC was known to many of the potential participant as a source for even handed facilitation and respectful interactions. More persistent personal communication was used to assemble the participants of the first workshop, which probably accounts for the high percentage of invitation acceptances—more than half. As we expanded the invitation list of potential participants, the acceptance rate dropped. A core group of stakeholders remained interested and engaged throughout the process. This core group consisted mainly of people whose professional positions involved management of water or related resources. An attempt was made to keep the project in the forefront of participants' minds by frequent communications about project progress and requests for input, but the development of high quality communications necessitated some time gaps. It was also important to consider the possibility of stakeholder fatigue.

A small group of key stakeholders can be useful as advisors when projects are linking stakeholders with research efforts. Their feedback can shape the design of stakeholder participation events and products meant to communicate results to the wider stakeholder group [44]. The choice of these key stakeholders may be based on existing relationships and may be made to ensure participation of key organizations and/or representatives with important content knowledge [56].

The choice of Advisory Committee members was based both on existing relationships and on their possession of key expertise. The extent of their engagement was directly related to their involvement in water management in the case study region and previous experience with the modeling framework. Their guidance remained important throughout the project, although emails tended to replace face-to-face and telephone meetings as the project progressed. Their assistance was also critical in framing discussions of transferability.

Organizations and agencies that are represented in a co-production process are committing resources and therefore they should see in the process a connection to their missions. Institutional support through the creation and communication of information and through participation in co-production dialogues can be instrumental in the success of these efforts. Without such support, the process may be jeopardized. The engagement of both ADWR and the City of Nogales was instrumental in linking GCASE to practical groundwater management considerations. They contributed staff time and expertise to the advisory committee, workshops, and other stakeholder communications. Their involvement could be considered intrinsic to the project, as both were already vested in development and application of the modeling framework at the inception of the GCASE project.

Communication between scientists and research users can be challenging when the parties are not accustomed to dialogue. Plain language should be used to communicate complex information [53]. In the GCASE project, communication issues were specifically addressed in workshop discussions. The meaning of key terms and their translations were agreed on through a consensus process. Trained facilitators led group discussions with a goal of promoting general understanding. At one point an online dialogue capability was suggested, so that the conversation could continue between workshops. Such a capability was developed, and all the participants were invited to join. However, it remained unused, perhaps because it required registration, but more likely because stakeholders were too busy. The capability was passive and not designed to push information to the stakeholders and no one was generating content regularly to spur responses. Workshops provided time to focus on GCASE related issues, engendering discussion.

People who are skilled and trusted as translators between experts and non-experts can be needed to close the gaps in understanding [43,53]. Translation proved an important skill for making sure the presentations of project scientists communicated effectively to other stakeholders. GCASE explained and promoted understanding among participants of climate uncertainty, likely future climate conditions, and the operation of the microbasin system as simulated by the modeling framework. It was important that everyone understand how model analyses were produced and what they meant. By pre-screening presentations, it was possible to improve their ability to communicate. Perhaps surprisingly, there was no resistance among the presenters to critiques that focused on communication to stakeholders. There was, however, in-depth discussion on how to translate the technical content without losing scientific integrity. An informal evaluation indicated that participants were satisfied with results.

Organizations that facilitate translation of science to use typically provide expertise that neither the research user nor the research provider possesses [57]. Thus, convening and nurturing stakeholder engagement is often best carried out by such an organization, which can tailor the processes and tools to specifics of the issue context [43]. Tailoring engagement processes and tools 3has been demonstrated to improve participation and outcomes [54]. The University of Arizona Water Resources Research Center provided these functions as a bridging organization. The WRRC brought together expertise in the modeling framework and climate science with expertise in the groundwater management context and the knowledge of concerned stakeholders for open exchange. Efforts to engage stakeholders and to keep them engaged involved frequent individual communication. Stakeholder comments and suggestions, not just from the advisory committee guided design of communications and workshop agendas. Facilitation during workshops afforded balance and direction to discussions.

In the co-production process, certain types of exchanges between scientists and other actors have been associated with successful stakeholder engagement. For example, Phillipson *et al.* found research was more likely to have an impact on stakeholders who participated in events, such as workshops, where they provided feedback on research findings [24].

Workshops were designed with the intension of providing in-depth background information first, in order for stakeholders to be able to discuss use of the model on the basis of shared knowledge. The focus of the background presentations necessarily shaped the following discussions. In addition, facilitators were prepared with questions to prompt discussion that also focused attention on project goals. These factors may have limited the range of topics discussed, but helped elicit information about stakeholder concerns specifically relevant to model scenario development and analysis. Facilitation also encouraged participation.

5. Conclusions

The GCASE case study provides one example of how knowledge co-production can be achieved through use of a hydrological model when two-way interactions are focused on development of usable science. Best practices in stakeholder engagement were employed in the GCASE project, which sought to enable decisions relating to the unique SCAMA groundwater management goals. The modeling framework, which was the focus of stakeholder engagement, provided the capability to integrate climate-related uncertainty in groundwater management decision processes. Stakeholders possessed context specific knowledge necessary for model analyses, and scientists acknowledged the need to educate stakeholders and to learn from them. A bridging organization ensured that stakeholders and scientists engaged in two-way communication regarding the use of the modeling framework.

The hydrologic modeling framework was developed in previous work for the SCAMA involving collaboration with ADWR, which has its own public participation policies, but without a well-defined process for stakeholder engagement. Its selection, enhancement, and use with stakeholders were based on the previously perceived need to account for climate uncertainty in simulations of groundwater behavior. The use of an existing modeling framework specifically developed for the study region focused stakeholder involvement on defining issues for analysis and scenarios for model simulation.

A statewide moratorium on rulemaking created a disconnection between the GCASE results and groundwater management decision making. Thus, stakeholders focused their attention of developing an understanding of model capabilities, climate change implications, and groundwater management. Although workshops in which scientists and stakeholders co-produced system knowledge advanced common understanding of these issues, rulemaking was not directly affected. Intimate involvement of ADWR personnel, however, aided communication among scientists, stakeholders and decision makers that focused interest on the potential of the project approach. The web-based data service being developed through a follow-on project resulted directly from knowledge co-production with influential stakeholders, specifically members of the SCAMA GUAC and ADWR modelers.

Workshops presented positive evidence that stakeholders and scientists who engage in facilitated two-way conversations can co-produce useful knowledge if they adhere to principles of respectful relations, dialogue, and focus. It was evident from discussions that stakeholder grasped the implications for groundwater managements of model outputs and the scientists grasped the implications for their research of stakeholder input. Transferability discussions produced suggestions for modifications of the modeling framework to fit new conditions and management issues.

The GCASE case study demonstrated the usefulness of best practices as offered by the literature on stakeholder engagement. Pre-existing knowledge of the stakeholders and their context can help tailor general guidance to a specific case. Personal contacts are useful in engaging key stakeholders and should be cultivated over time. Bridging organizations with a history of cultivating stakeholder contacts can supply the context knowledge and personal contact. Such organizations also can monitor and improve communications that are relevant to a co-production project.

The case study also illustrated that construction of stakeholder engagement around published strategies for successful co-production of knowledge can produce joint learning. Seeking the advice of key stakeholders, attention to stakeholder networks, and ensuring that diverse knowledge are represented all contributed positively to GCASE outputs. Important activities that expanded the toolbox of stakeholder communication included negotiating meaning and previewing workshop presentations. The productivity of workshop dialogue on the model and its use was enhanced by skilled facilitation, which ensured that model analyses were responsive to stakeholder queries. Finally, transferability assessments established the potential for future GCASE style projects.

Over the long term, however, efforts at knowledge co-production may be limited by resources and time constraints. Bridging organizations with established connections to the stakeholder community can maintain relationships and trust built up over the course of successful co-production processes as a basis for other projects. Results of individual projects ideally have a long-lasting impact on the resolution of groundwater management issues; however, widespread understanding and acceptance still depends on maintaining communication with a cast of interested parties likely to change over time. Maintaining communication may depend on structural support for long-term engagement mechanisms.

Acknowledgments: The authors would like to acknowledge the support of the NOAA Climate and Societal Interactions Sectoral Applications Research Program (NOAA Award NA12OAR4310099) for the GCASE project. Funding for this effort was also provided by the University of Arizona Technology and Research Initiative Fund. Additional support comes from the Arizona Department of Water Resources (Contract No: 2015-2994IGA, Title: Optimization of the water resources management in the Upper Santa Cruz River) Santa Cruz AMA Augmentation and Conservation Assistance Fund. The authors would also like to thank the members on the Project Advisory Committee: Alejandro Barcenas, Greg Kornrumph, Michael Lacey, and Jim Leenhouts, for their contributions and commitment, as well as all the stakeholders who participated in the GCASE project. Special thanks go to Jacob Prietto, who as the GCASE Graduate Outreach Assistant was instrumental to the success of the project's stakeholder engagement component, and to Nathaniel Delano, WRRC Research Analyst, for editorial assistance.

Author Contributions: Sharon B. Megdal and Eylon Shamir conceived and designed the project upon which this paper is based; Sharon B. Megdal, Eylon Shamir, Susanna Eden, and Karletta Chief carried out the project; Sharon B. Megdal provided project oversight; Eylon Shamir led the modeling effort; Susanna Eden implemented the stakeholder engagement; Karletta Chief contributed to performance of the stakeholder engagement; Susanna Eden wrote the paper with significant contributions from Kelly Mott Lacroix, Eylon Shamir, Sharon B. Megdal, and Karletta Chief.

Conflicts of Interest: The authors declare no conflicts of interest. The founding sponsors had no role in the design of the study; in the collection, analyses, or interpretation of data; in the writing of the manuscript, and in the decision to publish the results.

Abbreviations

The following abbreviations are used in this article:

ADWR	Arizona Department of Water Resources
AMA	Active Management Areas
GCASE	Groundwater, Climate, and Stakeholder Engagement
GUAC	Groundwater Users Advisory Councils
HRC	Hydrologic Research Center
NIWTP	Nogales International Wastewater Treatment Plant
SCAMA	Santa Cruz Active Management Area
WRRC	University of Arizona Water Resources Research Center

References

1. Tortajada, C. Water governance: Some critical issues. *Int. J. Water Resour. Dev.* **2010**, *26*, 297–307. [CrossRef]
2. Megdal, S.B.; Gerlak, A.K.; Varady, R.B.; Huang, L.-Y. Groundwater governance in the United States: Common priorities and challenges. *Groundwater* **2014**, *53*, 677–684. [CrossRef] [PubMed]

3. Peters, G.; Pierre, J. Governance without government? Rethinking public administration. *J. Public Adm. Res. Theory* **1998**, *8*, 223–243. [CrossRef]
4. Bingham, L.B.; Nabatchi, T.; O'Leary, R. The new governance: Practices and processes for stakeholder and citizen participation in the work of government. *Public Adm. Rev.* **2005**, *65*, 547–558. [CrossRef]
5. Pahl-Wostl, C.; Mostert, E.; Tàbara, D. The growing importance of social learning in water resources management and sustainability science. *Ecol. Soc.* **2008**, *13*, 24.
6. Lautze, J.; De Silva, S.; Giordano, M.; Sanford, L. Putting the cart before the horse: Water governance and IWRM. *Nat. Resour. Forum* **2011**, *35*, 1–8. [CrossRef]
7. Groundwater Governance. Global Framework for Action to Achieve the Vision on Groundwater Governance. 6 March 2015. Available online: http://www.groundwatergovernance.org/fileadmin/user_upload/groundwatergovernance/docs/general/GWG_FRAMEWORK.pdf (accessed on 9 May 2016).
8. Jolibert, C.; Wesselink, A. Research impacts and impact on research in biodiversity conservation: The influence of stakeholder engagement. *Environ. Sci. Policy* **2012**, *22*, 100–111. [CrossRef]
9. Voinov, A.; Brown Gaddis, E.J. Lessons for successful participatory watershed modeling: A perspective from modeling practitioners. *Ecol. Model.* **2008**, *216*, 197–207. [CrossRef]
10. Pahl-Wostl, C.; Craps, M.; Dewulf, A.; Mostert, E.; Tabara, D.; Taillieu, T. Social learning and water resources management. *Ecol. Soc.* **2007**, *12*, 5.
11. Antunes, P.; Giorgos Kallis, N.V.; Santos, R. Participation and evaluation for sustainable river basin governance. *Ecol. Econ.* **2009**, *68*, 931–939. [CrossRef]
12. Organization for Economic Cooperation and Development. Stakeholder Engagement for Inclusive Water Governance. OECD Studies on Water. 2015. Available online: http://www.oecd-ilibrary.org/governance/stakeholder-engagement-for-inclusive-water-governance_9789264231122-en (accessed on 9 May 2016).
13. Focht, W.; Trachtenbuerg, Z. A trust based guide to stakeholder participation. In *Swimming Upstream: Collaborative Approaches to Watershed Management*; Sabatier, P., Focht, W., Lubell, M., Trachtenberg, Z., Vedlitz, A., Matlock, M., Eds.; MIT Press: Cambridge, MA, USA, 2005.
14. Shamir, E.; Megdal, S.B.; Carrillo, C.; Castro, C.L.; Chang, H.; Chief, K.; Corkhill, F.; Eden, S.; Georgakakos, K.P.; Nelson, K.M. Climate change and water resources management in the upper Santa Cruz River, Arizona. *J. Hydrol.* **2015**, *521*, 18–33. [CrossRef]
15. Shamir, E.; Graham, N.E.; Meko, D.M.; Georgakakos, K.P. Hydrologic model for water resources planning in Santa-Cruz River, Southern Arizona. *J. Am. Water Resour. Assoc.* **2007**, *43*, 1155–1170. [CrossRef]
16. Sarewitz, D.; Pielke, R.A. The neglected heart of science policy: Reconciling supply of and demand for science. *Environ. Sci. Policy* **2007**, *10*, 5–16. [CrossRef]
17. McNie, E.C. Reconciling the supply of scientific information with user demands: An analysis of the problem and review of the literature. *Environ. Sci. Policy* **2007**, *10*, 17–38. [CrossRef]
18. Hegger, D.; Lamers, M.; van Zeijl-Rozema, A.; Dieperink, C. Conceptualising joint knowledge production in regional climate change adaptation projects: Success conditions and levers for action. *Environ. Sci. Policy* **2012**, *18*, 52–65. [CrossRef]
19. Michaels, S. Matching knowledge brokering strategies to environmental policy problems and settings. *Environ. Sci. Policy* **2009**, *12*, 994–1011. [CrossRef]
20. Roux, D.; Rogers, K.; Biggs, H.; Ashton, P.; Sergeant, A. Bridging the science–management divide: Moving from unidirectional knowledge transfer to knowledge interfacing and sharing. *Ecol. Soc.* **2006**, *11*, 4.
21. Lemos, M.C.; Morehouse, B.J. The co-production of science and policy in integrated climate assessments. *Glob. Environ. Chang.* **2005**, *15*, 57–68. [CrossRef]
22. Mitchell, M.; Curtis, A.; Sharp, E.; Mendham, E. Directions for social research to underpin improved groundwater management. *J. Hydrol.* **2012**, *448*, 223–231. [CrossRef]
23. Landry, R.; Amara, N.; Ouimet, M. Determinants of knowledge transfer: Evidence from Canadian University Researchers in Natural Sciences and Engineering. *J. Technol. Transf.* **2007**, *32*, 561–592. [CrossRef]
24. Phillipson, J.; Lowe, P.; Proctor, A.; Ruto, E. Stakeholder engagement and knowledge exchange in environmental research. *J. Environ. Manag.* **2012**, *95*, 56–65. [CrossRef] [PubMed]
25. Mitchell, M.; Curtis, A.; Sharp, E.; Mendham, E. *Social Research to Improve Groundwater Governance: Literature Review*; Institute for Land, Water and Society, Charles Stuart University: Albury, Australia, 2011; Available online: http://athene.riv.csu.edu.au/~{}acurtis/reports/66_Groundwater_lit_review_Nov_%202011.pdf (accessed on 9 May 2016).

26. Voinov, A.; Bousquet, F. Modelling with stakeholders. *Environ. Model. Softw.* **2011**, *25*, 1268–1281. [CrossRef]
27. Beall, A.M.; Ford, A. Reports from the field: Assessing the art and science of participatory environmental modeling. *Int. J. Inf. Syst. Soc. Chang.* **2010**, *1*, 72–89. [CrossRef]
28. Crona, B.I.; Parker, J.N. Learning in support of governance: Theories, methods, and a framework to assess how bridging organizations contribute to adaptive resource governance. *Ecol. Soc.* **2012**, *17*. [CrossRef]
29. Sternlieb, F.; Bixler, R.P.; Huber-Stearns, H.; Huayhuaca, C. A question of fit: Reflections on boundaries, organizations and social–ecological systems. *J. Environ. Manag.* **2013**, *130*, 117–125. [CrossRef] [PubMed]
30. Bartels, W.; Furman, C.A.; Royce, F.; Ortiz, B. Developing a Learning Community: Lessons from a Climate Working Group for Agriculture in the Southeast USA. 2012. Available online: http://www.seclimate.org/wp-content/uploads/2014/07/2012-bartels-et-al-secc-technical-series-developing-a-learning-community.pdf (accessed on 9 May 2016).
31. Biggs, R.; Westley, F.R.; Carpenter, S.R. Navigating the back loop: Fostering social innovation and transformation in ecosystem management. *Ecol. Soc.* **2010**, *15*, 9.
32. Guston, D.H. Boundary organizations in environmental policy and science: An introduction. *Sci. Technol. Hum. Values* **2001**, *26*, 399–408. [CrossRef]
33. Brown, L.D. Bridging organizations and sustainable development. *Hum. Relat.* **1991**, *44*, 807–831. [CrossRef]
34. Jacobs, K.; Lebel, L.; Buizer, B.; Addams, L.; Matson, P.; McCullough, E.; Garden, P.; Saliba, S.; Finan, T. Knowledge systems for sustainable development special feature sackler colloquium: Linking knowledge with action in the pursuit of sustainable water-resources management. *Proc. Natl. Acad. Sci. USA* **2010**. [CrossRef]
35. Miller, C. Hybrid management: Boundary organizations, science policy, and environmental governance in the climate regime. *Sci. Technol. Hum. Values* **2001**, *26*, 478–500. [CrossRef]
36. Carr, A.; Wilkinson, R. Beyond participation: Boundary organizations as a new space for farmers and scientists to interact. *Soc. Nat. Resour.* **2005**, *18*, 255–265. [CrossRef]
37. Berkes, F. Evolution of co-management: Role of knowledge generation, bridging organizations and social learning. *J. Environ. Manag.* **2009**, *90*, 1692–1702. [CrossRef] [PubMed]
38. Meadow, A.M.; Ferguson, D.F.; Guido, Z.; Horangic, A.; Owen, G.; Wall, T. Moving toward the deliberate coproduction of climate science knowledge. *Weather Clim. Soc.* **2015**, *7*, 179–191. [CrossRef]
39. Nelson, K. Risk Analysis of Pumping Impacts on Simulated Groundwater Flow in the Santa Cruz Active Management Area. Modeling Report No. 21. 2010. Available online: http://www.azwater.gov/AzDWR/Hydrology/Library/documents/Modeling_Report_21.pdf (accessed on 9 May 2016).
40. Shamir, E.; Georgakakos, K.P.; Graham, N.E.; Wang, J.; Meko, D.M. Generation and Analysis of Likely Hydrologic Scenarios for the Southern Santa Cruz River. Hydrologic Research Center Technical Report No. 4. 2005. Available online: http://www.hrc-lab.org/projects/projectpdfs/SANTACRUZ_REPORTS/TR4_01-10-06.pdf (accessed on 20 May 2016).
41. Shamir, E.; Wang, J.; Georgakakos, K.P. Probabilistic streamflow generation model for data sparse arid watersheds. *J. Am. Water Resour. Assoc.* **2007**, *43*, 1142–1154. [CrossRef]
42. Langsdale, S.; Beall, A.; Bourget, E.; Hagen, E.; Kudlas, S.; Palmer, R.; Tate, D.; Werick, W. Collaborative modeling for decision support in water resources: Principles and best practices. *J. Am. Water Resour. Assoc.* **2013**, *49*, 629–638. [CrossRef]
43. Glicken, J. Getting stakeholder participation 'right': A discussion of participatory processes and possible pitfalls. *Environ. Sci. Policy* **2000**, *3*, 305–310. [CrossRef]
44. Lacroix, K.E.M.; Megdal, S.B. Explore, synthesize, and repeat: Unraveling complex water issues through the stakeholder engagement wheel. *Water* **2016**, *8*. [CrossRef]
45. Arizona Department of Water Resources. Demand and Supply Assessment 1985–2025 Santa Cruz AMA. 2006. Available online: http://www.azwater.gov/AzDWR/WaterManagement/Assessments/documents/SCAMA_AssessmentSummarySheet.pdf (accessed on 9 May 2016).
46. Norman, L.M.; Villarreal, M.L.; Niraula, R.; Meixner, T.; Frisvold, G.; Labiosa, W. Framing scenarios of binational water policy with a tool to visualize, quantify and valuate changes in ecosystem services. *Water* **2013**, *5*, 852–874. [CrossRef]
47. Arizona Revised Statutes § 45-562.

48. Corkhill, F.; Dubas, L. Analysis of Historic Water Level Data Related to Proposed Assured Water Supply Physical Availability Criteria for the Santa Cruz Active Management Area Santa Cruz and Pima Counties, Arizona. Modeling Report No. 18. 2007. Available online: http://www.azwater.gov/AzDWR/Hydrology/Modeling/documents/Modeling_Report_18.pdf (accessed on 9 May 2016).
49. Arizona Department of Water Resources. Santa Cruz AMA. 2015. Available online: http://www.azwater.gov/AzDWR/WaterManagement/AMAs/SantaCruzAMA/default.htm (accessed on 1 June 2015).
50. Erwin, G. Groundwater Flow Model of the Santa Cruz Active Management Area Microbasins International Boundary to Nogales International Wastewater Treatment Plant Santa Cruz County, Arizona. Modeling Report #15. 2007. Available online: http://www.azwater.gov/AzDWR/Hydrology/Modeling/documents/Modeling_Report_15.pdf (accessed on 22 February 2016).
51. Nelson, K. Groundwater Flow Model of the Santa Cruz Active Management Area along the Effluent-Dominated Santa Cruz River, Santa Cruz and Pima Counties, Arizona. Modeling Report No. 14. 2007. Available online: http://www.azwater.gov/AzDWR/Hydrology/Modeling/documents/Modeling_Report_15.pdf (accessed on 22 February 2016).
52. Fischer, A.R.H.; Wentholt, M.T.A.; Rowe, G.; Frewer, L.J. Expert involvement in policy development: A systematic review of current practice. *Sci. Public Policy* **2014**, *41*, 332–343. [CrossRef]
53. Gardner, J.; Dowd, A.; Mason, C.; Ashworth, P. A Framework for Stakeholder Engagement on Climate Adaptation. CSIRO Climate Adaptation Flagship Working Paper No. 3. 2009. Available online: http://ipcc-wg2.gov/njlite_download2.php?id=8149 (accessed on 9 May 2016).
54. Baldwin, C.; Tan, P.; White, I.; Hoverman, S.; Burry, K. How scientific knowledge informs community understanding of groundwater. *J. Hydrol.* **2012**, *474*, 74–83. [CrossRef]
55. Fung, A. Varieties of participation in complex governance. *Public Adm. Rev.* **2006**, *66*, 66–75. [CrossRef]
56. Keown, K.; Van Eerd, D.; Irvin, E. Stakeholder engagement opportunities in systematic reviews: Knowledge transfer for policy and practice. *J. Contin. Educ. Health Prof.* **2008**, *28*, 67–72. [CrossRef] [PubMed]
57. Eden, S. Lessons on the generation of usable science from an assessment of decision support practices. *Environ. Sci. Policy* **2011**, *14*, 11–19. [CrossRef]

water

MDPI

Article

Exploring the Potential Impact of Serious Games on Social Learning and Stakeholder Collaborations for Transboundary Watershed Management of the St. Lawrence River Basin

Wietske Medema [1,*], Alison Furber [1], Jan Adamowski [1], Qiqi Zhou [2] and Igor Mayer [3]

[1] Department of Bioresource Engineering, McGill University, 21 111 Lakeshore, Ste Anne de Bellevue, QC H9X3V9, Canada; alison.furber@mail.mcgill.ca (A.F.); jan.adamowski@mcgill.ca (J.A.)

[2] TIAS School for Business and Society, Kromme Nieuwegracht 39, Utrecht 3512 HD, The Netherlands; qiqi.zhou@tias.edu

[3] Academy for Digital Entertainment, NHTV Breda University of Applied Sciences, Monseigneur Hopmansstraat 1, Breda 4817 JT, The Netherlands; i.s.mayer@hotmail.com

* Correspondence: wietske.medema@mcgill.ca; Tel.: +1-514-398-7786

Academic Editors: Sharon B. Megdal, Susanna Eden and Eylon Shamir
Received: 20 February 2016; Accepted: 7 April 2016; Published: 28 April 2016

Abstract: The meaningful participation of stakeholders in decision-making is now widely recognized as a crucial element of effective water resource management, particularly with regards to adapting to climate and environmental change. Social learning is increasingly being cited as an important component of engagement if meaningful participation is to be achieved. The exact definition of social learning is still a matter under debate, but is taken to be a process in which individuals experience a change in understanding that is brought about by social interaction. Social learning has been identified as particularly important in transboundary contexts, where it is necessary to reframe problems from a local to a basin-wide perspective. In this study, social learning is explored in the context of transboundary water resource management in the St. Lawrence River Basin. The overarching goal of this paper is to explore the potential role of serious games to improve social learning in the St. Lawrence River. To achieve this end, a two-pronged approach is followed: (1) Assessing whether social learning is currently occurring and identifying what the barriers to social learning are through interviews with the region's water resource managers; (2) Undertaking a literature review to understand the mechanisms through which serious games enhance social learning to understand which barriers serious games can break down. Interview questions were designed to explore the relevance of social learning in the St. Lawrence River basin context, and to identify the practices currently employed that impact on social learning. While examples of social learning that is occurring have been identified, preliminary results suggest that these examples are exceptions rather than the rule, and that on the whole, social learning is not occurring to its full potential. The literature review of serious games offers an assessment of such collaborative mechanisms in terms of design principles, modes of play, and their potential impact on social learning for transboundary watershed management. Serious game simulations provide new opportunities for multidirectional collaborative processes by bringing diverse stakeholders to the table, providing more equal access to a virtual negotiation or learning space to develop and share knowledge, integrating different knowledge domains, and providing opportunities to test and analyze the outcomes of novel management solutions. This paper concludes with a discussion of how serious games can address specific barriers and weaknesses to social learning in the transboundary watershed context of the St. Lawrence River Basin.

Keywords: social learning; stakeholder collaborations; transboundary water governance; watershed management; serious games

1. Introduction

A significant portion of the world's watersheds straddle jurisdictional boundaries where political and hydrological boundaries often do not correspond [1]. Particular challenges relating to resource management in such transboundary contexts include differing legal and regulatory frameworks, languages, economic conditions, political history and antecedents with public participation, and stakeholder organizations [2]. When looking at the current state of Canada's governing structure, as an example, it shows a division and fragmentation between jurisdictions of federal, provincial, territorial and the First Nations governments [3]. In Canada, provincial governments have jurisdiction over water as a resource, while the federal government mainly has jurisdictional authority over water issues and conflicts with the US [4]. This fragmentation has resulted in overlapping resource management policies that do not work together while creating transboundary issues, as well as a decreased water quality of Canadian watersheds [5]. In addition to this challenge of fragmentation, many environmental government authorities are experiencing various budgetary reductions, which is devolving the national- and provincial-level capacity to address water resource issues [4].

To ensure that the water governance system in Canada can cope given the challenges it is facing, including increasing pressures on water resources from highly urbanized and agricultural developments, the contamination of both surface and groundwater bodies, and climate change, a transition from current management regimes to a coordinated sharing of power and responsibilities, and partnerships between public, private and civil society stakeholders is necessary [6]. Overcoming transboundary water governance differences requires problems to be reframed from a local to a basin-wide perspective [7–9]. It also requires more inclusive forms of watershed management that seeks to increase the participation and engagement of a more diverse group of stakeholders [10]. Strong platforms for collaboration and interaction that enhance communications, information sharing, negotiations and knowledge co-creation, are mechanisms essential to facilitating positive relationships and dynamics between diverse stakeholders [7].

Social learning is becoming a "normative goal" for water governance and a key for addressing transboundary water issues [7–9]. According to Reed *et al.* [11], social learning refers to learning achieved by the social group as a whole, and involves a process in which individuals experience a change in understanding that has a social dimension in that it is brought about by social interaction and is linked to wider communities of practice. Mostert *et al.* [8] identify a number of components of collaboration that follow a social learning approach. In particular, they highlight the importance of recognizing mutual dependence and the role that trust plays. They emphasize the need to interact, share problem perceptions and develop alternative solutions. Lastly, they talk of the need for stakeholders to engage in collaborative decision making processes and jointly implement agreed actions. Mostert *et al.* [8] takes a broader perspective on social learning than Reed *et al.* [11] by recognizing mutual dependence and trust as prerequisites, while conceptualizing stakeholder interactions, the sharing of problem perceptions, and the development of alternative solutions as actions indicative of underlying social learning processes. Collaborative decision-making processes and the joint implementation of agreed actions are both prerequisites to, and potential desirable outputs of, social learning processes [12,13]. The value of this broader perspective is that, while assessing the extent to which internal change in understanding has taken place is complex and difficult to measure, external factors such as the extent of interaction and the degree to which problem perspectives are shared offer more tangible substitutes.

The goal of this paper is to assess the effectiveness of social learning practices for transboundary management within the St. Lawrence River watershed in North America and identify opportunities for enhancing social learning. This watershed faces a particularly complex set of transboundary challenges: jurisdiction is shared between two countries, several Canadian provinces, and numerous local governments [3], which leads to the involvement of a diverse range of stakeholders (e.g., the general public, industry, farmers, small business owners, civil servants and politicians of various levels). Such challenges require novel tools and methods to encourage dialogue, facilitate learning and enable systemic institutional change, so that decisions can be made from a position of shared knowledge and understanding. Serious games offer innovative solutions for enhancing learning and collaborations to overcome governance challenges [8,12,13]. These types of games combine computer simulation with role-play as an integrated method for complex policy making while triggering discussion and learning among stakeholders [14]. Serious games involve simulations of real-world events or processes designed for the purpose of solving contemporary societal challenges, and these games are therefore designed for a purpose beyond entertainment. Serious games have been successfully used in other sectors (e.g., education, military and health) and are now beginning to be explored in the water sector [15,16].

The following section will provide a literature review of the conditions and factors that are required for the facilitation of social learning and multi-stakeholder collaborations in a transboundary watershed context. Additionally, a review is offered through which serious games are assessed in terms of design principles, modes of play, and their potential impact on social learning for transboundary watershed governance. Following this, the methodology is presented, to achieve the research goal, in-depth interviews have been conducted with 10 of the region's key water resources managers. This paper concludes with a discussion of the study results and findings.

2. Social Learning and Serious Gaming

2.1. Social Learning for Transboundary Watershed Governance

A management paradigm called Adaptive Management was proposed in the 1970s to guide the management of resources and bring learning to the fore as a central tenant of effective management and to encourage resource managers to take steps to learn about the processes which govern a certain system [17–20]. With the need for more participatory forms of water resources management over the past decade, the Adaptive Management approach has been broadened with learning now commonly taken to be an essential social process [17]. As an example, Stringer *et al.* [21] describe Adaptive Management as a paradigm that "treats knowledge about ecosystems as both uncertain and pluralistic while recognizing that in order to create more sustainable management strategies, stakeholders must forge new relationships to enhance multi-directional information flows, learn from each other, and together develop flexible ways of managing their environment".

Multi-faceted issues associated with transboundary watershed governance are problematic due to diverse management regimes and potential asymmetries in resources, political structures, government agencies and institutions [22]. In order to achieve more adaptive and integrated forms of resource management, it is important to acknowledge that boundaries are always present and that collaborations across such boundaries are essential [8,23–25]. Such boundaries not only involve physical boundaries (e.g., between surface and groundwater, water quantity and quality, freshwater and coastal waters, water resources and land resources, different geographical scales) or jurisdictional boundaries (e.g., between different countries, government levels, policy sectors), but also boundaries of a social (between different social and economic groups, and between these groups and government) as well as a cognitive nature (between different disciplines and expertise). Social learning is recognized as a mechanism to support the facilitation of collaboration and interaction between stakeholders across these boundaries [8,12,13]. While there is not one stakeholder that carries all legal competencies, funds, information and other required resources to manage water issues, it is important for these parties to pool resources and "learn together to manage together" [8].

Such social learning processes take place in both a social context (*i.e.*, the governance system, economy and culture) as well as a natural context (*i.e.*, the hydrological and geographical conditions) that together play a crucial role in determining who the key stakeholders are and what they see as the key management issues. Social learning implies that these different stakeholders learn to resolve these issues through social interactions and relationships in order to come to a shared understanding of the problems at stake and the system to be managed, as well as to agree on a solution while ensuring that this solution is implemented [8]. The term "stakeholders" for this study encompasses parties who hold power and knowledge to influence the outcome of resource management decisions, and those affected by the actions taken [26]. Stakeholder participation is considered a critical component for transboundary watershed governance and the facilitation of social learning, and for this study it is conceptualized for this study according to typologies that distinguish between the degree to which stakeholders are engaged [27]. In this light, Arnstein [28] recognizes that stakeholder participation efforts vary in the amount of control given to stakeholders and identified different "rungs" on a "ladder of stakeholder participation" characterized by increasing stakeholder power [28]. This author developed different typologies based on these "rungs" to distinguish between the degree of stakeholder engagement. More recently, Chase *et al.* [29] outlined a similar range of approaches to stakeholder participation for the context of natural resources management, and argued that these approaches form a logical continuum over which the relative influence of citizens and agencies over management varies—from total agency control under the expert authority approach to broad power-sharing under co-management [29]. The degree of stakeholder engagement is conceptualized by these authors according to five typologies [30]:

1. *Expert authority*—agencies retain full responsibility for decision-making and do not consider stakeholder input during decision-making;
2. *Passive-receptive*—agencies consider stakeholder input, but do not actively seek it out;
3. *Inquisitive*—agencies make systematic attempts to gather stakeholder input while retaining authority for deciding how to weigh this information;
4. *Transactional*—agencies facilitate a process in which stakeholders work together to try to reach agreement on the best management decision, and;
5. *Co-management*—agencies work with stakeholders in partnership and involve them throughout the management process.

Co-management has been emphasized by Raadgever *et al.* [31] as being crucial for effective transboundary watershed governance where strong stakeholder networks must be in place that foster stakeholder participation and collaboration. Edelenbos *et al.* [32] highlight how such stakeholder networks are subsequently key in facilitating: the generation, acquisition, and diffusion of different knowledge types; multi-directional dialogue and interaction; shared understanding and sense making; mobilization and allocation of key resources for effective governance; commitment to common rules; willingness to engage in collaborative processes; and mutual trust and conflict resolution. Muro and Jeffrey [33] also provide a summary of conditions for promoting social learning based on many studies exploring the relationship between stakeholder engagement processes and outcomes. These conditions, as well as those identified by a number of other scholars that have studied different categories and types of collaborative network conditions in the context of water governance [32–43], are synthesized for this study into four key categories: (1) *properties of stakeholders*: the characteristics of stakeholders who are participating in collaborative networks; (2) *properties of collaborative processes*: the way interactions, communication and collaboration are organized; (3) *properties of relationships*: the quality of the relationships between members of a network; and (4) *properties of knowledge*: the properties of knowledge and understanding that is developed and shared within and between networks.

In assessing the effectiveness of social learning processes, it is complex and difficult to assess the extent to which *internal* changes in stakeholder understanding, norms and values have taken place as a result of social learning processes. For this reason, this paper focuses primarily on studying more tangible and *external* elements that effective social learning processes have in common. This study will look primarily at the characteristics of stakeholders participating in social learning processes (such as stakeholder preparedness to participate, and their available resources), the extent of stakeholder interactions and the way this engagement process is organized (such as stakeholders' recognition of their interdependence, and ongoing critical reflection), the quality of relationships between stakeholders and their attitudes towards one another (involving repeated interaction, mutual trust, respecting diversity, and the degree to which problem perspectives are shared). These factors are summarized in Table 1 below and form the foundation for the acquisition and analysis of research data for this study.

Table 1. Key elements of social learning processes.

Elements of Social Learning	References
Characteristics of stakeholders	
Stakeholder preparedness to participate	[35,37,39,44]
Available resources (e.g., facilities, organizational forms and competences)	[36,37,41]
Critical self-reflection on positions and goals	[8]
Stakeholder interactions and the way the engagement process is organized	
Involvement of crucial stakeholders	[32,34,37,39,44]
Including and respecting diverse interests, views and information	[8,32–34,36,37,42,43,45]
Development and assessment of potential solutions	[8]
Joint decision-making and implementation through open communication and interaction	[8,33,45]
Exchange of problem perspectives, knowledge and information	[8,33,36,37,46]
Ongoing reflection on positions, perspectives and goals	[8,33,34,36,47,48]
Quality of stakeholder relationships	
Development of strong network ties through prolonged and frequent interactions	[8,33,38,39,42,48,49]
Recognition of interdependence and shared goals	[8,34,42]
Mutual trust and commitment between stakeholders	[8,33,35,42]

2.2. *Serious Game Play for Social Learning*

Social learning may be viewed as an approach for analyzing and promoting stakeholder collaboration [8]. While social learning involves the crossing of boundaries and bridging of differences, it often requires some form of external facilitation that supports the social process and constructive interactions between stakeholders [8,12,13,50]. Mostert *et al.* [8] highlight that framing and re-framing processes are central to social learning, and emphasize the importance of re-framing from a common frame in transboundary watershed management of "sharing scarce water resources" to an alternative frame of "benefit sharing" that focuses on the benefits that may be derived from water resources, while turning transboundary watershed management into a win-win game [51,52]. Numerous methods to foster learning through collaborative and multi-stakeholder participation have been developed and implemented, from modeling and computer simulation to policy interventions such as panels, workshops and process management [14]. One promising category involves serious games that have the ability to encourage social learning. The concept of simultaneously addressing the techno-physical complexities of a system (*i.e.*, the underlying physical elements of the system) and the socio-political complexities (*i.e.*, the non-linear agencies of the stakeholder network) via an integrated simulation, or serious game, is very appealing [16,53,54].

Transboundary watershed governance involves a dynamic and interactive arena with a complex multi-stakeholder setting and often-conflicting interests that resembles a strategic and messy game [16]. These complex multi-stakeholder settings require methodologies and tools that are able to support managers in their ability to deal with the technical-physical as well as the social-political complexity of transboundary watershed governance. It is essential to integrate these two types of complexities, and serious games offer a mechanism through which stakeholders become a more intrinsic part of a computer model, not just as digital agents but as actual game players with real stakes, tacit knowledge, emotions, intuitions and interests [14]. A number of authors [16,18,55–59] highlight that the learning (both individually as well as socially) that occurs through such participation in a game simulation may be transferred to the real world outside the game, while at the same time providing a low-risk and safe environment in which to creatively experiment. Serious games may therefore be seen as a form of intervention within a multi-stakeholder network setting that involves learning and changing of stakeholders' mental frames of transboundary water issues, while at the same time offering opportunities to learn about and change the social-political structure (*i.e.*, by building trust, forming coalitions, and power plays in which the stakeholder network is embedded) [14].

Stakeholder participation in serious game simulations may provide significant support in the formation of new or stronger coalitions and collaborative partnerships while addressing existing power plays and building trust with other stakeholders [14,60,61]. Serious game simulations are particularly well suited for knowledge co-creation for situations where the underlying systems are vast in scale, interconnected or complex [62], which is often the case when it comes to transboundary river watersheds. Knowledge co-creation is an important element to social learning, while the sustainable governance of water resources relies greatly on diverse and multi-faceted knowledge systems, through which knowledge is continuously updated to reflect current understanding and needs [32]. Serious game simulation events provide creative platforms that allow stakeholders to share and transfer knowledge, understandings and perspectives through face-to-face interactions and discussions [14]. Such events offer opportunities and space for stakeholder communication, knowledge diffusion processes and the systemization of knowledge [32]. Interactive multi-player game formats that facilitate step-wise, round-based interactions allow participants to develop a greater understanding of different perspectives while also providing a platform for interactions between upstream and downstream contexts of a river watershed [53,62,63]. Role play characteristics of serious games may support the development of empathy and trust between stakeholders, as they develop a much greater understanding of what is required for effective transboundary governance.

Overall, it can be argued that serious games hold advantages over more conventional collaborative methodologies and tools owing to their competitive and entertainment aspects, as well as their role-play-game characteristics and feedback mechanisms [61]. A number of authors [13–15,53,54,60–67] have discussed key characteristics of serious games that are in support of social learning and stakeholder collaborations, which have been summarized in Table 2. Serious games can provide immersive learning opportunities, although the argument can be made that generally engagement goals go beyond just the desire to learn, *i.e.*, what is the problem to be solved that would bring key stakeholders together and want to make them play serious games? In this light, it should be noted that learning cannot remain restricted to acquiring knowledge of specific content matter, but also has to deal with selecting and using this knowledge for certain problem situations in a specific context—in this case, that of transboundary watersheds. Hummel *et al.* [68] highlight that social learning is about the acquisition of competences such as information skills, media literacy, problem-solving, communication and collaboration, as well as critical reflection about complex problems [68]. These authors also emphasize that serious games as virtual learning environments with scripted collaborative interactions have the potential to increase the quality of learning output. Although solitary (single-user) games may not provide the stakeholder interactions and collaborative effects that are required for social learning, multiplayer and multi-role games can enable and provoke social learning and collaborative task activity [53,62,63,68].

In serious games, stakeholders can interact both virtually as well as in reality. Allowing stakeholders to play different roles, not only provides opportunities for interactive learning, but provides players with great opportunities to take conflicts or specific water issues as a starting point for learning while discovering multiple aspects and perspectives of a problem during the game-play [68].

Such conflicts when solving a water management problem can be exchanged, reflected upon and integrated through participation in serious games by both taking an ecological as well as a governance perspective of the case [69]. While social learning does not occur due to one-off engagement but repeat engagement, it is essential that serious game events offer mechanisms that become a part of and contribute to an ongoing process of stakeholder interactions (through, e.g., game design and development, facilitated interactions, game play, post-game discussion, *etc.*), and that provide a type and format of knowledge and information that players can directly use and apply in their day-to-day activities and institutions [14,60]. When serious games meet these requirements, they contribute to social learning through both the "cognitive enhancement" of participants (*i.e.*, the acquisition and integration of knowledge) and "moral development", focusing on the interactive, inter-personal dimension of appraisal [69,70].

Table 2. Characteristics of serious games in support of social learning.

Serious Game Characteristics	Supporting Support Social Learning and Multi-Stakeholder Collaborations	References
Challenging	Facilitating deeper learning by including underlying competitive forces that affect decision-making at a variety of levels, while challenging participants to do better and to compete with others, themselves or a system	[14,60,61,66,67]
Entertaining and engaging	The immersive and competitive aspects of serious games can engage and entertain stakeholders that normally do not interact while providing incentives and enhancing motivation for coordinated action	[13,54,60,65]
Experiential	Experiential learning as an intended design element takes place as a result of the game play participation and relies on rounds of actions, trial and error, and feedback	[14,15,60,61]
Experimental	Learning occurs in a step-wise exploratory manner with actors experimenting with successive rounds of innovation in order to continually assess and improve upon the existing situation	[14,53,60,71]
Providing feedback	Exploring cause-effect relationships through feedback loops between outcomes and subsequent decisions is essential for players to develop knowledge and a deeper understanding of the system they are embedded in, while also encouraging collective sense making and critical self-reflection	[14,53,60,66,67]
Immersive	By creating a platform that stimulates creativity and innovation by exhibiting some degree of immersion that involves the replication of certain real-world elements (*i.e.*, through stories, visuals, a 3D world and levelling) to create a feeling of excitement and flow	[14,60,66,67,72]
Dynamic and interactive	Exhibiting various degrees of interaction with other players, with computers, with game paraphernalia and with facilitators, and thereby increasing stakeholders' capacity to communicate and collaborate	[14,60,66,67]
Realistic	A serious game must have a certain degree of realism allowing for participants to develop a deeper and richer understanding about the larger system that they are a part of	[15,54,64,73]
Low risk and safe environment	Offering risk-free opportunities for experimentation that carry no direct consequences for the outside world while supporting the integration of knowledge about the system	[14,60,66,67]
Multi-player role-play	Allowing participants to play different roles with the aim to develop a much deeper understanding of varying stakeholder interests and perspectives, stakeholder dynamics and power plays	[53,62,63]

2.3. Existing Serious Games for Water Management and Water Spatial Planning

There are many examples in non-water-related sectors of the adoption of advanced technology to foster interactive learning. The water sector has only begun exploring the application of computerized game simulations over the past decade [13,15]. This growth trend has, not surprisingly, coincided with an increasing implementation of integrated and adaptive concepts for water resources management involving principles such as bottom-up, multi-stakeholder participation and holistic systems-wide analysis [74]. In the context of water governance, the serious games developed to date are primarily aimed at building awareness and developing a shared understanding of common problems and trade-offs [53]. Examples of recent serious game simulations developed for water resources management and water-spatial planning include: Aqua Republica; CauxOperation; the Climate Game; EMOVER; Marine Spatial Planning Challenge; Shariva, and; the UVA Bay Game [14,15,73,75,76]. Some of these games involved stakeholders directly in the game design, including underlying systems and assumptions, while others have involved users only in the end-stage game play.

The Marine Spatial Planning Challenge is an example of an end-stage game play that offers an interactive game simulation involving stakeholders from different sectors to address complex, transboundary spatial development and water problems [14]. An example that involves stakeholders in the game design is CauxOperation, an agent-based multi-player role play game where users collectively designed and played a game to assess runoff and soil erosion issues in a French watershed [77]. Using the Companion Modeling ("CoMod") approach, Souchere *et al.* [77] combined scientific expertise and socio-political interactions in the CauxOperation game to enable users to understand complex bio-physical relationships and trade-offs. Another CoMod implementation is called Shariva (Thai for "shared river") and involves a serious game designed and implemented by researchers with the aim to create awareness, upgrade knowledge, build capacity and promote cooperation amongst stakeholders of an imaginary basin in the transboundary Mekong River. Game play was preceded with several pre-game training workshops designed to orientate players to the game context, teach negotiation and Alternative Dispute Resolution (ADR) skills, and provide training on tools to address and resolve transboundary conflicts. These workshops proved instrumental in achieving the game goals while also creating a positive learning experience. One of the objectives of the game was to develop an understanding amongst players of the viewpoints and interests of other players. The researchers designed the game to cause participants to switch roles during the game. For example, the same participants investigated two issues during the same game—one from the perspective of a fictional Flood Assessment Group and the other from the perspective of a fictional Shariva Expert Group. Despite the positive learning that occurred during this exercise, researchers observed that role-switching caused a degree of confusion amongst some of the participants [78].

The UVa Bay Game is an example of a participative computerized game that uses facilitated, round-based role play among a number of players who make decisions affecting their economy and their watershed [76]. Depending on the game, players assume the role of a stakeholder in a watershed or catchment. In each game, players must choose between only a few decisions during any single round. When the underlying models combine these decisions with those made by other players, the consequences can often be unexpected. Key to each of these games are the facilitated interactions, information sharing and negotiation that occur amongst players before, during and/or after rounds. Another game simulation called AquaRepublica is technically a one-player game simulation, although it can be played in the form of a competition between participants while allowing to monitor and track learning progress, as well as engage participants' interest and discussions [15]. The UVa Bay Game and AquaRepublica differ also in how they use computing technology. The UVa Bay Game uses a series of 51,000 differential equations to model inter-relationships between system variables with the primary purpose to scale up the rapid calculation of a larger set of variables affecting the watershed and its stakeholders—and not necessarily to create a more visually immersive environment. In its current form, the UVa Bay Game's game user interface is more akin to a computer "dashboard" than the virtual 3D world found in many medical and healthcare sector serious games [79]. Aqua

Republica, on the other hand, uses technology to create a visually engaging web-based interface and game play environment in addition to processing the calculations of the underlying physical and social models [15].

As mentioned, the Marine Spatial Planning Challenge involves considerable social interaction between stakeholders and is supported by a simulation model running in the background and a feedback system for measuring performance and enhancing learning [14]. Another interactive simulation, the Climate Game, integrates role play and scientific modelling and was developed to reveal and offer access to knowledge relating to climate, water and spatial planning, with the capability of calculating the measures taken by players in real time; this makes it easier for water managers and area developers to re-develop deltas and river areas in response to climate change (Ibid.). Zhou [14] found that although some game artefacts are more simplistic, involving no role playing, gaming interface or an immersive 3D virtual reality, these games may in some cases be more effective in facilitating real-world stakeholder collaborations than more sophisticated simulations. This study of several game simulations concluded that certain critical conditions relating to stakeholder commitment, consent, and equality were more effectively supported through simpler game artefacts in the establishment of a playful environment. It is essential to study in more depth how the involvement of stakeholders in the game design may impact the effectiveness of such games in general.

Several authors [14,16,77] emphasize that more research is needed on design principles, exploring appropriate serious games for complex water issues, and assessing the impact of these platforms on social learning and stakeholder collaboration for sustainable governance. It should be noted that various factors may work to undermine the beneficial impact of interactive processes, including: (1) power imbalances amongst players; (2) hidden agendas; (3) unequal stakeholder representation; (4) the inability to deal with fundamental value differences; (5) an overly simplified simulation that fails to offer a realistic and believable game environment; and (6) the occasionally perverse outcomes of consensus rules [80–82]. Maas [80] also comments that the quality of the stakeholder participation must be considered when designing, implementing and facilitating an interactive process as part of a serious game. It is clear, however, that the combination of declining cost and rapid growth of more powerful computing technologies (*i.e.*, artificial intelligence, virtual worlds, more sophisticated 3D engines incorporating real-world Geological Information Systems (GIS) data feeds, haptic devices, mobile computing, crowd-sourcing *etc.*) will continue to spur innovations in serious gaming—both within and outside the water sector.

3. Research Methodology

The objective of this study is to identify and assess the effectiveness of current social learning practices for transboundary management within the St. Lawrence River watershed. The study forms part of a much larger project that will explore the role and value of serious games to enhance social learning and collaboration processes for sustainable governance. In particular, this paper focuses on the following research themes:

1. *Stakeholder characteristics and institutional setting*, including stakeholder preparedness to participate, available resources, as well as their ability and willingness for critical self-reflection on positions and goals.
2. *Stakeholder interactions* and the way such processes are organized, while looking specifically at whether key stakeholders are involved, diverse interests, views and information are respected, potential solutions are developed and assessed, decisions are taken and implemented jointly, problem perspectives are exchanged, and ongoing reflection is facilitated on positions and goals.
3. *Stakeholder relationships* and their attitudes towards each other, involving strong relationships through repeated interactions, mutual trust, recognition of interdependence, and sharing of goals.

Interviews were conducted in the summer of 2014 with a total of 10 key stakeholders involved in water governance of the St. Lawrence River. These key individuals were identified in consultation with the coordinator of the Regional Round Table for the Upper St. Lawrence and Greater Montreal area. Although a snowball sampling method was used to identify a much more extensive list of relevant stakeholders, practical constraints to this first study demanded that the researchers focus primarily on a smaller group of interviewees with the most significant experience in watershed management of the St. Lawrence River context, while at the same time representing some of the key stakeholder groups that operate at different management levels. Due to this constraint in time and resources, the researchers focused primarily on stakeholders in Ontario and Québec. In future research, the researchers will include a much wider and more diverse group of stakeholders in their research activities. While this study forms a first step in a larger three-year partnership development project, this paper reports findings of the first wave of data collection and, as such has as one of its aims, to identify possible avenues for subsequent waves of data collection. A profile summary of the research participants is provided in Table 3. Names have been omitted to respect the anonymity of the participants, but details of the principal relevant geographic location in which they work and organization type have been included.

Table 3. Overview of profiles for research participants.

Participant Number	Geographic Location	Organization Type
1	Ontario	Binational
2	Ontario	Non-governmental organization
3	Québec	Governmental/regulatory body
4	Québec	National, not-for-profit
5	Québec	Not-for-profit
6	Québec	Higher education
7	Québec	Non-governmental organization
8	Québec	Not-for-profit, government-recognized
9	Vermont	Non-governmental organization
10	Québec	Binational coalition

The interviews, on the whole, followed a semi-structured interview schedule using probing and follow-up questions as required to add clarity to responses. The questions that were asked were aimed at developing an understanding of elements that are required for effective social learning. A "template approach" [83] was used in which relevant text segments from transcripts of the interview recordings were categorized under "codes" and "themes" that were selected *a priori* based on the key elements of social learning processes that were identified in Section 2.1 and Table 1. With the exception of one interview that was conducted via Skype, the interviews were conducted face-to-face and tape-recorded with the consent of the interviewees. Most interviews were conducted in English, although two of the interviews were conducted primarily in French. Interviews were transcribed in the language they were conducted in and then translated when required. Subsequently, responses were coded to group these responses according to the elements of social learning, as well as for any emergent themes. Responses were identified and analyzed using direct content analysis, and a discursive construction was used to assess the degree to which required conditions and elements for social learning are in place for the effective transboundary management of the St. Lawrence River.

4. Research Context

In the context of transboundary water issues, both the strength and weakness in dealing with such issues lies within the great potential to help in negotiation between governments and non-governmental stakeholders, and existing incentives for stakeholder cooperation [22]. Although the United States and Canada are renowned internationally for their ability to cooperate and resolve transboundary water issues as they arise [84], it should be noted that the structure of the Canadian political system relating to water resources management is provincially dominant, creating a point of conflict between the United States delegations and Canada, as the US system relies on the dominance of the federal government [85]. In other words, the states must cede negotiation in transboundary discussions regarding water issues to the federal level, whereas the provinces in Canada do not. There are over 300 transboundary lakes and rivers along the United States–Canada border, and water has been a defining issue in relations between the two countries. There have been many treaties and agreements regarding the usage and passage of water between the United States and Canada. The first of particular importance here was the 1909 Boundary Waters Treaty, which established the International Joint Commission (IJC) as the mechanism for conflict resolution should disputes arise over the use of transboundary waters.

In addition to the establishment of the IJC, which functions as a centralized commission to address transboundary water issues and manage conflict resolution, a number of legislative instruments and agreements have also been passed specifically to regulate use of waters from the Great Lakes. These include the Great Lakes Basin Compact of 1968, the Great Lakes Water Quality Agreement, which was first passed in 1972 and has since been renewed, and the Great Lakes Charter of 1985. The 1987 renewal of the Water Quality Agreement established 43 areas of particular environmental concern (AOCs) in the Great Lakes, five of which are transboundary. These agreements cover the Great Lakes and the international stretch of the St. Lawrence. As the St. Lawrence River leaves Ontario, just downstream of Cornwall, it becomes solely the jurisdiction of Québec and is no longer covered by these international agreements. This means that the transboundary lakes, rivers and streams that feed the St. Lawrence downstream of Cornwall are not covered by the Great Lakes Agreements. Bakker *et al.* [3,86] also emphasize that complications may arise in the transboundary management of the St. Lawrence River between Ontario and Québec because of a lack of coordination between the various governing bodies and the multitude of stakeholders involved.

In 2005, the Great Lakes–St. Lawrence River Basin Sustainable Water Resources Agreement was signed by the states and provinces surrounding the Great Lakes and St. Lawrence River. This agreement aims to strengthen participation of, and cooperation between, stakeholders in the region (with particular reference to data collection and sharing), and aims to adapt management models to changing climate conditions [87]. This agreement is enacted through laws brought in by each jurisdiction separately. The United States has implemented the agreement through the Great Lakes Compact, which was signed in 2008. The agreement was enacted in Québec in 2009 and in Ontario in 2007, though the regulations were only implemented in Ontario as of 1 January 2015 (Great Lakes-St. Lawrence River Basin regional body 2015). The past three decades have seen a shift from national to regional management of water resources in North America through the decentralization of governance [88,89]. There are now thousands of local, regional and special-purpose governing bodies who have the responsibility to govern some aspect of the regions water resources [90]. In Québec, for example, 40 watershed organizations have been established, each charged with planning and facilitating the implementation of watershed management plans within their own jurisdiction [12]. Ontario has put in place a similar governance structure, with 36 Conservation Authorities overseeing the water resource management across the province. Also of relevance here is Stratégies Saint-Laurent, a not-for-profit organization that oversees the *comités ZIP*, the committees of 13 priority intervention zones along the St. Lawrence that act to solve local ecosystem problems.

5. Results

In this section each of the research aims as identified in Section 2, are considered in turn with the goal of identifying insights that can be gained from the analysis of the interview transcripts.

5.1. Characteristics of Stakeholders and Institutional Setting

Although all interviewees recognized the importance of active participation between key stakeholders, the **fragmentation of jurisdiction and laws** has been highlighted as a key barrier to the effective involvement of diverse stakeholders in the transboundary management of the St. Lawrence River. Participant 2 (non-governmental organization), for example, identified fragmentation of jurisdictions as a key barrier to effective transboundary collaborations and consequently social learning. The Great Lakes Water Quality Agreement covers only the international part of the St. Lawrence River and the part of the St. Lawrence River that is solely in Québec is excluded from this agreement. Where policies differ, efficiency is lost and issues can arise between the stakeholders in the regions. This participant also provided an example of the American eel to illustrate this point. The eel has a unique physiology and biology, and spawns in the ocean, but its normal habitat is upstream in Ontario. To make this journey, the eel must pass through two dams and eel ladders are provided for this purpose. Since 2008, the eel has been listed as an endangered species in Ontario, where restoration efforts are ongoing. The eel is not listed as endangered in Québec, however, and the eels are often caught along with fish. With the lack of a unified and common framework and the corresponding structure for collaboration, inefficiencies and disputes can arise in the region. Participant 4 (national, not-for-profit) emphasized, *"There should be a transboundary working group in charge of developing priorities, objectives, goals on specific issues related to transboundary water issues in the St. Lawrence River"*.

In addition to the challenge of fragmentation, participants 1 (binational), 2 (non-governmental organization), 3 (governmental/regulatory body), 4 (national, not-for-profit), 5 (not-for-profit) and 9 (non-governmental organization) brought up **political will and commitment** as an essential element that forms either a barrier or driving force to the facilitation of transboundary collaborations. As participant 1 described, *"You've got folks who have absolutely no interest or connection or investment in the resource to folks who are very committed and invested and that extends politically as well in terms of those that are engaged and those that aren't"*. This political will is emphasized as depending on the perceived seriousness and gravity of water issues at play, but is also often limited by a lack of civic engagement in water issues relating to the Great Lakes and St. Lawrence River. Participant 9 (non-governmental) highlighted, as an example, a record-breaking flood in Lake Champlain that prompted sudden stakeholder interest and involvement. *"When we talk about transboundary water governance"*, explained Participant 4, *"we are not talking about day to day needs or, you know, a quick return on whatever you do"*. This same participant suggested that the transboundary management of the St. Lawrence River requires a long-term vision and commitment that often do not coincide with the shorter term objectives and planning cycles of local or provincial government officials and organizations.

All participants cited **limited time, funding and resources** as barriers to working in greater coordination with their stakeholder counterparts. Participants 4 (national, not-for-profit) and 5 (not-for-profit) referred to a lack of resources to implement face-to-face meetings as inhibiting communication. Participant 5 explained that *"we have so much to take care of that we don't necessarily have time to sit around the table with others. There is not enough funding, not enough resources, to better our management practices"*. Participant 4 spoke specifically of *"an erosion of scientific capacity"*. All participants clearly recognized the importance of pooling resources and sharing knowledge between stakeholders in order to develop more sustainable solutions and outcomes. Participant 4 identified current high staff turnover in government agencies as a barrier to building the long-term relations needed for successful collaboration and social learning. This participant noted that *"while workers used to stay in jobs for life, that is much rarer now"*. Participant 10 (binational coalition) spoke of concerns that many knowledgeable and well-connected people within governmental organizations are approaching retirement age, and that this may present challenges for further transboundary collaboration in the future. Participants 5 (not-for-profit), 6 (higher education) and 8 (not-for-profit, government-recognized) described the impact that language barriers are having on relations between Québec and neighboring provinces and states. According to Participant 5, the lack of bilingual people and resources to translate project documentations impedes the effectiveness of communication in the region. All participants discussed the problem of enforcing existing laws and regulations, and of implementing joint decision-making, both limited in large by a lack of funding and resources as well.

5.2. Stakeholder Interactions and the Way this Engagement Process is Organized

All participants referred to the **level of interaction and communication** between the federal governments as adequate to good. None of the participants highlighted communication between federal governments as a particular issue. Participant 4 (national, not-for-profit) went on to add that communication between federal governments was formal and largely directed by the IJC. At lower levels of government, however, the majority of participants highlighted deficiencies in communication between provincial and state departments. Participants 4 and 8 (not-for-profit, government-recognized) described relations between the states and provinces as operating in "silo mode". Participant 4 provided the example of the St. Lawrence River area of concern at Massena, New York, and Cornwall, Ontario, and stated, "*Whilst the region has been designated as a focus for transboundary management by the Great Lakes Water Quality Agreement* (i.e., *a federal agreement), in practice the State of New York and Province of Ontario have separate remediation plans and proceed towards the overarching goals independently*". Participant 3 (governmental/regulatory body), in contrast, described interaction as good at all levels of government and between stakeholders from non-governmental organizations. Participant 1 (binational) mentioned that, "*The larger the scale of the planning effort, the more difficult it is to get stakeholders engaged and the more difficult it is for them to relate their own particular interests or objectives to the issues being discussed at the larger scale*". This same participant emphasized that many collaborative processes at such larger scales require adequate human resources and change makers at different levels communicating effectively, while developing a clear shared vision for a process and outcome.

A number of participants highlighted *the issue of engagement of the public* in transboundary water management. Public participation was widely regarded as desirable, but participants 4 (national, not-for-profit) and 5 (not-for-profit) spoke of the way it was generally lacking in practice. The preoccupation with what was going on in their people's own back yards and individualistic values were cited as possible reasons for this. Participant 1 (binational) was of the opinion that grass roots passion has declined in recent years but also spoke of some knowledgeable, committed and professional stakeholders that do engage in the processes. This participant explained that, "*It is difficult for citizens and the public to have access to information on what is planned and what has been done and by whom*". Participants 4 (national, not-for-profit) and 7 (non-governmental organization) spoke of the way that the general public loses interest when scientists and politicians present differing perspectives of problems. Several participants referred to the mandate of the IJC to engage citizens in transboundary water resource management processes. In fact, the only vehicle for hearing public views and opinions on transboundary matters identified during the interviews was the IJC's calls for public consultations. According to participant 4, however, "*Last year there was an opportunity for the IJC to consult on a new regulation plan for Lake Ontario—St. Lawrence River. So you are talking about half a day, and with a dozen people in the room. So that's not much of a consult really*". Conversely, Participant 10 argued that the IJC did make an effort to be inclusive. Many of the participants referred to a need for increasing direct communication with communities.

During the interviews, participants were asked who, in their opinion, are the most powerful stakeholders. They were also asked to comment upon the **nature of the power held by different stakeholders**. Most participants agreed that decisions are made primarily at the federal, provincial and state levels (who hold the ultimate regulatory power), and that regional and local stakeholders are much less involved in decision-making processes. The balance of decision-making power with regards to treaties and negotiations in transboundary discussions is with the provinces in Canada and the federal government in the United States. It was also widely noted that the IJC holds a particular regulatory power on the specific topic of water levels and flows. Power over funding and resources was also widely attributed to governmental organizations. Participant 5 (not-for-profit) specifically discussed power in terms of expertise and knowledge, and attributed this to the ground based organizations who have been collecting data and building relationships with local stakeholders over the years. This participant explained that these grassroots level organizations are put in place to represent local stakeholders and the public in general. They shared that these organizations often feel like their expertise, knowledge and perspectives are not sufficiently taken into account at higher levels and scales, and explained that to an extent, "*The citizen's voice isn't taken into account because large stakeholders have so much power and influence*". Participant 4 (national, not-for-profit) also noted that, "*First Nations communities, and their traditional knowledge, are not getting the responsibility and recognition that they deserve for transboundary water management*".

There was widespread agreement among the research participants that the **decision-making process is largely top-down**, although the participants presented a variety of different ideas regarding the basis upon which these decisions are taken. Participant 1 (binational) described how, "*Agreements are more formal and comprehensive, and resources are more transparent at the bi-national, federal, provincial and state levels, but when you move to provincial, regional and local levels, collaboration is much less comprehensive or ongoing, and more on a one-off basis through a specific project*". Participant 1 argued that decisions were driven by the commitments made by federal-federal and federal-state/province agreements. Participant 2 (non-governmental organization) spoke of the role of scientific evidence and public opinion in decision-making. They provided the example of the Water Level Plan of 2014, in which an early proposal based on scientific evidence was adjusted following public concerns about water levels and erosion in the Port of Montréal. Participant 5 (not-for-profit) argued that decisions are weighted too significantly in favor of large stakeholders (for example, large hydropower companies). In their opinion, large stakeholder concerns are the primary basis on which decisions are taken. Participants 6 (higher education) and 10 (binational coalition) also identified stakeholder concerns as a key factor in decision making processes, with Participant 10 saying that scientific knowledge plays a role in the background but that decisions are taken through negotiation by stakeholders.

Opinions also varied on the **extent to which decisions are taken collaboratively**. Participant 5 (not-for-profit) said categorically that they did not think that a collaborative decision-making structure exists. Coming from an organization with strong connections to the general public, this participant lamented the fact that, "*Decisions are made under the table and then when we present them to people because we want to consult them, we are doing it on the basis of the decision is already made*". Participant 8 (not-for-profit, government recognized) felt that their organization was not consulted as much as it should be. Participants 1 (binational), 2 (non-governmental organization) and 7 (non-governmental organization) expressed the view that decisions were made at the federal, state and provincial level, but that input was sought from local actors. Participant 1 said, in reference to individual citizens, they did not always get to make decisions but that there was an opportunity to be heard. Nobody argued that local actors and members of the public had any role in decision-making beyond presenting opinion, and everyone agreed that the role of the public and local actors could beneficially be strengthened.

5.3. Quality of Stakeholder Relationships

As the research data reveals, there are limited examples of social learning occurring through the majority of the St. Lawrence River Basin. With the exception of the highest levels of government (*i.e.*, federal government) and isolated localities, all but one participant described how there is **little interaction between stakeholders** involved in transboundary water resources management. Even where stretches of the river do come under the purview of the Great Lakes Agreement (*i.e.*, the international stretch between Lake Ontario and Cornwall, Ontario), little transboundary interaction occurs between the states and provinces. In this case, participants reported individual agencies operating in "silo mode", both pulling towards the goals as set out in the high level strategic plans but doing so independently. The overarching agreements help align the objectives of the transboundary parties, but without implementing significant processes for interaction and collaboration, little social learning can occur. Participant 9 (non-governmental organization) explained that "*every agency of government should really connect at the local level with organizations and people who are living in the watershed area*". They emphasized that their particular basin program has been successful due to regular meetings of their steering committee over the years, which allowed for different stakeholders to get to know each other, develop stronger relationships and ferment trust.

Participant 9 (non-governmental organization) presented a number of examples with clear implications for successful social learning in the region. In the Lake Champlain Basin, which is a shared resource of Québec, New York and Vermont, Canada and the United States had been individually collecting their own water quality data in their own territory. Each of the jurisdictions had developed a very rigorous testing program, but the problem was that the protocols for testing differed and as a result the data was not comparable, which limited a complete understanding of what was going on with regards to water quality in the lake. The three jurisdictions have collaborated effectively to find a solution to this problem by **aligning data collection methods** in Lake Champlain. Now, when data are collected, the samples are split between the jurisdictions and each runs tests on samples from across the lake following the same protocol. As a result of a change in problem framing from a local (*i.e.*, the state or province's section of the lake) to a global (*i.e.*, whole lake) perspective, a more complete understanding of water quality issues has been gained. This change in problem framing is a clear example of social learning occurring. The experience of such a positive collaboration is likely to have strengthened relations between the jurisdictions, enabling further collaboration in the future.

In another example, Participant 9 (non-governmental organization) described the reaction of journalists from Québec and the United States following a flood event in Lake Champlain. Both sides of the border held the impression that water level management decisions of the other jurisdiction had been the cause of the flooding and subsequent damages in their own territory. Underlying this issue were widely held misconceptions about the locations of dams in the region and the actual impact that operation of the dams has on water levels. In fact, neither jurisdiction had taken decisions, which negatively impacted the other. **Workshops were held to increase understanding** of the Lake Champlain dam system and management and its consequent impact on flooding. In this example, the learning is clear; members of the public increased their understanding of water level management in the basin. In contrast to the previous example, which occurred at the level of government agencies, this example involves learning between citizens. There is a social dimension to the problem due to the fact that the problem was causing ill feeling between communities and their neighbors. The strength of the social dimension with respect to learning for this second example is presently unclear. It is not clear from the interview transcript exactly how the workshops were conducted. It is important to explore this further, to establish whether citizens were brought together from the different jurisdictions, and to determine the process and extent to which social learning actually occurred.

All the interview participants **clearly recognized their interdependence** with other transboundary stakeholders in the region. Despite this recognition, the participants agreed that most stakeholders function too much in silo mode. Participant 4 (national, not-for-profit) provided an example of the Province of Québec that has developed the St. Lawrence plan which stimulates an effort for collaboration between the federal and provincial levels. When this St. Lawrence plan is matched with the Canada-Ontario Agreement program for the Great Lakes, however, these two plans are not at all integrated, while only matching certain priorities. Despite this lack of a current integration of plans and actions across different jurisdictions and boundaries, this participant is convinced that "*If the means and resources would be there for us to collaborate and interact, it would be much more ongoing and constant*". Participant 9 (non-governmental organization) stressed that "*If you want to get real about solving your water issues, you've got to solve the problem where it happens and you've got to be looking on both sides of the border*". This participant stated that this would be achieved not through new formal structures, but through the development and facilitation of new collaborative partnerships between stakeholders at different levels and scales. Another issue highlighted by two of the participants was the perception that the general public lacks understanding of their dependence on each other, both in their own provinces and states, and outside. Participant 4 (national, not-for-profit) said explicitly that they were unsure whether the public in their area of work knew where their water came from or of the impact that upstream states and provinces had on their water.

All of the participants described **reasonable working relations and trust** between government agencies at the federal, state and provincial level. Participant 10 (binational coalition) described trust between individuals involved in transboundary collaboration as stronger than the resulting trust between the organizations they represent. Participant 9 (non-governmental organization) noted that, in their experience, the trust between Québec and Canada is not as strong as that between the other provincial and state parties, citing historical events as the probable cause. This participant also highlighted that trust of citizens in their government is very low due to the unwillingness of those government organizations to make tough decisions, stating that *"many citizens don't trust their state or their province to be doing the best thing for water quality"*. Participant 4 (national, no-for-profit), referring to attitudes between government agencies, expressed the feeling that *"the doors are open for collaboration"*. When it comes to developing mutual trust, Participant 3 explained, *"Trust is long to build and quick to disappear which is why you have to be very careful when addressing issues. When all stakeholders are around the table it is very important that there is no denial of certain issues or perspectives"*. Three of the participants made references to tensions in the relationships between First Nations communities in the region, with Participant 2 (non-governmental organization) explicitly describing the way levels of trust held by members of the Akwesaske community in government officials varies.

6. Discussion

Although certain examples of social learning occurring have been identified during the interviews, preliminary results strongly suggest that these examples are the exception rather than the rule, and that on the whole social learning and transboundary stakeholder collaborations are not occurring to their full potential. The positive examples brought up by some of the interview participants all refer to events that occurred in the Lake Champlain area. Although these examples clearly highlight the fact that social learning is possible in a transboundary watershed context, and that really positive benefits can be realized, it should be noted that Lake Champlain is exceptional in the region, in that it has been designated as one of seven demonstration watersheds worldwide within the UNESCO Hydrology, Environment, Life, and Policy program. For these reasons, the examples of social learning that were identified are hypothesized to be exceptions rather than an indication of broader social learning routinely occurring throughout the region. Further work is proposed, employing a case study approach, to look in more detail at the successes achieved in the Lake Champlain basin. The goal of this further work will be to analyze these successes to identify the mechanisms that made them possible, and to explore the extent to which these successes may be replicable in other basins in the region.

It may come as somewhat of a surprise that more stakeholder interactions do not occur in locations within the St. Lawrence where the objectives of the transboundary agencies are under the purview of and in alignment with overarching strategies and objectives of the Great Lakes Agreement. Clearly where two agencies are working towards different goals, communication will be impeded by differing agendas. When objectives are aligned, however, it seems intuitive that it is in the best interests of both parties to collaborate. Table 4 below provides a summary for each category of analysis, indicating the key barriers to social learning and stakeholder collaborations that have been identified through the interviews. The findings are discussed, as well as their implications for transboundary watershed governance of the St. Lawrence, while also exploring potential solutions that serious game simulations may offer to address the barriers to social learning and stakeholder collaborations that have been identified within the study context.

Table 4. Summary of barriers to social learning for transboundary water governance.

Research Themes	Barriers to Social Learning
Characteristics of stakeholders and Institutional Setting	• Fragmentation of jurisdiction and laws • Varying levels of political will and commitment • Limited time, funding and resources • Erosion of scientific capacity • Loss of knowledge and expertise due to high staff turnover
Stakeholder interactions and the way this engagement process is organized	• Lack of a unified and common framework and corresponding structure for collaboration • Top-down approach to decision-making • Limited civic engagement that prompts political will and stakeholder interests • Lack of involvement of First Nations • Limited participation of non-governmental stakeholders
Quality of stakeholder relationships	• Limited stakeholder interactions • Power imbalance between stakeholders at different levels • Language barriers

6.1. Characteristics of Stakeholders and Institutional Setting

Overall, participants cited limited time and financial resources as barriers to working in greater coordination with their counterparts. Participants agreed that the challenges they face in their own areas often leave them with limited time for transboundary communication and collaboration. Although all participants recognized that increased collaborations may provide opportunities to pool resources, share knowledge, and work more effectively, a fragmentation of jurisdiction and laws, as well as varying levels of political will and commitment impact the extent to which transboundary stakeholder collaborations take place. It may be argued that there is a high degree of initial effort involved in setting up inter-agency networks and stakeholder collaboration processes, and that this barrier must be overcome before interactions can be fruitful and benefits reaped. Exploring and analyzing innovative mechanisms and interventions that support the facilitation of transboundary stakeholder collaborations and support social learning are therefore essential.

Social simulation and computer-based role-playing games can help stakeholders build their capacities by exploring skills, methods, and concepts rapidly within an engaging nonthreatening environment ripe with experiential and behavioral learning components [62]. Serious games can also be viewed as powerful new means of communication, and an even more powerful means of persuasion while offering an opportunity for innovation to policy makers [14]. In light of this study, it is also important to explore in more detail how such games may offer mechanisms to effect change in social behavior as well as in political ideas. Due to the mention of limited time, resources and funding highlighted by all research participants, it seems essential in this case to offer a serious game simulation that is not only interactive, but also provides immediate effects and output based on participants' input [14,60].

6.2. Stakeholder Interactions and the Way This Engagement Process Is Organized

Transboundary engagement in the region, with the exception of isolated localities, is limited to top levels of the federal government and occurs in a formal manner, coordinated by the International Joint Commission. With such little transboundary agency interactions and communication, it is no surprise that collaborative decision making is not occurring on the whole at the state/provincial level. In those regions which are covered by the Great Lakes Agreement, this is likely to lead to inefficiencies in management strategies. In the majority of the St. Lawrence River Basin where the influence of over-arching agreements does not appear to be felt as yet, this is likely to lead to a situation where the different jurisdictions enter a cycle whereby they take decisions separately, which leads to a lack of co-ordination of objectives, creating barriers to effective communication, thus further reducing the ability of the agencies to take decisions collaboratively. While there is still a very top-down approach to water governance of the St. Lawrence River with limited interaction and communication at lower levels of government, and with stakeholders from non-governmental organizations, it is important to find ways to scale up collaborations and learning across the basin. Serious games present a promising, bottom-up tool through which stakeholders at different levels may gain an appreciation of the larger context in which they are operating [32]. Serious games have been found to be ideally suited to applications in contexts in which complex, interconnected problems require reconciliation of divergent viewpoints [62]. It has been argued that they allow players to co-develop knowledge through experimentation but without real world consequences [15]. Although serious game simulations may offer support in the development of initial interactions and connections between stakeholders that do not collaborate in real life, it must be studied further whether such interactions will be able to establish longer term learning partnerships and stronger stakeholder relationships.

The social learning paradigm is concerned with the extent to which knowledge is able to flow in multiple directions. Such learning is meant to occur between a broader range of stakeholders than just governmental and non-governmental agencies. The interview data suggest that the relationships required to enable social learning are very limited between the general public and governmental agencies, and the First Nations and governmental agencies, and also to an extent between non-governmental and governmental agencies. During the interviews participants highlighted a number of issues with regards to the participatory role of the general public in transboundary water resource management. Despite some examples of citizen engagement, the public on the whole was characterized as disengaged. Correspondingly, the degree of transboundary collaboration in decision making for the St. Lawrence River may be considered low, falling somewhere between an "Expert Authority Approach" and an "Inquisitive Approach" on Lauber and Knuth's [30] participation scale. In order to develop increased stakeholder interactions, it is crucial to understand what mechanisms and tools may support this in addition to more conventional ways of stakeholder interactions (*i.e.*, meetings, emails, phone calls, *etc.*). One way to increase stakeholder interactions is through the involvement of stakeholders not in only the game play, but also the design and development of a game simulation. Gurung *et al.* [91], as an example, involved their study participants in the process of developing the game they would later play, much like a collaborative modelling process. This not only helped to increase interactions, but also encouraged consensus building regarding the facts of the underlying system, as well as the development of a shared vision and goals.

6.3. Quality of Stakeholder Relationships

Mostert *et al.* [8] identified the recognition of mutual dependence and trust between stakeholders as important prerequisites to social learning. The results of this study show that interviewees recognize the mutual dependence between key stakeholder groups. Interview participants also spoke of high levels of trust between governmental and non-governmental agencies at all levels. For effective transboundary collaborations, such good relations and trust must be developed with a much wider group of stakeholders at all levels. While the interviews have been conducted primarily with representatives from governmental and non-governmental agencies, it is important in future research to explore the relations and dynamics with diverse stakeholder groups further to gain a greater understanding of their perspectives on the issues at hand. Additionally, there was one mention of strained relations between Québec and Canada, which is a well-known historical problem that could be further explored in the context of transboundary water resources management in the region.

Serious games offer the opportunity for stakeholders to not only voice their perspectives, but have other stakeholders experience it themselves through multi-player role-game-play. Hagen [92] discuss how useful serious game play can be in developing trust and empathy amongst stakeholders for the problems faced by other stakeholder groups. Their study indicated not only a strengthening of relationships between diverse stakeholders through increased mutual trust and empathy, but also an improved understanding of the system and complex issues at play. This was also found by Gurung *et al.* [91] who asked farmers from villages in conflict over water allocations to play a simulation game in which the farmers swapped roles for one round of the game. They found that, "the scenario with swapped roles, during which upper villagers played the role of lower villagers and vice versa, was very effective at sharing different points of view".

7. Conclusions

Generally good relations between Canada and the United States create a context that is ideal for social learning to occur in the St. Lawrence River Basin. The findings from this preliminary study show that only limited examples of successful social learning and transboundary stakeholder engagement have been taking place in the context of the St. Lawrence River Basin. Transboundary stakeholder engagement in the region, with the exception of isolated localities, is limited to the top levels of federal government and occurs in a formal manner, coordinated by the International Joint Commission.

As there is little interaction at lower levels of government, the lack of social learning at these levels is not surprising. The isolated examples of social learning occurring demonstrate its potential for increasing the effectiveness of resource management in the region. Further work is proposed to analyze the mechanisms that enabled the identified examples of social learning and those which create barriers in other cases. In this way, recommendations can be made as to how social learning can be scaled up across the basin. Serious games present a promising, bottom-up tool or intervention platform through which learning partnerships and networks may be created that support the development of stronger stakeholder relationships, as well as increase interactions and communications between these diverse stakeholder groups.

This study is the first phase of a larger three-year project that is aimed at exploring the role and value of serious games to enhance social learning and collaboration processes for sustainable governance. With this research, the authors have identified a series of required elements for social learning (see Table 1) as a way to "test" for the utility of serious games in advancing social learning in the St. Lawrence River system. The second phase of this project entails the actual testing of existing game simulations with youth and students, as well as with stakeholders that operate in the St. Lawrence River through the Regional Round Table for the Upper St. Lawrence and Greater Montréal area. A unique game simulation will be development for the St. Lawrence River context that will be tested during an AquaHacking event that is to take place in October 2016 in Montréal, Canada.

Acknowledgments: The research team gives thanks to Genevieve Grenon (a PhD student at McGill), who conducted the interviews with each of the research participants, and to Meetu Vijay (an MSc student at McGill), who was involved in the early stages of the research. This study was funded by a Social Sciences and Humanities Research Council of Canada (SSHRC) Partnership Development Grant held by Jan Adamowski and contributed to by Nguyen from the McGill University Brace Centre for Water Resources Management.

Author Contributions: Wietske Medema and Jan Adamowski conceived and designed the experiments; Alison Furber and Wietske Medema analyzed the data; Wietske Medema, Alison Furber, Igor Mayer and Qiqi Zhou wrote the paper.

Conflicts of Interest: The authors declare no conflict of interest.

References

1. Uitto, J.; Duda, A. Management of transboundary water resources: Lessons from international cooperation for conflict prevention. *Geogr. J.* **2002**, *168*, 365–378. [CrossRef]
2. Pahl Wostl, C.; Bouwen, R.; Craps, M.; Maurel, P.; Mostert, E.; Ridder, D.; Thallieu, T. The importance of processes of social learning for transboundary water management—Perspectives from the haminicop project. In *Conference on Integrated Water Management of Transboundary Catchments*; TRANSCAT: Venice, Italy, 2004.
3. Bakker, K.; Cook, C. Water governance in Canada: Innovation and fragmentation. *Water Resour. Dev.* **2011**, *27*, 275–289. [CrossRef]
4. Morris, T.J.; Boyd, D.R.; Brandes, O.M.; Bruce, J.P.; Hudon, M.; Lucas, B.; Maas, T.; Nowlan, L.; Pentland, R.; Phare, M. *Changing the Flow: A Blueprint for Federal Action on Freshwater*; The Gordon Water Group of Concerned Scientists and Citizens: Toronto, ON, Canada, 2007.
5. Lagacé, E. *Shared Water, One Framework: What Canada Can Learn from EU Water Governance*; Gordon Water Policy Fellow, Forum for Leadership on Water (FLOW): Toronto, ON, Canada, 2011.
6. Adamowski, J.; Zyla, C.; Cuenca, E.; Medema, W.; Clamen, M.; Reig, P. *Integrated and Adaptive Water Resources Planning, Management, and Governance*; Water Resources Publications LLC: Littleton, CO, USA, 2013.
7. Timmerman, J.; Bernardini, F. Adapting to Climate Change in Transboundary Water Management. In Proceedings of the 5th World Water Forum, Istanbul, Turkey, 16–22 March 2009.
8. Mostert, E.; Craps, M.; Pahl-Wostl, C. Social learning: The key to integrated water resources management? *Water Int.* **2008**, *33*, 293–304. [CrossRef]
9. Hughes, T.P.; Gunderson, L.; Folke, C. Adaptive management of the great barrier reef and the grand canyon world heritage areas. *AMBIO J. Hum. Environ.* **2007**, *36*, 586–592. [CrossRef]
10. Huitema, D.; Mostert, E.; Egas, W.; Moellenkamp, S.; Pahl-Wostl, C.; Yalcin, R. Adaptive water governance: Assessing the institutional prescriptions of adaptive (co-) management from a governance perspective and defining a research agenda. *Ecol. Soc.* **2009**, *14*, 26.
11. Reed, M.S.; Evely, A.C.; Cundill, G.; Fazey, I.; Glass, J.; Laing, A.; Newig, J.; Parrish, B.; Prell, C.; Raymond, C.; *et al.* What is social learning? *Ecol. Soc.* **2010**, *15*, r1.
12. Medema, W.; Adamowski, J.; Orr, C.; Wals, A.; Milot, N. Towards sustainable water governance: Examining water governance issues in Québec through the lens of multi-loop social learning. *Can. Water Resour. J.* **2015**, *40*, 373–391. [CrossRef]
13. Medema, W.; Wals, A.; Adamowski, J. Multi-loop social learning for sustainable land and water governance: Towards a research agenda on the potential of virtual learning platforms. *Wagening. J. Life Sci.* **2014**, *69*, 23–38. [CrossRef]
14. Zhou, Q. The Princess in the Castle: Challenging Serious Game Play for Integrated Policy Analysis and Planning. Ph.D. Thesis, Delft University, Delft, The Netherlands, 15 September 2014.
15. Chew, C.; Lioyd, G.J.; Knudsen, E. An Interactive Capacity Building Experience—An Approach with Serious Games. 2013. Available online: http://www.dhigroup.com/upload/ publications/mikebasin/Chew_2013.pdf (accessed on 28 February 2016).
16. Mayer, I.S. The gaming of policy and the politics of gaming: A review. *Simul. Gaming* **2009**, *40*, 825–862. [CrossRef]
17. Fabricius, C.; Cundill, G. Learning in adaptive management: Insights from published practice. *Ecol. Soc.* **2014**, *19*, 29. [CrossRef]
18. Medema, W. *Integrated Water Resources Management and Adaptive Management: Shaping Science and Practice*; Cranfield University: Cranfield, UK, 2009.
19. Shea, K. Management of populations in conservation, harvesting, and control. *Trends Ecol. Evol.* **1998**, *13*, 371–375. [CrossRef]
20. Holling, C.S. *Adaptive Environmental Assessment and Management*; John Wiley: New York, NY, USA, 1978.
21. Stringer, L.C.; Dougill, A.J.; Fraser, E.; Hubacek, K.; Prell, C.; Reed, M.S. Unpacking "participation" in the adaptive management of social-ecological systems: A critical review. *Ecol. Soc.* **2006**, *11*, 39.
22. Timmerman, J.G. *Adaptation to Climate Change: Challenges for Transboundary Water Management*; Springer Berlin Heidelberg: Berlin, Germany, 2011.

23. Prins, S.; Craps, M.; Taillieu, T. Boundary dynamics in natural resources management: The ambiguity of stakeholder inclusion. *Rev. Gouv.* **2006**, *2*, 2–12.
24. Global Water Partnership. Toolbox integrated water resources management. In *Internet-Only Publication*; Global Water Partnership: Stockholm, Sweden, 2005.
25. Bouwen, R.; Taillieu, T. Multi-party collaboration as social learning for interdependence: Developing relational knowing for sustainable natural resource management. *J. Community Appl. Soc. Psychol.* **2004**, *14*, 137–153. [CrossRef]
26. Freeman, R.E. *Strategic Management: A Stakeholder Approach*; Pitman: Boston, MA, USA, 1984.
27. Reed, M.S. Stakeholder participation for environmental management: A literature review. *Biol. Conserv.* **2008**, *141*, 2417–2431. [CrossRef]
28. Arnstein, A. A ladder of citizenship participation. *J. Am. Inst. Plan.* **1969**, *26*, 216–233. [CrossRef]
29. Chase, L.C.; Decker, D.J.; Lauber, T.B. Public participation in wildlife management: What do stakeholders want? *Soc. Nat. Resour.* **2004**, *17*, 629–639. [CrossRef]
30. Lauber, T.B.; Knuth, B.A. *Citizen Participation in Natural Resource Mangement: A Synthesis of Hdru Research*; Cornell University: Ithaca, NY, USA, 2000.
31. Raadgever, G.T. Assessing management regimes in transboundary river basins: Do they support adaptive management. *Ecol. Soc.* **2008**, *13*, 14.
32. Edelenbos, J.; Van Buuren, A.; Van Schie, N. Co-producing knowledge: Joint knowledge production between experts, bureaucrats and stakeholders in dutch water management projects. *Env. Sci. Policy* **2011**, *14*, 675–684. [CrossRef]
33. Muro, M.; Jeffrey, P. Time to talk? How the structure of dialog processes shapes stakeholder learning in participatory water resources management. *Ecol. Soc.* **2012**, *17*, 3. [CrossRef]
34. Reed, M.S.; Stringer, L.C.; Fazey, I.; Evely, A.C.; Kruijsen, J.H.J. Five principles for the practice of knowledge exchange in environment management. *J. Environ. Manag.* **2014**, *146*, 337–345. [CrossRef] [PubMed]
35. Sol, J.; Beers, P.J.; Wals, A.E.J. Social learning in regional innovation networks: Trust, commitment and reframing as emergent properties of interaction. *J. Clean. Prod.* **2012**, *49*, 35–43. [CrossRef]
36. Crona, B.I.; Parker, J.N. Learning in support of governance: Theories, methods, and a framework to assess how bridging organizations contribute to adaptive resource governance. *Ecol. Soc.* **2012**, *17*, 32. [CrossRef]
37. Hegger, D.; Lamers, M.; Van Zeijl-Rozema, A.; Dieperink, C. Conceptualising joint knowledge production in regional climate change adaptation projects: Success conditions and levers for action. *Environ. Sci. Policy* **2012**, *18*, 52–65. [CrossRef]
38. Weichselgartner, J.; Kasperson, R. Barriers in the science-policy-practice interface: Toward a knowledge-actionsystem in global environmental change research. *Glob. Environ. Chang.* **2010**, *20*, 266–277. [CrossRef]
39. Bodin, O.; Crona, B.I. The role of social networks in natural resource governance: What relational patterns make a difference? *Glob. Environ. Chang.* **2009**, *19*, 366–374. [CrossRef]
40. Olsson, P.; Gunderson, L.H.; Carpenter, S.R.; Ryan, P.; Lebel, L.; Folke, C.; Holling, C.S. Shooting the rapids: Navigating transitions to adaptive governance of social-ecological systems. *Ecol. Soc.* **2006**, *11*, 1–21.
41. Roux, D.J.; Rogers, K.H.; Biggs, H.C.; Ashton, P.J.; Sergeant, A. Bridging the science-management divide: Moving from unidirectional knowledge transfer to knowledge interfacing and sharing. *Ecol. Soc.* **2006**, *11*, 4–23.
42. Inkpen, A.C.; Tsang, E.W.K. Social capital, networks, and knowledge transfer. *Acad. Manag. Rev.* **2005**, *30*, 146–165. [CrossRef]
43. Argote, L.; Mcevily, B.; Reagans, R. Managing knowledge in organizations: An integrative framework and review of emerging themes. *Manag. Sci.* **2003**, *49*, 571–583. [CrossRef]
44. Lamers, M.; Ottow, B.; Francois, G.; Korff, Y.V. Beyond dry feet? Experiences from a participatory water-management planning case in the Netherlands. *Ecol. Soc.* **2010**, *15*, 14.
45. Mostert, E.C.; Pahl-Wostl, Y.; Rees, Y.; Searle, B.; Tabara, D.; Tippett, J. Social learning in European river-basin management: Barriers and fostering mechanisms from 10 river basins. *Ecol. Soc.* **2007**, *12*, 19.
46. Folke, C.; Hahn, T.; Olsson, P.; Norberg, J. Adaptive governance of social-ecological systems. *Annu. Rev. Environ. Resour.* **2005**, *30*, 441–473. [CrossRef]
47. Fernandez-Gimenez, M.E.; Ballard, H.L.; Sturtevant, V.E. Adaptive management and social learning in collaborative and community-based monitoring: A study of five community-based forestry organizations in the western USA. *Ecol. Soc.* **2008**, *13*, 4.

48. Dyball, R.; Brown, V.; Keen, M. Towards sustainability: Five strands of social learning. In *Social Learning towards a Sustainable World; Principles, Perspectives, and Praxis*; Wals, A.E.J., Ed.; Wageningen Academic Publishers: Wageningen, The Netherlands, 2007; pp. 181–195.
49. Keen, M.; Bruck, T.; Dyball, R. Social learning: A new approach to environmental management. In *Social Learning in Environmental Management: Towards a sustainable Future*; Keen, M., Brown, V., Dyball, R., Eds.; Earthscan: London, UK, 2005; pp. 3–21.
50. Ridder, D.; Mostert, E.; Wolters, H.A. *Learning Together to Manage Together; Improving Participation in Water Management*; University of Osnabrück, USF: Osnabrück, Germany, 2005.
51. Klaphake, A. *Kooperation an Internationalen Flüssen aus Ökonomischer Perspektive: Das Konzept des Benefit Sharing*; Deutsches Institut für Entwicklungspolitik: Bonn, Germany, 2005.
52. Sadoff, C.W.; Grey, D. Beyond the river: The benefits of cooperation on international rivers. *Water Policy* **2002**, *4*, 389–403. [CrossRef]
53. Hoekstra, A.Y. Computer-supported games and role plays in teaching water management. *Hydrol. Earth Syst. Sci.* **2012**, *16*, 2985–2994. [CrossRef]
54. Van Bilsen, A.; Bekebrede, G.; Mayer, I. Understanding complex adaptive systems by playing games. *Inform. Educ.* **2010**, *9*, 1–18.
55. Mayer, I. Gaming for policy analysis. Learning about complex multi-actor systems. In *Why do Games Work? In Search of the Active Substance*; de Caluwé, L., Hofstede, G.J., Peters, V., Eds.; Kluwer: Dordrecht, The Netherlands, 2008; pp. 31–40.
56. Geurts, J.L.A.; Duke, R.D.; Vermeulen, P.A.M. Policy gaming for strategy and change. *Long Range Plan.* **2007**, *40*, 535–558. [CrossRef]
57. Juul, J. *Half-Real—Video Games between Real Rules and Fictional Worlds*; MIT Press: Boston, MA, USA, 2005; p. 233.
58. Duke, R.D.; Geurts, J.L.A. Designing the policy exercise. In *Policy Games for Strategic Management: Pathways into the Unknown*; Dutch University Press: Amsterdam, The Netherlands, 2004; pp. 269–305.
59. Mayer, I.; Veeneman, W. *Games in a World of Infrastructures. Simulation-Games for Research, Learning and Intervention*; Eburon: Delft, The Netherlands, 2002.
60. Mayer, I.; Bekebrede, G.; Harteveld, C.; Warmelink, H.; Zhou, Q.; van Ruijven, T.; Lo, J.; Kortmann, R.; Wenzler, I. The research and evaluation of serious games: Toward a comprehensive methodology. *Br. J. Educ. Technol.* **2013**, *45*, 502–527. [CrossRef]
61. Graafland, M.; Schraagen, J.M.; Schijven, M.P. Systematic review of serious games for medical education and surgical skills training. *Br. J. Surg.* **2012**, *99*, 1322–1330. [CrossRef] [PubMed]
62. Reckien, D.; Eisenack, K. Climate change gaming on board and screen: A review. *Simul. Gaming* **2013**, *44*, 253–271. [CrossRef]
63. Schusler, T.M.; Decker, D.J.; Pfeffer, M.J. Social learning for collaborative natural resource management. *Soc. Nat. Resour.* **2003**, *16*, 309–326. [CrossRef]
64. Seibert, J.; Vis, M.J.P. Irrigania—A web-based game about sharing water resources. *Hydrol. Earth Syst. Sci.* **2012**, *16*, 2523–2530. [CrossRef]
65. Wiecha, J.; Heyden, R.; Sternthal, E.; Merialdi, M. Learning in a virtual world: Experience with using second life for medical education. *J. Med. Internet Res.* **2010**, *12*, e1. [CrossRef] [PubMed]
66. Mayer, I.S.; van Bueren, E.M.; Bots, P.W.G.; van der Voort, H.; Seijdel, R. Decisionmaking for sustainable urban renewal projects: A simulation—Gaming approach. *Environ. Plan. B Plan. Des.* **2005**, *32*, 403–423. [CrossRef]
67. Mayer, I.; van Daalen, C.; Bots, P. Perspectives on policy analysis: A framework for under-standing and design. *Int. J. Technol. Policy Des.* **2004**, *4*, 169–191. [CrossRef]
68. Hummel, H.G.K.; van Houcke, J.; Nadolski, R.J.; van der Hiele, T.; Kurvers, H.; Lohr, A. Scripted collaboration in serious gaming for complex learning: Effects of multiple perspectives when acquiring water management skills. *Br. J. Educ. Technol.* **2011**, *42*, 1029–1041. [CrossRef]
69. Haug, C.; Huitema, D. Learning through games? Evaluating the learning effect of a policy exercise on european climate policy. In Proceedings of Amsterdam Conference on the Human Dimensions of Global Environmental Change, Volendam, The Netherlands, 2–4 December 2009.

70. Webler, A.; Kastenholz, H.; Renn, O. Public participation in impact assessment: A social learning perspective. *Environ. Impact Assess. Rev.* **1995**, *15*, 443–463. [CrossRef]
71. Pahl-Wostl, C. A conceptual framework for analysing adaptive capacity and multi-level learning processes in resource governance regimes. *Glob. Environ. Chang.* **2009**, *19*, 354–365. [CrossRef]
72. Lucke, B. Vbs2 Introduction. Online presentation. 2011. Available online: http://www.slideshare.net/BrianLucke/vbs2-introduction-v15-details-8415168?from_search=1 (accessed on 4 October 2013).
73. Mayer, I.S.; Zhou, Q.; Lo, J.; Abspoel, L.; Keijser, X.; Olsen, E.; Kannen, A. Integrated, ecosystem-based marine spatial planning: First results from international simulation-game experiment. In Proceedings of the Third International Engineering Systems Symposium (IEEE), CESUN, Delft University of Technology, Delft, The Netherlands, 18–20 June 2012; pp. 18–20.
74. Rusca, M.; Heun, J.; Schwartz, K. Water management simulation games and the construction of knowledge. *Hydrol. Earth Syst. Sci.* **2012**, *16*, 2749–2757. [CrossRef]
75. Islam, S.; Susskind, L. Water diplomacy: Creating value and building trust in transboundary water negotiations. *Sci. Dipl.* **2012**, *1*, 1–7.
76. Learmonth, G.P.; Smith, D.E.; Sherman, W.H.; White, M.A.; Plank, J. A practical approach to the complex problem of environmental sustainability: The UVa bay game. *Public Sect. Innov. J.* **2011**, *16*, 1–8.
77. Souchère, V.; Millair, L.; Echeverria, J.; Bousquet, F.; Le Page, C.; Etienne, M. Co-constructing with stakeholders a role-playing game to initiate collective management of erosive runoff risks at the watershed scale. *Environ. Model. Softw.* **2010**, *25*, 1359–1370. [CrossRef]
78. Douven, W.; Mul, M.L.; Son, L.; Bakker, N.; Radosevich, G.; Hendriks, A. Games to create awareness and design policies for transboundary cooperation in river basins: Lessons from the shariva game of the Mekong river commission. *Water Resour. Manag.* **2014**, *28*, 1431–1447. [CrossRef]
79. Learmonth, G.P., Sr.; Bobko, R. Informing water policy with large-scale, high fidelity simulation. In Proceedings of the Association for Public Policy Analysis and Management (APPAM) Annual Meeting, University of Virginia, Charlottesville, VA, USA, 1 December 2011.
80. Maas, A. Water, Governance and Sustainability: A case Study of Water Allocation in Whiteman's Creek Watershed, Ontario. Master's Thesis, University of Waterloo, Waterloo, ON, Canada, 30 April 2011.
81. Dray, A.; Perez, P.; Le Page, C.; D'Aquino, P.; White, I. Who wants to terminate the game? The role of vested interests and metaplayers in the atollgame experience. *Simul. Gaming* **2007**, *38*, 494–511. [CrossRef]
82. Gunton, T.I.; Day, J.C.; Williams, P.W. The role of collaborative planning in environmental management: The north American experience. *Environments* **2003**, *31*, 1.
83. Robson, C. *Real World Research: A Resource for Social Scientists and Practitioner-Researchers*; Wiley: Hoboken, NJ, USA, 2002.
84. Parrish, A. Mixed blessings: The great lakes compact and agreement, the IJC, and international dispute resolution. *Mich. State Law Rev.* **2006**, *5*, 1299–1321.
85. Norman, E.S.; Bakker, K. Transgressing scales: Water governance across the Canada-US borderland. *Ann. Assoc. Am. Geogr.* **2009**, *99*, 99–117. [CrossRef]
86. Bakker, K.; Norman, E. Local Canada-US transboundary water governance: Issues, drivers, and barriers. In *Briefing Note*; University of British Columbia: Vancouver, BC, Canada, 2007.
87. Schulte, P. The great lakes water agreements. In *The World's Water Volume 7: The Biennial Report on Freshwater Resources*; Gleick, P.H., Ed.; Island Press: Washington, DC, USA, 2011.
88. Norman, E.S. *Governing Transboundary Waters: Canada, the United States and Indigenous Communities*; Routledge: New York, NY, USA, 2015.
89. De Loë, R.; Kreutzwiser, R. Challenging the status quo: The evolution of water governance in Canada. In *Eau Canada:* The Future of Canada's Water; Bakker, K., Ed.; UBC Press: Vancouver, BC, Canada, 2007; pp. 85–103.
90. Clamen, M.; Macfarlane, D. The international joint commission, water levels, and transboundary governance in the great lakes. *Rev. Policy Res.* **2015**, *32*, 40–59. [CrossRef]
91. Gurung, T.R.; Bousquet, F.; Trebuil, G. Companion modelling, conflict resolution, and institution building: Sharing irrigation water in the lingmuteychu watershed, bhutan. *Ecol. Soc.* **2006**, *11*, 36.

92. Hagen, E. New approaches in the potomac river basin and beyond—Pioneering work in the development of shared vision planning. In *Converging Waters: Integrating Collaborative Modeling with Participatory Processes to Make Water Resources Decisions*; Bourget, E., Ed.; Institute for Water Resources, U.S. Army Corps of Engineers: Alexandria, VA, USA, 2011; pp. 35–58.

water

MDPI

Article

International Severe Weather and Flash Flood Hazard Early Warning Systems—Leveraging Coordination, Cooperation, and Partnerships through a Hydrometeorological Project in Southern Africa

Robert Jubach [1,*] and A. Sezin Tokar [2]

1 Hydrologic Research Center, 12555 High Bluff Drive, San Diego, CA 92130, USA
2 U.S. Agency for International Development, Office of U.S. Foreign Disaster Assistance, Ronald Reagan Building, Washington, DC 20523, USA; stokar@usaid.gov
* Correspondence: rjubach@hrcwater.org; Tel.: +1-858-461-4560; Fax: +1-858-704-4955

Academic Editors: Sharon B. Megdal and Susanna Eden
Received: 20 February 2016; Accepted: 12 June 2016; Published: 20 June 2016

Abstract: Climate, weather and water hazards do not recognize national boundaries. Transboundary/regional programs and cooperation are essential to reduce the loss of lives and damage to livelihoods when facing these hazards. The development and implementation of systems to provide early warnings for severe weather events such as cyclones and flash floods requires data and information sharing in real time, and coordination among the government agencies at all levels. Within a country, this includes local, municipal, provincial-to-national levels as well as regional and international entities involved in hydrometeorological services and Disaster Risk Reduction (DRR). Of key importance are the National Meteorological and Hydrologic Services (NMHSs). The NMHS is generally the authority solely responsible for issuing warnings for these hazards. However, in many regions of the world, the linkages and interfaces between the NMHS and other agencies are weak or non-existent. Therefore, there is a critical need to assess, strengthen, and formalize collaborations when addressing the concept of reducing risk and impacts from severe weather and floods. The U.S. Agency for International Development/Office of U.S. Foreign Disaster Assistance; the United Nations World Meteorological Organization (WMO); the WMO Southern Africa Regional Specialized Meteorological Center, hosted by the South African Weather Service; the U.S. National Oceanic and Atmospheric Administration/National Weather Service and the Hydrologic Research Center (a non-profit corporation) are currently implementing a project working with Southern Africa NMHSs on addressing this gap. The project aims to strengthen coordination and collaboration mechanisms from national to local levels. The project partners are working with the NMHSs to apply and implement appropriate tools and infrastructure to enhance currently operational severe weather and flash flood early warning systems in each country in support of delivery and communication of warnings for the DRR entities at the regional, national and local levels in order to reduce the loss of life and property.

Keywords: early warning; hydrometeorological hazards; severe weather; flash flood guidance; disaster risk reduction

1. Introduction

The following article overviews the framework of a project currently underway in Southern Africa designed to reduce the risks from severe weather and flash floods in the region. The stated goal of the project is to improve accuracy, lead time, communication and dissemination of severe weather and flash flood warnings to communities. The project aims to (1) strengthen the capacity of National

Meteorological and Hydrologic Services (NMHSs) to provide timely and effective early warnings of extreme hydrometeorological events (with appropriate lead times for taking action); and (2) develop products and information that address needs of users such as National Disaster Management Agencies, Non-Governmental Organizations and the general public to reduce the loss of lives and livelihoods and to protect property and the environment in the Southern Africa region.

At this point the project is not mature enough to measure its effectiveness and so we do not provide any detailed statistics or information in this regard. There are plans to evaluate the project as well as the impact of improved forecasts and early warnings—these plans are discussed in Section 5. The project is presented here at this time because the authors believe that this project is a unique approach to improving early warnings on a regional basis through the application of existing operational systems that provide data and information for extreme weather and flash floods to all countries in the region. We believe that a regional approach to using actual, proven operating systems as a catalyst for the NMHS and supporting governmental agencies to fully develop their early warning potential and address the critical end-to-end early warning system components and requirements will prove most successful.

2. Background

Southern Africa is among the most vulnerable regions of the world to hydrometeorological hazards including floods, cyclones, droughts and extreme temperatures. According to the Center for Research on the Epidemiology of Disasters [1], 72% of natural disasters reported in the countries in the Southern Africa region were severe weather and hydrometeorological in nature (e.g., floods, flash floods, storms, droughts, extreme temperatures). As a result of a United Nations World Meteorological Organization (WMO) survey [2] and from discussions at various regional hydrometeorological hazard workshops conducted by the Hydrologic Research Center (HRC), it was noted that all countries in the region report being impacted in some way by the effects of these hazards. In a recent workshop organized by the WMO, the HRC and the U.S. Agency for International Development/Office of U.S. Foreign Disaster Assistance (USAID/OFDA), meteorological and hydrologic forecasters and disaster managers from Botswana, Lesotho, Malawi, Mozambique, Namibia, South Africa, Swaziland, Zambia, and Zimbabwe stated that flash floods/floods, heavy precipitation, strong winds, and thunderstorms were the most prevalent hydrometeorological hazards affecting their countries [3].

A recent example includes the major flash flooding and riverine flooding that occurred in Malawi, Mozambique, Madagascar, and Zimbabwe from December 2014 to January 2015. This flooding affected more than 135,000 people in the four countries. Malawi was especially hit hard, with flooding in over half of the country's districts [4].

The Sendai Framework for Disaster Risk Reduction: 2015–2030 (SFDRR) that was adopted by 187 nations aimed at the substantial reduction of disaster risk and losses in lives, livelihoods and health and in the economic, physical, social, cultural and environmental assets of persons, businesses, communities and countries. As outlined in the principles and priorities under SFDRR, a critical component of Disaster Risk Reduction (DRR) is to enable early action. One way achieve this is through a well-functioning early warning system that delivers accurate, reliable and understandable warnings, in a timely manner, to authorized disaster managers and populations at risk. Effective early warning systems involve four operational components requiring close cooperation and coordination among the all agencies involved. The United Nations International Strategy on Disaster Reduction (UNISDR) [5] defined the four components including:

1. Monitoring and Warning Service—Develop hazard monitoring and early warning services (observing, detecting, monitoring and forecasting hazards, and developing warning messages);
2. Risk Knowledge—Systematically collect data on hazards and vulnerability and undertake risk assessments (assessing the potential risks and integrating risk information in the warning messages);
3. Dissemination and Communication—Communicate risk information and early warnings (distributing, rapidly and reliably, understandable warnings to authorities, disaster managers and

the population at risk, with levels of warning that are linked to levels of preparedness, readiness and emergency operations); and,

4. Response Capability—Build national and community response capabilities (community-based emergency planning, preparedness and training programs to ensure effective response to warnings to reduce the potential impacts).

Of primary importance for early warning systems is that they are truly end-to-end in nature. An end-to-end system consists of a warning and response system where the components are interconnected. Each component in this process is critical to reduce the impacts of hydrometeorological extremes and provide essential lead times to aid decisions. Failure of one component will lead to failure of the entire system to save lives and livelihoods. State-of-the-art technology and a perfect forecast will not save lives if the populations at risk are not informed in a timely manner or do not have plans and policies in place in advance to reduce impacts. Well-prepared communities remain vulnerable to these hazards if they do not have access to and understand information that provides the lead time needed to take necessary actions. Communications and warning/information dissemination are important attributes of a successful warning system as well as investment into building the capacity of forecasters and user knowledge. In addition, close coordination must occur between all sectors and between national and local governments for systems to function properly with clear lines of roles and responsibilities to avoid confusion and chaos during disasters. Important parts of this process are the systems that enable user feedback to periodically improve to address the needs of decision makers.

As our focus in this article is on severe weather and hydrologic/hydrometeorological hazards, the primary agencies we are concerned with that must deal with early warnings for these hazards are the NMHSs. The role of the NMHSs is critical; the majority of natural disasters are weather related as discussed during the previously referenced workshop [3]. Flash floods, floods, heavy precipitation, drought, strong winds, thunderstorms (including lightning and hail), tropical storms, extreme temperatures, heat/cold waves, wildfires, fog, hazardous seas/lakes, snow, and storm surges are common hazards that are a threat to lives and livelihoods in Southern African countries.

Rogers and Tsirkunov [6] outlined the importance of sustainable partnerships and commitment to cooperation and information exchange between NMHSs with disaster management agencies for early warning systems. These partnerships must be supported by legal authority and institutional arrangements for the successful development and sustainability of effective early warning systems. As they have noted, partnerships are important in two ways: early warning tasks in an end-to-end system are not usually managed by a single entity, and they provide the ability to leverage resources and provide opportunities to share costs, knowledge, and lessons learned. Most significantly, effective partnerships can ensure a consistent message from multiple sources (e.g., severe weather and flash flood warnings and information). To be effective partners, the NMHSs must understand the decision-making process made by all sectors affected by the hazard to ensure that information is tailored to the specific needs of the user. This includes:

- Providing and communicating efficient, clear and timely weather- and warning-related data and information;
- Understanding the hazard's effect on the user operations and needs that vary from user to user and event to event at different temporal and spatial scales;
- Having a quantitative understanding of the social and economic cost and benefit of the warnings;
- Understanding the decisions that depend on the warning and the lead time; and,
- Understanding and conveying the uncertainty in the warning.

Over the years, international and regional entities have worked on strengthening the different components of early warning systems and DRR. This has included the development of specialized technical capacities for monitoring, detecting, modeling and forecasting for different natural hazards, legal and institutional aspects, emergency preparedness and response planning for communities, coordination of humanitarian response, and public education. In some countries, hazard-specific early warning systems have been developed. Yet the success of some of these programs has been mixed. Often what are needed to integrate all the various components of early warning systems are projects which focus on actual operational systems that act as a catalyst for the NMHS and supporting governmental agencies to fully develop their early warning potential and address the critical early warning system components and requirements as noted above. One such project, now fully underway in Southern Africa, deals with both severe weather-related hazards (a meteorological issue) and flash floods (a hydrometeorological issue). This project, known as the Delivery of Warnings of Hydrometeorological Hazards—Southern Africa Region is discussed in the following sections.

3. Southern Africa Region Project Overview

USAID/OFDA along with the WMO, RSMC Pretoria (WMO Regional Specialized Meteorological Center, managed by the South African Weather Service), the U.S. National Oceanic and Atmospheric Administration/National Weather Service (NOAA/NWS) and the HRC have partnered to implement a project to link and enhance the operational Southern Africa Severe Weather Forecast Demonstration Project (SWFDP) and the Southern Africa Region Flash Flood Guidance (SARFFG) systems with the objective of leveraging those systems to provide improved service delivery of forecasts, warnings and information to the DRR agencies/organizations at the regional, national and local levels in Southern Africa. The project includes the Southern African countries of Botswana, Lesotho, Malawi, Mozambique, Namibia, South Africa, Swaziland, Zambia, and Zimbabwe. It was kicked off with workshops in Pretoria, South Africa, conducted by the project partners and attended by representatives from all nine countries [3].

Both the SWFDP and the SARFFG are regional-based systems that have been designed to provide forecasters at the NMHSs the necessary data and information to provide timely and effective warnings of severe hydrometeorological events, including conditions for rainfall that could lead to flash floods. Because they are regional-based systems hosted by a Regional Center (RSMC Pretoria), data, information and forecast products can be generated efficiently and cost-effectively, thereby complementing the limited capacity available at national levels. Both systems encompass end-to-end early warning system concepts by including the elements of (1) observing, detecting and assisting forecasters to develop hazard forecasts and warnings; (2) assessing the potential risks and integrating risk information in the warning messages; and (3) distributing, rapidly and reliably, understandable warnings to authorities, risk managers and the population at risk as well as many other users, such as the agricultural sector. Linking the two systems both programmatically and technically enhances their capabilities to address a fourth element of an end-to-end early warning system—emergency preparedness and response to warnings at all relevant levels (national to local) to minimize the potential impacts of extreme hydrometeorological events. (Additional discussion regarding both systems is provided below.)

This project: (1) integrates data from both the operational SARFFG and SWFDP systems; (2) improves forecast operations at RSMC Pretoria and the nine Southern African countries that have access to both systems; (3) builds capacity of the hydrological and meteorological operations of each NMHS and RSMC Pretoria; (4) improves service delivery to the National and Regional Disaster Management agencies and DRR users through improved warning dissemination (e.g., lead time, message content and impact information); and, (5) enhances severe weather disaster awareness, preparedness and response. For the DRR sector in each country, the degree of outreach and readiness is assessed. This includes the legal frameworks in effect to assure effective use of improved warnings and forecasts.

An essential aspect of this project is having the DRR community clearly define the information and products they need to maximize response and minimize loss of life and property. This is best done through prototyping and feedback, as the DRR community must understand the forecast uncertainties well in order to provide useful information back to the NMHSs. This is because of the high uncertainties of the forecasts in small regions that requires risk management conditional on the local forecast uncertainties (rather than based only on a priori knowledge).

For this project to deliver improved services, it integrates/links and enhances the two fully operational systems, builds capacity through coordinated training and develops and implements a severe weather/flash flood Concept of Operations (CONOPS) and Roadmap for each country for warning services to connect to disaster response operations and users.

Linkages between the systems will lead to improvements in the capabilities of both initiatives and create closer collaborations between NMHSs, leading to improved service delivery. In essence these two systems have been implemented in parallel with similar concepts of operation, basic approach, and region (southern Africa). At the same time, they have distinct differences in relation to both the temporal and spatial nature of each extreme hydrometeorological event and on the emphasis of the hazards, respectively, which require different approaches. Each separate system could benefit from programmatic and technical linkages, including, for example:

- Shared outputs from each program's planning/steering committee meetings;
- Joint workshops that focus on using products from either system for DRR;
- A common format for alert/warning protocols;
- Joint validation programs;
- Joint access to system products (e.g., rainfall estimates and forecasts);
- Hypertext linked access to each system; and,
- Collaborations between system forecasters for the development of consistent forecast and outlook discussions relevant to both systems.

A particular emphasis will be placed on "reaching the last mile", making sure warnings developed from both systems reach the affected population and users in a timely and accurate manner and are in a format that can be easily understood and acted upon to achieve maximum value in resulting disaster response actions. Improved linkage between NMHS warning products and services to national and community disaster response activities is the principal goal for integrating these two successful forecast and warning systems.

3.1. Severe Weather Forecasting Demonstration Project (SWFDP)

In essence the SWFDP is a process whereby scientists from global and regional centers work with severe weather forecasters at the national level to identify services that would assist the national disaster response and risk reduction efforts, and which can be implemented almost immediately by tailoring numerical weather prediction model outputs and other forecasting tools that exist in the most advanced centers, making them routinely available at the national level. The majority of NMHSs are not able to develop or run the weather forecast models due to limited capacity and resources. The SWFDP employs a 'Cascading Forecasting Process' whereby outputs from forecast systems that are available and free in advanced global centers are cascaded to NMHSs through a designated regional center which provides interpretation and guidance on severe weather from the next few hours (nowcasting) to the next five days (short- to medium-range forecasting). This allows forecasters at the NMHSs to focus limited resources on considering the impact of this weather in their country, service delivery, and communicating the message to users in their countries to ensure timely and effective early warnings and protective responses. The SWFDP project includes building the capacity of national meteorologists in the application of the cascaded information and in the development of services to meet the disaster management communities' needs. Opportunities are taken to involve disaster managers and the news media in preparation of the user requirements.

The SWFDP has improved the lead time and reliability for alerts about high-impact events such as heavy precipitation, strong winds, and damaging waves. It has strengthened NMHS interaction with disaster management and civil protection agencies, local communities and media. The SWFDP is making a major contribution to SFDRR and is contributing to the Sustainable Development Goals [7] as well as climate change adaptation. More broadly, the project is benefiting society in terms of safety of both the population and operations, including DRR, and in the performance of key socio-economic areas, including agriculture, fisheries, aviation, water resource management, health and marine transportation, where meteorological prediction is crucial.

3.2. *Southern Africa Region Flash Flood Guidance System (SARFFG)*

The SARFFG was designed and developed for use by meteorological and hydrologic forecasters in the Southern Africa region (specifically the countries of Botswana, Lesotho, Malawi, Mozambique, Namibia, South Africa, Swaziland, Zambia and Zimbabwe). The primary purpose of the SARFFG is to provide operational forecasters and disaster management agencies with real-time informational guidance products pertaining to the threat of small-scale flash flooding throughout the region. The SARFFG provides the necessary products to support the development of warnings for flash floods from rainfall events through the use of satellite-based rainfall estimates and hydrologic models.

The SARFFG outputs are made available to users (forecasters) to support their analysis of weather-related events that can initiate flash floods (e.g., heavy rainfall, rainfall on saturated soils) and then to make a rapid evaluation of the potential for a flash flood at a location. To assess the threat of a local flash flood, the SARFFG is designed to allow product adjustments based on the forecaster's experience with local conditions, the incorporation of other information (e.g., Numerical Weather Prediction output) and any last-minute local observations (e.g., non-traditional rain gauge data) or local observer reports. The system can be used in its real-time mode or in a forecast mode when outputs are used along with nowcasting and Numerical Weather Prediction precipitation forecasts. The system supports evaluations of the threat of flash flooding over hourly to six-hourly time scales.

Important technical elements of the SARFFG system are the development and use of a bias-corrected satellite precipitation estimate field and the use of land-surface hydrologic modeling. The system then provides information on rainfall and hydrologic response, the two important factors in determining the potential for a flash flood. The system is based on the concept of flash flood guidance and flash flood threat. Both indices provide the user with the information needed to evaluate the potential for a flash flood, including assessing the uncertainty associated with the data.

The flash flood guidance approach to developing flash flood warnings rests on the real-time comparison of observed or forecasted rainfall volume of a given duration and over a given catchment to a characteristic volume of rainfall for that duration and catchment that generates bank full flow conditions at the catchment outlet. If the observed or forecasted rainfall volume is greater than the characteristic rainfall volume then flooding in the catchment is likely (a flash flood threat). The characteristic rainfall volume for a particular catchment and duration, called flash flood guidance, depends on the catchment and drainage network characteristics and the soil water deficit determined by antecedent rainfall, evapotranspiration and groundwater loss [8–10].

The SARFFG system was an important tool in developing and issuing warnings for the previously mentioned flash flooding in Malawi in January 2015. The Malawi Department of Climate Change and Meteorological Services (DCCMS) applied the system to issue timely and accurate warnings to communities impacted by the flooding. The DCCMS used the flash flood guidance and rainfall forecasts from the system to develop the warnings [11]. Lessons learned from events such as the Malawi flooding event plus evaluations of the addition of data and information from the SWFDP system during such events will be factored in to the Southern Africa Region Project in order to further improve the development and dissemination of warnings.

4. Roadmap and CONOPS Development

On the operational front, effective early warning systems require the establishment of Roadmap and CONOPS documents by the NMHS to ensure coordination and operational cooperation among the responsible agencies and, for this project, a focus on the SWFDP and SARFFG linked systems. The Roadmap will outline objectives, deliverables, work plans, and organizational makeup. The CONOPS will provide a clear outline of the flow of activities necessary to produce routine and non-routine operations, data/information flow, responsibilities, and ongoing training programs for the linked systems; included are linkages to all aspects of the end-to-end processes that comprise EWS. The development of these two documents by the countries will be accomplished under the guidance of the project team—USAID/OFDA, WMO, RSMC, NOAA/NWS, and the HRC. The Roadmap and CONOPS development is a critical element of this project as these documents will be used as input to country-specific implementation plans. The implementation plan then provides a mechanism to track progress in the full integration of the SWFDP and SARFFG systems into NMHS operations and the development of the end-to-end early warning systems for hydrometeorological hazards, specifically for severe weather and flash floods.

4.1. Roadmap Development

One of the important aspects of this project is the development of a Roadmap (framework) outlining the actions required to help meet the objective to provide improved service delivery of forecasts, warnings and information to the DRR agencies/organizations at the regional, national and local levels in Southern Africa. This Roadmap is a tool to recognize existing and needed relationships between these assets and the skills, technologies and requirements to meet future needs and goals. This Roadmap integrates product development and training outcomes from this project and accounts for the end-to-end early warning system process.

The various steps and information used in the development of the country roadmaps are as outlined below.

1. Current NMHS Capabilities—Provide a clear and unambiguous understanding of the current conditions of the capabilities to support an end-to-end early warning system for severe weather and flash floods such that appropriate timely and accurate warnings and information can be provided to the affected population.
2. Required Outcomes—In evaluating the required outcomes, the following are to be considered [6]:
    - The governance and institutional arrangements necessary to support the development and sustainability of effective early warning systems;
    - Appropriate and available technical and operational capacities for observing, detecting, monitoring, forecasting and warning of hazards;
    - What constitutes effective partnerships between stakeholders;
    - Efficient, clear and timely communication of weather- and warning-related data and information;
    - A clear understanding of the decision-making process made by all sectors affected by the hazard to ensure that information is tailored to the specific needs of the user, including:
        - Understanding the effect of weather and warning-related data and information on the user operations and needs
        - Knowing the operations and needs for each stakeholder for a given event, which could vary widely
        - Quantitative understanding of the social and economic cost and benefit of the warnings
        - An understanding of and ability to convey the uncertainty in the warning
        - Understanding the decisions that depend on the warning
        - Level of acceptance of false alarms (positives or negatives)
    - Good communication is important for an effective warning system.
3. Gap Analysis—An important step is to undertake a gap analysis to identify areas needing improvement between the current capabilities and the required outcomes. Critical systems to include in the analysis are observation networks, data/information tools, technical

operating procedures/protocols (e.g., SWFDP/SARFFG operations manuals), meteorological and hydrologic forecasting infrastructure, communication infrastructure to ensure data and information reach the target users, and internal and user capacity building.

Roadmap Activities—Table 1 provides examples of roadmap activities for the NMHSs.

Table 1. Roadmap activities (NMHS responsibilities).

Review, Develop Disaster Management Policies
❖ Analyze existing national- to community-level disaster management plans, procedures and protocols for emergency situations.
❖ Analyze existing linkages/collaborations between providers and users of severe weather and flash flood warnings.
❖ Evaluate mandates/roles of the National Meteorological and Hydrologic Service (NMHS) and other national to community disaster management agencies/organizations (what can be sent and to whom by the NMHS).
❖ Develop partnerships between all stakeholders.
Review, Develop Operational Capacities for Observing, Forecasting, and Warning
❖ Develop technical and funding plans to optimize monitoring and communication networks identified as deficient in the gap analysis.
❖ Implement top priority upgrades to monitoring and communication systems when funding becomes available.
❖ Integrate the application of the Severe Weather Forecast Demonstration Project (SWFDP) and Southern Africa Region Flash Flood Guidance (SARFFG) systems, with updated products, into current and future operations.
❖ Develop technical capacity-building and training programs (including funding plans) to expand forecasting and infrastructure maintenance resources.
❖ Develop a Concept of Operations document outlining NMHS operations for the development and dissemination of severe weather and flash flood warnings.
Develop Guidance Products and Protocols
❖ Assess the extent to which existing severe weather and flash flood warning services are able to meet user demands and requirements.
❖ Create a working platform to bring together providers and users of severe weather and flash flood warnings to promote better understanding of the diverse and evolving needs of the user groups, especially the last mile, to jointly identify, design, and implement strategies and actions to improve responsiveness to warning services. This platform (which could include stakeholder workshops) will address content, format, delivery, lead time, communication of technical limitations (including uncertainty), literacy and gender.
❖ Develop appropriate products and data and information packages for users; utilize existing and new products from the SWFDP and SARFFG systems as applicable.
❖ Create capacity-building and training programs for users of severe weather and flash flood warnings to improve user product development, delivery, usability, evaluation and interpretation.
❖ Perform continuous assessments of issues (potential or actual) that could affect the NMHS warning services; test feasible options to mitigate the issues.
Develop Communication and Dissemination Mechanisms
❖ Develop technical and funding plans to optimize communication infrastructure identified as deficient in the gap analysis.
❖ Develop and implement a disaster managers' web-based portal.
❖ Develop protocols for the dissemination of severe weather and flash flood warnings and other specialized forecasts and NMHS products exchanges with target users.
❖ Develop protocols for improving warning dissemination through formal and informal channels, including feedback mechanisms before, during, and after an event.
❖ Implement simulation exercises annually with users to test the communication and dissemination systems for severe weather and flash flood warnings, including feedback mechanisms.

4.2. CONOPS Development

A CONOPS document is critical and provides a clear outline of the flow of activities necessary to produce routine and non-routine operations, data/information flow, responsibilities, and ongoing training. A CONOPS describes the likely operation of a future or existing system in the terminology of its users [12]. The countries are in the process of developing an integrated CONOPS document for the operations of both the SWFDP and SARFFG systems for their application to an end-to-end early warning system, including accounting for the linkages between the two systems. Included are references to all aspects of the end-to-end processes that comprise an early warning system and ensure integration into current and future NMHS forecast operations. The CONOPS documents provide the top-level technical specifications and functionality of the early warning system, as well as its look and feel.

Key elements in the Southern Africa country CONOPS documents include:

1. Description of the early warning system;
2. Operations of the early warning system
   a integrating products from the SWFDP and SARFFG systems
   b integrating local data and information
   c information flow;
3. Responsibilities of operational components
   a forecast centers
   b national, provincial, local national disaster management centers;
4. Responsibilities of support infrastructure—IT and other technical support components;
5. NMHS routine forecast and warning operations
   a SARFFG and SWFDP product overview
   b meteorological, hydrologic, and precipitation forecast evaluations
   c SARFFG and SWFDP product evaluations
   d routine bulletins—e.g., guidance, discussions, dissemination protocols
   e alerts/warnings—e.g., disaster management, public, dissemination protocols
   f validation;
6. Staff training;
7. NMHS non-routine operations—e.g., operations during unusual events; and
8. Outreach
   a public
   b disaster managers.

5. Project Evaluation

The project described in this article is underway and it will take several years to determine its effectiveness. During this time the project team will be working with the countries to monitor and measure its effectiveness, especially with regards to improved service delivery to NMHS users such as DRR organizations. Future activities will involve the review and monitoring of outputs from the application of the Roadmap and Concept of Operation activities to determine if improvements have been made with early warnings for extreme hydrometeorological events. This evaluation will also include a determination of the need for future improvements and/or modifications in operational procedures. In addition, approaches to forecast and warning verification will be documented so they can be applied consistently by participating NMHSs. Warning verifications are especially critical and, with the cooperation of the countries, statistics can be derived for accuracy, timeliness, and warning lead times. For forecast verifications, the countries will be trained on the application of a WMO document which will be revised to accommodate SARFFG products under this project [13]. The evaluation will also include assessing the effectiveness and accuracy of the SWFDP and SARFFG systems as tools to support NMHS operations.

Current plans are to have these verifications presented, reviewed and evaluated at a series of steering committee meetings and/or SWFDP/SARFFG technical workshops for discussion, leading to recommendations for further improvements to the end-to-end early warning system including emergency preparedness and response measures.

Of particular interest to the project team, other than the forecast and warning verification statistics, will be feedback to assess the following success metrics:

- Receive positive user responses as to the timeliness and effectiveness of severe weather and flash flood warning products developed by the NMHS using the SWFDP and SARFFG systems and other available (e.g., local) data.
- Success stories that demonstrate reduction in economic losses and lives saved through either the mitigation or aversion of severe weather or flash flood disasters as a direct consequence of improvements in warnings delivered through the NMHS.

6. Conclusions

In the developing or least developed countries in Southern Africa, many National Meteorological and Hydrologic Services have limited or no capacity to provide early warning services for extreme hydrometeorological events, including severe weather and flash flood alerts and warnings. There are some NMHSs which have a limited warning capacity but are facing challenges in implementing the currently existing systems available to them. The currently implemented project, Delivery of Warnings of Hydrometeorological Hazards—Southern Africa Region, looks to link and leverage the operational Southern Africa Flash Flood Guidance System and the Severe Weather Forecast Demonstration Project to provide the catalyst for the NMHS and supporting governmental agencies needed to integrate all the various components of early warning systems to fully develop their early warning potential. A key element of the project is the development of Roadmap and CONOPS documents by each country as input to country-specific project implementation plans. These documents are used to track progress in the full integration of the SWFDP and SARFFG systems into NMHS operations and the development of the end-to-end early warning systems for hydrometeorological hazards, specifically for severe weather and flash floods.

Acknowledgments: The authors would like to thank Paul Pilon, Ayhan Sayin, and Abdoulaye Harou of the World Meteorological Organization, Curt Barrett of the U.S. Agency for International Development/Office of U.S. Foreign Disaster Assistance, and Eugene Poolman of the South African Weather Service for their involvement in the development and implementation of this important project for Southern Africa. As part of the Project Management Team, each was involved in the development of the project concept and implementation and work plans critical for a successful project. A special thanks to Konstantine Georgakakos and Eylon Shamir of the Hydrologic Research Center for their review and comments on the article.

Conflicts of Interest: The authors declare no conflict of interest.

Disclaimer: The program was made possible through support provided by the Office of U.S. Foreign Disaster Assistance, U.S. Agency for International Development, under the terms of Award No. AID-OFDA-A-11-00016. The opinions expressed in this publication are those of the authors and do not necessarily reflect the views of the U.S. Agency for International Development.

References

1. Guha-Sapir, D.; Below, R.; Hoyois, P. *EM-DAT: International Disaster Database*; Université Catholique de Louvain: Brussels, Belgium, 2016.
2. United Nations World Meteorological Organization. *Capacity Assessment of National Meteorological and Hydrological Services in Support of Disaster Risk Reduction—Analysis of the 2006 WMO Disaster Risk Reduction Country-Level Survey*; World Meteorological Organization: Geneva, Switzerland, 2008.

3. United Nations World Meteorological Organization. *Delivery of Warnings of Hydrometeorological Hazards—Southern Africa Region—Regional Technical Meeting/CONOPS Workshop, Pretoria, South Africa, 26–30 October 2015, Final Report*; World Meteorological Organization: Geneva, Switzerland, 2015.
4. ReliefWeb. *Southern Africa Floods—January 2015*; U.N. Office for the Coordination of Humanitarian Affairs: Geneva, Switzerland, 2015.
5. United Nations International Strategy for Disaster Reduction. *Global Survey of Early Warning Systems*; Third International Conference on Early Warning: Bonn, Germany, 2006.
6. Rogers, D.; Tsirkunov, V. *Implementing Hazard Early Warning Systems*; World Bank Global Facility for Disaster Reduction and Recovery: Washington, DC, USA, 2011.
7. United Nations. Transforming Our World: The 2030 Agenda for Sustainable Development. Available online: http://www.un.org/pga/wp-content/uploads/sites/3/2015/08/120815_outcome-document-of-Summit-for-adoption-of-the-post-2015-development-agenda.pdf (accessed on 17 June 2016).
8. Carpenter, T.; Sperfslage, J.; Georgakakos, K.; Sweeney, T.; Fread, D. National Threshold Runoff Estimation Utilizing GIS in Support of Operational Flash Flood Warning Systems. *J. Hydrol.* **1999**, *224*, 21–44. [CrossRef]
9. Georgakakos, K. Analytical Results for Operational Flash Flood Guidance. *J. Hydrol.* **2006**, *317*, 81–103. [CrossRef]
10. Georgakakos, K.; Graham, R.; Jubach, R.; Modrick, T.; Shamir, E.; Spencer, C.; Sperfslage, J. *Global Flash Flood Guidance System Phase I*; HRC Technical Report No. 9; Hydrologic Research Center: San Diego, CA, USA, 2013.
11. Graham, R., Ed.; *Flash Flood Guidance Gazette*; Hydrologic Research Center: San Diego, CA, USA, 2015; Volume 5.
12. University Corporation for Atmospheric Research. *Flash Flood Early Warning System Reference Guide*; ISBN: 978-0-615-37421-5. University Corporation for Atmospheric Research: Boulder, CO, USA, 2010.
13. World Meteorological Organization. *Forecast Verification for the African Severe Weather Forecasting Demonstration Projects*; WMO Doc-No. 1132; World Meteorological Organization: Geneva, Switzerland, 2014.

water

MDPI

Article

Undermining Demand Management with Supply Management: Moral Hazard in Israeli Water Policies

David Katz

Department of Geography and Environmental Studies, University of Haifa, Haifa 39105, Israel; katzd@geo.haifa.ac.il; Tel.: +972-4-824-8106

Academic Editors: Sharon B. Megdal, Susanna Eden and Eylon Shamir
Received: 21 February 2016; Accepted: 14 April 2016; Published: 20 April 2016

Abstract: Most water managers use a mixture of both supply-side and demand-side policies, seeking to capitalize on the relative advantages of each. However, supply augmentation undertaken to avoid overdrafts can reduce the effectiveness of demand management policies if the two strategies are not carefully integrated. Such a result can stem from a type of moral hazard phenomenon by which consumers, aware of the increases in potential supply, discount the importance of conservation. This is illustrated by the case of Israel. Initial national-scale water-supply projects were followed by over-extraction, which, in turn, compelled implementation of wide-spread demand management measures to reduce consumption. With the recent advent of large-scale desalination in Israel, public perception regarding the importance of conservation has diminished and consumption has increased—this, despite periodic drought conditions and critically low levels of water reserves.

Keywords: conservation; demand management; supply management; moral hazard; water policy

1. Introduction

Water management options have typically been categorized as either supply management or demand management. The former focuses on enlarging the amount of resources available, while the second focuses on reducing the amount of needed for consumptive purposes. Historically, civil and water engineers have focused on large-scale supply augmentation infrastructure projects, while economists and environmentalists have tended to advocate for efficiency improvements and conservation oriented policies typically associated with demand management (e.g., [1–4]). Each approach has its relative merits. Supply-side policies enlarge the pie, promoting possibilities for increased economic activity and avoiding the difficult social and political obstacles involved in such demand-side options as cutting water quotas or increasing prices. Demand management options are often cheaper, more economically efficient, and have less negative environmental impacts than supply augmentation [2–5].

Most countries (or municipalities or other water governing bodies) employ a mix of supply-side and demand-side management strategies. The relative emphasis placed on each shifts over time and according to circumstances. For instance, most developing countries tend to focus first on supply management, building water and sanitation infrastructure. Indeed, access to "improved" water supplies providing safe drinking water is one of the first priorities of many developing countries, even at low levels of economic development [6–8]. More advanced economies often shift their focus to demand management options, after basic infrastructure is in place and relatively low cost supply-side options have been exhausted. This is especially the case in countries or areas that have fully tapped readily available renewable resources. Demand management options are also more likely to be emphasized when dealing with immediate and/or short term issues such as drought adaptation, as supply-side options often necessitate longer planning horizons.

Water scarcity already affects billions of people around the globe [9]. This situation is expected by many to worsen in the near future due to population growth, economic growth and climate change which is predicted to alter the quantity, timing, intensity, and duration of precipitation events, as well as increased evapotranspiration [10–12]. In order to address such challenges, many water managers and policymakers are actively pursuing both supply-side and demand-side management strategies concomitantly, hoping to capitalize on the relative advantages of each.

Simultaneous pursuit of both types of management strategies is often recommended within the extensive literature on Integrated Water Resource Management (IWRM) (e.g., [13–15]). There can be, however, tradeoffs between water management goals, whereby attempts to implement one type of policy, inadvertently reduces the effectiveness of the other. For instance, increases in technical efficiency can lead to "rebound effects", which can, at least partially, cause increases in consumption due to the reduced cost per unit of consumption or associated output [16,17].

Another possible tradeoff, and the focus of this paper, is the risk that investing in significant expansion of supplies can inadvertently undermine various demand management policies, especially if the two strategies are not specifically developed in an integrated and mutually reinforcing manner. This can stem from a type of moral hazard phenomenon by which consumers, aware of the increases in potential supply, discount the importance of conservation. Such a result is likelier the more the government or water utilities emphasize the scale and/or merits of the supply augmentation projects. Thus, though water managers may be tempted to implement both types of policies under the assumption that they are complementary, they should be aware that in some cases they may work at cross-purposes, and thus, the effect of joint implementation may be less than the sum of the effects of each policy implemented in isolation.

To illustrate this potential conflict, this article demonstrates how Israel's large-scale adoption of seawater desalination has weakened public perception of water scarcity and therefore has undone much of the efforts invested over the past decades in public awareness regarding the importance of conservation. The public's perception of a reduction in water scarcity and, as a result, an attenuation in its conservation habits, come despite continued periodic droughts and critically low levels of water reserves. To support the article's argument, this case study presents both publically available data on water consumption as well as results of a randomized survey of the Israeli public's perceptions and knowledge regarding the water sector and the importance of water conservation.

The article proceeds as follows. The following section provides an overview of available renewable water resources in Israel and of indicators of national water scarcity. Section 3 provides a chronological review of shifts in Israeli water policy, from an emphasis on supply-side management strategies, such as large scale water pumping and conveyance projects, cloud-seeding, and large-scale wastewater reuse, to a heavier reliance on demand management policies such as reduction in water quotas, pricing mechanisms, and conservation campaigns. It then shows how the combination of increasing demand for water within Israel and by its neighbors, together with decreasing reliability of precipitation, led the government to pursue large-scale seawater desalination. Section 4 presents data on changes in water consumption over the period since the inauguration of desalination and results of a randomized survey of public perception of water scarcity and the importance of water conservation in order to show how conservation habits have changed since the advent of desalination. Section 5 presents a discussion of the case study results, and Section 6 presents concluding remarks.

2. Overview of Available Water Resources in Israel

Israel is a small (20,770 square kilometers, official pre-1967 boundaries), semi-arid country. Over 60% of its landmass is desert. It is densely populated (roughly 8.4 million inhabitants as of 2016 [18]), especially in the non-desert areas. Average annual renewable freshwater resources were long estimated at approximately 1600 million cubic meters (mcm) [19]; however, this estimate has been revised in recent years to roughly 1200 mcm [20]. The non-desert portion of Israel is characterized by a Mediterranean climate, with the majority of the rainfall concentrated in the winter months, with long dry and hot summers, and high rates of evaporation. Because of geographical and topographic conditions—the non-desert area is long, narrow and hilly—much of the rainfall is lost to runoff to the Mediterranean Sea. Approximately 30% of the natural water supply is from surface water (primarily the Sea of Galilee—the country's only lake), with almost all of the remainder coming from groundwater. Nearly all of the country's natural freshwater resources are transboundary in nature, being shared with the Palestinian Authority and Jordan, and to a lesser extent with Syria and Lebanon.

The Falkenmark index is a commonly referenced measure of national water scarcity. According to this index, countries with annual renewable water resources of less than 1000 cubic meters (m^3) per capita suffer from water scarcity and countries with less than 500 m^3 per capita suffer from chronic or "absolute" water scarcity [21]. Using the upper-bound estimate of average annual renewable freshwater listed above (1600 mcm), national per capita availability is at 190 m^3, well below the threshold for chronic water scarcity. Using the revised estimate of water availability (1200 mcm), the situation is even more dire, at less than 150 m^3. As will be elaborated more fully in the following section, Israel has invested heavily in non-conventional water resource augmentation, such as cloud-seeding, treated brackish water, wastewater reuse, and, most recently, large-scale seawater desalination. Together, these non-conventional sources add roughly 1000 mcm to the national annual water budget. Taking these water sources into consideration, using the two estimates above, annual per capita availability increases to approximately 260–310 m^3; still well within the Falkenmark standard for chronic water scarcity.

Climate change models largely predict additional stress on the country's water supplies. Climate scientists have already identified trends of warming temperatures and decreasing precipitation over the last several decades. According to Alpert *et al.*, "average temperature over the Mediterranean area has increased by 1.5–4 °C in the last 100 years" and "The precipitation above most of the Mediterranean shows a dominant negative trend in the last 50 years" ([22], p. 163). Other research has already identified increasing variability and uncertainty in temporal and spatial precipitation patterns in the region, and predicts a continuation of such trends [23].

Predictions regarding expected rainfall in the region differ by model. The combined results from nine climate models, using the A2 scenario of the International Panel on Climate Change (IPCC), for instance, predict a reduction of 2.5%–9.5% for the period 2045–2065 relative to the period 1961–1990 [24]. Results from the B2 scenario, however, are unclear [19]. Temperature is expected to increase (by 1.9–2.7 °C in the A2 scenario for the period 2045–2065 [20]). The region is also expected to have more frequent and longer duration heatwaves [25]. This may lead to an associated increase in evaporation, though recent research from the region shows that the relationship between temperature and surface-water evaporation is not linear, and that temperature increases may be offset by changes in wind patterns [26]. Another likely impact is salinization of the coastal aquifer due to sea-level rise [27,28]. One estimate predicts that each 50 cm rise in sea level would reduce the storage capacity of the coastal aquifer by 16.3 mcm per kilometer of coast [29]. In summary, climate change is already evident and is expected to further reduce naturally available water resources.

3. A Brief Review of Historical Supply and Demand Management in Israel

Feitelson [30] divides Israeli water management into four distinct periods. The first, from the founding of the state in 1948 until 1964, he termed the "Hydraulic Mission Period". This was characterized by several large-scale water supply projects. It was during this period that the nation developed the Yarkon–Negev water project to bring water from the coastal plain to the Negev desert in the south. This was the largest water conveyance project of its kind in the world at the time [31]. The capstone of the period was the building of the National Water Carrier (NWC), completed in 1964, which brought water from the Sea of Galilee to the coastal plain, where the majority of the population lived. The NWC essentially connected almost the entire country into one centralized water delivery system. During this period the country also began heavy pumping of groundwater, which was eventually also incorporated into the NWC delivery system.

During the second period, which lasted until roughly 1990, policy was also largely focused on supply augmentation, but this time by means of non-conventional sources. By the mid-1960s, shortly following the inauguration of the NWC, Israel had essentially utilized 100% of available renewable fresh-water supplies. The environmental toll of such intensive development of water supplies was tremendous. The flow of the Jordan River was reduced to less than 10% of its natural flow. The level of the Dead Sea, a terminal lake at the end of the Jordan River, began declining by over a meter per year. Many of the coastal streams were desiccated and their springs dried up completely [32]. In addition, saltwater intrusion began along the coastal aquifer.

The capture of the West Bank, Gaza, and the Golan Heights in 1967 did not result in substantial increases in water consumption [33]. Most of the relevant water resources were transboundary in nature and Israel had already been utilizing them prior to the war. By controlling this territory, however, Israel was able to regulate withdrawals from these areas in order to ensure that the water it was already using continued to be available.

Given the limited available resources, in order to address growing water demand due to a growing population and an increasing standard of living, Israel began to develop non-conventional water sources. A plan to develop large-scale seawater desalination did not come to fruition during this period due to high costs. Other technologies did take off, however, and substantially contributed to the countries available water supplies. The first was a cloud-seeding program, primarily in the north of the country. It is claimed that this program has added approximately 13% to annual rainfall [34,35], though these figures have been questioned by some [36].

Another major contributor to increased water supplies that occurred during this second period of water development was reuse of treated wastewater for irrigation purposes. Israel has become a world leader in wastewater reuse. As of the time of the writing of this article, over 80% of municipal sewage is treated and reused—by far, the highest rate in the world [37]. This adds over 400 mcm to the national water budget, roughly a third of natural renewable supplies.

Some initial advancement was made on demand management during this period as well. Drip irrigation was developed, which had the effect of substantially reducing the water needs per unit of crop. Despite widespread adoption of this potential demand management technology, however, it did not reduce agricultural water consumption. Rather, with water for the agricultural sector still priced at below its market value (shadow price), it remained a quantity-rationed good. Thus, rather than leading to a decline in consumption with stable yields, the increase in efficiency produced significant gains in yields with the existing agricultural water allocations [38]—that is, it resulted in "more crop per drop" but not in less drops dedicated to crops.

The third period in Israel's water history, beginning in the 1990s and lasting until the advent of wide-scale desalination beginning in 2005, saw a shift from the supply-side dominated policies of the first two periods, to a greater emphasis on demand management. Facing recurring droughts, depleted aquifers and levels of the Sea of Galilee at record lows, often below the "red line" set by authorities as the minimum level necessary to avoid permanent ecological damage (see Figure 1), Israeli policy changed its focus to reducing consumption. The first and most significant area of reduced consumption was allocations of freshwater to the agricultural sector, which the government cut by nearly half during this period (Figure 2a). Some of this was compensated for by increased allocations of treated wastewater (Figure 2b), but, overall, freshwater consumption declined by roughly 25% to approximately the revised annual renewable rates of 1200 mcm.

With the decline in agricultural consumption, municipal consumption overtook agricultural consumption as the primary consuming sector of freshwater in the country (see Figure 2) and thus became a more prominent focus of demand management policies. Periodic restrictions on municipal use were put in place, such as on lawn watering, especially during drought periods. Low-flow toilets were mandated for new construction. However, unlike the agricultural and industrial sectors in Israel, water supplies to the municipal sector are not allocated by administrative quotas, and thus, not as amenable to command and control type regulation. The authorities, therefore, made use of other demand management tools. (For a review of Israeli demand management policies, see [39].) Prices on municipal water were gradually raised and several high-profile awareness raising campaigns were advanced by the water authority. Campaigns emphasized that "every drop counts", and newspapers and other popular media outlets began publishing the levels of the Sea of Galilee as a means of highlighting the water crisis.

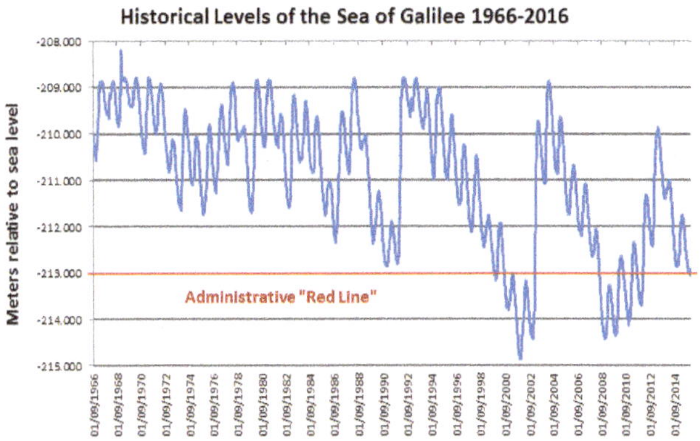

Figure 1. Historical levels of the Sea of Galilee (09/1966–01/2016). Based on Data from [40].

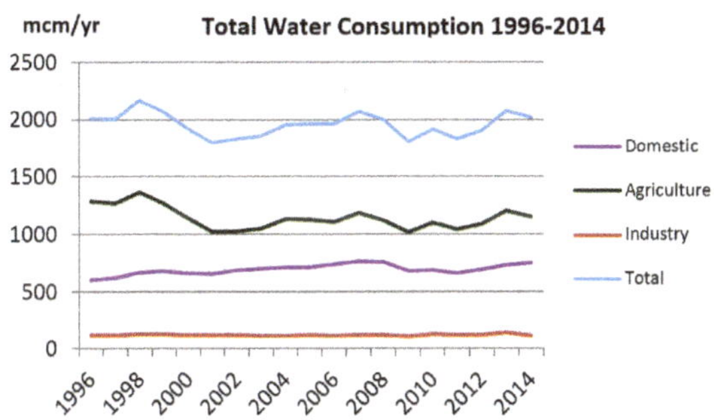

(a)

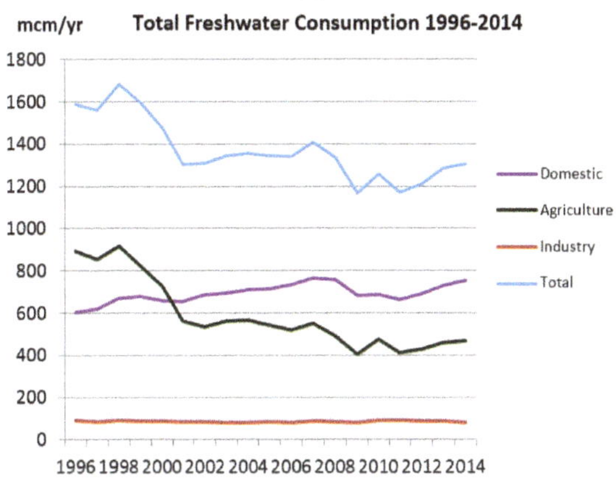

(b)

Figure 2. *Cont.*

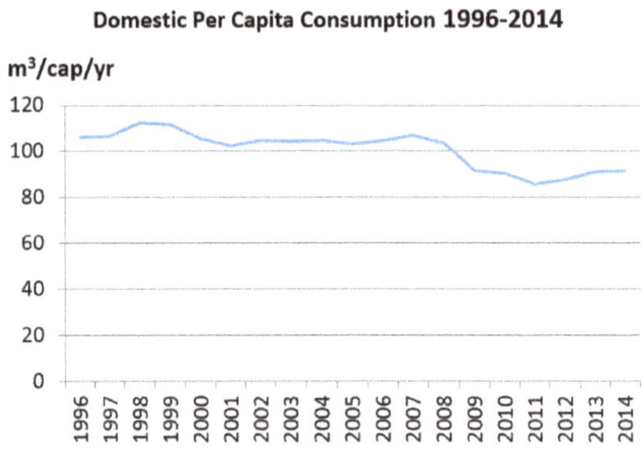

(c)

Figure 2. (**a**) total water consumption 1996–2014; (**b**) freshwater consumption 1996–2014; (**c**) domestic per capita water consumption 1996–2014. Based on data from [41].

During this period, however, new demands on freshwater were being made. The country was absorbing over a million new immigrants, primarily those coming from the former Soviet Union. Peace agreements with the Palestinians and Jordanians both contained clauses committing Israel to transfer more water to its neighbors. In addition, growing environmental awareness resulted in new demands for in-stream water allocations for ecological rehabilitation purposes [32]. During this period, Israel also experienced serious droughts. In order to accommodate the increased demands on scarce water supplies, concomitantly with the implementation of demand management, the government committed to long-term supply augmentation via seawater desalination.

The year 2005 marked the beginning of the fourth period in Israel's water management according to Feitelson, with the inauguration of the country's first large-scale desalination unit, adding 120 mcm per year of freshwater. By 2013, the country had four large desalination plants operating, producing nearly 500 mcm, annually, and had commissioned yet another plant, due to come on line in 2016 and provide an additional 100 mcm per year. Within less than a decade, collectively, desalination plants had become the country's single largest source of freshwater, providing more than the Sea of Galilee or any of the country's aquifers.

During the first few years of this period, when desalination was just coming online, water reserves were at historical lows and the country was experiencing a severe drought. Therefore, the authorities continued implementing demand management policies. Subsidies for water were reduced or eliminated across sectors. The average price of both agricultural and municipal water more than doubled between 2000 and 2010 [42,43], with most of the price increase coming in the post-desalination years (see Figure 3) (note: much of the spike in 2010 was a result of consolidation of water supply and sewage treatment into one payment, rather than pure price increases).

The cost of desalination was important (though not the sole) factor leading to price increases. According to the Israeli Water Authority, as of early 2016 desalination costs represented 16% of the cost of municipal water supply and treatment [43]. Other factors contributing to price increases included a removal of subsidies as part of a move towards full-cost pricing and a restructuring of municipal water supply from public to private water utilities. However, desalination was indirectly responsible for a greater share of the price increases than its mere share of the overall costs would indicate. This is because overall elimination of subsidies and promotion of full-cost recovery were among the measures put forth by the Ministry of Finance as conditions in order to gain its support for the costly desalination projects.

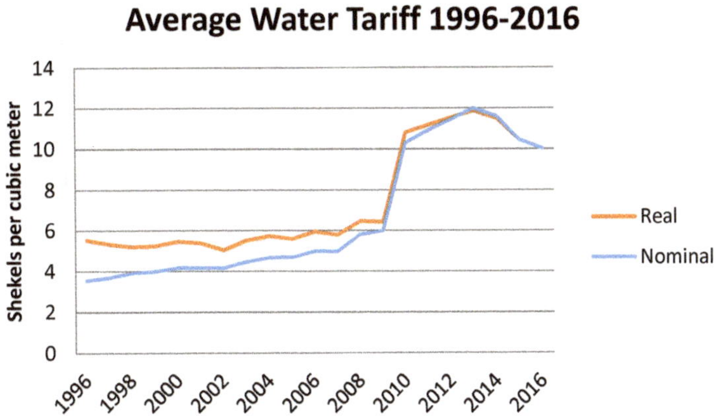

Figure 3. Average municipal water tariffs 1996–2016. Based on data from [43]. Real prices in 2016 shekels.

The Water Authority, the governmental body responsible for water management, also intensified its campaigns for conservation, enlisting several celebrities in its efforts. Assessments of the campaigns determined them to be effective [44], though, because they were implemented together with other demand management policies, and it is difficult to isolate the exact impact of the campaigns alone [45]. Other conservation measures were also implemented, for instance, distribution of low-flow faucet filters. The result of the combination of demand management policies was a nearly 20% drop in municipal water consumption between 2007 and 2009 (see Figure 2). Already by 2008, consumption of natural sources of freshwater had dropped to its lowest levels since 1967 [46].

4. Tradeoffs between Supply and Demand Management Policies

Seemingly, Israel had succeeded in both significantly reducing demand and increasing supplies. However, several events and issues have presented challenges to this mutual coexistence. As of 2011, Israel was no longer depleting its aquifers and the level of the Sea of Galilee had risen to well above the red-line (though still well below its full capacity). Government officials and water managers, proud of their achievements in rapidly increasing the amount of freshwater supplies, and also wanting to justify the cost increases to the population, came out with statements to the effect that Israel was no longer suffering from a water shortage. For instance, Uri Shani, the former head of the Water Authority, was quoted as saying in 2011 that Israel was already overcoming "*the critical water shortages*" it had been facing. Despite comments by other officials that "*the hydrological crisis that the State of Israel has been in over the past years is far from over. The country's citizens must still continue to conserve*", the headlines of the paper read "*The Departing Director of the Water Authority: Israel's Water Crisis is Over*" [47].

In the same vein, in 2013, Uzi Landau, then serving as Minister of National Infrastructures, which oversees the Water Authority, stated in an interview "*Today, it can be claimed with confidence that the water crisis is behind us*" [48]. He said that a combination of both additional supplies from desalination and reduced demand due to campaigns and pricing had achieved "stability" in the water sector. According to one observer, by 2013 the Israeli government had declared "water independence from weather" [49].

Statements such as these, together with a de-emphasis of demand management measures, caused many people to conclude that water in Israel was no longer scarce. Public protest at the rapid and dramatic increase in tariffs for water, led to pressure on policymakers to reduce prices. A conservation fee imposed on the highest tier of a three-tiered municipal water tariff (*i.e.*, on the biggest water consumers) was ceased in 2009 only several months after it was initiated. Between 2013 and 2016, both average and marginal real prices for municipal water were reduced by nearly 20% (based on data from [43]); this, despite the fact that expenditures for water represent only 1% of household disposable income (based on [50]) and the fact that, as of 2013, average tariff rates for municipal water in Israel were lower than average for OECD countries [51].

In addition, the government, weary of "conservation fatigue" on the part of consumers, changed the nature of their campaigns. In 2010 and 2011, it asked consumers to "hold out" and keep conserving for another bit until further desalination came online. By 2012, it had stopped major conservation campaigns altogether.

While, in essence, the additional desalinated water had merely allowed Israel to cease withdrawing beyond natural recharge rates. The statements of policymakers confidently singing the praises of desalination and declaring an end to the water crisis, were seen by many as evidence that water was no longer scarce, and even that Israel now had water surpluses. For instance, a writer of a recently published book on Israeli water management stated that Israel was *"not just water secure but even abundant"* [52]. Similar views were shared by water officials and the media. For instance, in 2014, Avraham Tenne, head of the desalination division of Israel's Water Authority, stated "*We have all the water we need, even in the year which was the worst year ever regarding precipitation*" [53]. An article on "the end of Israel's water shortage" stated *"Remember all the years of being told to conserve 'every drop'? Well, times have changed: Today, Israel has so much affordable water, it can offer to export it."* [54].

A survey conducted for the purposes of this article in early 2016 asked a random sample of 70 Israeli adults a number of questions about their perception of water resources in Israel. When posed the question of whether or not Israel faced water scarcity, less than 38% indicated that they believed Israel faced water scarcity (Table 1). Nearly half of respondents indicated that they believed that Israel had faced scarcity in the past, but no longer did, due to desalination, while almost 15% believed that Israel had never faced scarcity.

Table 1. Public perceptions of water scarcity ($n = 70$).

Attitude towards National Water Scarcity	Percent of Respondents
Israel faces water scarcity	37.5%
Israel did face water scarcity in the past, but no longer does	47.8%
Israel has never faced water scarcity	14.5%

As a result of the decline in the implementation of demand management measures, together with the decrease in public perception of scarcity due to desalination, both overall water consumption and consumption of freshwater have increased each year since 2011 (see Figure 2b). Municipal consumption also increased, both in aggregate and per capita terms, each year since 2011 (Figure 2c). Per capita municipal consumption in 2014 was more than 7% higher than 2011 levels (based on data from [41]). Thus, much of the effect of the previous demand management measures was essentially eroded during this period.

The increased consumption would not be problematic if supplies were, in fact, greater than demand. However, that was not the case. Israel and its neighbors, which also suffer from chronic water scarcity, have growing populations. Jordan also has the additional burden of supplying water for hundreds of thousands of Syrian refugees, and Israel has obliged by increasing water transfers to Jordan. Furthermore, Israel and the region are still prone to drought conditions. As a result of a combination of scant rainfall in the north of the country in 2014 and 2015, and increased consumption, the level of Sea of Galilee was once again below the red-line as of late 2015 and early 2016 (Figure 1), this despite a significant reduction in withdrawals from the lake in recent years.

The impression that water is no longer scarce is in stark contrast to the fact, that, as mentioned earlier, even with the addition of desalination, according to the Falkenmark index, Israel is still solidly in the category of countries suffering from chronic water scarcity. The additional desalination capacity has brought annual freshwater supplies to only slightly higher than the original estimate of 1600 mcm that had served water planning in Israel for decades. In addition, the country still has huge historical deficits of water, depleted aquifers, and a legacy of desiccated streams. According to the former Minister of National Infrastructures Uzi Landau, historical over-withdrawals meant that Israel faced a cumulative water debt of 1500–2000 mcm that needed to be restored to nature [55]. This is three to four times the total amount of annual desalination capacity, and is likely to take years to replenish. These facts seem clearly at odds with the public perception, often promoted by government officials, that Israel no longer faces water scarcity.

5. Discussion

From the data presented in the previous section, it seems that the supply-side policies served to diminish the once successful demand-side management measures that were in place. The dynamic at play appears to be more than simply a market reaction to increased supply. In theory, the increase in consumption could simply be due to the lowering of tariff rates over the past two years. Even if this were the case, it still could support the claim that expanded supplies undermined demand management, since much of the rationale for these price decreases was itself due to pressure from the public that was convinced that water was no longer scarce, and thus, high prices no longer justified. However, it does not appear that the increased consumption was primarily in response to price decreases. The increase in consumption has been continuous since 2011, while prices began to decline only in 2013 (see Figures 2c and 3). In addition, the price decline in the last few years did not get nearly the media attention that price increases did. Thus, most of the public is unaware of the price decrease, and so is unlikely to have changed behavior in response.

In the survey conducted for this study in 2016, when given ranges of prices for a cubic meter of water, only 17% of those surveyed correctly identified the range covering the actual price range over the last eight years, while an identical percentage chose price ranges higher than the prices at their peak level in 2013, and over 62% answered that they had no idea. More importantly, when asked about the change in water prices over the last two and half years, only 13% correctly indicated that they had declined (all indicating a "slight decline"), while over 65% answered that they had increased, with over 26% replying that they had increased substantially (Table 2). This lack of public awareness of water prices confirms similar findings from earlier surveys (e.g., [56]).

Table 2. Public perception of price changes over the past two years (n = 70).

Perception of Price Change	Percent of Respondents
Decreased substantially	0.0%
Decreased slightly	13.0%
Remained the same	21.7%
Increased slightly	39.1%
Increased substantially	26.1%

A few conclusions can be inferred from such findings. One is that price decreases have not been widely felt by the public, and, thus, are unlikely to explain much of the increase in consumption over the last years. The increase in consumption began during a period following the cessation of public campaigns, during which prices were actually increasing. This seems to indicate both that prices have been, at best, a crude tool for regulating consumption over the past years, and that public campaigns and perceptions seem to play an important role in impacting consumption.

Following official government committee investigations into the causes of the water management crisis in 2010 [57], policymakers were eager both to demonstrate that they had taken significant steps to alleviate the situation, as well as to justify the costs of their primary solution, desalination. Their response was to tout the merits of desalination as a permanent solution to the nation's water shortages. While in the interim, desalination had merely obviated the need for further overdrafts, a common perception was that the country now had surpluses and had solved the issue of water scarcity completely. Given this perception, it will be difficult to once again convince the public to conserve.

As such, an effective demand management tool is essentially no longer available, even at a time when reserves are at critically low levels, even below the so-called red-lines.

6. Conclusions

Water managers are always likely to seek a mix of supply and demand management strategies. Ideally, these tools would be complementary and their impact additive. This study aimed to show a potential trade-off between supply and demand-side policies, namely, the risk of moral hazard, by which consumers, knowing that additional supplies are available or are soon to be available, discount the need for conservation.

One should be careful not to infer from this case study that increased supplies will always detract from conservation efforts. In the case examined, policymakers, for various reasons, perhaps overstated the contribution of desalination to alleviating the country's water shortages. Arguably, the media also contributed by oversimplifying and misrepresenting the situation.

It is possible that better planned policy could have avoided much of the undesired impacts. For instance, the rapid increase in prices caused public outrage. In order to justify the costs of desalination, policymakers responded by emphasizing that desalination would solve the country's water shortages. A more gradual increase of costs may have reduced some of the need to tout desalination's benefits, and, as a result, also may have avoided some of the public backlash. This also may have avoided the need for an eventual decrease in tariffs that came as a response to the backlash.

Continuation of conservation campaigns, even as new supplies were coming online, could also have prevented the perception that water was no longer scarce. Finally, while policymakers are largely not responsible for media coverage of their actions, better care could have been taken to ensure more accurate and nuanced coverage that emphasized that water scarcity was still an ever-present issue. While there was responsible journalism, which attempted to remind the public that shortages were not a thing of the past and that conservation was still warranted (e.g., [58,59]), it was largely reacting to the already ingrained perception that water was now abundant.

Challenges in achieving coexistence in Israel seem also to apply to the field of water policy options. Expanding available supplies by means of desalination is seen as the bulwark of Israel's long-term water plan [20]. These supply augmentation efforts, however, have, intentionally or not, eroded the effectiveness of various demand management policies implemented over the course of the last several years. The lesson for water managers is that special care needs be taken to ensure that provision of increased supplies does not undermine conservation objectives. This is likely to entail coordinated pricing policies as well as specific strategies to communicate to the public the importance of continued conservation.

Conflicts of Interest: The author declares no conflict of interest.

Abbreviations

The following abbreviations are used in this manuscript:

IWRM	Integrated Water Resource Management
IPCC	International Panel on Climate Change
MCM	Million cubic meters
NWC	National Water Carrier
OECD	Organization of Economic Cooperation and Development

References

1. Dziegielewski, B. Strategies for Managing Water Demand. *Water Resour. Update* **2003**, *126*, 29–39.
2. Dolatyar, M.; Gray, T. *Water Politics in the Middle East: A Context for Conflict or Cooperation?* Palgrave Macmillan: Basingstroke, Hampshire, UK; New York, NY, USA, 2000.

3. Gleick, P.H. Global Freshwater Resources: Soft-Path Solutions for the 21st Century. *Science* **2003**, *302*, 1524–1528. [CrossRef] [PubMed]
4. Brandes, O.; Brooks, D. The Soft-Path for Water in a Nutshell. In *Friends of the Earth Canada*; POLIS Project on Ecological Governance; University of Victoria: Victoria, BC, Canada, 2007.
5. Butler, D.; Memon, F.A. (Eds.) *Water Demand Management*; IWA Press: London, UK; Seattle, WA, USA, 2006.
6. Shafik, N.; Bandyopadhyay, S. *Economic Growth and Environmental Quality: Time Series and Cross-Country Evidence*; World Bank: Washington, DC, USA, 1992.
7. Gleick, P. (Ed.) *The World's Water Volume 8: The Biennial Report on Freshwater Resources*; Island Press: Washington, DC, USA, 2014.
8. The United Nations Children's Emergency Fund (UNICEF); World Health Organization. *Progress on Sanitation and Drinking Water—2015 Update and MDG Assessment*; WHO Press: Geneva, Switzerland, 2015.
9. Mekonnen, M.M.; Hoekstra, A.Y. Four billion people facing severe water scarcity. *Sci. Adv.* **2016**, 2. [CrossRef] [PubMed]
10. Alavian, V.; Qaddumi, H.M.; Dickson, E.; Diez, S.M.; Danilenko, A.V.; Hirji, R.F.; Puz, G.; Pizarro, C.; Jacobsen, M.; Blankespoor, B. *Water and Climate Change: Understanding the Risks and Making Climate-Smart Investment Decisions*; World Bank: Washington, DC, USA, 2009.
11. Cooley, H. Water Management in a Changing Climate. In *The World's Water Volume 6: The Biennial Report on Freshwater Resources*; Gleick, P., Ed.; Island Press: Washington, DC, USA, 2008.
12. Miller, K.A. Climate Change and Water Resources: The Challenges Ahead. *J. Int. Aff.* **2008**, *61*, 35–50.
13. Jønch-Clausen, T.; Fugl, J. Firming up the Conceptual Basis of Integrated Water Resources Management. *Int. J. Water Res. Dev.* **2001**, *17*, 501–510. [CrossRef]
14. Global Water Partnership. *Sharing Knowledge for Equitable, Efficient and Sustainable Water Management*; Global Water Partnership: Stockholm, Sweden, 2003.
15. Borchardt, D., Ibisch, R., Eds.; *Integrated Water Resources Management in a Changing World: Lessons Learnt and Innovative Perspectives*; IWA Publishing: London, UK, 2013.
16. Dixon, A.M.; McManus, M. An Introduction to Life-Cycle and Rebound Effects in Water Systems. In *Water Demand Management*; Butler, D., Memon, F.A., Eds.; IWA Press: London, UK; Seattle, WA, USA, 2006.
17. Berbel, J.; Gutiérrez-Martín, C.; Rodríguez-Díaz, J.A.; Camacho, E.; Montesinos, P. Literature Review on Rebound Effect of Water Saving Measures and Analysis of a Spanish Case Study. *Water Resour. Manag.* **2014**, *29*, 663–678. [CrossRef]
18. Central Bureau of Statistics (Israel). Population. Table B/1.-Population, By Population Group. Available online: http://www1.cbs.gov.il/webpub/pub/text_page_eng.html?publ=93#2 (accessed on 8 February 2016).
19. Yakobovitz, M. *Water in Israel*; Shikmona Publishing Company: Haifa, Israel, 1971. (In Hebrew)
20. Israel Water Authority. Long-Term Master Plan for the National Water Sector Part A—Policy Document Version 4. Available online: http://www.water.gov.il/Hebrew/Planning-and-Development/Planning/MasterPlan/DocLib4/MasterPlan-en-v.4.pdf (accessed on 30 December 2015).
21. Falkenmark, M.; Lindh, G. *Water for a Starving World*; Westview Press: Boulder, CO, USA, 1976.
22. Alpert, P.; Krichak, S.O.; Shafir, H.; Haim, D.; Osetinsky, I. Climatic trends to extremes employing regional modeling and statistical interpretation over the E. Mediterranean. *Glob. Planet. Change* **2008**, *63*, 163–170. [CrossRef]
23. Paz, S.; Kutiel, H. Rainfall Regime Uncertainty (RRU) in an Eastern Mediterranean region—A methodological approach. *Israel J. Earth Sci.* **2003**, *52*, 47–63. [CrossRef]
24. The World Bank. *The Little Data Book on Climate Change 11*; The World Bank: Washington, DC, USA, 2012.
25. Ziv, B.; Saaroni, H.; Baharad, A.; Yekutieli, D.; Alpert, P. Indications for aggravation in summer heat conditions over the Mediterranean basin. *Geophys. Res. Lett.* **2005**, *32*, L12706. [CrossRef]
26. Shilo, E.; Ziv, B.; Shamir, E.; Rimmer, A. Evaporation from Lake Kinneret, Israel, during Hot Summer Days. *J. Hydrol.* **2015**, *528*, 264–275. [CrossRef]
27. Golan-Engelko, I.; Bar-Or, Y. Israel's Preparation for Global Climatic Changes, Phase I—The Implications of Climate Change for Israel: Ministry of Environmental Protection. 2008. Available online: http://www.sviva.gov.il/Enviroment/Static/Binaries/ModulKvatzim/adaptation_report_to_yeshayahu_revised_2010-_remarks_2.pdf (accessed on 14 January 2015).

28. Sowers, J.; Vengosh, A.; Weinthal, E. Climate change, water resources and the politics of adaptation in the Middle East and North Africa. *Clim. Chang.* **2011**, *104*, 599–627. [CrossRef]
29. Melloul, A.J.; Collins, M.L. Hydrological changes in coastal aquifers due to sea level rise. *Ocean Coast. Manag.* **2006**, *49*, 281–297. [CrossRef]
30. Feitelson, E. The Four Eras of Israeli Water Policies. In *Water Policy in Israel*; Becker, N., Ed.; Springer: Dordrecht, The Netherland, 2013.
31. Tal, A. *Pollution in a Promised Land: An Environmental History of Israel*; University of California Press: Berkeley, CA, USA, 2002.
32. Tal, A.; Katz, D. Rehabilitating Israel's Streams and Rivers. *Int. J. River Basin Manag.* **2012**, *10*, 317–330. [CrossRef]
33. Israel Water Commission. *The Water in Israel—Consumption and Production 1962–1970*; Ministry of Agriculture: Tel Aviv, Israel, 1972.
34. Gagin, A.; Neumann, J. The Second Israeli Randomized Cloud Seeding Experiment: Evaluation of the Results. *J. Appl. Meteorol.* **1981**, *20*, 1301–1311. [CrossRef]
35. Gvirtzman, H. *Water Resources in Israel*; Yad Ben-Zvi: Jerusalem, Israel, 2002. (In Hebrew)
36. Rangno, A.L.; Robbs, P.V. A New Look at the Israeli Cloud Seeding Experiments. *J. Appl. Meteorol.* **1995**, *34*, 1169–1193. [CrossRef]
37. Israel Water Authority. The Wastewater and Treated Effluents Infrastructure Development in Israel. Available online: http://water.gov.il/Hebrew/ProfessionalInfoAndData/2012/03-The%20Wastewater%20and%20Treated%20Effluents%20Infrastructure%20Development%20in%20Israel.pdf (accessed on 26 March 2015).
38. Kislev, Y. *The Water Economy of Israel*, Policy Paper No. 2011.15; The Taub Center for Social Policy Studies in Israel: Jerusalem, Israel, 2011.
39. Katz, D. Policies for Water Demand Management in Israel. In *Water Policy in Israel: Context, Issues and Options*; Becker, N., Ed.; Springer Press: Dordrecht, The Netherland, 2013.
40. Israel Water Authority. Kinneret Water Levels. Available online: http://water.gov.il/Hebrew/WaterResources/Kinneret-Basin/Pages/default.aspx (accessed on 8 February 2016). (In Hebrew)
41. Israel Water Authority. Consumption Survey 2014. Available online: http://water.gov.il/Hebrew/ProfessionalInfoAndData/Allocation-Consumption-and-production/Pages/Consumer-survey.aspx (accessed on 8 February 2016). (In Hebrew)
42. Kislev, Y. Water in Agriculture. In *Water Policy in Israel: Context, Issues and Options*; Becker, N., Ed.; Springer Press: Dordrecht, The Netherland, 2013.
43. Israel Water Authority. Water Tariffs. Available online: http://water.gov.il/Hebrew/Rates/Pages/Rates.aspx (accessed on 8 February 2016). (In Hebrew)
44. Lavee, D.; Danieli, Y.; Beniad, G.; Shvartzman, T.; Ash, T. Examining the Effectiveness of Residential Water Demand-side Management Policies in Israel. *Water Policy* **2013**, *15*, 585–597. [CrossRef]
45. Katz, D.; Grinstein, A.; Kronrod, A.; Nisan, U. Comparing marketing and price and mechanisms for water conservation. *J. Environ. Manag.* **2016**, in press.
46. Rinat, T. Water Consumption from Natural Sources the Lowest Since '67. *Haaretz*, 6 April 2008. Available online: http://www.haaretz.co.il/misc/1.1316400 (accessed on 6 April 2008). (In Hebrew)
47. Cohen, A. The Departing Director of the Water Authority: Israel's Water Crisis is Over. Available online: http://www.themarker.com/news/1.643631 (accessed on 18 May 2011). (In Hebrew)
48. Landau, U. Today, it can be claimed with confidence that the water crisis is behind us. *Mayim V'Hashkaya (Water and Irrigation)* **2013**, 8–12. (In Hebrew)
49. Siegel, S. *Let There Be Water: Israel's Solution for a Water-Starved World*; Thomas Dunne Books/St. Martin's Press: New York, NY, USA, 2015.
50. Central Bureau of Statistics (Israel). The 2013 Household Expenditure Survey. Table 1.1.-Monthly Income And Consumption Expenditure. Available online: http://cbs.gov.il/publications15/1613/pdf/t01_01.pdf (accessed on 2 April 2016).
51. OECD. *Environment at a Glance 2015: OECD Indicators*; OECD Publishing: Paris, France, 2015.
52. Siegel, S. Water Blog—A Cinderella Moment. Available online: http://www.sethmsiegel.com/a-cinderella-moment/ (accessed on 8 February 2016).

53. The Associated Press. Israel's Desalination Program Averts Future Water Crises. *Haaretz*, Available online: http://www.haaretz.com/israel-news/science/1.596270 (accessed on 31 May 2014). (In Hebrew)
54. Elizur, Y. Over and Drought: Why the End of Israel's Water Shortage Is a Secret. *Haaretz*, Available online: http://www.haaretz.com/israel-news/1.570374 (accessed on 24 January 2014). (In Hebrew)
55. Trilnik. Uzi Landau: By the End of the Decade Israel's Water Problems Will be Solved. *The Marker*, Available online: http://www.themarker.com/misc/article-print-page/1.1740828 (accessed on 26 May 2012). (In Hebrew)
56. National Investigative Committee. *Report of the National Investigative Committee for the Management of the Water Sector in Israel*; National Investigative Committee: Haifa, Israel, March, 2010. (In Hebrew)
57. Peled, M. Survey: 77% of the Public do not Know How Much They Will Pay for Water. *Calcalist*, Available online: http://www.calcalist.co.il/local/articles/0,7340,L-3380543,00.html. (accessed on 30 December 2009). (In Hebrew)
58. Porat, T. Waiting for the Desalination Tidings. *Ynet*, Available online: http://www.ynet.co.il/articles/0,7340, L-4137270,00.html (accessed on 22 October 2011). (In Hebrew)
59. Lavee, A. The Water Crisis, Remember? NRG. Available online: http://www.nrg.co.il/online/1/ART2/285/878.html (accessed on 15 September 2011). (In Hebrew)

water

MDPI

Article

Adaptive Governance of Water Resources Shared with Indigenous Peoples: The Role of Law

Barbara Cosens [1,*] and Brian C. Chaffin [2]

1 College of Law and Waters of the West Program, University of Idaho, 875 Perimeter Dr., Moscow, ID 83844-2321, USA
2 College of Forestry and Conservation, University of Montana, 32 Campus Drive, Missoula, MT 59812, USA; brian.chaffin@umontana.edu
* Correspondence: bcosens@uidaho.edu; Tel.: +1-208-885-6298

Academic Editors: Sharon B. Megdal, Susanna Eden and Eylon Shamir
Received: 7 January 2016; Accepted: 3 March 2016; Published: 11 March 2016

Abstract: Adaptive governance is an emergent phenomenon resulting from the interaction of locally driven collaborative efforts with a hierarchy of governmental regulation and management and is thought to be capable of navigating social–ecological change as society responds to the effects of climate change. The assertion of Native American water rights on highly developed water systems in North America has triggered governance innovations that resemble certain aspects of adaptive governance, and have emerged to accommodate the need for Indigenous water development and restoration of cultural and ecological resources. Similar innovations are observed in the assertion of Indigenous voices in Australia. This presents an opportunity to analyze the emergence of adaptive processes within complex legal systems. We explore the role of law in locally driven innovation in this context, concluding that any system of governance that requires greater flexibility will only be viewed as legitimate, and thus succeed, if attention is given not only to adaptive capacity, but also to aspects of good governance. Through examples of the assertion of Indigenous rights, we illustrate critical links between adaptive capacity in water management, good governance, and law.

Keywords: adaptive governance; legitimacy; water law; water management; stakeholder processes; Native American rights; Aboriginal rights; Indigenous rights

1. Introduction

The provision of water in the face of climate change and increasing scarcity is complicated by the fact that it is essential for life and has no substitute. As a result, solutions to water scarcity are as much issues of governance as they are issues of technology. Complicating governance efforts is the fact that water ignores jurisdictional, cultural, and economic boundaries. Climate change has introduced a level of uncertainty that challenges traditional bright line approaches to managing shared resources across jurisdictional boundaries. New approaches to governance are necessary to reconcile our obligations as stewards of water resources with our ability to extract its benefits, and to assure that those benefits are equitably distributed.

Much of the current discussion focuses on the need for greater adaptive capacity to cope with uncertainty and complexity associated with social and ecological change [1,2]. Scholarship also suggests an increasing awareness that these new forms of governance will not gain acceptance without the normative overlay of aspects of legitimacy, transparency, accountability, and increased public participation reflected in the term "good governance" [3–7]. In recent efforts to understand the role of law in facilitation of adaptive governance, it has become clear that in setting the stage, law provides an avenue to facilitate both adaptive capacity and "good governance" and that while the first is essential to adaptive governance, the second is essential to its implementation [7]. The importance of understanding the role of law in enhancing the ability of water basin societies to adapt and solve problems cannot be overstated. Governments cannot force participation or dictate self-organization and innovation. They can only influence how the stage is set. Thus, it is in attention to the law and formal institutional arrangements that society can provide the tools for the emergence of adaptive processes and legitimacy in their outcomes.

Over the past three decades in the United States and Australia, the assertion of Indigenous rights has triggered the emergence of processes with robust public participation, enhanced local adaptive capacity, and strong local legitimacy. On the ground, these processes appear self-organizing, locally driven by stakeholders with close ties to the water resources and therefore resemble the emergent process described in the adaptive governance literature. In reality, none would emerge without a legal and institutional framework capable of re-balancing interests and thus triggering dialog, providing support and a forum for local adaptive capacity building, and institutionalization of solutions, thus illustrating the normative overlay of good governance.

We begin by defining adaptive governance as used in this article and discussing the increasing recognition that attention to aspects of good governance is necessary for its emergence. We then describe the intersection of law, good governance and increased adaptive capacity in collaborative efforts to share and co-manage water resources across jurisdictional boundaries with Indigenous communities. We conclude with a discussion of the critical role of law and good governance in the emergence of adaptive governance.

2. Background

The term "governance" is used to describe the means, including processes and institutions, through which collective goals are chosen, decisions are made, and action is taken to achieve the chosen goals. Government, which is in part a reflection of the law, is merely a subset of governance [6–10]. "Adaptive governance" is *governance* that allows adaptive processes to emerge and much of the literature focuses on the non-governmental emergent aspects that are reflected in enhanced adaptive capacity [1,4,11–13]. Strong adaptive capacity is also observed in efforts described as co-management and collaborative management [6,8,14–20] and we use the term "adaptive governance" to include these types of processes [11]. While there is no exact formula to define adaptive governance, its attributes are thought to include: the capacity to learn, reflect and adapt; the distribution of authority over multiple entities capable of acting in the same sphere (referred to as polycentricity); the capacity to integrate management across jurisdictions and sectors that affect the resource in question through arrangements such as co-management; and the formal and informal engagement of those affected by decision making [6,11].

Substantial overlap exists between the attributes described as adaptive governance and what is referred to as "new governance" [3,21]. This overlap is important, because the new governance literature observes that the move to new governance may be a product of the rise of informal, non-governmental aspects of governance, the increase in interdependency and interaction of actors, the pressure for a greater citizen voice in natural resources management by those affected, and the rise of collaborative governance [3]. The focus of the adaptive governance literature is on managing change and uncertainty in ecological systems as the driving force behind the emergence of new approaches. Thus, new governance approaches emergent, locally driven collaboration from the perspective of a dynamic society and adaptive governance approaches it from the perspective a dynamic ecological system. The two systems are interrelated in any water basin, thus we draw on both perspectives.

Both new governance and adaptive governance call for devolution of governance to local levels, and adaptive governance has as its key focus increased flexibility in environmental governance. Yet it is recognized that major barriers to these approaches include lack of local capacity and practices to assure accountability and legitimacy, lack of attention to the broader public good [1,3–6], and loss of legitimacy from the instability resulting from flexible management [22]. Searching for a way to address these issues has led us to the literature on good governance [3,8,23,24].

Good or "effective" water governance is intended to address social justice and stability [24]. The means include clear definitions of rights and obligations [24], as well as legitimacy; transparency; accountability, inclusiveness; fairness; integration; capability, and adaptability [3,8]. Rather than engage in the problematic discussions of how is "good" governance measured and who decides, we use the overlap between long-tested approaches to governmental process to achieve these same goals of legitimacy; transparency; accountability; inclusiveness; fairness; and integration that is, or can be, imbedded in administrative law [22,25].

The Adaptive Water Governance Project in which the authors are participants seeks to integrate the approaches to good governance with adaptive governance principles to develop guidelines for the role of law in achieving flexible water management [7,26]. Law is thus not only the mechanism to set the stage for the emergence of adaptive governance, but in the viewpoint brought by legal scholarship, flexible, local governance will fail without attention to attributes of law and the reflection of good governance in the legal process. In particular, implementation of adaptive governance will fail to achieve its objectives without attention to: the role of governance in a just and equitable society; the need for resources and capacity building for local participation and decision making; the need to balance flexibility with stability; and the importance of legitimacy, transparency and accountability in any devolution of decision making to the local level. Thus, it is through this lens that we explore the intersection of law, good governance and increased adaptive capacity in collaborative efforts to share and co-manage water resources across jurisdictional boundaries with Indigenous communities.

3. Indigenous Rights to Water

While adaptive governance itself is a product of self-organization of social actors including organizations, it may emerge within complex institutional and legal settings [7,27]. In the U.S., Native American water settlements have become one vehicle for holistic resolution of multiple water resource issues on shared watersheds while at the same time improving equitable distribution of water. Many have done so through collaborative processes that resemble "adaptive governance" (see e.g., [28,29]). Yet, as settlement of litigation generally involving historic treaty rights, these processes are emerging within a heavily constrained institutional setting. The question we ask in this paper is: What is the role of law and good governance in the emergence of adaptive water governance in the context of conflict over scarce water resources? We believe that the applicability goes beyond the setting of Indigenous rights and that these merely represent contexts in which water scarcity has recently come to a head. We address this question through examination and discussion of four legal processes for determining Indigenous rights to water: (1) the state of Montana, USA process for settlement of reserved water rights; (2) a water settlement process in the Klamath River basin, USA; (3) the federal process for participation in Native American water right settlements; and (4) the co-management efforts of the Ngarrindjeri Nation in the Lower Murray River, Australia. In each example we seek to identify: (1) the role of law in creating a trigger or disturbance that may result in the emergence of collaborative processes; (2) the role of law in establishing a process to facilitate adaptive governance; and (3) the role of law in institutionalizing the resulting solutions. Through each of the examinations of the role of law we will trace the thread of good governance and identify any relation to the emergence of adaptive governance.

3.1. Montana and the Settlement of the Water Rights of the Chippewa Cree Tribe of the Rocky Boy's Reservation

Under U.S. law, Native American Reservations hold water rights as necessary for the purpose of the reservation with a priority date of the date of establishment of the reservation [30,31]. The existence of what are generally senior water rights poses a threat to other water uses on a shared source. The unquantified nature of the right poses an impediment to protection and development of the rights for Native American tribes. Over the past five decades, numerous states have engaged in massive general stream adjudications, in part to quantify these vaguely defined rights [32]. The filing of litigation combined with the serious threat to junior developed water rights and the need for water development on reservations has triggered innovative settlements of water rights for over 30 reservations [33]. The process of settling Native American water rights in the U.S. state of Montana has led to both creative holistic watershed-scale solutions to scarcity and to dialog between tribes and their non-Indian neighbors that have increased adaptive capacity within shared watersheds. The Rocky Boy's Reservation Settlement in Montana provides an example of emergence of a collaborative process.

Established on 7 September 1916 "for Rocky Boy's Band of Chippewas and ... other homeless Indians" [34], the Rocky Boy's Reservation in the Bearpaw Mountains of north-central Montana is home to over 3000 tribal members. Water right settlement negotiations began in 1992 among the Chippewa-Cree Tribe of the Rocky Boy's Reservation, the state of Montana, and the United States as part of the state-wide adjudication of water rights. At the time, the reservation had high unemployment and half the population lived below the poverty line. Although serving an area economically dependent on agriculture and ranching, the grass lands on the reservation receive approximately twelve inches of precipitation per year. Snowpack in the Bearpaw Mountains provides substantial spring runoff. The two major drainages arising on the reservation are shared with downstream farmers and ranchers [35].

The process chosen by the state of Montana has been held as a model of good governance in settlement negotiations [36], and facilitated emergence of adaptive processes that crossed jurisdictional divides to take a bioregional approach, developed polycentric co-management at the local level to develop a regional tribal/rural community drinking water supply, and incorporated local knowledge into the decision-making process. As part of the 1979 amendments to the Montana Water Use Act, the Montana legislature established the Montana Reserved Water Rights Compact Commission ("commission"), to negotiate water rights settlements "for the equitable division and apportionment of waters between the state and its people and the several Indian tribes claiming reserved water rights within the state" [37]. The bipartisan commission acts on behalf of the state and its citizens as a whole. It does not represent the interests of individual water users [38]. The policy of the state of Montana is to conduct negotiations with Indian tribes on a government-to-government basis [35].

By establishing a clear policy in favor of negotiation, the state of Montana provides a forum to resolve conflicts with practical solutions. Negotiations of Indian reserved water rights in Montana involve three parties: the tribe, the commission, and the United States as trustee for the tribe. Each party is governed by its own laws and rules for participation in a proceeding. The initial step in negotiation is to discuss the basic elements of the process and attempt to integrate the constraints each party brings to the table. Of greatest importance in the emergence of adaptive governance in the Montana process are the agreements on: (1) exchange/development of information; and (2) public participation.

Quantification of Indian water rights requires collection and analysis of technical data on subjects ranging from soil composition, water supply, and climate, to the economics of irrigation, and may involve complex hydrologic models. The focus in negotiation is interest based; thus, rather than discuss the possible litigated outcomes, dialog focuses on the needs and future plans of a tribe for sustained development or resource preservation and the needs of nearby water users to protect their investments. Each technical variable has a range of possible values. If each party were to collect and analyze its own data, negotiation would become mired in efforts to resolve technical issues rather than focus on the issues of policy that negotiators must address. As an illustration of good governance, the commission tries to encourage joint efforts at technical work among the parties. Negotiators then discuss issues of policy from a common database and a more common understanding among parties of how the water system functions [35].

The commission uses a number of approaches to ensure that public involvement is a two-way dialog. Among the approaches are: (1) public meetings are used to inform the public, identify issues and identify leadership in the community; (2) commission staff attend local meetings of various water-related entities, also for purposes of providing information and to identify leadership; (3) commission staff identify interested persons that have not been present at meetings and contact them specifically to provide information and to seek their engagement; (4) through leadership that either emerges in the process or is identified through attending local meetings, commission staff work more closely with local leadership to try to address their interests in positions taken in negotiations; and (5) in instances where a small group of people or entities will be affected by a particular issue or position, commission staff meet individually and will even do so by going to the particular person's home [29,35].

The locally focused process that involves continual contact throughout negotiations results in tailored settlements that address many local issues that state-level representatives would not be aware of. This process also allows the use of local and traditional knowledge to inform the development of solutions. The process recognizes that there must be a high level of local understanding, acceptance, and even ownership for smooth implementation of a settlement and that the individuals who live within a watershed have the greatest knowledge of their watershed. The lens of local knowledge is essential to the design and evaluation of solutions. The high degree of public involvement has generally led to the commission filling the role of facilitator between a tribe and their neighbors, while the solutions themselves are developed in the emergent local process [35]. The law assures that this emergent process will continue as appointments to the commission change, a factor in providing a stable overlay to local innovation. This robust process of not merely public involvement but also local capacity building takes far more time and resources during the phase of developing an agreement. Nevertheless, as illustrated in the following paragraphs by the emergent collaboration among Rocky Boy's and its neighbors, it pays off in the long term.

At the time of settlement negotiations, the existing domestic water supply on the Rocky Boy's Reservation developed by the Indian Health Service was deficient in both quality and supply [35]. Due to the limited and unreliable nature of the on-reservation water supply, the tribe and the state became convinced that alternative sources of supply must be considered if the tribe was to receive adequate and safe drinking water, despite the fact that off-reservation supplies were beyond the scope of the tribe's legal claims. Many communities near the reservation faced similar drinking water problems [35]. Nine municipal systems and fifteen rural water systems in the area serve populations ranging from seventy to 10,500 people [35]. As a result of changes in the Safe Drinking Water Act standards [39], and obsolescence, in addition to increases in population of the service area, each system faced the need for major modifications or repairs in the coming decades [35]. An off-reservation irrigator suggested importing domestic water through construction of a rural water supply system as a solution. Triggered by dialog arising out of the water rights settlement, efforts to solve this problem evolved beyond the water right negotiations and are now being addressed in an ongoing collaboration between the Rocky Boy's Tribe and 22 rural water entities in North Central Montana [40]. That process is without precedent in both its character and level of tribal-non-tribal cooperation [35]. It is emerging as a polycentric and bioregional local process.

The settlement of water rights on the reservation allocated 10,000 acre-feet of water from surface and groundwater sources to the tribe. The allocation included water to maintain and enhance fish and wildlife habitat, water for snow making at the tribe's ski area, and water for irrigation. An off-reservation source was allocated for the domestic water supply. Funding was made available for the expansion of two reservoirs on the reservation. In an arid region, these measures are the difference between viable and unviable economic development [41,42].

Once completed, the settlement was memorialized in the laws of the state [41], the tribe [43], and the United States [42], and reviewed by the court in which the litigation was filed. This institutionalization of a settlement in law, including funding for its implementation and authorization for any cooperative state/tribal entity or cooperation needed for ongoing resolution of water disputes assures that the emergence of collaborative behavior is not a one-off phenomenon. Instead, it becomes imbedded in a new norm for relations between the state and the tribe and between the tribe and its neighbors.

3.2. Water Settlement in the Klamath River Basin, USA

Rights to appropriate water and associated resources in the Klamath River basin of southcentral Oregon and northern California, USA have been contested for the better part of the last century [44]. At the center of this conflict are four federally recognized Native American tribes spread across the Klamath River basin: the Hoopa Valley Tribe; the Karuk Tribe; the Klamath Tribes; and the Yurok Tribe. Similar to the common narrative across the western United States, the tribes of the Klamath basin were displaced by Euro-American settlers in the late-19th century [45]. Despite a treaty and executive action creating reservations in the basin, tribes were subject to an almost complete loss of access to culturally important lands, waters, and species, such as Klamath River salmon populations in the lower basin and populations of sucker fish in upper basin lakes and tributaries—both of which served as critical food sources for basin tribes [46]. In addition, widespread development of the Klamath River system for agriculture and hydropower and the over-allocation of state-based water rights in the upper basin significantly increased the complexity, and ultimately the difficulty, for tribes to regain access to culturally significant resources (e.g., water) even considering the rise of favorable policy and processes for renegotiating tribal water rights at the federal level.

Unlike the state of Montana, USA, the states of California and Oregon do not have a formal state entity authorized to negotiate on their behalf. With regards to water and other natural resources, federally recognized tribes in both California and Oregon are treated similarly to tribes in Montana and elsewhere in the United States, as sovereigns, with the ability to negotiate as such with both state and federal governments (see Section 3.3 below). This negotiating position, combined with previously favorable litigation by the Klamath Tribes, the unique geography of the basin, and the state of Oregon process for adjudicating water rights, set the stage for the emergence of adaptive governance in the Klamath basin [28].

The Klamath River basin is roughly divided in half: the upper basin representing a mostly arid, high-elevation landscape, with intermittent lakes, marshes, and swamps, subsequently drained for agriculture; and the lower basin representing a rain-driven, mountainous region with numerous tributaries providing salmonid spawning and rearing habitat [46]. The basin looks like an hourglass tilted on its side, and historically, the two halves were segregated by the Oregon—California state line and the Klamath Hydroelectric Project—a series of river impoundments in the middle sections of the basin that divert, dampen, and regulate the Klamath River flow to the lower basin and completely cut off fish passage from the lower to upper portions of the basin [44].

In 1976, the state of Oregon initiated a general stream adjudication in the upper Klamath Basin (Oregon half only) to determine who had water rights in the basin and what the order of use or priority dates each of these water rights retained. The Klamath Tribes filed claims for water rights in the upper basin, based on the boundary of their historical reservation created in the Treaty of 1864 (this reservation was "terminated" under the federal policy of the same name in 1954; although the tribe's status as a federally recognized tribe was reestablished in 1986, the reservation was not). At the same time, the Klamath Tribes of the upper basin and the Hoopa Valley, Karuk, and Yurok Tribes of the lower basin (along with environmental groups and public scientists) worked diligently to establish three species of fish as federally listed threatened (Coho salmon) and endangered (Lost River and shortnose sucker) species under the U.S. Endangered Species Act (ESA). This process helped the tribes establish critical relationships with state and federal agencies as consulted co-managers of fisheries and other natural resources in the basin.

In the spring and summer of 2001, the legal requirement to leave water in stream for ESA-listed threatened and endangered fish in the Klamath basin, combined with the continuation of a crippling drought, dictated a curtailment of water deliveries to a large, federal irrigation project (210,000 acres; 1400 family farms) in the upper basin resulting in public protest, civil disobedience, and renewed racial tension in the basin between mostly Euro-American farmers and Native Americans [44]. In the fall of the following year, a reversed interpretation of the ESA provided more water for the irrigation project than left in stream at a critical time for migrating fall Chinook salmon, causing a rapid spread of disease killing over 30,000 salmon along the banks of the Yurok Reservation [46]. These two devastating losses for both the farming and tribal communities in the basin set the stage for years of protest, heated media conflict, contentious litigation, and a virtual stalemate in collaborative relations in the basin [44]. At the same time, these events—triggered by law—ripened the social contexts for the emergence of adaptive governance by first exhausting adversarial options and opening conflicting parties to the idea of collaborative dialog [28].

The expiration of the Federal Energy Regulatory Commission (FERC) hydropower license for the Klamath Hydroelectric Project in 2006 represented a window of opportunity for a more complete emergence of adaptive governance processes in the Klamath basin. As a result of amendments to the U.S. Federal Power Act, private licensees of hydroelectric projects could request an "alternative process" instead of applying to relicense projects. This process had only recently been employed, generally in situations of high conflict over water uses and ecological resources, many of which involved unsettled tribal claims to water [47]. The alternative process created a venue for the project's owner, PacifiCorp, to sponsor a series of facilitated negotiations between vested stakeholders in water and power use in the Klamath basin (tribes, irrigators, environmental groups, federal agencies, both land and species management agencies, commercial fishing groups, and others)—as opposed to forcing PacifiCorp to either apply to relicense the dams amidst significant stakeholder opposition, or alternatively, to abdicate their license (upon relicensing, PacifiCorp would face legally mandated installation of fish passage at four of its mainstem Klamath River dams, rendering the prospect of relicensing a financial loss) [48]. Once engaged in negotiation towards a settlement of hydropower operations on the Klamath River, PacifiCorp, the federal government, and other involved stakeholders recognized that discussions about governing water in the Klamath were bigger than the dams alone [49]. This prompted a spin-off negotiation from the alternative relicensing talks with a goal of settling Native American water rights in the upper basin and facilitating agreements between tribes and irrigators that conserved threatened and endangered species while preserving the basin's agricultural economies [28,49]. The process—while closed to only those private parties and public agencies with a legal stake in state-based water rights negotiations, threatened and endangered species management, or those organizations with valid litigation potential under law—included aspects of good governance and emergent adaptive governance at the outset including: space for structured conflict resolution-based dialog; trust building among stakeholder leadership; joint fact finding; and co-production of basin modeling [50].

In 2010, the non-federal stakeholders involved in this process signed two agreements, the Klamath Basin Restoration Agreement (KRBA) [51], and the Klamath Hydroelectric Settlement Agreement (KHSA) [52]. Legally, these agreements comprised both a FERC alternative to relicensing settlement and an Indian water rights settlement, and practically, these agreements represented an effort toward holistic, basin-wide restoration of Klamath River ecosystems, communities, and economies through habitat restoration, fishery reintroduction, dam removal, and preparation for climate change (increasing aridity and drought). The agreements resembled a significant step toward adaptive governance in the basin: opportunities for public participation in a federal review of the dam removal action (under the National Environmental Policy Act); local governance of the agreements via public meetings of a committee of stakeholder leadership representatives; and mediated conflict resolution processes for resolving conflicts with the agreements in the future [50].

Law has played a critical role in the changing nature of Klamath basin water conflicts, both in creating space for the emergence of collaborative processes as well as in creating windows of opportunity for key aspects of adaptive governance to grow and mature. The role of law in institutionalizing adaptive governance in the Klamath basin, however, is still unclear. The state of Oregon-led Klamath basin water rights adjudication process finally concluded in 2013, and although the Klamath Tribes had already entered into agreements with upper basin irrigators as to the future of many of their water claims, the tribes' legal rights to water were finally and clearly identified, quantified, and dated—many of those rights now holding a priority date of "time immemorial" giving them a significant negotiating position among upper basin agricultural interests. This position was put to the test immediately when they called on their rights ("called the river") to have water left instream to bolster aquatic habitat during a dry year in 2013. This action brought the furthest upstream irrigators, many of which who did not participate in or sign the Klamath Agreements (or broke off after 2010), back to the table to negotiate a third agreement for comprehensive water rights settlement in the upper basin [28]. For this suite of three Klamath Agreements to become binding law, and for federal agencies to have legal authorization to participate in the programs designed under the negotiations, Congress must approve legislation authorizing the agreements as well as approve funding to carry them out. The process of seeking Congressional approval and authorization of funding is one embedded in the U.S. system of passing laws. At the same time, the role of Congress is not merely to rubber stamp a local solution. Its role in setting priorities for use of federal funding and in assuring that all interests have been accounted for is of equal importance in providing the stability that allows for local innovation. It remains to be seen how this will play out in the recent emergence of new avenues for both accomplishing and funding aspects of the settlement while leaving other aspects behind.

3.3. Federal Participation in Native American Water Right Settlements

The emergence of adaptive governance in the settlement processes in Montana and in the Klamath River basin were not only triggered by state and federal law, but could not have emerged without a supportive federal process. Native American land and resources, including water, are held by the United States in trust for tribes. While many have criticized the federal process (see e.g., [35]), had the United States chosen a litigation path rather than articulating a preference for negotiation and providing the resources for it to proceed, collaborative settlements would never have been possible.

The federal government's policy to negotiate settlement of Indian water rights claims was declared by President George H.W. Bush in 1989 [53]. The president's statements committed the administration to "establishing criteria and procedures to guide future Indian land and water claim settlement negotiations" [53]. The Department of the Interior responded with publication of the criteria and procedures for federal participation in Indian water rights settlements in the Federal Register [54].

The criteria and procedures provide guidelines for federal participation in negotiations that assure coordination of federal interests. A federal team is assigned to each negotiation and is composed of representatives of various bureaus within the Department of the Interior with interests in the watersheds involved as well as a representative of the Department of Justice. All decisions are made through the Working Group on Indian Water Rights Settlements, composed of assistant secretaries and a counselor to the Secretary of the Interior in Washington D.C. [54].

Although the process of formulating a federal position only after the state and tribal parties have identified a solution has been criticized as being in direct conflict with the dynamic necessary for negotiation [35], the presence of people knowledgeable about the various federal interests has proven invaluable in reaching resolution [55]. In addition, institutional stability in the program since its inception means that a wealth of knowledge in creative problem solving accompanies each federal team to the table.

3.4. Ngarrindjeri Nation in the Lower Murray Basin in South Australia

Unlike the other case studies, the rise of the Ngarrindjeri Nation to a role of co-management on the lower Murry River, the possibility of legal recognition of Indigenous rights played only a minor role in creating a trigger for collaboration. Instead, governance capacity building and diplomacy through a combination of self-organization by an Aboriginal community, assistance from external entities in that capacity building, and water reform initiated during the Millennium Drought in Australia led to a process of collaboration. Here we analyze the Ngarrindjeri Nation's path to obtaining a stake in water governance, as a non-U.S. analog to the previous examples, demonstrating that law can be a vehicle not only for emergence, but also for institutionalization of the resulting co-management.

Australia recognized no rights to land or water of Aboriginal people until 1992 when the High Court of Australia rejected the long-standing doctrine of *terra nullis*. The doctrine stood for the proposition that on discovery by European nations the land belonged to no one, despite long inhabitation by Indigenous people [56]. Even the ruling in 1992 recognized only limited rights and did not recognize Aboriginal peoples as sovereign. The Commonwealth followed the ruling with passage of the Native Title Act of 1993, with the purpose of setting up a process for the establishment of native title to land and water, and limiting its future extinguishment [57]. Under the act, native title must be established through application to the Federal Court, but the court may refer the process to mediation [57]. In addition, authority to determine claims may be delegated to a state or territorial body under certain conditions [57]. Compensation is provided where actions such as private settlement prior to 1996 are inconsistent with native title that would otherwise exist [57]. Importantly, prior granting of exclusive right to land or water to a private person or entity extinguishes native title, leaving compensation as the only legal remedy for Indigenous people in the more populated areas of Australia. [57].

While the possibility of establishment of native title provided some leverage for Aboriginal communities, the possibility of extinguishment due to settlement and development dampened the threat in the populated parts of Australia. Since the passage of the Native Title Act of 1993, some progress has been made in Australia establishing native title to land in the less populated interior and northeast. However, even in these situations there is a disparity between recognition of rights to land and rights to water. Jackson and Langston assert that:

> *[t]he current distribution of water entitlements in Australia is transparently inequitable. A comparison of indigenous land ownership and water entitlements demonstrates the extent of the inequity: Indigenous people own almost 20 per cent of the country's land mass/ while Indigenous specific water entitlements are at present such a minuscule figure that is estimated at less than 0.01 per cent of Australian water diversions* ([58] at page 110).

Water allocation is generally a matter of state law in Australia [59]. However, in 2004, the Council of Australian Governments (COAG) with membership including *the Prime Minister, State and Territory Premiers and Chief Ministers, and the President of the Australian Local Government Association* [60], developed the Intergovernmental Agreement on a National Water Initiative, with the governments of New South Wales, Victoria, South Australia, Queensland, Northern Territory, and Australian Capital Territory signing in 2004, Tasmania signing in 2005, and Western Australia signing in 2006 [59]. The NWI recognizes the right of Indigenous access to water and calls for Indigenous representation in water planning and accounting of water allocated to traditional cultural purposes ([61] at section 52). The focus on cultural purposes rather than the broad spectrum of economic and environmental uses that a community of people might seek to make of water is thought to be the limitation on the recognition of Indigenous water rights [58]. Nevertheless, some states have taken a broader view of Indigenous involvement. The 2009 "Water for Good" water plan for South Australia calls for collaboration with Indigenous communities including incorporation of Indigenous knowledge and attention to the water supply needs of the communities [62].

The combination of these legal changes and the crisis presented by the impact of the Millennium Drought (lasting from 1996 to 2010) on the Murray River and the Coorong (a series of lakes, wetlands, saline lagoon and estuary at the mouth of the Murray River) provided the impetus for self-organization by the Ngarrindjeri people—Aboriginal peoples with deep ties to the region at the mouth of the Murray River. Through internal organization the Ngarrindjeri articulated their vision of "caring for country" through development of the 2006 Ngarrindjeri Yarluwar-Ruwe (Sea-Country) Plan for the region of the lower Murray River, the Coorang and associated wetlands [63]. The plan notes that the "river, lakes, wetlands/nurseries, Coorong estuary and sea have sustained us culturally and economically for tens of thousands of years" and seeks to work with others "for the benefit of our shared future and for our land and waters upon which we all depend" ([63] at page 6).

With assistance from organizations such as the Yunggorendi First Nation Centre at Flinders University the community built governance capacity and the Ngarrindjeri Nation Authority was formed and incorporated in 2007 to represent the interests of the people. While law only provided some leverage in emergence of this local stakeholder-driven effort, it provided the means to institutionalize that effort through incorporation of the authority. Following incorporation, the Ngarrindjeri Nation Authority approached local, state and interstate governments to enter into co-management agreements using private contracts as the legal means for rendering the agreements binding ([64] at page 4). The use of legally binding agreements gave the parties greater protection than reliance on "consultation," thus filling the current void in state and federal definition of Indigenous rights [65]. Similar to the water settlement process in the United States, the contract approach allows Indigenous communities to represent themselves and to enter agreements or refuse them, voluntarily [65].

The sophisticated approach to organization and diplomacy taken by the Ngarrindjeri Nation is unprecedented and has been recognized as such. On 22 September 2015, the Ngarrindjeri Regional Authority in partnership with the South Australia Department of Environment, Water and Natural Resources, accepted the prestigious Australian River Prize from the International River Foundation for its work on the Murray River [66]. The following is from the press release on the much-deserved award to the Ngarrindjeri Regional Authority:

> *The Ngarrindjeri's Kungun Ngarrindjeri Yunnan Agreement (KNYA) established a new and positive relationship between the Ngarrindjeri and the South Australian Government, which has seen an innovative and integrated approach to river basin management for the Murray. In particular, the Agreement emphasised a participatory approach with the land's traditional custodians—moving past historical barriers to Aboriginal involvement in integrated river basin management—which has led to opportunities to develop Aboriginal-led wetland management plans for land owned by the Ngarrindjeri people* ([66]).

4. Discussion

We set out in our initial paragraphs to ask the question: What is the role of law and good governance in the emergence of adaptive water governance in the context of conflict over scarce water resources? In the complex social–ecological system of a water basin, it would likely be impossible to conclude that a specific law or governmental action directly led to a particular collaborative process or to its success—there are too many interacting factors acting upon or within these contexts. What we can do is observe the correlation between emergence of collaborative processes and enabling law, and identify the governmental actions that likely improved accountability, transparency, legitimacy, inclusion, stability, and attention to the broader public good and what role they may have played in assuring a successful collaboration.

First, without the legal forum for the assertion of Native American water rights and the relatively recent recognition of native title in Australia, along with a growing policy in favor of consultation, the development of capacity and entree into water discussions of these formally marginalized populations may not have been possible. In fact, it is possible that the destabilizing effect of recognition of formerly marginalized rights was sufficient to trigger a willingness to seek solutions on the part of those currently in command of the water resource in the United States thus leading to emergence of bioregional processes. Even in the Australia example in which legal rights of the Ngarrindjeri had not been recognized and capacity building was done through self-organization and assistance from non-governmental entities, had the governments of Australia and the state of South Australia not been undergoing a policy shift toward recognition of the need for consultation and co-management, the effort would not have been successful.

Second, without the support for capacity building from the federal government in the United States and from external interests such as higher education in Australia, it is possible that the leadership, problem-solving capacity, and political strategy that Native American communities in the United States and the Ngarrindjeri in Australia have now taken to a highly sophisticated level, would have been much slower in coming. The legal and policy shift of the United States from a paternal approach to governmental assistance to Native American tribes to the fostering of self-determination [67], as well as the UN recognition of the importance of self-determination [68], has led to a greater focus on

Indigenous capacity building worldwide. Without this increased capacity, the sophisticated diplomacy and emergent innovations illustrated in our examples would have been much less likely.

Third, without the legally mandated transparent process in U.S. states like Montana, and vetting of results through legislative and court proceedings, the legitimacy of settlement resulting from local collaborative problem solving might have been questioned by both the tribal communities and external constituencies, and would likely never have led to an ongoing collaboration between a tribe and its neighbours. Similarly, the institutionalization of self-organization of the Njarrindjeri through formation of an authority to speak for their interests enhances accountability and reduces the possibility of claimed representation by a special faction. The attention to a process for determining who to recognize under the Australian Native Title Act and the U.S. federal process for recognition of tribes, while potentially subject to use for exclusion, increases the likelihood of a transparent and accountable process to determine who speaks for an Indigenous people.

Fourth, the formal process and institutional memory of the U.S. federal government in its participation in settlements provides an avenue to consider the broader public good and even-handed treatment across U.S. Indian Country. The federal use of a systematic process memorialized in the Federal Register (*i.e.*, the criteria and procedures referred to above) also provides an avenue for overall stability while leaving open the possibility of local innovation. Similarly, the participation of the state government of South Australia in formalizing the agreement on co-management with the Ngarrindjeri Regional Authority may, over time, result in a systemized approach to similar efforts that will assure consideration of the public good and maintain regional stability as local and bioregional processes are given room to innovate.

Fifth, without the state, tribal and federal processes in the United States used to institutionalize and fund the results of settlement, and the use of a legal contract in Australia to recognize co-management, the good feelings and handshake agreements from a collaborative watershed process could quickly be undermined by powerful special interests seeking a better deal. While the difficulties the Klamath settlement experienced in the U.S. Congress illustrates that these legal processes themselves can be captured by special interests and larger-scale political forces, without the ability to obtain legal recognition and funding, no settlement would be implemented. It is also important that a body such as congress consider settlements in light of the broader public good to assure that they have not merely become a vehicle to fund special-interest projects and that the limited funds available are well spent.

5. Conclusions

The past four decades of settlement of Native American water rights in the United States offers a window on what happens when the law alters the relative power among communities sharing a scarce water source and a chance to study the emergence of adaptation and innovation as communities respond to this new paradigm. It provides a means to consider the role of the legal process in assuring the accountability, transparency, fairness, and inclusiveness in these local emergent collaborations. The example of the Ngarrindjeri Nation in Australia offers a chance to view emergence of adaptive processes from a completely different starting point—internal self-organization and the building of governance capacity in the absence of a clear legal avenue to establish rights. Yet law became the avenue to institutionalize the emerging solution and the willingness of the Ngarrindjeri to formalize their own emergence in law through creation of the Ngarrindjeri Regional Authority provides the basis for legitimacy as they move forward. Adaptive, collaborative, and new governance mechanisms hold promise for the innovation and adaptation needed to respond to the challenges of water scarcity in the current millennium, but it is clear that they do not emerge in a vacuum. Not only may legal and institutional tools be used to trigger these processes and to finalize their results, but without attention to aspects of good governance through legal process, local innovation will also not have the legitimacy necessary to succeed.

Acknowledgments: The authors would like to acknowledge the support of the National Socio-Environmental Synthesis Center (SESYNC) under funding from the National Science Foundation, DBI-1052875, for the Adaptive Water Governance Project and the hard work and inspiration of the AWG team on which this work relies. We would also like to express a deep gratitude for the Native and non-Native water users of the Klamath basin and the many watersheds in Montana who provided us the opportunity to personally witness the emergence of new approaches to governance that offer promise for our collective water future. Co-author Barbara Cosens would like to acknowledge the support of the Goyder Institute for Water Research and Flinders University in South Australia for providing her the opportunity to learn about Australian water and indigenous law as a Visiting Professor in Public Sector Policy and Management in the spring of 2015. Finally, co-author Barbara Cosens would like to thank the people of the Ngarrindjeri Nation and the Yunggorendi First Nation Centre at Flinders University for the discussion and opportunity to observe a truly novel approach to co-management, and the people of the Rocky Boy's Reservation and their neighbors who taught her the power of collaboration and the importance of local involvement.

Author Contributions: Lead author Cosens provided material on the Montana, Federal and Australian processes. Co-author Chaffin provided material on the Klamath settlements. The authors collaborated on the introductory material and lead author Cosens provided the discussion and conclusion.

Conflicts of Interest: The authors declare no conflict of interest, but believe it important to note that co-author Cosens was lead negotiator for the state of Montana in the water rights negotiations with the Chippewa-Cree Tribe of the Rocky Boy's Reservation.

References

1. Folke, C.; Hahn, T.; Olsson, P.; Norberg, J. Adaptive governance of social-ecological systems. *Annu. Rev. Environm. Resour.* **2005**, *30*, 441–473. [CrossRef]
2. Pelling, M.; High, C.; Dearing, J.; Smith, D. Shadow spaces for social learning: A relational understanding of adaptive capacity to climate change within organisations. *Environ. Plan. A* **2008**, *40*, 867–884. [CrossRef]
3. Lockwood, M.; Davidson, J.; Curtis, A.; Stratford, E.; Griffith, R. Governance Principles for Natural Resource Management. *Soc. Nat. Resour.* **2010**, *23*, 986–1001. [CrossRef]
4. Dietz, T.; Ostrom, E.; Stern, P.C. The Struggle to Govern the Commons. *Science* **2003**, *302*, 1907–1912. [CrossRef] [PubMed]
5. Lebel, L.; Anderies, J.M.; Cambell, B.; Folke, C.; Hatfield-Dodds, S.; Hughes, T.P.; Wilson, J. Governance and the capacity to manage resilience in regional social-ecological systems. *Ecol. Soc.* **2006**, *11*, 19.
6. Huitema, D.; Mostert, E.; Egas, W.; Moellenkamp, S.; Pahl-Wostl, C.; Yalcin, R. Adaptive water governance: Assessing the institutional prescriptions of adaptive (co) management from a governance perspective and defining a research agenda. *Ecol. Soc.* **2009**, *14*, 26.
7. Cosens, B.; Craig, R.K.; Hirsch, S.; Allen, C.; Arnold, A.; Benson, M.; DeCaro, D.; Fremier, A.; Garmestani, A.; Ruhl, J.B.; *et al.* The Role of Law in the Emergence of Adaptive Governance. *Ecol. Soc.* **2016**. under review.
8. Rogers, P.; Hall, A. *Effective Water Governance*; Global Water Partnership Technical Committee (TEC) Background Paper 7.2003; Global Water Partnership: Stockholm, Sweden, 2003; Available online: http://www.gwp.org/Global/ToolBox/Publications/Background%20papers/07%20Effective%20Water %20Governance%20%282003%29%20English.pdf (accessed on 30 December 2015).
9. Lemos, M.C.; Agrawal, A. Environmental governance. *Annu. Rev. Environ. Resour.* **2006**, *31*, 297–325. [CrossRef]
10. UN System Task Team on the Post-2015 UN Development Agenda: Governance and Development. 2012. Available online: http://www.un.org/millenniumgoals/pdf/Think%20Pieces/7_governance.pdf (accessed on 30 December 2015).
11. Chaffin, B.C.; Gosnell, H.; Cosens, B. A decade of adaptive governance scholarship: Synthesis and future directions. *Ecol. Soc.* **2014**, *19*, 56. [CrossRef]
12. Cosens, B.; Gunderson, L.; Chaffin, B. The Adaptive Water Governance Project: Assessing Law, Resilience and Governance in Regional Socio-Ecological Water Systems Facing a Changing Climate. *Nat. Resour. Environ. Law Ed. Idaho Law Rev.* **2014**, *51*, 1–27.
13. Schultz, L.; Folke, C.; Österblom, H.; Olsson, P. Adaptive governance, ecosystem management, and natural capital. *PNAS* **2015**, *112*, 7369–7374. [CrossRef] [PubMed]
14. Hahn, T.; Olsson, P.; Folke, C.; Johansson, K. Trustbuilding, knowledge generation and organizational innovations: The role of a bridging organization for adaptive comanagement of a wetland landscape around Kristianstad, Sweden. *Hum. Ecol.* **2006**, *34*, 573–592. [CrossRef]

15. Olsson, P.; Folke, C.; Berkes, F. Adaptive comanagement for building resilience in social-ecological systems. *Environ. Manag.* **2004**, *34*, 75–90. [CrossRef] [PubMed]
16. Olsson, P.; Folke, C.; Hahn, T. Social-ecological transformation for ecosystem management: The development of adaptive co-management of a wetland landscape in southern Sweden. *Ecol. Soc.* **2004**, *9*, 2.
17. Olsson, P.; Gunderson, L.H.; Carpenter, S.R.; Ryan, P.; Lebel, L.; Folke, C.; Holling, C.S. Shooting the rapids: Navigating transitions to adaptive governance of social-ecological systems. *Ecol. Soc.* **2006**, *11*, 18.
18. Olsson, P.; Folke, C.; Galaz, V.; Hahn, T.; Schultz, L. Enhancing the fit through adaptive co-management: Creating and maintaining bridging functions for matching scales in the Kristianstads Vattenrike biosphere reserve, Sweden. *Ecol. Soc.* **2007**, *12*, 28.
19. Plummer, R.; Armitage, D.; de Loë, R.C. Adaptive Comanagement and its relationship to environmental governance. *Ecol. Soc.* **2013**, *18*, 21. [CrossRef]
20. Brunner, R.D.; Steelman, T.A.; Coe-Juell, L.; Cromley, C.M.; Edwards, C.M.; Tucker, D.W. *Adaptive Governance: Integrating Science, Policy, and Decision Making*; Columbia University Press: New York, NY, USA, 2005.
21. Lee, M. Conceptualizing the new governance: A new institution of social coordination. In Proceedings of the Institutional Analysis and Development Mini-Conference, Bloomington, IN, USA, 3–5 May 2003.
22. Cosens, B.A. Legitimacy, adaptation, and resilience in ecosystem management. *Ecol. Soc.* **2013**, *18*, 3. [CrossRef]
23. Dublin Principles. International Conference on Water and Environment. 1992. Available online: http://www.wmo.int/pages/prog/hwrp/documents/english/icwedece.html#principles (accessed on 31 December 2015).
24. UN WWAP, United Nations. Water for People, Water for Life. UN World Water Development Report 1. World Water Assessment Program. 2003. Available online: http://www.unesco.org/new/en/natural-sciences/environment/water/wwap/wwdr/wwdr1-2003/ (accessed on 31 December 2015).
25. Esty, D.C. Good governance at the supranational scale: Globalizing administrative law. *Yale Law J.* **2006**, *115*, 1490–1562. [CrossRef]
26. Cosens, B. Application of the Adaptive Water Governance Project to the management of the Lake Eyre Basin and Its connections to the Great Artesian Basin. Report to Flinders University and the Goyder Institution. 2015. Available online: http://www.goyderinstitute.org/uploads/FU%20LEB%20GAP%20Report-WEB.pdf (accessed on 31 December 2015).
27. Chaffin, B.C.; Gunderson, L.H. Emergence, Institutionalization and Renewal: Rhythms of Adaptive Governance in Complex Social-Ecological Systems. *J. Environ. Manag.* **2016**, *165*, 81–87. [CrossRef] [PubMed]
28. Chaffin, B.C.; Craig, R.K.; Gosnell, H. Resilience, Adaptation, and Transformation in the Klamath River Basin Social-Ecological System. *Nat. Resour. Environ. Law Ed. Idaho Law Rev.* **2014**, *51*, 157–193.
29. Cosens, B. Water Dispute Resolution in the West: Process Elements for the Modern Era in Basin-wide Problem Solving. *Environ. Law* **2003**, *33*, 949–1018.
30. Winters v. United States. 207 U.S. 564, 1908.
31. Cosens, B. The Legacy of Winters v. United States and the Winters Doctrine, One Hundred Year Later. In *The Future of Federal and Indian Reserved Water Rights: The Winters Centennial*; Cosens, B., Royster, J., Eds.; University of New Mexico Press: Albuquerque, NW, USA, 2012.
32. Thorson, J.E. Reflections on Western General Stream Adjudications upon the Signing of Wyoming's Big Horn River Adjudication Final Decree. *Wyo. Law Rev.* **2015**, *15*, 383–411.
33. Native American Water Right Settlement Electronic Repository (NAWRS), a Project of the American Indian Law Center, University of New Mexico and University of Idaho College of Law. Available online: http://repository.unm.edu/handle/1928/21727 (accessed on 31 December 2015).
34. Act Providing for the Opening of the Fort Assinniboire Military Reservation. Public Law No. 261, United States Statutes at Large Volume 39, page 739, 1916.
35. Cosens, B. The 1997 Water Rights Settlement between the State of Montana and the Chippewa Cree Tribe of the Rocky Boy's Reservation—The Role of Community and of the Trustee. *UCLA J. Environ. Law Policy* **1998**, *16*, 255–295.
36. Cosens, B., Royster, J., Eds.; *The Future of Federal and Indian Reserved Water Rights: The Winters Centennial*; University of New Mexico Press: Albuquerque, NW, USA, 2012.
37. Montana Code Annotated §§ 85-2-701(2) and 702. 1995.
38. Montana Code Annotated § 2-15-212. 1995.

39. Safe Drinking Water Act, United States Code, Volume 42, Chapter 6A, Subchapter XII §§ 300f to 300j-26. Available online: https://www.gpo.gov/fdsys/pkg/USCODE-2014-title42/pdf/USCODE-2014-title42-chap6A-subchapXII.pdf (accessed on 10 March 2016).
40. Rocky Boy's North Central Montana Regional Water Supply Project. Available online: http://www.rockyboynorthcentral.com/ (accessed on 31 December 2015).
41. Chippewa-Cree Tribe—Montana Compact. In Montana Code Annotated 85-20-601. Available online: http://leg.mt.gov/bills/2014/mca/85/20/85-20-601.htm (accessed on 10 March 2016).
42. Chippewa-Cree Tribe of the Rocky Boy Reservation Indian Reserved Water Rights Settlement and Water Supply Enhancement Act of 1999. Public Law 106–163, United States Statutes at Large Volume 113 page 1778, 1999.
43. Resolution of the Chippewa-Cree Tribe of the Rocky Boy Reservation of 9 January 1997, approving the water rights compact between the Tribe and the State of Montana.
44. Doremus, H.; Tarlock, A.D. *Water War in the Klamath Basin: Macho Law, Combat Biology, and Dirty Politics*; Island Press: Washington, DC, USA, 2008.
45. Most, S. *River of Renewal: Myth and History in the Klamath Basin*; Oregon Historical Society Press: Portland, OR, USA, 2006.
46. National Research Council. *Endangered and Threatened Fishes in the Klamath River Basin: Causes of Decline and Strategies for Recovery*; National Academies of Sciences: Washington, DC, USA, 2004.
47. Blumm, M.C.; Erickson, A.B. Dam removal in the Pacific Northwest: Lessons for the nation. *Environ. Law* **2012**, *42*, 1043–1100. [CrossRef]
48. Spain, G. Dams, water reforms, and endangered species in the Klamath basin. *J. Environ. Law and Litig.* **2007**, *22*, 49–129.
49. Gosnell, H.; Kelly, E.C. Peace on the river? Social-ecological restoration and large dam removal in the Klamath Basin, USA. *Water Altern.* **2010**, *3*, 361–383.
50. Chaffin, B.C. Reallocating Resources, Rebuilding Community: The Klamath Basin Restoration Agreements and the Emergence of Adaptive Governance. Ph.D. Thesis, Oregon State University, Corvallis, OR, USA, 2014.
51. Klamath Basin Restoration Agreement (KRBA). Available online: http://klamathriverrestoration.org/index.php/kbra-summary (accessed on 30 December 2015).
52. Klamath Hydroelectric Settlement Agreement (KHSA). Available online: http://www.klamathrestoration.org/index.php/klamath-hydroelectric-settlement-agreement (accessed on 30 December 2015).
53. Bush, G. 1 Statement on Signing the Puyallup Tribe of Indians Settlement Act of 1989, Public Papers 771, 772 (21 June 1989). Available online: http://www.presidency.ucsb.edu/ws/index.php?pid=17186 (accessed on 8 March 2016).
54. Criteria and procedures for participation of the federal government in negotiations for the settlement of Indian water right claims. 55 Federal Register Volume 55, No. 48, page 9, 223. 1990. Available online: http://www.usbr.gov/native/policy/12mar1990_fedreg_indianwaterrights.pdf (accessed on 10 March 2016).
55. Cosens, B. 2005 Indian Water Rights Settlement Conference Keynote Address. *Denver Water Law Rev.* **2006**, *9*, 285–297.
56. Mabo v Queensland (No 2) (Mabo) [1992] HCA 23; (1992) 175 CLR 1 (3 June 1992). 1992.
57. Native Title Act 1993 as amended. Available online: http://www.comlaw.gov.au/Series/C2004A04665 (accessed on 31 December 2015).
58. Jackson, S.; Langton, M. Trends in the Recognition of Indigenous Water Needs in Australian Water Reform: The Limitations of "Cultural" Entitlements in Achieving Water Equity. *J. Water Law* **2011**, *22*, 109–123.
59. Stoeckel, K.; Webb, R.; Woodward, L.; Hankinson, A. *Australian Water Law*; Thomson Reuters: Sydney, Australia, 2012.
60. Council of Australian Governments (COAG). Standing Council on Primary Industries. Available online: https://www.coag.gov.au/ (accessed on 31 December 2015).
61. National Water Initiative of 2004 (NWI). Available online: http://nwc.gov.au/__data/assets/pdf_file/0008/24749/Intergovernmental-Agreement-on-a-national-water-initiative.pdf (accessed on 31 December 2015).
62. South Australia Department of Environment, Water, and Natural Resources. *Water for Good Plan*; Department of Environment, Water, and Natural Resources: Adelaide, Australia, 2009.

63. Ngarrindjeri Yarluwar-Ruwe (Sea-Country) Plan: Caring for Ngarrindjeri Sea Country and Culture. 2006. Available online: http://www.environment.gov.au/indigenous/publications/pubs/ngarrindjeri-scp-2006-1.pdf (accessed on 31 December 2015).
64. Hemming, S.; Rigney, D. *Indigenous Engagement in Environmental Water Planning, Research and Management: Innovations in South Australia's Murray-Darling Basin Region*; Technical Report Series No. XX/x; Goyder Institute for Water Research: Adelaide, Australia, 2014.
65. Hemming, S.; Rigney, D.; Berg, S. Chapter 6: Ngarrindjeri Futures: Negotiation, Governance and Environmental Management. In *Unsettling the Settler State: Creativity and Resistance in Indigenous Settler-State Governance*; Maddison, S., Brigg, M., Eds.; The Federation Press: Annandale, Australia, 2011.
66. Ngarrindjeri regional authority accepts prestigious Australian Riverprize for Murray River basin management. Available online: http://www.riverfoundation.org.au/admin/structured/strc__resource_3219_1___MEDIA_RELEASE___2015_Australian_Riverprize_winner_announced___FINAL.pdf?1442963108 (accessed on 31 December 2015).
67. Indian Self-Determination and Education Assistance Act of 1975. Public Law 93-638, 1975.
68. United Nations Declaration on the Rights of Indigenous Peoples, 13 September 2007. Available online: http://www.un.org/esa/socdev/unpfii/documents/DRIPS_en.pdf (accessed on 1 January 2016).

water

MDPI

Article

Groundwater Governance: The Role of Legal Frameworks at the Local and National Level—Established Practice and Emerging Trends

Kerstin Mechlem

International Natural Resources Law and Human Rights Consultant, 69120 Heidelberg, Germany; k.mechlem@gmail.com; Tel.: +49-6221-4208-179

Academic Editors: Sharon B. Megdal, Susanna Eden and Eylon Shamir
Received: 4 May 2016; Accepted: 10 August 2016; Published: 17 August 2016

Abstract: Legal frameworks play a crucial role for effective groundwater governance. They flank and support water policy and provide users and the administration with rights and obligations to use, manage, and protect vital resources in order to achieve the overall goal of equitable and sustainable water use. This paper discusses key challenges that have to be addressed in water law to manage and protect groundwater effectively. It will provide an overview of established practice in groundwater legislation and discuss recent trends and developments in light of current challenges. It focuses on permit-based systems of administrative water rights but will to a limited extent also deal with customary, community-based, and informal arrangements. It will show that increasingly domestic groundwater legislation is strengthened and ranked on a par with surface water regimes, ideally by dealing with all water resources in an integrated manner.

Keywords: water law; groundwater law; legislation; governance; water management; institutions; human rights; abstraction control; pollution control; environmental law

1. Introduction

> *"[T]he existence, origin, movement and course of [underground] waters, and the causes which govern and direct their movements, are so secret, occult and concealed, that an attempt to administer any set of legal rules in respect to them would be involved in hopeless uncertainty, and would be, therefore, practically impossible."*
>
> ([1])

Although groundwater has traditionally received less attention than surface water in policy and law, views as expressed in this quote from *Frazier v. Brown* (1861), which were not uncommon in the 19th century, are by now fortunately obsolete. Nonetheless, a certain lack of attention to groundwater still manifests itself in piecemeal legislative approaches, inadequate institutional set-ups and insufficient implementation of groundwater law in many parts of the world. Increasingly, however, it is acknowledged that legal frameworks play a crucial role for effective groundwater governance, which need to flank and complement science and policy. Modern water laws take an integrative approach to surface water and groundwater resources, define clear rights and obligations, set up management tools, protect quantity and quality, involve stakeholders and are matched with robust institutions in charge of their implementation. They regulate, amongst other issues, the conditions for access to groundwater, the protection of aquifers against depletion and pollution, the use of monitoring and planning tools, the way private and public interests have to be balanced and the involvement of stakeholders in decision-making and management processes. Typically, formal legislation addresses these issues. In some parts of the world, however, informal and customary rules continue to play a significant role that has to be acknowledged.

This paper will provide an overview of established practices as well as recent trends in groundwater legislation. It will discuss key challenges that legislation has to address to manage and protect groundwater and to effectively counter the unsettling effects of groundwater depletion and pollution. The latter are on the rise globally due to groundwater overexploitation and are likely to increase in times of climate change. The article focuses on a comparative analysis of domestic legislation. The examples provided from legislation worldwide are exemplary and, due to the breadth of the subject matter, in no way comprehensive. Where one of the more than 592 transboundary aquifers (including transboundary groundwater bodies as defined in the EU Water Framework Directive) is at stake, the domestic legal frameworks have to take account of and reflect international water law obligations and obligations from other areas of international law, such as human rights laws [2,3]. These will be addressed to the extent necessary without going into the details that have been discussed elsewhere [4–7].

Modern technologies allow using the subsurface for a range of newer activities, which may contaminate groundwater. Among them are the storage of hazardous waste such as nuclear waste, nuclear testing, the injection of fluids for soluble mineral extraction, the storage and recovery of heat and hydrocarbons, carbon capture and sequestration (CCS), hydraulic fracturing ('fracking'), and the injection of residual geothermal fluids. These activities pose new challenges, and to a large extent require specialized laws and regulations which overall still need to catch up with technological developments. Their regulation is beyond the scope of this paper, while their impacts on groundwater make groundwater protection rules, particularly with respect to groundwater quality, ever more relevant.

2. Recognizing the Importance of Human Rights

Until a few years ago water laws made no mention of the human right to water. Indeed, the right to water itself—an implicit element of the right to an adequate standard of living recognized most prominently in the 1966 International Covenant on Economic, Social and Cultural Rights—received very little attention [8]. This situation changed when the United Nations Committee on Economic, Social and Cultural Rights General adopted General Comment No. 15 in 2002 [9]. The latter marked the beginning of a quickly developing debate on the right to water, which resulted, inter alia, in the recognition of the right to water in United Nations General Assembly and Human Rights Council resolutions, numerous constitutions and an increasing number of water laws [10].

The right to water entitles everyone to sufficient, safe, acceptable, physically accessible and affordable water for personal and domestic uses [9]. The right to water poses an obligation upon the state to ensure such access in a non-discriminatory manner. It has implications for water resources legislation and for the regulation of the water industry.

In water resources legislation drinking and domestic uses have to take priority over other uses. Whereas such priority has traditionally been accorded to drinking water needs in the allocation of water to different uses, framing a water law in human rights terms adds a new dimension by elevating drinking water needs to requirements as of right instead of policy choice, i.e., to entitlements. Due to the generally superior quality of groundwater compared to surface water groundwater is often ideally suited to the satisfaction of the right to water. Human rights considerations also play a role in the protection of water resources from contamination, the regulation of abstraction to prevent dropping water tables and drying up of wells and many other issues. With respect to the regulation of the water industry, these include appropriate pricing policies, freedom from arbitrary or unjustified disconnections from the supply network, and non-discriminatory access to sufficient and continuous water supply. In many countries, the quantities and uses of water protected by the right to water fall into the category of permit-free de minimis uses (see below, No. 5.), the protection of which therefore may not be compromised unless access to a different source of water is guaranteed. The human right to water requires a look at all water laws, policies, and strategies through a human rights lens and to measure them against human rights criteria, particularly with respect to non-discrimination.

3. Groundwater as a Public Good

As much as access to water is deemed a human right, water as a resource is by now overwhelmingly regarded as a public good.

Historically, the opposite was the case for groundwater. Being a "hidden resource" it was often treated as a private good accessible to owners of land and out of the purview of public regulation. Water rights were essentially a subsidiary component of land tenure rights, i.e., the right to use groundwater was conferred on the owner of the overlying land. In many non-western systems, groundwater could not be privately owned. In Islamic law groundwater is considered a public good, but the ownership of a well entails ownership of a certain amount of adjacent land called *harim* or forbidden area. It varies in size according to different schools [11]. Customary regimes in many parts of the world view groundwater resources as belonging to the community and reject the concept of individual rights over water.

Better understanding of the characteristics and nature of groundwater and increasing pressure on the resource have instilled a predominant trend to vest ownership and control over all water resources in the state or to recognize the state's superior right to the management of water resources [12]. The state becomes the guardian or trustee of groundwater resources. The right to access and use of groundwater is independent of the regime of the overlying land. Any user who wants to abstract groundwater must apply for a permit (also styled license, authorization, or concession) in order to obtain a "right to use". The separation between ownership and usufructuary rights is one of the cornerstones of modern formal water rights regimes and allows the government to manage and protect groundwater resources in the interest of the public.

In some parts of the world, rules granting private ownership of groundwater persevere. For instance, in Texas, India, or Pakistan the rule of capture is still dominant, allowing landowners to extract groundwater freely on their land [13]. As a result, regulation of access to and extraction of groundwater in the interest of the public is very weak in those countries.

More importantly, even where groundwater is formally a public good and users have only usufructuary rights, perceptions of it being "private" often linger on. They interfere with compliance with government regulation or generate transition problems from an unregulated to a regulated regime. The legal notion or the perception of groundwater being private property can be a strong driver for overexploitation. Socially more equitable access to groundwater and sustainable management depends thus on severing the link between land ownership and control over groundwater and on supporting such change by educational and awareness raising measures.

Customary laws have their own approaches to ownership issues. Customary regimes in many parts of the world regard groundwater resources as belonging to the community and reject the concept of individual rights over water. In much of Africa and Asia, customary water rights are intrinsically linked to land and embedded in land tenure [14]. Formal water legislation might provide for the recognition of customary water rights, which is an emerging trend reflected in recent water legislation from countries as diverse as Bhutan, Namibia, Peru, Tanzania, and Zambia [15,16]. States resort to a variety of approaches [17]. Mozambique (Water Licensing Regulations, 2007), for instance, grants priority of allocation to customary water rights. The same applies to the customary rights of traditional communities in Paraguay (Law on Water Resources, 2007). Another option is to order or allow the transformation of a customary right into a statutory right (Water Law of Tucuman Province, Argentina, 2001; Tanzania's Water Resources Management Act, 2009). Others, e.g., Namibia (Water Management Act, 2013), place the government under a duty to have regard to customary rights and practices when considering applications for a borehole license or a license to abstract water). Unfortunately, a large number of water laws protect customary water rights but do not provide the necessary details on the interface between customary and statutory rights, thereby creating legal uncertainty as to their de facto status and protection [17].

4. Groundwater Exploration

Borehole digging or drilling and well construction for exploration and exploitation are the first steps towards groundwater use. To protect groundwater, to retain control of and information over access to it, and to prevent conflict among users, groundwater exploration and borehole digging or drilling tend to be subject to prior notification, a permit, or registration, as legislation from countries as diverse as Kenya, Kyrgyzstan, Namibia, the Northern Territory of Australia, Oman, the Philippines, or Costa Rica (with a piece of special purpose legislation) [18] demonstrates. In addition, there are rules on borehole construction, well maintenance, and the sealing of abandoned wells. Obligations to sample and to file drilling reports ensure that drillers supply groundwater data to the administration [19]. Legislation may impose drilling fees may and the maintenance of boreholes registers.

In view of the skills required, legislation may subject the exercise of the profession of commercial well digging or drilling to registration or licensing requirements to ensure that the driller is appropriately qualified and that borehole construction standards are maintained [20]. Where this is the case, registries of drillers are often kept. In many cases, however, the drilling industry is not regulated and opportunities for getting away with poor constructions standards abound.

5. Groundwater Quantity Protection: Regulating Abstraction and Recharge

Where water demand outstrips supply, significant negative environmental impacts are likely to occur. These can be felt on groundwater through (1) diminishing water tables and eventual depletion of an aquifer; (2) migration of low quality water; or (3) salinization in coastal areas. Equally it can impact negatively on springs, seepage zones, and the baseflow of rivers. Often, social and economic repercussions ensue. Since 2011 the ongoing drought in California, for instance, has caused tremendous aquifer overdraft resulting in significantly declining groundwater levels (often below historic lows), land-subsidence in part of the Central Valley, and a rush for digging ever deeper wells [21].

Water rights are the central element of most current water laws to control water demand. Under a system of water rights, users have to acquire a permit before (ground-)water can legally be abstracted. This obligation is incumbent upon whoever the user is, including state agencies such as those involved in the development of irrigation schemes. Small-scale or de minimis uses are usually exempt (see below). Permit-based systems have gained prominence, in particular since more powerful pumps, population growth, and economic development have driven demand for groundwater, often in excess of supply. They enable the water administration to allocate water to different uses ranging from domestic, agricultural, and industrial uses to environmental ones such as sustaining wetlands and the baseflow of rivers.

Although permit-based systems have long been in place, in many countries, including western ones, the regulation of groundwater still lags behind the regulation of surface water. An example is the Canadian province of British Columbia where as late as 29 February 2016, with the entry into force of the new Water Sustainability Act, nondomestic groundwater use was licensed and made subject to fees and rentals [22]. Prior to this change even large industrial users like the bottling company Nestlé Waters Canada that bottles 265 million liters per annum could take limitless amounts of groundwater for free even in situations of drought [23]. Under new regulations (Water Sustainability Act) fees will be charged, which are, however, heavily criticized as being too low for industrial users and which might undergo further revision [24].

Permits are typically granted for a renewable time period, which is short enough to provide the state flexibility in the resource management, on the one hand, and provide a stable basis for the user for planning and investment decisions, on the other hand. The length of this time period varies considerably ranging from very short (e.g., 5 years) to very long (e.g., 50 years) and depends on local context, the purpose of abstraction, and the state of the aquifer.

Permits provide for annual or seasonal volumetric allocations or are based on an area quota. They also state the abstraction purpose, water protection measures, and the amount of fees or charges due, among other things. Permits require that use be made of the water right failing which the right will lapse ("use-it-or-lose-it" rule). An exception is Chile which imposes a tax in case of nonuse [25]. The metering of wells is often imposed in order to verify compliance with the conditions attached to a permit and to measure the amount of water abstracted. Permit holders have the obligation to report regularly (normally on an annual basis) how much water has been abstracted. Where the transaction costs of metering would be too high, estimates are sometimes used. Permits are usually recorded in a registry, which serves as a tool for planning purposes as much as a record and legal evidence of abstraction rights against the claims of third parties and as a source of information to government and fellow or prospective water users.

In some instances, legislation determines the overall amount of permitted withdrawals per year. One example is the Edwards Aquifer Authority Act (Texas), which precludes the Edwards Aquifer Authority from authorizing withdrawals from the entire aquifer exceeding 572,000 acre-feet (approximately 705,550 cubic meters) of water annually [26]. Another example is the Namibian Water Resources Management Act (2013), which empowers the Namibian Water Minister to restrict or limit groundwater abstraction to ensure that the total abstraction of the water permitted from an aquifer does not exceed the volume of water, which can be abstracted over a period without adversely affecting the quantity and quality of the water or the environment above [27]. In the same direction goes Tanzania's Water Resources Management Act (2009).

In some countries, e.g., Australia, Chile, and the US, users can trade groundwater abstraction permits, subject to some form of prior involvement of the water administration to protect both private and public interests and to mitigate negative impacts of such trades [28]. Other countries, especially in South Asia, have informal water trading schemes. Worldwide there is no clear trend on the sensitive issue of trading water permits, with recent water legislation in as many cases allowing as prohibiting it [15,29].

Administrating and implementing a permit system is a costly, administratively challenging, and time-consuming process, especially in countries with very high numbers of small-scale users. The introduction of a permit system is likely to fail where it is not well designed and tailored to the local context and administrative capacity.

For administrative and user convenience and cost it is standard practice to exclude small-scale or de minimis uses—which are particularly important in developing countries—from permit requirements. Such de minimis uses can be defined with regard to volumes, area, or purpose and need to take into account local conditions. Widely accepted de minimis uses include the use of water for drinking, watering domestic animals and poultry, recreational uses such as bathing, meeting of basic household needs, the watering of garden plots, and firefighting [30]. Defining de minimis uses is a tricky task that requires taking into account local circumstances. While for instance in the Murray–Darling basin small-scale users are allowed to extract water without a permit for domestic or livestock needs or for irrigating small plots of 2 ha or less, the same rule would exempt over 95% of current groundwater irrigators if applied in South Asia or on the North China plains [31].

In terms of water tenure, users face the disadvantage that even where de minimis use is described in terms of a right to abstract water and use it for specific purposes, it may be difficult to assert an unquantified right against the state or other users in case of drying up wells or polluted water. Insofar individual de minimis rights lack security and often also equitable mechanisms for allocation, even where legislation protects de minimis uses as a category [30,32].

The protection of the significant social and public health benefits derived from small-scale uses requires effective aquifer planning and protection measures to prevent de minimis sources from running dry or becoming unfit for use. A permit can be made a prerequisite for drilling so information about the location of wells for de minimis uses is available. Cumulatively, or alternatively, a reporting regime whereby users are required to periodically provide details of their abstraction and use of groundwater can help to keep the scope of de minimis uses under review. Community management approaches may provide further options for groundwater allocation and protection for de minimis uses.

The protection of groundwater from overdraft requires regulation not only of abstraction but protection, and in some cases artificial enhancement, of recharge.

Natural groundwater recharge processes have to be protected from land-based interference including paving, development, or logging. Land surface zoning is a tool often applied to protect recharge areas and to establish zones where abstraction is restricted or prohibited. Also, other land management measures such as changing cropping or irrigation practices offer scope to regulate recharge to improve groundwater quantity. Under certain conditions, recharge rates under cropland may increase by one to two orders of magnitude compared with native perennial vegetation [33]. Equally, reducing or changing highly water-intensive cropping practices preserve precious groundwater resources. An example is the Indian States of Punjab and Haryana, which prohibited to sow paddy nursery and to transplant paddy before certain dates in May and June to avoid the high evaporation rates in early summer [34,35]. Robust effects on reducing groundwater depletion have been reported for the Punjab [36].

Recharge can be enhanced through traditional water conservation measures and artificial direct surface or subsurface as well as indirect recharge techniques. Managed aquifer recharge is the intentional recharge and storage of water (river water, reclaimed water, desalinated seawater, rainwater, or groundwater from other aquifers) into an aquifer for subsequent recovery or for environmental benefits using spreading and infiltration techniques or injection wells. It is used for a number of purposes, including replenishing an aquifer to increase the water table and storing water for drought and emergency supplies. Managed aquifer recharge schemes require prior authorization. Licenses are required for the abstraction of the source water, storing of water, aquifer recharging, and re-abstracting water [37]. Legislation also requires that certain conditions are met as to the qualifications of the operators in the sector and to water quality.

Some legal instruments, such as the 2011 Indian Draft Groundwater Model Bill, explicitly call for a number of measures to protect and enhance recharge, namely water harvesting, including rooftop rainwater harvesting, catchment conservation using appropriate groundwater structures or pits, the creation of protection zones in natural recharge areas and in areas that require special attention with regard to the artificial recharge of groundwater, programs for the recharge of aquifers, setting up artificial recharge structures, afforestation, and reforestation [38]. The Andhra Pradesh (India) Land, Water and Trees Act and Rules (2002) explicitly enables the relevant authority to issue guidelines for constructing appropriate rainwater-harvesting structures in all residential, commercial, and other premises and larger open spaces [39].

6. Groundwater Quality Protection: Pollution Control

Groundwater quality protection—with a view to protecting public health and achieving social, economic, and increasingly environmental objectives such as minimizing harm to dependent ecosystems—is the second main goal of legal frameworks for groundwater. A plethora of activities and a much wider set of actors than groundwater abstractors causes groundwater pollution and degradation, which are under certain conditions irreversible. Causes of pollution are inappropriate drilling, abstraction, and well maintenance practices resulting in contamination from the well, inter-aquifer leakages or saline water intrusion, pollution caused by polluted surface (river, lake) water, and, in particular, pollution caused by land-based activities.

Groundwater quantity and quality protection are interrelated. Poorly performed drilling operations, inadequate well construction and maintenance, and poor well-casing may result in contamination from the well or inter-aquifer leakage and groundwater degradation by mixing water from different aquifers/layers of aquifers of different quality. Drilling and pumping can mobilize naturally occurring pollutants such as arsenic or fluoride or induce saline intrusion or the migration of lower quality or brackish waters. In many small island states such as The Bahamas and Barbados or Malta, salinization from overpumping of aquifers and thus salinity intrusion into the islands' underlying freshwater lens is a matter of great concern [40,41]. The regulation of the profession of drillers, groundwater exploration, and groundwater abstraction (discussed above) serve groundwater quality protection purposes, as do regulations on borehole construction, well maintenance, and the sealing of abandoned wells.

Land-based activities causing pollution fall broadly into two categories, each triggering distinct legal answers. Point-sources of pollution relate to pollution whose entry point into an aquifer can be established with sufficient certainty such as landfills, waste dumps, industrial accidents, mines, and the underground storage of substances. In contrast, non-point source pollution originates from diffuse or indistinct sources, often spread over relatively large areas, whose origins, entry point into groundwater, and impact are difficult or impossible to determine with accuracy. The prime example is agricultural runoff, but stormwater runoff in urban areas and airborne pollution transmitted by precipitation also play a role.

Point-source pollution is addressed by absolute prohibitions or limitations on emitting certain substances. Polluting activities may be made conditional upon a wastewater or pollution discharge permit, prior treatment of the substance, and compliance with effluent standards [11]. The regulation of potential causes of point-source pollution includes provisions aimed at preventing harm to groundwater resources, such as regulations on mining activities or the sitting and operation of landfills or storage tanks and the use of best available technologies. Non-point source pollution prevention measures include the regulation of land uses giving rise to diffuse discharge and the imposition of best agricultural and environmental practices. They include prohibiting or limiting certain polluting and water-using activities; limiting the use pesticides, herbicides, and fertilizers, especially to reduce nitrogen and phosphorus built up; restricting certain cropping patterns; reducing animal-grazing intensity; land reclamation; and drainage [19]. The whole array of actors involved in land use and pollution management—including urban water supply utility managers and agricultural and environmental agencies seeking to regulate the application of fertilizers (e.g., nitrates) and pesticides—have to be involved in groundwater quality protection and addressed by legislation. Given growing pressures on aquifers it becomes increasingly acute to develop and coordinate land use plans, in both urban and rural areas, with river basin or aquifer development plans.

Water quality standards and threshold values are a related mechanism to protect groundwater quality. For the members of the European Union (and for accession countries) the EU Water Framework Directive and its daughter directive on the Protection of Groundwater against Pollution and Deterioration provide a particularly stringent system of groundwater quality protection [3,42].

Land surface zoning is used for critical areas of high vulnerability, including recharge areas, to restrict or prohibit activities with adverse impact on groundwater, such as industrial chemical handling or effluent discharge to the ground, mining, or certain agricultural land use practices. The protection of drinking water sources is a particularly important case in point: the capture zones of the main areas of potable water-supply abstraction require to be designated as protected areas. Successful zoning depends on sufficient knowledge about the characteristics of the aquifer, on land use planning processes, which take groundwater issues appropriately into account, and on a set of accompanying measures that support the restrictions brought about by zoning.

7. Responding to Environmental Concerns and Climate Change

As groundwater resources are becoming overly used and polluted, environmental concerns play a more and more prominent role in groundwater legislation. The principle of sustainability has emerged as a key crosscutting principle and cornerstone of modern groundwater legislation. Equally, other principles of environmental law such as the precautionary principle and the preventive principle increasingly feature in water law. European Union directives and—for transboundary groundwater resources—European international legal instruments such as the 1992 European Convention on International Rivers and Lakes and its accompanying 2012 Model Provisions on Transboundary

Groundwaters [43,44] most clearly bear out this tendency, but it can equally be observed in legislation from other regions of the world such as from Bhutan (Water Act) [45], Uruguay (National Water Policy Law) and Venezuela (Water Law) [46,47]. The obligation to use and manage aquifers sustainably preserves the integrity of aquifers and protects the demands of the environment and of water-dependent ecosystems such as wetlands in the discharge areas of aquifers as much as human needs.

The principle of sustainability is expressed in a variety of ways [48]. It plays a role in allocation decisions and planning processes where the recharge capacity of an aquifer has to be taken into account. Environmental objectives support the allocation of groundwater to groundwater-dependent ecosystems and the baseflow of rivers. The viability of aquatic ecosystem and protected area sustenance may take precedence over other demands competing for water at times of scarcity except those protected by the right to water. Abstraction and pollution discharge permits must meet environmental criteria, and if environmental concerns so require then abstraction rights may be curtailed or revoked, which may raise intricate issues of compensation. Where transfers of an abstraction right are permitted, the environment may not be harmed. Environmental concerns can be taken into account in determining the charges for water abstraction and wastewater disposal and they play a significant role in zoning and planning [15,29]. For instance, zones may be established in recharge areas to protect the aquifer as well as in discharge areas to protect wetlands.

The obligation to carry out environmental impact assessments is becoming a prerequisite for the granting of abstraction or discharge permits above a certain quantity of water or before allowing projects of industrial, agricultural or other type with potentially negative impact on aquifers. The water legislation of a large number of countries—among them jurisdictions as diverse as Cameroon, Mexico, Kenya and Paraguay [48]—gives evidence of this trend.

The allocation of water to an "environmental reserve" as in Namibia, South Africa, Tanzania, or Zambia aims at promoting sustainable use of water resources and preserving dependent ecosystems such as wetlands [49–52].

An environmental concern of particular salience is climate change, one of the big challenges of the future with pervasive impacts on the water sector, including on groundwater availability and quality. The interactions between groundwater and climate change are diverse, depending on the context, and still largely uncertain. It is, however, certain that there will be direct and indirect impacts.

Among the direct impacts are changes in precipitation and evapotranspiration, which may reduce groundwater recharge and result in falling water tables. Climate change is projected to reduce renewable surface and groundwater resources in most dry subtropical regions [53]. Increases in the frequency and severity of floods may result in higher surface runoff and soil erosion, possibly also reducing recharge. Preliminary evidence suggests that changes in snowmelt regimes will reduce the seasonal duration and magnitude of recharge [33]. In other regions recharge may increase and groundwater levels rise. Sea level rise and storm surges will put coastal and island aquifers at risk of saline intrusion [33].

Indirect effects of climate change on groundwater occur through increases in demand on groundwater in a warming climate, in particular irrigation demand, and changes in the availability of different water sources. In adaption to increases in precipitation variability, the importance of groundwater is likely to increase.

It is necessary to assess continually the impact of climate change on water resources and, where necessary, to take measures, such as the possibility to revoke abstraction permits, as a consequence of climate-induced deterioration of the state of an aquifer. With few exceptions climate change is not yet a specific concern in modern water policy and law. While scarcity and sustainability are addressed, the broader connections between climate, water, forests, and land use are often not yet tackled in mitigation attempts. Where environmental legislation addresses climate change it often remains at a generic level. Especially land use and its effects on groundwater have to be addressed as interlinked issues.

National action plans on climate change tend to address inadequately impacts on water resources, and action plans on climate change and water are not sufficiently interlinked. Positive examples of the contrary are the 2012 Murray Darling Basin Plan (Australia), which is based on the 2007 Water Act (a Commonwealth statute amended in 2008) and deals with the nexus between water resources, including groundwater, and climate change (see Sections 4.03, 5.03, 6.06, 8.04, 8.07, 10.51) [54]. Also, the Draft National Water Framework Bill of India (2013) refers to climate change in its sections on project planning and management, and on coordination and policy support mechanisms [55]. In contrast, Indian state-level groundwater laws, the Indian 2005 Model Bill to Regulate and Control the Development and Management of Groundwater and its planned successor, the 2011 Draft Model Bill for the Conservation, Protection and Regulation of Groundwater, only refer to environmental protection but do not address climate change directly [38,56].

8. Monitoring

The main piece of water legislation tends to contain obligations to monitor groundwater use and status, while subsidiary legislation or technical guidelines sets out detailed parameters. Groundwater quantity and quality monitoring provide the basis for planning, allocation, and conservation decisions. Legislation may (and should) require groundwater-related institutions to coordinate data-gathering, interpretation, and storing.

Information from modeling and monitoring enables scientifically-based management decisions to ensure that in case of aquifers under pressure water remains available and of sufficient quality. Water rights remain secure and are not threatened by falling water tables putting water out of reach or deteriorating quality attributes making water unfit for the purposes a right was granted for. Once baseline data of the characteristics of an aquifer system is available, regular monitoring of changes in flow, storage, and water quality informs about the impact of abstraction and pollution activities. Inventories of wells (sometimes limited to certain locations or to wells showing specific characteristics), registers of abstraction and wastewater discharge permits, the information provided by well owners on their abstraction, and the data made available through monitoring together enable the water administration to recognize critical situations and to intervene accordingly. Monitoring also helps to establish liability, for instance for polluting incidents, although the concealed and inaccessible nature of groundwater and the slow changes in quantity and quality cause difficulties in practice.

Monitoring aquifers is a technically demanding and costly exercise so that information about many aquifers is highly incomplete. The costs for setting up a monitoring network, collecting data regularly, processing, interpreting, and storing it in databases, including unified databases at the national level, are one of the reasons why in developed and developing countries alike water rights and monitoring obligations, especially with respect to groundwater, may remain unimplemented.

9. Planning

Water legislation regularly requires resource management plans, for instance, in Morocco, South Africa, Uganda, South Australia, and Texas and by the European Water Framework Directive [32]. Water management plans are developed to promote rational, effective, and fair water management and decision-making. In some countries they may even have a legally binding nature.

They typically contain the characteristics of a river basin or aquifer, a review of the impact of human activity, objectives with respect to quantity and quality of the resource and measures to meet the objectives. In the case of nonrenewable aquifers, or of aquifers for which a policy choice in favor of mining has been made, the plan may need to address depletion. Where the water legislation does not establish development and management priorities, a management plan may include them.

The type of plan the respective legal and institutional framework provides for depends on the hydrogeological setting, the degree of surface–groundwater linkage and the significance of the aquifer. For significant aquifers or fossil aquifers, management plans can be developed to complement integrated water resource planning at the national level. An example is the Groundwater Management

Plan for the important Edwards Aquifer in Texas, which guides the activities of the Edwards Aquifer Authority. Other examples are the Draft Model Groundwater Model Bill of India that suggests aquifer-based use and protection plans [38] and the water legislation of Algeria, which establishes important aquifers as units for planning [57].

Alternatively, as all recharging aquifers occur within river basin management units, groundwater can be integrated into river basin planning. Where legal and institutional frameworks are weak or do not adequately cover institutional responsibility for groundwater resources or groundwater user groups are not represented in a river basin agency or the agency's personnel lacks groundwater knowledge, such an approach might, however, be premature [58].

There is a clear trend to involve stakeholders in plan preparation and revision. Participation in planning processes under the EU Water Framework Directive is a case in point: under Art. 14 of the EU Water Framework Directive, Member States shall encourage the active involvement of all interested parties in the implementation of the EU Water Framework Directive and in particular in the production, review, and updating of river basin management plans, which are mandatory for all basins under the directive [3].

10. Designing Effective Multilevel Institutional Set-Ups

Institutional set-ups to manage aquifers and to administer water rights vary highly across countries. As discussed in depth elsewhere, in most countries a plethora of multilevel water resources administration institutions exists [11]. As intensive development of groundwater is recent, institutional responses tend to lag behind the tasks created by legislation. In federal countries, federal, state and local institutions may be vested with functions and powers in respect of groundwater. Only a few countries have attempted to set up dedicated groundwater management agencies.

The highest and ultimate responsibility for water resources management is often conferred upon one or more ministers, usually acting through a statutory Director or Director-General of water resources, such as the Department of Water and Sanitation in South Africa, or some other statutory body such as an authority (e.g., the Jamaica Water Authority) or agency (e.g., the Environment Agency in England) or Directorate of Waters (e.g., the Direccción General de Aguas in Chile) [30,59]. Ideally, the entire range of groundwater management issues is placed in the hands of a single water resources institution, which would also be in charge of surface water [60]. While such a set-up, which reflects allegiance to the principle of integrated water resources management, is still more exception than rule, recent water legislation of a number of countries, among them Honduras, Peru, Tanzania and Zambia, point to a consolidating trend in this direction [15]. An inter-ministerial/interagency mechanism such as a council, commission or committee usually has the task of coordinating those ministries and agencies that have a stake in aquifer management [19].

The main tasks of a water authority comprise planning and modeling future demands and impacts on water resources, organizing stakeholder fora, monitoring water quality and quantity, issuing and administrating water rights, including the maintenance of registers, and implementing and enforcing the water law and water rights regimes, e.g., by meting out administrative fines in cases of non-respect for legal obligations or the terms and conditions of a permit [59]. Water authorities often fall short of delivery on these tasks due to institutional weaknesses, lack of empowerment and finances, gaps in mandate, and a number of other factors. Institutional deficits inevitably result in implementation deficits, which erode the contribution a legal framework can make to protecting and managing an aquifer.

There is a trend to complement the water resources institution, which bears overall responsibility throughout the jurisdiction with coordinating/decision-making institutions at the drainage basin or aquifer level. The scope of institutions set up at the river basin often includes groundwater management, even where aquifer boundaries do not follow the boundaries of the river basin. E.g., under the Art. 3 Water Framework Directive where groundwater does not follow a particular river basin it is assigned to the nearest or most appropriate river basin [3]. Less frequently, a mechanism is set up specifically at the aquifer level.

In federal countries, large aquifers spanning several states may call for interstate institutional mechanisms. This is the case of the Great Artesian Basin (GAB) shared by Queensland, New South Wales, South Australia, and the Northern Territory, one of the largest basins in the world covering 22% of Australia, for which the Australian and state/territory governments in consultation with the GAB Coordinating Committee (community and sector representatives), national organizations, landholders, and water users develop a Strategic Management Plan.

Where the scale of groundwater issues is rather small, local solutions may be the most appropriate given the localized nature of groundwater access and use. In such circumstances local level groundwater management by self-organized user groups may be a promising and appropriate approach.

The legislation of a number of countries provides for the establishment of (ground-)water user associations or aquifer management organizations, for instance Chile, Spain, Mexico, Australia, and the western states of the US [61]. They have been established especially where aquifers are at risk of being degraded or depleted. The Spanish Water Law of 1985 (revised in 2001) makes the establishment of groundwater user organizations compulsory in overexploited aquifers [62]. In a number of western states of the U.S. groundwater management or conservation districts, a form of water users association, have been established in respect to about 89% of groundwater resources. Local users control them and may set limits on pumping and wells, adopt groundwater management and development policies and programs, and propose water allocations criteria. In Mexico, where groundwater resources are severely overexploited, COTAS (Comités técnicos de aguas subterráneas—technical groundwater committees) have been created. They are civil society organizations, which are supported financially by the public hand. The National Water Commission carefully facilitated their set-up. Inter alia, the COTAS support the implementation of groundwater management plans, support the government in groundwater rights administration, provide services to groundwater users, support consensus-building for future integrated water resources management, establish dialogue with groundwater users, and help improve data (e.g., by helping the water administration to validate, update, and correct databases on wells) [63].

Also, the Expert Group on Ground Water Management and Ownership of the Indian National Planning Commission suggested a shift in focus from state control to community management by user groups because of difficulties of implementing a permit-based system. It recommended that user groups be responsible for planning the use of groundwater within groundwater management units. Based on the goal of sustainable-yield management, withdrawal should not exceed long-term recharge. The Central Ground Water Board and the State Ground Water Board would be responsible for scientific monitoring of groundwater levels and for estimating a sustainable level of groundwater use. In case water levels fell below the replenishable level, the Central government could declare an area as *"environmentally threatened"* and, in consultation with stakeholders, a strategy for addressing this problem would be developed [64].

Stakeholder self-organization depends on a number of conditions. It can occur where strong community values and norms exist, particularly if they are embedded in and facilitated by a conducive larger regime [65]. A well-informed community of groundwater users may agree on maximum acceptable drawdowns in pumped wells or ban the application of pesticides across an aquifer that furnishes potable water supplies [66]. Local self-management will need technical support when problems occur, and preferably before, for realistic appreciations of the state of the resource. Management of groundwater resources by user groups also does not replace state regulation, implementation and enforcement, and economic mechanisms, rather the three work as a tripod on which to base aquifer governance [67]. There are relatively few examples of self-organization in relation to the scale and intensity of groundwater development [68].

The legal framework has to define clearly the role and responsibilities of groundwater user groups on the one hand and the water administration and other public institutions on the other. For large-scale groundwater users, such as Coca-Cola in the case of India, a permit and registration approach with clear indications of conditions should exist in parallel to community approaches [69]. Among possible organizational forms are water users associations, groundwater user groups, village water supply groups, aquifer management organizations (AMORs), or Non-governmental organizations (NGOs). AMORs have been suggested for larger high-yielding aquifers and should include all local water user associations, groundwater user groups, village water-supply councils etc., and also representatives of national and/or local groundwater resource agencies and of the corresponding local government [70]. The governance arrangements that may work across a small aquifer may not work in large aquifers supplying water to a range of different activities in larger urban, agricultural, industrial, and mining areas [66].

11. Conclusions

Modern water legislation based on administrative water rights and a range of planning and monitoring tools has been introduced worldwide over the last decades. It offers mechanisms to strike a balance between growing demand and the need to protect and preserve groundwater resources for current and future generations.

A number of trends in groundwater legislation can be distilled. The human right to water is increasingly recognized not only at the constitutional level but also in water resources legislation. There is a strong trend to treat groundwater as a public and no longer as private good attached to land rights. Administrative permit-based water rights systems are becoming a norm. They tend to work best in highly formalized water economies and always have to take into account local conditions, the existence of customary rules—where applicable—and existing administrative capacity. In developing countries with limited administrative capacity and high numbers of small-scale users they face implementation challenges. Environmental concerns and principles continue to gain importance in (ground-)water legislation. The array of mechanisms apt to deal with the uncertainties generated by climate variability are becoming ever more important as the latter exacerbates the need to strike a balance between the unpredictability of water resources availability and quality, the security of water rights tenure sought by investors and environmental allocations. Stakeholder involvement plays an increasing role in planning and decision-making processes, and management approaches at the community or aquifer level are developing.

All in all, more and more countries strengthen domestic groundwater legislation and rank it on a par with surface water regimes, often by dealing with all water resources in an integrated manner based on the principle of integrated water resources management. These encouraging developments with respect to law, institutions, and administration have to be supported and embedded in supportive policies in other sectors, such as agriculture and energy, to achieve effective and sustainable improvements of groundwater governance and reverse the global trend of groundwater depletion and degradation. Legal frameworks for groundwater have to form an integral part of context-based groundwater governance, which must be adapted to each country's financial, technological and institutional capabilities and be appropriate to its geography and environment, customs, cultures, political systems, and prevailing practices [71].

Acknowledgments: This paper builds on a thematic paper on legal and institutional frameworks prepared for a GEF project on Groundwater Governance. The author was the lead consultant for the legal and institutional work carried out with respect to this output under the project [72].

Conflicts of Interest: The author declares no conflict of interest.

References

1. *Frazier* v. *Brown*; 12 Ohio St. 294 (1861).
2. IGRAC. Transboundary Aquifers of the World, 2015. Available online: http://www.un-igrac.org/sites/default/files/resources/files/TBAmap_2015.pdf (accessed on 7 April 2016).

3. Directive 2000/60/EC of the European Parliament and of the Council of 23 October 2000 Establishing a Framework for Community Action in the Field of Water Policy. Available online: http://eur-lex.europa.eu/legal-content/en/TXT/?uri=CELEX:32000L0060 (accessed on 15 August 2016).
4. Eckstein, G. Managing Buried Treasure across Frontiers: The International Law of Transboundary Aquifers. *Water Int.* **2011**, *36*, 573–583. [CrossRef]
5. McCaffrey, S. The International Law Commission Adopts Draft Articles on Transboundary Aquifers. *Am. J. Int. Law* **2009**, *103*, 272–293. [CrossRef]
6. McIntyre, O. International Water Resources Law and the International Law Commission Draft Articles on Transboundary Aquifers: A Missed Opportunity for Cross-fertilisation? *Am. J. Int. Law* **2011**, *105*, 237–254.
7. Mechlem, K. Past, Present and Future of International Law of Transboundary Aquifers. *Int. Community Law Rev.* **2011**, *13*, 209–222. [CrossRef]
8. International Covenant on Economic, Social and Cultural Rights. Available online: http://www.refworld.org/docid/3ae6b36c0.html (accessed on 15 August 2016).
9. General Comment No. 15. The Right to Water. Available online: http://www2.ohchr.org/english/issues/water/docs/CESCR_GC_15.pdf (accessed on 15 August 2016).
10. UN General Assembly. Resolution 64/292 of 28 July 2010. UN Doc. A/RES/64/292 of 3 August 2010. For References to Further Resolutions, Information on the Right to Water in Constitutions and Domestic Legislation See. Available online: http://www.righttowater.info/progress-so-far/ (accessed on 12 April 2016).
11. Caponera, D. *Principles of Water Law and Administration, National and International*, 2nd ed.; Nanni, M., Ed.; Taylor & Francis: London, UK, 2007; p. 11.
12. Burchi, S.; Nanni, M. How Groundwater Ownership and Rights Influence Groundwater Intensive Use Management. In *Intensive Use of Groundwater: Challenges and Opportunities*; Llamas, R., Custodio, E., Eds.; Balkema: Lisse, The Netherlands, 2003; pp. 227–240.
13. Edwards Aquifer Authority V. Day. Supreme Court of Texas. Available online: http://caselaw.findlaw.com/tx-supreme-court/1595644.html (accessed on 10 August 2016).
14. Meinzen-Dick, R.; Nkonya, L. Understanding Legal Pluralism in Water and Land Rights: Lessons from Africa and Asia. In *Community-Based Water Law and Water Resource Management Reform in Developing Countries*; van Koppen, B., Giordano, M., Butterworth, J., Eds.; Cabi: Wallingford, UK, 2007; pp. 12–27.
15. Burchi, S. 2009–2011 Year-End Review. *Water Law* **2011**, *22*, 248–265.
16. Burchi, S. 2005 Year-End Review. *Water Law* **2005**, *16*, 155–179.
17. Burchi, S. A Comparative Review of Contemporary Water Resources Legislation: Trends, Developments and an Agenda for Reform. *Water Int.* **2012**, *37*, 613–627. [CrossRef]
18. Reglamento de Perforación del Subsuelo para la Exploración y Aprovechamiento de Aguas Subterráneas. Available online: http://www.da.go.cr/textos/documentos/35884-.pdf (accessed on 10 August 2016).
19. Nanni, M.; Burchi, S.; Mechlem, K.; Stephan, R.M. Sustainable Utilization of Groundwater Resources: Legal and Institutional Aspects. In *Non-Renewable Groundwater Resources—A Guidebook on Socially Sustainable Management for Water Policy-Makers*; Foster, S., Loucks, D.P., Eds.; United Nations Educational, Scientific and Cultural Organization (UNESCO): Paris, France, 2006; pp. 49–58.
20. Preparing National Regulations for Water Resources Management—Principles and Practice. Available online: http://www.fao.org/docrep/006/y5051e/y5051e00.htm (accessed on 20 April 2016).
21. Assessing CA Groundwater Management One Year after Landmark Legislation. Available online: http://waterinthewest.stanford.edu/resources/forum/assessing-ca-groundwater-management-one-year-after-landmark-legislation (accessed on 13 April 2016).
22. Water Sustainability Act, British Columbia, Canada. Available online: https://www.leg.bc.ca/Pages/BCLASS-Legacy.aspx#%2Fcontent%2Flegacy%2Fweb%2F40th2nd%2F3rd_read%2Fgov18-3.htm (accessed on 7 April 2016).
23. CBC News. Nestlé Faces Renewed Criticism as B.C. Drought Continues. Available online: http://www.cbc.ca/news/canada/british-columbia/nestlé-faces-renewed-criticism-as-b-c-drought-continues-1.3145929 (accessed on 8 April 2016).
24. Nestlé is about to Suck BC Dry—For $2.25 per Million Litres to be Exact. Available online: http://action.sumofus.org/a/bc-bottled-water/13/2/?sub=homepage (accessed on 8 April 2016).

25. Ley N° 20.099—Introduce Modificaciones a la Ley N° 20.017, que Modifica el Código de Aguas. Available online: http://faolex.fao.org/cgi-bin/faolex.exe?rec_id=045571&database=faolex&search_type=link&table=result&lang=eng&format_name=@ERALL (accessed on 10 August 2016).
26. Edwards Aquifer Authority Act, 1993. Available online: http://www.edwardsaquifer.org/legislation-and-policy/the-eaa-act (accessed on 10 August 2016).
27. Promulgation of Water Resources Management Act, 2013. Available online: http://docplayer.net/11376739-Government-gazette-of-the-republic-of-namibia-n-36-40-windhoek-19-december-2013-no-5367.html (accessed on 10 August 2016).
28. Solanes, M. Institutional and Legal Issues Relevant to the Implementation of Water Markets. In *Groundwater: Legal and Policy Perspectives*; Salman, S., Ed.; World Bank Technical Paper No. 456; World Bank: Washingon, DC, USA, 1999; pp. 69–90.
29. Burchi, S. 2008 Year-End Review. *Water Law* **2008**, *19*, 223–228.
30. Hodgson, S. Exploring the Concept of Water Tenure. FAO, 2016. Available online: http://www.fao.org/3/I5435E (accessed on 8 April 2016).
31. Shah, T.; Roy, A.D.; Qureshi, A.S.; Wang, J. Sustaining Asia's Groundwater Boom: An Overview of Issues and Evidence. *Nat. Resour. Forum* **2003**, *27*, 130–141. [CrossRef]
32. Van Koppen, B. Dispossession at the Interface of Community-Based Water Law and Permit Systems. In *Community-Based Water Law and Water Resource Management Reform in Developing Countries*; van Koppen, B., Giordano, M., Butterworth, J., Eds.; Cabi: Wallingford, UK, 2007; pp. 46–64.
33. Taylor, R.G.; Scanlon, B.; Doll, P.; Rodell, M.; van Beek, R.; Wada, Y.; Longuevergne, L.; Leblanc, M.; Famiglietti, J.S.; Edmunds, M.; et al. Ground Water and Climate Change. *Nat. Clim. Chang.* **2013**, *3*, 322–329. [CrossRef]
34. The Punjab Preservation of Sub-Soil Water Act, 2009. Punjab Act No. 6 of 2009. Available online: http://agripb.gov.in/abt_deptt/pdf/Pb%20preservation%20of%20Subsoil%20Act,2009.pdf (accessed on 10 August 2016).
35. The Haryana Preservation of Sub-Soil Water Act, 2009. Available online: http://www.cseindia.org/userfiles/haryana_sub-soil_wateract.pdf (accessed on 10 August 2016).
36. Tripathi, A.; Mishra, A.K.; Verma, G. Impact of Preservation of Subsoil Water Act on Groundwater Depletion: The Case of Punjab, India. *Environ. Manag.* **2016**, *58*, 48–59. [CrossRef] [PubMed]
37. Burchi, S. Study on the Legislative Framework Regulating the Recharge of Aquifers with Adequately Treated Wastewater. 2014. Available online: http://www.swim-sm.eu/files/Legislative_framework_regulating_the_recharge_of_aquifers_EN.pdf (accessed on 14 April 2016).
38. India, Draft Model Bill for the Conservation, Protection and Regulation of Groundwater 2011, in Planning Commission, Report of the Steering Committee on Water Resources and Sanitation for Twelfth Five Year Plan (Government of India 2012) 154. Available online: http://planningcommission.gov.in/aboutus/committee/strgrp12/str_sani.pdf (accessed on 16 April 2016).
39. Andhra Pradesh (India). Water, Land and Trees Act and Rules (2002), s. 17. Available online: http://faolex.fao.org/docs/pdf/ind41989.pdf (accessed on 27 July 2016).
40. Intergovernmental Panel on Climate Change (IPCC). Small Island States. Climate Change and Variability. Available online: http://www.grida.no/climate/ipcc_tar/wg2/pdf/wg2TARchap17.pdf (accessed on 14 April 2016).
41. FAO. Malta Water Resources Report. 2006. Available online: ftp://ftp.fao.org/docrep/fao/009/a0994e/a0994e.pdf (accessed on 14 April 2016).
42. Directive 2006/118/EC of the European Parliament and of the Council of 12 December 2006 on the Protection of Groundwater against Pollution and Deterioration. Available online: http://eur-lex.europa.eu/legal-content/HR/ALL/?uri=CELEX:32006L0118 (accessed on 15 August 2016).
43. United Nations Treaty Series. *1936 United Nations Treaty Series U.N.T.S. 269*; United Nations Treaty Series: New York, NY, USA.
44. Model Provisions on Transboundary Groundwaters. Available online: https://www.unece.org/fileadmin/DAM/env/water/publications/WAT_model_provisions/ece_mp.wat_40_eng.pdf (accessed on 10 August 2016).
45. The Water Act of Buthan 2011. Art. 43. Available online: http://faolex.fao.org/docs/pdf/bhu106322.pdf (accessed on 10 August 2016).

46. Ley N° 18.610 Política Nacional de Aguas. Available online: http://www.ose.com.uy/descargas/documentos/leyes/ley_18_610.pdf (accessed on 10 August 2016).
47. Ley de Aguas. Available online: http://www.leyesvenezolanas.com/aguas.html (accessed on 10 August 2016).
48. The Greening of Water Law: Managing Freshwater Resources for People and the Environment. UNEP. Available online: http://www.unep.org/delc/Portals/119/UNEP_Greening_water_law.pdf (accessed on 16 April 2016).
49. Namibia. Water Resources Management Act, No. 11 of 2013. Available online: http://docplayer.net/11376739-Government-gazette-of-the-republic-of-namibia-n-36-40-windhoek-19-december-2013-no-5367.html (accessed on 10 August 2016).
50. South Africa. National Water Act, No. 36 of 1998, Part 3. Available online: https://www.dwa.gov.za/Documents/Legislature/nw_act/NWA.pdf (accessed on 10 August 2016).
51. Tanzania. The Water Resources Management Act, No. 11 of 2009, Art. 33. Available online: http://faolex.fao.org/docs/pdf/tan96340.pdf (accessed on 10 August 2016).
52. Zambia. Water Resources Management Act, No. 21 of 2011. Art. 44. Available online: http://faolex.fao.org/docs/pdf/zam117433.pdf (accessed on 10 August 2016).
53. IPCC. Climate Change 2014: Synthesis Report. Contribution of Working Groups I, II and III to the Fifth Assessment Report of the Intergovernmental Panel on Climate Change. Available online: https://www.ipcc.ch/pdf/assessment-report/ar5/syr/SYR_AR5_FINAL_full_wcover.pdf (accessed on 20 April 2016).
54. Available online: https://www.legislation.gov.au/Details/F2012L02240 (accessed on 21 April 2016).
55. Available online: http://www.indiaenvironmentportal.org.in/files/file/Draft%20national%20framework%20bill%202013.pdf (accessed on 17 April 2016).
56. Available online: http://www.indiawaterportal.org/sites/indiawaterportal.org/files/Model%20bill%20to%20regulate%20and%20control%20development%20of%20groundwater%20by%20Ministry%20of%20Water%20Resources%20%282005%29_0.pdf (accessed on 20 April 2016).
57. Loi No. 05-12 du 28 Joumada Ethania 1426 correspondant au 4 août 2005 relative à l'eau. Art. 3, 56, 57. Available online: http://www.droit-afrique.com/upload/doc/algerie/Algerie-Loi-2005-12-eau.pdf (accessed on 10 August 2016).
58. Garduño, H.; Foster, S.; Nanni, M.; Kemper, K.; Tuinhof, A.; Koundouri, P. *Groundwater Dimensions of National Water Resource and River Basin Planning*; GW Mate Briefing Notes Series; World Bank: Washington, DC, USA, 2006.
59. Hodgson, S. *Modern Water Rights*; Food and Agriculture Organization (FAO): Rome, Italy, 2006.
60. India Groundwater Governance Case Study. Available online: http://www.groundwatergovernance.org/downloads/GWGovernanceIndia.pdf (accessed on 7 April 2016).
61. Lopez-Gunn, E.; Martinez Cortina, L. Is Self-Regulation a Myth? Case Study on Spanish Groundwater User Associations and the Role of Higher Level Authorities. *Hydrogeol. J.* **2006**, *14*, 361–379. [CrossRef]
62. Real Decreto Legislativo 1/2001, de 20 de julio, por el que se aprueba el texto refundido de la Ley de Aguas. Available online: https://www.boe.es/diario_boe/txt.php?id=BOE-A-2001-14276 (accessed on 10 August 2016).
63. Foster, S.; Garduño, H.; Kemper, K. *The "COTAS": Progress with Stakeholder Participation in Groundwater Management in Guanajuato, Mexico*; GW-Mate Case Profile Collection; World Bank: Washington, DC, USA, 2004.
64. Expert Group on Ground Water Management and Ownership of the National Planning Commission of the Government of India, 2007. Available online: http://www.planningcommission.nic.in/reports/genrep/rep_grndwat.pdf (accessed on 16 April 2016).
65. Ostrom, E. Reformulation the Commons. Ambiente. & Sociedade, Ano V No. 10, 1o Semestre de 2002. Available online: http://www.scielo.br/pdf/asoc/n10/16883.pdf (accessed on 16 April 2016).
66. GEF Groundwater Governance Project. Synthesis of Thematic Papers/Case Studies, 2012. Available online: http://www.groundwatergovernance.org/fileadmin/user_upload/groundwatergovernance/docs/Thematic_papers/GWG_Synthesis_report_update.pdf (accessed on 16 April 2016).
67. Meinzen-Dick, R. Beyond Panaceas in Water Institutions. *Proc. Natl. Acad. Sci. USA* **2007**, *104*, 15200–15205. [CrossRef] [PubMed]

68. Moench, M.; Burke, J.; Kulkarni, H. Trends in Local Groundwater Management Institutions & User Partnerships. GEF Groundwater Governance Project, Thematic Paper No. 7, 2012. Available online: http://www.groundwatergovernance.org/fileadmin/user_upload/groundwatergovernance/docs/Thematic_papers/GWG_Thematic_Paper_7.pdf (accessed on 16 April 2016).
69. Nanni, M. *Comments on the Report of the Expert Group on Ground Water Management and Ownership of the National Planning Commission of the Government of India*. Unpublished.
70. Garduño, H.; van Steenbergen, F.; Foster, S. *Stakeholder Participation in Groundwater Management*; GW Mate Briefing Note Series; World Bank: Washington, DC, USA, 2010.
71. Groundwater Policy and Governance. GEF Groundwater Governance Project, Thematic Paper No. 5, 2012. Available online: http://www.groundwatergovernance.org/resources/thematic-papers/en/ (accessed on 25 July 2016).
72. Legal and Institional Frameworks. GEF Groundwater Governance Project, Thematic Paper No. 6, 2012. Available online: http://www.groundwatergovernance.org/resources/thematic-papers/en/ (accessed on 7 April 2016).

water

MDPI

Review

Modes and Approaches of Groundwater Governance: A Survey of Lessons Learned from Selected Cases across the Globe

Robert G. Varady [1], Adriana A. Zuniga-Teran [1,*], Andrea K. Gerlak [1,2] and Sharon B. Megdal [3]

1 Udall Center for Studies in Public Policy, University of Arizona, Tucson, AZ 85719, USA; rvarady@email.arizona.edu (R.G.V.); agerlak@email.arizona.edu (A.K.G.)

2 School of Geography and Development, University of Arizona, Tucson, AZ 85719, USA

3 Water Resources Research Center, College of Agriculture and Life Sciences, University of Arizona, Tucson, AZ 85719, USA; smegdal@email.arizona.edu

* Correspondence: aazuniga@email.arizona.edu; Tel.: +1-520-626-0693

Academic Editor: Arjen Y. Hoekstra

Received: 12 May 2016; Accepted: 18 September 2016; Published: 23 September 2016

Abstract: The crucial role of groundwater and the centrality of water governance in accommodating growing water demands sustainably are becoming well recognized. We review 10 case studies of groundwater governance—representing diverse global regions and local contexts—from the perspective of four well-established elements: (1) institutional setting; (2) availability and access to information and science; (3) robustness of civil society; and (4) economic and regulatory frameworks. For institutional setting, we find that governing is often a thankless task that paradoxically requires popularity; legislation does not always translate to implementation; conflict resolution is central to governance; and funding is critical for governance. In terms of information access, we see: a need for research for natural systems, social systems, and institutions; trust as an essential element in research; and that urbanized landscapes are critical components of groundwater governance. Looking at civil society robustness, we observe that equity is an essential element for governance; community-based governance requires intention; and leaders can play a powerful role in uniting stakeholders. As for frameworks, the cases suggest that economic incentives sometimes yield unintended results; "indirect" management should be used cautiously; and economic incentives' effectiveness depends on the system employed. Collectively, the lessons speak to the need for shared governance capacities on the part of governments at multiple levels and civil society actors.

Keywords: groundwater governance; water; stakeholder participation; equity; water security; case studies

---

1. Introduction

Two global programs underscore the value of examining groundwater-governance practices through case study analysis. The Global Environment Facility's (GEF) comprehensive project, "Groundwater Governance—A Global Framework for Action" (2011–2015), aims to raise awareness of the "*paramount importance of groundwater resources and their sustainable management in averting the impending water crisis*" [1]. A parallel global effort to examine water governance (not specifically groundwater), initiated in 2013 by the Organization for Economic Co-operation and Development (OECD), acknowledges that "*managing and securing access to water for all is not only a question of money, but equally a matter of good governance*" [2].

At the core of the GEF et al. (2016) [1] Groundwater Governance project are 12 thematic papers, followed by five continental-scale regional consultations held around the world, and a "Global Groundwater Diagnostic." These steps surveyed groundwater-governance practices and identified relevant opportunities, constraints, and challenges. In this article—which has its origins in GEF-commissioned series' Thematic Paper Number 5, "Groundwater policy and governance" [3]—we revisit the findings of that 2013 report by analyzing five of its most cogent case studies and complementing those sources with five more recent studies.

The objective of this article is—through an analysis of selected case studies—to demonstrate how the modes and approaches of groundwater governance depend on four governance elements: (1) institutional setting; (2) availability and access to information and science; (3) robustness of civil society; and (4) economic and regulatory frameworks.

1.1. Institutional Setting

If we accept a working definition of groundwater governance as *"the overarching framework of groundwater use laws, regulations, and customs, as well as the processes of engaging the public sector, the private sector, and civil society"* (p. 678) [4], the role and influence of institutions is the principal pillar upon which governance rests. The institutional setting of a nation's or a region's groundwater sector comprises the governmental, nongovernmental, and private sector agencies, organizations, and decision-making and managerial practices. A thorough appreciation of the elements, facets, approaches, rules, financial arrangements, actors, and power relationships within this setting is critical to achieving responsible groundwater use.

In particular, over the past two decades—with growing emphasis on sustainable approaches to managing resources—many societies have recognized the importance of bottom-up decision-making [5–8]. This has led to better alignment of actions and responsibilities among sometime-competing agencies and actors involved in governance [9,10], and to more efficient vertical integration among those entities [11].

Going beyond small-scale local institutional arrangements, observers have noted the importance of looking at the "big picture"—that is, linking groundwater governance to larger, macro-level, multisectoral policies [12,13]. Such a view permits linking groundwater governance to water governance more generally, and to the governance of intricately related sectors such as food, energy, environment, and trade—often characterized as the "nexus" [14–16].

At both small-scale and large-scale levels, the enhanced focus on the significance of institutions has permitted greater attention to issues of water security and its components of ethics, equity, access, and social justice. These notions, sometimes embodied in the term "hydrosolidarity"—an ethical concept intended to encompass sustainable aspects of water management that go beyond political and technical considerations—have begun to penetrate national and even transnational policies and practice [17] (e.g., the Water Integrity Network that aims to fight corruption in water governance worldwide; www.waterintegrity.org). While water politics continue to be driven by powerful interests at all levels (local to international), the growing prominence of bottom-up governance and its sibling attribute, transparency, has helped manage and transform water conflicts [18].

1.2. Availability and Access to Information and Science

Effective groundwater governance requires availability and access to information and science. Possessing information is very important for monitoring and assessment [3], and for negotiations over water allocation and management [19]. Accordingly, acquiring and analyzing data and information (both natural and social), and conducting research are foundational to good governance.

Among the needs are quantitative data—such as indicators—particularly for the development of early warning systems to increase disaster-preparedness [3]. In addition, collecting sociopolitical information is strongly recommended because of the significance of poverty, health, and vulnerability—all related to governance [14,19,20]. Furthermore, a key strategy in governance is to increase the co-production of knowledge, where the public actively participates in science—and this can be achieved via bridging organizations that act as intermediaries and can function at different scales [21–23].

However, the collection of sometimes sparse, dispersed, less-than-reliable data and information can be challenging and expensive, particularly for developing countries. Use of remotely-sensed data may be a cost-effective, reliable, and impartial way to skirt difficult on-the-ground searches [3]. Also, developing countries often have access to financial help from the international donor community to conduct research on groundwater resources. Multilateral and bilateral aid mechanisms can support data and information collection and management [24]. Moreover, collaboration and sharing among organizations can reduce research costs and overcome institutional asymmetries [25].

The benefits of better and more reliable data, more information, and more equitable access are self-evident, but there can be too much emphasis on this issue. At times, demands for those commodities may become action-averting strategies. So while calls for ever more authoritative information may be legitimate, they can also mask such weaknesses as lack of will, administrative and scientific inertia, bureaucratic bottlenecks, political meddling, insufficiency of trained personnel, or lack of resources.

1.3. Robustness of Civil Society

Engaging stakeholders in environmental decision-making was not always a prevalent mode of governance. However, over the past three to four decades, the procedure has gained acceptance across the world, especially in democratic societies. Involving a broad, representative, and active collection of stakeholders is essential to governance because that process is about influencing behavior [12]. To engage stakeholders effectively, it may be necessary to organize community and civil servants, who assist in public-participation processes where participants learn about relevant issues, actors, roles, and responsibilities, and become familiar with modes of complying with policies and regulations [3,14]. Strong leadership facilitates communication between stakeholders to ensure implementation [20].

However, enlisting and mobilizing stakeholders is not without its limitations. An important constraint that applies to all stakeholder processes is the difficulty of sustaining participant interest in a process that requires serious commitment over an extended period. Given the slowness of the decision-making process, even highly committed stakeholders become frustrated at the pace of responsiveness [26]. Stakeholder endurance palpably limits the empowerment of people to participate in policymaking.

Transparent, open, and fair mechanisms can help overcome this limitation [3]. The emerging social-media arsenal of Internet-based platforms and information technologies can help link and energize stakeholders [23]. Such connective tools facilitate the development of communities of practice that can exchange learning experiences from different regions [3]. With suitable technical and policymaking training of stakeholders, social-media tools hold potential to enhance substantially groundwater decision-making and environmental policymaking [3]. Nonetheless, even with the best tools possible, substantive stakeholder engagement may require redistribution of power and resources to sufficiently engage and empower disadvantaged and less organized stakeholders [3]. Further, even if civil society engagement is robust, implementation of particular policies or regulation may be uneven or may impact stakeholders in different, or perhaps inequitable ways.

1.4. Economic and Regulatory Frameworks

Regulations regarding groundwater use, quality, and monitoring determine how economic incentives and responses to those incentives influence groundwater-use practices and usability. Groundwater-users—like users of all goods and services—respond to price signals and other economic incentives. If water is available to the user cost-free, the potential for over-extraction or unsustainable groundwater use rises, especially where natural replenishment rates are slow.

Establishing water rights and measuring water use by individual users allow for providing behavior-changing economic incentives through pricing [11]. It is possible to recognize the necessity of water for living and still incorporate water scarcity into pricing through carefully crafted water-rate structures. However, even where water is metered and priced, water prices reflect the cost of extracting, treating, and delivering the water as incurred by the entity establishing these prices. There are often third-party impacts that should be incorporated into water rates. For example, water extraction by one user may cause higher energy consumption (from water pumps) and higher economic cost to a neighbor [3]. Establishing and sharing best practices for measuring third-party impacts can improve the policies governing groundwater management [3]. In addition, economic factors enter water-user decision-making via financing and paying for water infrastructure.

Water recharge and banking programs—where water is stored in the aquifer for future use—can optimize groundwater use based on intertemporal availability of surface-water supplies [27]. In addition, a carefully designed water-banking system may include mechanisms for marketing the right to pump water stored underground [27]. More generally, water marketing (whose frameworks might be varied) can permit market-based optimization of water allocation. Because water is necessary for basic living and since market control can lead to exploitation, there is a need for governmental oversight of water markets [13]. Therefore, analysis of the potential for water markets to improve groundwater allocation will necessarily need to consider equity and societal goals. While the expert community can offer insights on the scope of equity issues and societal impacts, it will be up to decision-makers to determine how to incorporate such considerations into groundwater governance approaches.

2. Materials and Methods—Case Study Approach and Analytical Structure

We review and compare observations from 10 studies that cover an expansive terrain—including the perspectives of geography, thematic content, problematics, and instructiveness. We selected these studies based on their location, quality, authoritativeness, and most of all, their direct, targeted attention to groundwater governance. We also chose them to represent diverse regions of the world—including developed and developing nations—and because they are rich in data, observations, conclusions, and recommendations.

We rely on experiences, observations, and findings steeped in real-world explorations of groundwater governance. Through records of earlier published cases, we overlay our own composite observations, which highlight context while seeking to identify commonalities and generalizable lessons.

The first five studies (published between 2000 and 2011) were important references in Thematic Paper Number 5 by Varady et al. (2013) [3] and helped provide a basis for that report's findings. To reflect more current understandings—from a limited pool of post-2011 groundwater case-study publications with an emphasis on governance—we selected five studies that are of comparable breadth and quality to the other five (See Figure 1 and Table 1). We elected to limit our sample to 10 studies to permit in-depth review and analysis of the cases, while identifying broad trends and patterns in groundwater-governance case research.

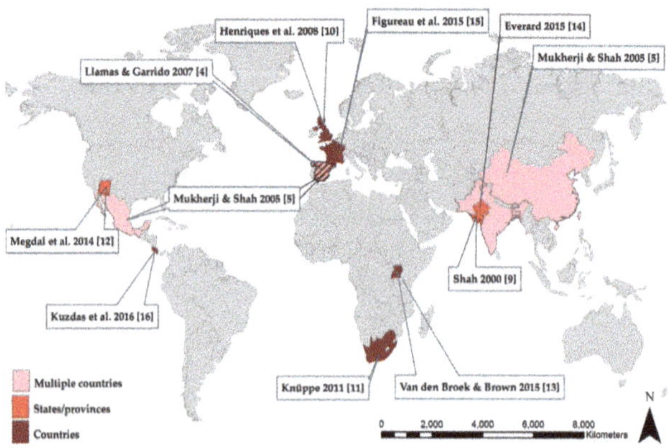

Figure 1. Location of case studies. This map illustrates the diversity of the geographic location of the case studies analyzed in this paper.

Table 1. Selected case studies. The 10 case studies reviewed, given in order of year of publication.

Study	Location	Summary
Shah (2000) [28] [a]	Gujarat, India	Looks at a social movement in western India, where a groundwater-recharge initiative grew to a large social phenomenon.
Mukherji & Shah (2005) [14] [a]	India, Pakistan, Bangladesh, China, Spain, Mexico	Examines benefits & disadvantages of groundwater use, and recommends a change in paradigm: from management to governance.
Llamas & Garrido (2007) [29] [a]	Spain	Explores the "silent revolution" phenomenon, where developers use ground-water with little or no scientific, technological, or administrative control; examines stakeholder participation in groundwater management.
Henriques, Holman, Audsley & Pearn (2008) [30] [a]	East Anglia and North West England	Uses interactive Regional Impact Simulator tool to examine impacts of farm-level cropping decisions in regional water availability, including multi-scalar approach at field, farm, sub-catchment and catchment, and regional levels.
Knüppe (2011) [20] [a]	South Africa	Describes qualitative assessment of groundwater management that includes expert interviews from entities including national government, research and conservation organizations, and consultants.
Megdal, Dillon & Seasholes (2014) [27] [b]	Arizona, USA	Explores policy perspectives of managed aquifer recharge, or "water banking", in Arizona, USA, and the context in which this strategy is implemented.
Van den Broek & Brown (2015) [31] [b]	Uganda	Analyzes impacts of two decades of managing rural groundwater in Uganda, following the Community Based Management (CBM) paradigm.
Everard (2015) [32] [b]	Rajasthan, India	Studies the successful community-based groundwater-recharge program in a region of groundwater-level decline and related socioecological degradation.
Figureau, Montginoul & Rinaudo (2015) [33] [b]	France	Offers empirical assessment of opposing groundwater-management regulations strategies that may be implemented in five groundwater basins in France, identifying barriers and factors that can facilitate implementation.
Kuzdas, Warner, Weik & Vignola (2016) [34] [b]	Guanacaste, Costa Rica	Provides a transdisciplinary, multi-criteria sustainability assessment of groundwater resources, based on current management approach in five governance regimes in Guanacaste, Costa Rica.

Notes: [a] Papers from Thematic Paper No. 5 (Varady et al., 2013) [3]; [b] Papers from recent literature.

We approached our analysis by selecting four practical elements (or action steps) of governance employed in the thematic paper [3]. These elements—institutional setting, availability and access to information and science, robustness of civil society, and economic and regulatory frameworks—are instrumental to groundwater governance and policy. We also worked inductively to query systematically the updated case studies using these four governance elements and to identify commonalities in the cases.

Our aim is to report on the broader trends and patterns across the cases and to illustrate the most pertinent or interesting findings within the various cases. As such, we do not treat every case equally. While we acknowledge that other combinations of articles plausibly could entail such a review, our selection emphasizes diversity in location, quality, relevance, and authoritativeness, and in so doing, best addresses the four identified governance elements. This approach enables us to take the next step to see what animates governance at ground level in diverse settings.

3. Results—What Do the Case Studies Reveal?

3.1. Institutional Setting

As noted, effective groundwater governance relies on a strong institutional setting. Legitimate, durable, flexible, responsive, and above all, viably funded institutions are indispensable elements of a nation's or a region's ability to manage its groundwater resources. However, those attributes—important as they are—can be read as platitudes. In real-world settings, what are some the ingredients of a strong institutional base? Our 10 case studies offer useful insights into this question.

Virtually all the studies consider groundwater policymaking as strongly dependent on institutional factors. Unsurprisingly, the role of the state and its various agencies is a hub for achieving good governance. In this vein, many of our authors acknowledge the importance of state agencies and ask whether and how they might be more effective.

3.1.1. Role of Government

Governments sometimes have responsibilities that exceed their human and financial capacity. Governing groundwater is a highly complex and fraught activity; it depends upon both the physical configuration of groundwater structures (e.g., wells, pumps, etc.) and an area's reliance on groundwater resources. These considerations include the following variables: (i) per-annum volume of groundwater abstractions; (ii) number and type of groundwater structures; (iii) percentage of population whose livelihood depend on groundwater; and (iv) the prevailing political system [14]. To navigate such complications, rules and procedures are indispensable.

In order for governing institutions to have power over groundwater resources, they should be able to make policy. In Arizona, USA, the state water agency—the Arizona Department of Water Resources (ADWR)—is enabled to make policy. The regulatory water-rights framework of the region's major surface-water source, the Colorado River, provides ADWR with the legal right to store water via managed aquifer recharge [27]. The agency also has authority to limit urban growth through an Assured Water Supply program, which mandates that residential developments have a guaranteed 100-year water supply [27]. Yet even where favorable institutional conditions exist, they can be constrained budgetarily, politically, and managerially.

Elsewhere, writing broadly about countries in three continents, Mukherji and Shah (2005) [14] found that if agencies are to function effectively via rules and procedures, while devising policies and making laws may be feasible, enforcing them usually proves to be far more difficult. Knüppe (2011) [20], considering South Africa, confirms that legislation alone is insufficient to address groundwater-management problems; it is equally necessary to educate and raise awareness, induce cooperation among existing networks, and increase stakeholder engagement. But when state agencies are absent altogether or weak, Van den Broek and Brown (2015) [31] found in Uganda that this sort of lacuna (as promoted by Community Bases Management (CBM)) leads to unsuccessful groundwater-governance outcomes. However, there may be times when government's role can remain negligible, yet without harm. A case study in India [28] shows that communities by themselves may be able to improve the health of a watershed with complete autonomy—particularly when, with strong community leadership, positive externalities are rapidly internalized; in this case, the effort was tied to religious beliefs.

Van den Broek and Brown (2015) [31] suggest a middle ground: *the role of government should change, but not disappear*. Reluctance to change can be manifested in a country's legal framework, where in extreme cases, conservation actions can actually be illegal, underlining why change is essential and should be encouraged. In Rajasthan, India, Everard (2015) [32] illustrates reluctance to change; there, locally built water-harvesting structures are considered illegal, if nevertheless unenforced. Van den Broek and Brown's (2015) [31] work in Uganda shows that the state should not abandon groundwater-resources governance. The authors recognize that assigning to a community the full responsibility to manage groundwater resources is generally unsuccessful.

3.1.2. Matching Administrative and Hydrologic Units

In the past, water management—as with all forms of government-led management—was achieved within formal administrative units (nations, provinces, states, districts, municipalities, and other divisions). While this approach may have been efficient bureaucratically, it failed to account for landscape, natural processes, communications, and societal practices. In short, it disregarded hydrologic and cultural boundaries. Everard's (2015) [32] study in Rajasthan, while recognizing barriers to radical organizational transformation, exemplifies the need for administrative units to match as closely as possible hydrological ones. The techniques used to build local water-harvesting structures used traditional knowledge and social institutions. The Non-Governmental Organization (NGO) leading this effort, Tarun Bharat Sangh (TBS), responded to demands to maintain traditional village management institutions. One example of such an institution is the Gram Sabha, the water-management decision-making body. TBS created "water parliaments" that group villages within a catchment area to create unified policies [32].

In the United Kingdom, as well, matching functional units is important. Henriques et al. (2008) [30] point out that water is transferred over long distances, and that abstractions are not necessarily linked to points of use. For this reason, the authors model water supply and demand at the regional scale. However, sometimes there can be a mismatch between regional boundaries and catchment boundaries. In this case, the authors scaled catchment water availability in their hydrological model according to proportion of catchment area outside the regional boundary.

The Arizona study also addresses this issue by describing the flexibility of the recharge program that allows a recharge project to have a different location than the point where the groundwater withdrawal occurs. This locational disconnection has eased the use of this renewable resource, but increased the risk of localized overdraft. Still, Arizona has attempted to match hydrologic and administrative units. The regulatory regime established in the state's Groundwater Management Act (GMA) is aligned with Active Management Areas (AMAs), the units created for the state's most populated groundwater basins [27].

Management via traditional administrative units further complicates planning and coordination. Henriques et al. (2008) [30] observed that urban- and regional-planning policies affect landscape and hydrology, including evapotranspiration and surface sealing. More generally, Kuzdas et al. (2016) [34] found that in Guanacaste, Costa Rica, a lack of clear regional-planning objectives and poor coordination between national- and local-level decisions lead to mismatches in outcomes and local community needs and sustainability.

3.1.3. Financing Practices

Financial support for water agencies is always of the utmost importance, particularly in developing countries. In Costa Rica, Kuzdas et al. (2016) [34] reported that shortfalls can lead to corruption, sometimes manifested as the approval of development projects as *quid-pro-quos* for accepting developers' donations. Sometimes corruption practices can be subtler yet still damaging, when governments "do nothing" to stop the overexploitation of the aquifers [34]. One way to lessen such questionable practices is to assure that water professionals receive equitable salaries [34].

In India, corruption cases are also evident and often with seriously adverse results. Everard (2015) [32] observed that the Indian government granted fishing permits to contractors, a practice that allowed them to exploit restored ecosystems. This became a huge social conflict with local communities that were personally involved in restoration efforts. Those communities did not allow the contractor to access the river; and the angry contractor, in revenge, poured pesticides into the river killing fisheries, negatively affecting local populations [32].

In Uganda the situation is somewhat similar. Van den Broek and Brown (2015) [31] noted that lack of salaries for people collecting fees led to corrupt practices, because the collectors spent the fees for personal expenses, attributing their behavior to their lack of salaries. In addition to corruption, lack of funding can result in disengagement and lack of continuity. Trained Water Use Committee members in Uganda were unwilling to work without pay [31]. These individuals lacked capacity-building opportunities, forgot their original training, or simply decided to leave the community [31].

However, corruption cases are not exclusive to developing countries. In Spain, e.g., Llamas and Garrido (2007) [29] found that subsidies can become "perverse" by encouraging corrupt practices.

For example, European Union subsidies intended to decrease groundwater abstractions for 10 years failed to stop illegal pumping and drilling of new wells [29]. The intent of regulating groundwater is to reduce corruption but this practice, too, requires funding. Mukherji and Shah (2005) [14] recommend regulation through registration of wells and licensing, and pricing policies, particularly in developing countries. However, installing groundwater-monitoring networks represents a high cost and presents a serious obstacle for regulators.

Conflict resolution also requires financial investment. In France, inadequacies within agencies prompted a need for non-agency intermediaries, via "joint liability contracts," to solve social conflicts [33]. And similarly, in Costa Rica agencies brought in a social scientist to facilitate deliberation in a workshop that aimed to evaluate five governance approaches (scenarios) for groundwater governance [34].Without doubt, funding constitutes an overarching issue that affects many aspects of groundwater governance. The following sections include reference to issues and challenges that emerge when there is lack of enough funding.

3.1.4. Government as Supplier or as Custodian?

Whether agencies are amply or inadequately funded, and whether they are centrally controlled or locally managed, many observers have noted a shift in their role: *from "supplier" to "custodian"*. The Rajasthan study points to potential conflict when government does not act as a resource custodian, and instead grants rights such as fishing permits to contractors to exploit restored ecosystems. As explained above, conflict with local communities resulted in negative impacts for everyone—the contractor did not get access to the river, and the communities suffered adverse effects from the poisoning of the fish [32].

By contrast, some studies illustrate instances when a government's resource custodianship can be strong. For example, in England and Wales, according to Henriques et al. (2008) [30], the Environment Agency enforces control of abstractions through a system of licenses and drought management by water utilities. In South Africa, the government's custodianship is evidenced in the 1998 National Water Act, the country's principal water-resources-management legal instrument. In this law, the ecosystem's needs are included in the principle of "Ecological Reserve", according to which, environmental requirements must be "reserved" before any allocation for human use. However, even with ecological needs legally accounted for, sustainable groundwater governance remains challenging in South Africa [20]. In the case of Arizona, the state government became the resource custodian in 1980 when it passed the GMA after a period of overdraft that reached a critical point in the late 1970s. Megdal et al. (2014) [27] describe the GMA's function in requiring new urban developments to use renewable water supplies instead of groundwater.

In Costa Rica, the need for resource custodianship is evidenced by overexploitation of aquifers (a partial consequence of greedy practices), excessive reliance on supply-side approaches such as augmentation, a lack of flexible water management frameworks, and insufficient government oversight [34]. Yet, such government oversight is sometimes unwelcome, as shown in the periodic instances of resistance and discontent by French farmers when there was a shift from open access resource to a regulated one [33].

3.2. *Availability and Access to Information and Science*

Our case studies review showed that availability and access to information and science is critical to groundwater governance, pointing to a need to stimulate research on groundwater resources. However, a lack of funding, particularly in developing nations, limits availability and access to reliable information.

3.2.1. Research Needs and Transparency

The studies show that effective groundwater governance should be based on research on the resource itself and on interrelated sectors including agriculture, tourism, mining, forestry, and the environment—all at multiple scales. Effective research requires expertise on groundwater and its connected sectors. However, in some countries such expertise is often in short supply. In South Africa, for instance, Knüppe (2011) [20] describes some bureaucratic issues where a paucity of trained personnel and facilities limits research and progress on effective approaches to governance.

In Arizona, where research capacity is ample (both in terms of field investigations and university-like research), applied science on groundwater resources is fundamental for groundwater governance. There, water from the Colorado River is transported via canal by the Central Arizona Project (CAP) to recharge aquifers for storage (i.e., managed aquifer recharge, or "water banking"). However, in order to "evaluate hydrologic feasibility", an Arizona study on banking used quantitative data derived from groundwater-flow models. That evaluation process also considered water quality and potential damage to adjacent properties [27].

The U.K. experience reveals that groundwater-resources research also should consider a multi-scale approach (e.g., field, farm, sub-catchment and catchment, region) [30]. In addition, sustainable management requires not only access to information, but also sustained data collection for monitoring quality and quantity [20]. In Spain, Llamas and Garrido (2007) [29] similarly found that more and better hydrological data are needed to reduce uncertainty in groundwater management and to make better decisions in general.

Our review also revealed that research should include the effects of urban-planning policy. As large water-users, municipalities are becoming critical actors for groundwater governance in terms of land-use change that affects hydrologic processes. The UK study confirms that urban- and regional-planning policies affect landscape and hydrological effects, including evapotranspiration and surface sealing [30]. And lack of clear planning objectives can lead to a mismatch in outcomes that negatively affects community needs and sustainability [34].

However, research can be biased and manipulated. In Uganda, van der Broek and Brown (2015) [31] found that the government over-reported official figures of functioning pumps in order to meet sustainable development goals. In Costa Rica, Kuzdas et al. (2016) [34] observed that when the government does research without the participation of local communities, there is a risk of bias in studies that explore the feasibility of development projects. Transparency in data availability—as well as broader development planning—is necessary to avoid conflicts and suspicions [33,34]. Yet transparency, while desirable in theory, may trigger further conflict. In the case of France, although institutional stakeholders believed that transparency would enhance responsible behavior, most farmers did not believe that making data available would result in higher compliance of rules [33].

3.2.2. Funding for Research

The case studies suggest that the degree of financial resources required for research frequently impedes sustainable groundwater governance, particularly in disadvantaged, developing societies. There, the challenges to obtaining access to high-quality data are substantially greater than in developed nations [14,34]. NGOs have filled this gap to some degree. The Rajasthan study demonstrated that such organizations might sometimes function as mediators while also attracting funds from the international community [32]. Developed nations are not exempt from funding challenges. The South African case reveals that insufficient funding for groundwater-resources research and for training water-management personnel negatively affects pricing systems and legal provisions [20].

3.2.3. Information for Negotiation

Our review found that water allocation and management require reliable information and that the use of indicators promotes transparency. In Costa Rica, Kuzdas et al. (2016) [34] learned that judicious use of such indicators helped identify key water-governance challenges, and that transparency engaged stakeholders in the decision-making process.

Lack of reliability reported in the Uganda case study shows that official over-reporting of water-pump functionality is either a case of corruption or lack of capacity causing a serious impediment to development [31]. A study by the Uganda Ministry of Water and Environment reported that 64 percent of the rural population had access to water (through hand pumps) and that 85 percent of the pumps functioned properly. However, a study by Koestler et al. (2010; cited in van den Broek and Brown 2015 [31]) determined that in actuality only 40 percent of the pumps were functional. This finding suggests that government-funded projects may overestimate such indicators as pump functionality for self-serving purposes. In this instance, the government would like to report that they are in compliance with Uganda's constitution and national policy, which stipulates that that by the year 2015, 77 percent of the rural population would have access to water and 90 percent of the pumps would be functioning [31].

Access to information and scientific research is critical to governance because these commodities can be used to enhance political leverage. The Arizona case exemplifies how these capabilities can help in negotiations with neighboring governments. Because CAP water was "underutilized" (that is, used at less than permitted, allocated amounts) and because there was a risk of losing this water to a large and powerful neighboring state, Arizona decided to use CAP river water to recharge some of its aquifers. The different recharge projects range in scale, but infiltration rates are carefully monitored. Thanks to this information, Arizona has been able to take—and save—its entire allocation of Colorado River water. This full utilization, along with interstate managed aquifer recharge (discussed above), were important to interstate negotiations on sharing of Colorado River shortages [27].

3.2.4. Socioeconomic Data Needs

The case studies uniformly suggest that socioeconomic factors are critical for groundwater governance and that wherever possible, research should factor such considerations. In Spain, Llamas and Garrido (2007) [29] found that social factors are the primary cause of environmental problems, more so than natural factors. This was also true for South Africa, where Knüppe (2011) [20] describes that achieving sustainable groundwater-resources management requires integration of the socioeconomic and hydrogeological dimensions. For example, aquifer properties and their relationship to human wellbeing are of equal importance because of the linkages between aquifer properties and poverty, vulnerability, and health [20]. Similarly, in Costa Rica, Kuzdas et al. (2015) [34] concluded that water issues are linked to socioeconomic issues faced by communities that are proximate to ecotourism development sites. These communities incurred such negatives consequences as loss of land, low wages, and increased prices. In addition to tourism, other economic sectors that depend on groundwater resources include agriculture, and mining (20), therefore researching how much groundwater contributes to these economic activities is crucial. In the UK, Henriques et al. (2008) [30] state that water demand is linked to population, which is the dominant socioeconomic indicator. In India, Mukherji and Shah (2005) [14] concur, arguing that although hydrogeological data are greatly needed, having reliable socioeconomic data is as important. For example, in India, livelihoods are directly linked to water levels. Everard (2015) [32] found that water levels directly affected women because when levels were restored, women became able to engage in productive activities rather than spend entire days foraging water. Mukherji and Shah (2005) [14] include the percentage of population whose livelihoods depend on groundwater along with its political systems as one out of three factors that may explain the success of groundwater-demand management in some countries and not in others. More research is needed on the economic and social dimensions of groundwater use and on the institutions that govern this use [14].

3.3. *Robustness of Civil Society*

Our case studies concur that groundwater governance is a community process that requires participation by governmental and nongovernmental groups. However, participatory governance is often lacking, especially in places where the status quo favors certain groups that use the resource, or when it promotes inequities in access to groundwater or in the implementation and exercise of regulation. We also found that many stakeholders indicate that they would like to be involved in knowledge production. As a major reason, they cite the desire to reduce the potential bias of research organizations—particularly when these are sponsored by local government, which is frequently distrusted. The studies also revealed that educating groundwater-user communities and promoting learning in participation is a key to groundwater governance, as is wide communication of successful and unsuccessful strategies. Finally, the studies frequently highlighted that leadership is a key element in stakeholder engagement.

3.3.1. Participatory Governance

Our review reveals that groundwater governance is undertaken by diverse governmental and nongovernmental stakeholders, including policymakers, public-sector organizations, and private-sector organizations [20]. In Arizona, the CAP, which delivers water to the largest cities in the state (Phoenix and Tucson) and to sovereign Native Nations, is governed by a 15-member board of directors that represents each of the counties served by CAP [27]. In India, a local NGO created the "Rashitriya Jal Biradari" (national water brotherhood), which facilitated a community-led groundwater recharge effort in a semiarid region of Rajasthan, India [32]. The group includes water experts, farmers, NGOs, researchers, voluntary organizations, social groups, and social scientists who meet once a year to discuss management issues. The Llamas and Garrido (2007) [29] study of groundwater governance in Spain indicates that the timing of engagement is crucial. A community-based governance approach that engages stakeholders at the outset is a key factor for successful management of aquifers.

In some contexts, participation is notably lacking. An assessment of alternative governance regimes for Guanacaste Province, Costa Rica, highlighted a call by stakeholders for more participatory processes [34]. Mukherji and Shah (2005) [14] found reluctance in Mexican farmers to join governance institutions because the main institutional goal usually is to limit groundwater abstractions, which is something farmers are not willing to do.

3.3.2. Top Down or Bottom Up?

The case studies surveyed indicate general agreement that groundwater governance should be decentralized and must involve local communities. According to Knüppe (2011) [20], governments can support bottom-up approaches to water governance by funding technical services, infrastructure, capacity building, and coordinating initiatives. In Spain, Llamas & Garrido (2007) [29] also suggest such a balanced approach. They conclude that it would likely be beneficial to collectively and inclusively manage aquifers, with some control from water authorities.

In a recent study in France, Figureau et al. (2015) [33] explored the role of Groundwater Users Associations, which attempt to adapt rules to local contexts and facilitate social acceptability. The most desirable management mode among three evaluated approaches was the strategy that empowered local farmers while reducing government interference and promoting self-regulation.

The case studies showed that bottom-up, more community-based approaches require vertical integration. In Arizona, there is a kind of coherence across scales that integrates local actions with state policies. The state legislature passed the GMA creating an institution to administer it, ADWR. This institution has an administrative framework that regulates water rights [27]. In his study of groundwater in Rajasthan, Everard (2015) [32] demonstrates the advantages of vertical integration. He points out that (i) state and central governments have significantly different water-management perspectives than local communities; and (ii) that the government has more institutional and legal power than village councils, NGOs, and water parliaments. This unequal balance leads to suboptimal groundwater decision-making (and other resource governance). Comparably, in Uganda, Van den Broek and Brown (2015) [31] observed that Water User Committees, working without support from the national government, lack legal status and authority and were generally unsuccessful.

Everard's (2015) [32] study—which explores late colonial and early post-independence centralization of water governance in the state of Rajasthan—illustrates the need to decentralize groundwater governance and involve local communities. Everard found that centralized approaches resulted in a loss of local indigenous knowledge about water harvesting structures and a corresponding shift in the value of water, disengaging the local community in the maintenance of water structures. This abandonment of local water management led to unsustainable livelihoods, degraded ecosystems, and societal collapse. However, during the mid-1980s, an NGO—TBS—worked with local communities to restore water systems in what they called "community self-reliance through resource conservation". This permitted construction of traditional water-harvesting structures with successful results for both social and ecological systems [32].

Similarly, in Costa Rica, Kuzdas et al. (2016) [34] determined that top-down approaches to groundwater governance would not meet sustainability goals. In Uganda, van den Broek and Brown (2015) [31] noted the desirability of involving the community in groundwater governance, but their case study concludes that leaving the process entirely to the community does not work. In addition, the appropriate Water User Committee (WUC) lacked legal standing, resulting in its inability to secure operation and maintenance funds for water pumps. The authors also found that community meetings that bring up the need to pay fees discouraged participation—and those members of the WUC charged with collecting such fees are often seen as enemies.

In countries with highly centralized governments, such as China, governments are expected to enforce groundwater-use rules. However, there, implementation of regulations is not very transparent and has not been studied [14]. In South Africa, centralization of power has challenged the achievement of sustainable groundwater management [20]. Equity, effectiveness, and efficiency in groundwater management are best accomplished through a decentralized management system, according to the papers reviewed here.

3.3.3. Co-Production of Knowledge

As a body, the case studies evince a desire by stakeholders to be engaged in the production of knowledge for groundwater governance. Stakeholders in Guanacaste province in Costa Rica, for instance, believe that there is a need for research on the aquifer's capacity and recharge potential in order to improve governance. Although complex quantitative hydrogeological models may not be suitable for the coproduction of knowledge, other types of assessments can be feasible. For example, in western India, where a social movement harvested rainwater to recharge wells, a series of NGO-led technological experiments in water-conservation strategies (in-situ recharge and water-use reduction) used traditional water-harvesting methods. The group experimented until it obtained a simple, inexpensive way to direct rainwater to a tank that removes silt and allows clean water to recharge wells [28]. In Guanacaste province in Costa Rica, stakeholders believe that there is a need for research on the aquifer's capacity and recharge potential in order to improve governance. However, to minimize risk of bias and to ensure accuracy and legitimacy, government-led studies that explore the feasibility of development projects should include input from and contributions by local communities [34].

3.3.4. Awareness: Education and Communication

The cases underlined the importance of education in stakeholder engagement. Stakeholder participation in groundwater governance is essential for sustainability in the use of the resource, and to succeed, the process should include informed and educated groundwater-users [29]. Educating the public and raising awareness of the importance of groundwater resources is a fundamental principle of effective groundwater governance [20]. Education and awareness programs are likely to result in greater willingness for community participation when some stakeholders may be reluctant to engage [14].

The selected studies suggest that communication is a central determinant of successful groundwater governance. Shah (2000) [28] reported that audio and video recordings enhanced public participation in a western Indian social movement to recharge wells in India, particularly when the leader is the speaker in the recordings or in person. Another effective way to communicate successful experiences is through money. Shah (2000) [28] reports that demonstrating the attractive cost-benefit ratio of well-recharge expenses versus the economic benefits of a recovered watershed and increased groundwater levels served as anecdotal experience to convince other communities. In addition, the distribution of illustrative pictures of working well-recharge technologies helped the proliferation of India's water conservation movement [28].

3.3.5. Role of Leadership

The robustness of civil society remains highly dependent on the availability and quality of leadership, which determines the degree of stakeholder engagement. Our two case studies in India highlight this. In Gujarat, traditional Hindu spirituality and ethics played a role in a social movement to conserve water and recharge wells. There, charismatic and skillful leaders interpreted and narrated scriptures (*Katha-kar*), thereby demonstrating great power to direct community residents [28]. In the other Indian case study, an NGO leader, Rajendra Singh, advocated traditional knowledge and brought necessary funding to initiate water-conservation techniques. He embodied the role of the strong leader needed to manage groundwater effectively [32]. In South Africa, Knüppe (2011) [20] observed that strong leadership that improves guidance and communication is a vital element of governance. A leader, she noted, must be educated, open, flexible, and trustworthy.

3.4. *Economic and Regulatory Frameworks*

The diverse set of case studies provides evidence that water-users generally respond to economic incentives, be they charges or subsidies, although outcomes may not fully align with expectations. The studies also illustrate some of the complications associated with changes in economic policies, especially for the agricultural sector and in locations where water previously had not been priced. They also demonstrate that the efficacy of market-like mechanisms in achieving policy objectives depends on underlying water rights and other elements of the regulatory framework, along with regulatory oversight. This underscores the importance of economic policies related to electricity.

3.4.1. Economic Instruments

Figureau et al. (2015) [33] look at economic instruments to promote sustainable groundwater use in France; these can be positive (e.g., subsidies, payments) or negative (e.g., pricing, taxes, penalties). There, pro-social policies have been employed to promote fairness, accountability, trust, moral inclusion, and reputation—and a combination of both economic incentives and pro-social policies. Not surprisingly, the study found that farmers prefer pro-social policies over economic instruments, most likely because these are policies with which farmers are familiar. The authors found that when evaluating the effectiveness of a strategy based on economic instruments (payments and penalties), participants recommend aligning the economic instrument to local conditions (economic, hydrological, and climatic). They discussed the level of the economic instrument (penalty) that should avoid farm bankruptcy and be based on tradeoffs between long-term viability goals and efficiency. Their finding that most farmers would prefer to pay the penalty rather than change their practices suggests that penalty rates were not set to provide behavior-changing economic incentives. The study found that penalties to induce water conservation might have had the opposite effect. They found that farmers who pay the imposed penalty might feel that they have the right to exceed their water allocation. This points to the need for effective compliance mechanisms. Comparably, in Uganda, Van den Broek and Brown (2015) [31] identified the inability of WUC to collect water fees from users as a reason for the failure of CBM to manage groundwater. The feasibility of implementing economic policies and monitoring compliance needs to be considered.

3.4.2. Communal Ownership

Contributions toward infrastructure influence behavior. The case study in Rajasthan found that behavior is tied to economic cost. There, the NGO TBS requires that 30 percent of the cost of structures be provided by the community in order to ensure "communal ownership and continued maintenance" [32]. However, community contribution toward the cost of water pumps does not necessarily foster a sense of ownership [32]. The Ugandan case study failed to find that an upfront payment for pumps resulted in intention to maintain the installed pumps [31]. The study discovered that the reason for resistance to pay for operation and maintenance of pumps in Uganda was a general belief that water should be free of charge. People expect the government or NGOs to pay for infrastructure, and changing to a system that requires payment for previously free water services can be difficult [31].

3.4.3. Subsidies

Conceptually, subsidies to support desired behaviors can have the same economic impacts as economic charges to discourage groundwater over extraction. Yet, Llamas and Garrido (2007) [29] found that subsidies also can have perverse effects, as discussed in Section 3.1.3 above. They reported that European Union (EU) subsidies given to Spain to compensate the decrease in groundwater abstractions for 10 years did not stop the drilling and use of illegal wells. Another case study in Spain concluded that EU subsidies resulted in a temporary halt to groundwater abstractions, but not in positive collective action [14]. This study did find that income-compensation schemes have been effective in decreasing groundwater abstractions. The confidence that water-users have in delivery of promised economic support will influence future effectiveness of economic policies. Shah (2000) [28] found that when the government of India promised subsidies to the people who participated in the well-recharge movement, some people delayed their participation to wait for the subsidies, which never materialized.

In their case study of Mexico, Mukherji and Shah (2005) [14] found that substantial electricity subsidies have served well as a "carrot" to well-owners to register their wells. However, monitoring water volumes of abstraction remains challenging. They report on several strategies employed by agencies—with differing levels of success—to manage groundwater demand, including: regulations and enforcement, modification of property rights and new institutions such as water markets, and economic incentives. Likewise, Knüppe (2011) [20] found that economic instruments available to manage groundwater demand include a range of policies, such as subsidies to water-saving measures, subsidies to efficient irrigation technology, and subsidies for water-treatment technologies offered to industries and municipalities.

3.4.4. Changes in Crop Types

How agricultural water users respond to economic incentives can vary widely. Farm subsidies designed to influence the mix of crops produced has the potential to affect water use [30]. Henriques et al. (2008) [30] asserts that economic instruments to affect agricultural water demand should not only include water pricing, but also consider markets and food production priorities. Moreover, the study identifies the important role of the cost of inputs, such as fertilizers and labor, in agricultural production and associated water use. The authors recommend that regulatory restrictions on water use for irrigation, such as licensing, or economic instruments, such as volumetric abstraction costs, must be used with caution and should include the associated environmental costs. However, for developed countries, Mukherji and Shah (2005) [14] recommend regulation through registration of wells and licensing, and pricing policies.

In considering future scenarios, Henriques et al. (2008) [14] found that controlling water demand in agriculture through water pricing may result in the substitution of high-value crops for lower-value crops. While such behavior may be seen as a rational economic response by the irrigator to higher water rates, it could in fact result in higher volumes of water used.

3.4.5. Water Pricing

Water pricing may affect decisions to use surface water rather than groundwater. In the U.S. state of Arizona, the statutory recharge and recovery framework allows for farmers to partner with municipalities or with the Arizona Water Banking Authority so that irrigators use less groundwater and more surface water. The nonagricultural partner buys down or subsidizes the cost of surface water so that the irrigators will use surface water instead of lower-cost groundwater, to which the irrigators have extraction rights. The partner, such as the Banking Authority, accrues credits for future recovery, with a small "cut to the aquifer" not eligible for future recovery. The credits are granted by the state water agency through a carefully enforced permitting and accounting system. This regulatory framework has been effective in meeting statewide goals to store surface water in times of availability for future use during shortage [27].

3.4.6. Water-Energy-Food Nexus

Electricity is the most crucial operating cost for groundwater pumpers. In India, in particular, the instrument that is most commonly used to control groundwater extractions is electric supply. Shah (2000) [28] identifies a ban on electric supply as an effective means to reduce extractions. He reports that the electricity supplier—the state of Gujarat—is a monopoly that can play a significant role in "disciplining private pumpers by skillful use of electricity pricing and supply policies as levels for groundwater demand management" (p. 200).

Mukherji and Shah (2005) [14] clearly see the connections of food and energy policies. They identify a window of opportunity to manage India's groundwater economy through the rationalization of electricity supply. They also identify other control instruments—e.g., electricity subsidies, and policies for foodgrain production. Elsewhere, Henriques et al. (2008) [30] note that in the UK climate change mitigation policies could alter crop production strategies to favor non-food bioenergy crops that are linked to water use. However, while identifying such potential levers, Mukherji and Shah (2005) [14] suggest caution and discretion, acknowledging that practical, political reasons may pose a barrier to electric bans and subsidy reductions and changes in crop-subsidy policies. They also note that not all measures of water governance work everywhere, and that contextual differences play a major role in the success of groundwater governance.

3.4.7. Role of Markets

Market mechanisms are playing an increasing role in water allocation. Henriques et al. (2008) [30] describe water markets as an economic instrument that can replace drought regulations in the UK. However, the authors warn that this mechanism leads to changes in water use (e.g., away from agriculture and more towards municipal use and even leisure activities such as golf). Therefore, this economic mechanism does not necessarily decrease water demand. When Figureau et al. (2015) [33] studied scenarios of water-management strategies among stakeholders in France, they found that farmers believe that water markets would undermine the "collaborative spirit" that a pro-social strategy promotes. And Shah (2000) [28] attributes the worst groundwater catastrophe in the Gujarat region in India to water markets that promoted agricultural expansion and economic growth beyond the aquifer's carrying capacity. The study further identified a need to regulate water markets in order to restore environmental health.

In Arizona, the integrity of the managed aquifer recharge process is maintained by accountability including permits, monitoring, and reporting [27]. In addition, the flexibility of the water-banking process in Arizona—also known as "mass-balance approach," where recharge is done in one place and abstractions in another—is supported by extensive monitoring. Arizona's water banking strategy has been extended to the neighboring state of Nevada. This complex mechanism to store Colorado River water in aquifers for future use by another state requires an accounting system that tracks quantified rights, and permits and fees associated with storage and future retrieval [27].

In Spain, groundwater markets are difficult to assess. Llamas and Garrido (2007) [29] describe that the 1985 Water Act declared groundwater as a public domain; before 1985, groundwater was considered a private domain. Groundwater use prior to 1986 could continue to function as private property using the same amount as before, but new abstractions require a permit [29]. However, the government has not been able to get the private wells built before 1986 registered and the volumes extracted are still unknown [29]. Water markets were introduced in the late 1990s without much impact on water reallocation; but groundwater markets are not supposed to be pertinent because groundwater was somehow still considered a private asset, therefore it could be leased, bought, or sold like any other asset. There are informal/illegal groundwater markets in Spain, but data on these are not trusted [29]. So, groundwater rights in this country are unclearly established mainly because most of the wells are illegal, with the exception of the Canary Islands, which have a different water code because groundwater is the main water source [29]. So, basically all groundwater in this region is privately owned [29].

Mukherji and Shah (2005) [14] indicate that water markets work better in areas where water is not scarce. In water-short regions, water markets could actually result in resource depletion. Long-term water-storage credits under the Arizona system are in fact marketable. While there has been increased activity in the marketing of long-term water storage credits in water-scarce Arizona, analysis of the economic and water use implications of sales of storage credits awaits further analysis [27,35].

4. Discussion—Lessons Learned: Context and Commonality

In this section, we digest the rich materials consulted in 10 case studies in order to draw some lessons (Table 2). The papers range across five continents and treat developed and developing societies. It could not be clearer that—at all levels—contextual considerations are (and ought to continue to be) the primary determinant of governance policies based on traditions and practices. This reflects a growing recognition in water governance of the importance of mixed-governance strategies that appeal to multiple values and fit into local circumstances [36]. Context in water governance is recognized to shape collaboration, innovation, and on-the-ground implementation of projects and policies [37]. Yet many of the observations in our case studies converge. While we emphasize that there exists no single, all-purpose toolkit for sustainable groundwater management, we believe that the take-home messages from the ensemble of our sources do provide valuable tips for planners, decision-makers, managers, and stakeholders.

To discuss these messages, we follow the organization of the previous sections of the paper. Thus, we consider, in order, lessons learned regarding: institutional setting, availability and access to information and science, robustness of civil society, and economic and regulatory frameworks.

Table 2. Lessons learned from case studies across key governance elements.

Governance Elements	Lessons Learned
Institutional Setting	• Governing is often a thankless task, yet it requires popularity • Legislation does not always translate into implementation • Conflict resolution is central to groundwater governance • Sufficient funding is of the utmost significance for governance
Availability and Access to Information and Science	• Natural systems, social systems, and institutions all have been understudied and would benefit greatly from additional research • Trust is a necessary element for all research • Urbanized landscapes are critical components of groundwater governance
Robustness of Civil Society	• Equity is an essential ingredient of groundwater governance • Community-based governance requires deliberate, purposeful intention • Leaders can unite stakeholders
Economic and Regulatory Frameworks	• Economic incentives can be effective, but may sometimes yield unintended, even opposite results • "Indirect" management approaches may be suitable in certain settings, but they should be used cautiously • The effectiveness economic incentives as use-control mechanisms depends greatly on the system employed

4.1. Institutional Setting

Our review of what the case studies tell us about institutional issues suggests four observations: (a) governing is a thankless and often criticized task, yet paradoxically it requires popularity; (b) legislation does not translate to implementation; (c) conflict resolution is central to groundwater governance; and (d) sufficient funding is of the utmost significance for governance.

Members of the scientific community (and the authors of our case studies) have noted that the role of the government should transform from "resource supplier" to "resource custodian" [38]. However, the role of resource custodian is problematic for many reasons; in particular, it offers an uneasy fit for most government agencies. Water-users, for instance, often disdain or resist restrictions, while in democratic societies politicians intrinsically seek popular support to retain their offices. If by default government absents itself from the process, groundwater management typically suffers [31]. Where livelihoods depend on groundwater resources, as in many developing societies—which coincidentally need greater human and financial capacity—this tension complicates stewardship [14].

When governments do pass water laws, they may discover that lawmaking is easy; law enforcement is the chief challenge [14], particularly when such laws require a shift from open-access to regulated resource [33]. To function as intended, legislation should be accompanied by awareness-raising educational programs and enhanced stakeholder engagement [20].

Broad participation can be an effective antidote to the shortcomings of top-down legislation. For example, water conservation is more likely to occur when it stems from a bottom-up approach than when it is mandated but unenforced [20]. In a variation on this theme, in Rajasthan, we see an opposite phenomenon. There, groundwater recharge is occurring on a massive scale—even though it is deemed illegal [28,32].

Whether dealt with top-down or bottom-up, water problems are primarily socioeconomic and political in nature and therefore can engender potential conflict over a precious commodity. For this reason, it is important to include conflict-resolution techniques in groundwater management. Van den Broek and Brown (2015) [31] found that social divisions based on wealth and education are not easily overcome during participatory processes. Innovative, context-specific conflict-resolution techniques can help alleviate struggles inherent in water-management processes [29]. One approach cited in our case studies is to employ social-science-trained experts as intermediaries [33,34].

An overarching need is captured in a common dictum: "No funding, no governance". Insufficient financing forecloses opportunities for technical services, infrastructure development, use of bottom-up approaches, capacity building efforts, and coordination of initiatives [20]. Moreover, it can corrupt water agencies [34]. Other products of lack of funds include dearth of trained personnel [3], and paucity of scientific knowledge [20], and too few social scientists to facilitate stakeholder engagement and deal with conflict resolution [33]. All these factors impede groundwater governance.

4.2. Availability and Access to Information and Science

The 10 case studies point to three lessons around the availability and access to data, information, and science: (a) natural systems, social systems, and institutions all have been understudied and would benefit greatly from additional research; (b) trust is a necessary element for all research; and (c) urbanized landscapes are critical components of groundwater governance.

Natural systems, which may have the richest research base, nevertheless could use additional exploration. There are needs for basic hydrologic data [29] to monitor quality and quantity of groundwater resources. Relating groundwater to other sectors and to ecosystem services also requires additional information and study [20].

Studies increasingly have demonstrated that socioeconomic factors are a primary cause of environmental problems, more so than natural factors [29]. Yet the role and significance of social systems in groundwater governance remains poorly appreciated and requires further study [14].

Demographic forces, for example, are critical to groundwater management because resource use is directly related to population [30].

Finally, the place and function of institutions, which serve as the backbone of governance processes, are incompletely understood. Robust research on their role would measurably improve governance capacity [14].

In all of this, trust comprises an essential element of research. When government conducts or sponsors research, there is evidence that communities may be skeptical, suspicious, or even hostile—especially when such research explores the feasibility of development projects [34]. The co-production of knowledge—i.e., involving a local community in a research project—is widely recommended [7,28,39].

Urbanized landscapes, which are proliferating in the face of population shifts to cities and megacities, are sites for land-use changes that affect hydrologic processes as well as major users of groundwater. As a result, municipalities are becoming increasingly important agents of governance. Simultaneously, developers are becoming key players because they have the power to contribute to water conservation or to resist it. In one Indian region, for instance, developers played a large role by installing water-harvesting systems in their buildings and have helped propagate the national water-conservation movement in India [28]. However, in Costa Rica, by contrast, the economic power of large development interests corrupted water agencies, authorizing new, groundwater-dependent development even under water-scarce conditions [34].

Whether in village settings or in large urban agglomerations, multi-participatory, multi-sectoral regional planning appears to be the soundest way to assure community needs and implement sustainable practices—not only for groundwater governance, but for resources governance more generally.

4.3. Robustness of Civil Society

Three observations related to robust civil society emerge from our review: (a) equity is an essential ingredient of groundwater governance; (b) community-based governance requires intention; and (c) forceful, effective, and progressive leaders can unite stakeholders.

Equity is critical to sustainable water governance [34]. Access to resources nearly always is accompanied by equity issues. In some instances, wealthy landowners can afford to dig deeper wells and exploit restored groundwater resources, while poor farmers cannot do so. Further, equity should include gender issues because water availability directly affects women and their potential participation in other activities, including water management. When water levels were restored in areas in India, women became able to access water with simple water pumps. This allowed them more time to engage in productive activities rather than spending their day foraging for water [32].

Equity issues also can relate to transparency in governance. On the one hand, transparency around behavior and water use can trigger debate on ethical issues regarding private information. Some might argue that personal-information disclosure infringes individual liberty [33]. On the other hand, lack of transparency (self-monitoring) can sometimes cause a "climate of suspicion" that can generate social tension [33]. In Costa Rica, insufficient transparency in development planning was related to water conflicts because it promoted a loss of credibility: community members do not believe enough water exists to serve future development [34].

Ethical values in natural-resource management can serve to promote civil-society engagement [28]. Yet, implementing or executing principles of "human solidarity" in practice is often seen to be difficult [29]. So too are mitigating social divisions based on wealth and education in stakeholder participatory processes [31].

It has become widely accepted that a community-based governance approach is central to groundwater governance, but it requires deliberate, purposeful intention. Community-based governance needs to be initiated from the outset [29] and requires legal status [31]. Many examples of community-based management have been seen as relatively unsuccessful throughout the world, with a few successful cases [14]. Success is tied to intention around the purpose and legal basis of the governance mode, but it is also linked to the focus of the governance in terms of supply and demand. Soft-path approaches are generally gaining acceptance and these tend to favor demand-management strategies that are viewed as more sustainable than supply-side ones. However, in some instances, successes have been attributed to supply-side schemes. For example, community-based initiatives aimed at supply augmentation can appeal to stakeholders and raise engagement because the conversation is not so narrowly focused on the burdens of limiting water use [14].

Finally, an emerging lesson from the cases is the role of leaders in uniting diverse participants in governance. Leadership is necessary to initiate action and acquire funding [32]. Committed leaders can unite community members by integrating traditional knowledge [32], and also by embracing ethics and where appropriate, spirituality. They can empower and inspire people for action [28]. Flexibility, trustworthiness, and communication are considered critical leadership skills for effective civil-society engagement in groundwater governance [20].

4.4. Economic and Regulatory Frameworks

Economic and financial practices, pricing policies, and regulatory policies vary substantially across the globe and that characteristic certainly is manifest in our case studies. Nonetheless, we do find three common strands: (a) economic incentives can be effective, though at other times they may provide unintended, even opposite results; (b) "indirect" management approaches may be suitable in certain settings, but they should be used cautiously; and (c) the effectiveness of economic incentives as use-control mechanisms depends greatly on the system employed.

Llamas and Garrido (2007) [29] provide an example of such perverse outcome by examining subsidies in Spain. These actions stopped neither the drilling of illegal wells, nor their pumpage. Likewise, in India, subsidies delayed some communities' participation in a well-recharge movement [28]. In France, paying penalties encouraged farmers to feel they have a right to exceed their water allocations [33].

At a larger scale, changes to the economic framework through "indirect" management—i.e., management of water resources via other sectors such as energy, food, and trade—can be problematic [14]. For example, the Indian government tried to reduce groundwater pumpage by implementing bans to electric supply, but new political candidates promised the exact opposite mechanism: electric connections for water-users [28]. Still, not all indirect management necessarily fails. In Mexico, for instance, substantial electric subsidies have effectively incentivized well-owners to register their wells [14].

Changes to governance regimes that move from a no-pricing to a pricing system may be met by resistance. This is not surprising: no one is eager to pay for something they previously have had for free. The Uganda case shows that people believe that water should be free of charge, therefore they refused to pay for the operation and maintenance of community-owned water pumps [31].

There is a role for market mechanisms in water governance systems, but with oversight. Without such oversight, water markets can promote agricultural expansion and growth beyond the carrying capacity of aquifers [28]. Alternatively, such practices can lead to water-use changes that do not necessarily decrease demand [30]. In other instances, insufficient oversight could result in resource depletion, particularly in arid lands [14]. In Arizona, where groundwater regulation is relatively strong, emerging water markets have shown successful results [27].

Our review suggests that the effectiveness of groundwater-use control via economic incentives depends on the particular system of incentives and enforcement and its relevance to and viability within its context. Because not all societies possess the financial and human capacity to enforce compliance with economic instruments, this sort of governance strategy is highly context-specific and should be used conservatively, lest it lead to unintended results.

5. Conclusions

There is growing recognition within policy circles and in academic scholarship of the importance of groundwater governance. To better understand lessons in groundwater governance around the world today, we reviewed and analyzed carefully 10 selected groundwater-governance case studies representing diverse regions. The cases provide a wealth of experience and observations, draw cogent conclusions, and offer potentially useful recommendations. We integrated and synthesized across the cases to distill key lessons on what animates actual, on-the-ground governance. We approached the case studies through the lens of four practical elements (or action steps) of governance employed in an earlier thematic paper on global groundwater governance and policy [3]. These elements—institutional setting, availability and access to information and science, robustness of civil society, and economic and regulatory frameworks—are seen as instrumental to groundwater governance and policy [3].

In terms of institutional setting, we found that: governing is a trying and usually unappreciated task, yet it requires popular support; legislation does not automatically translate to implementation; and that sufficient and reliable funding is essential for governance. Lemos and Agrawal (2006) [6] and other scholars within the broader environmental governance community frequently speak to the importance of implementation and funding.

Regarding availability and access to information, the cases highlighted the need for research on natural systems, social systems, and institutions; trust is an essential element for all research; and urbanized landscapes are becoming critical components of groundwater governance. This observation was confirmed by water-governance research concerning surface water (e.g., [40–42]).

Apropos the robustness of civil society, we learned that equity is an indispensable element of governance; community-based governance requires planned, purposeful intention; and leaders can unite stakeholders. This view of stakeholder engagement and participatory processes to support the engagement of diverse stakeholders is echoed and well-articulated in water-governance literature (e.g., [22,38,43,44]). Knowledge co-production processes are increasingly seen as important in not just building products—but building relationships between stakeholders [45].

Finally, in terms of economic and regulatory frameworks, the case studies illustrate how economic incentives sometimes provide unintended results; "indirect" management should be used with caution; and the effectiveness of economic incentives depends on the system employed. These reflections are perhaps less commonly found in western, non-Marxist economic literature, which tends to favor market-based approaches while discounting such factors as context and difficult-to-value commodities such as streamflow and aquifer storage.

Of course, many of these findings are not unique to groundwater *per se*. In their review of conditions to support effective science–policy interaction in transboundary river basins, Armitage et al. (2015) [46] conclude that the social and institutional context in which actors create and utilize scientific knowledge is fundamental. Increasingly, a number of water scholars are calling for greater attention to issues of democratization, equity, and human welfare in water governance (e.g., [47–52]). Strands of collaborative-governance research in public policy fields and studies of adaptive capacity well articulate issues of participation, knowledge and learning, leadership, and resources in governance (for a review, see [53]).

In addition, there exist common groundwater governance challenges that are not well articulated in the case studies reviewed here; some of these are nonetheless important issues related to governance. In terms of participation, the cases did not really engage in limits to participation or failed strategies to participation which may in part heighten the promotion of participatory approaches as a panacea to water governance [54]. Particularly in the cases of developing countries and communities associated with disenfranchised and poor citizens, participation may be implemented in inequitable ways that do not well certain segments of a population fairly. Redistribution of power and resources may be needed to engage the poor and achieve substantive stakeholder participation [55]. In terms of availability and access to information and science, there is little attention in the case studies reviewed here that speaks to the communication of information and data to decision-makers and the complexities of adaptive governance and the science-policy interface in water governance (e.g., [56–58]).

Collectively, the lessons drawn from our case-study analysis suggest a need for dual governance capacities—on the part of governments at multiple levels as well as civil-society actors. As any other common pool resource, groundwater faces challenges of overexploitation and protection [58]. Good policy design (even if we can agree on what this ought to entail) alone is insufficient for effective groundwater governance. Rather, implementation requires sufficient investments, reliable science, good leadership, and equitable decision-making. Even then—as real-world experiences show—unanticipated and often undesirable and costly consequences can result.

In spite of a number of observed commonalities, our review emphasizes that contextual considerations—i.e., locational, temporal, sociocultural, and procedural—are primary determinants of governance policies. There is no universal toolkit for groundwater governance, so communities will need to learn from each other. Additionally, in this regard, the growing body of case study research on groundwater governance will do well to address some of the heretofore unexplored or insufficiently addressed aspects of the subject including challenges of participation, intra-community equity, and legitimacy of leadership. Finally, we believe that decision-makers, stakeholders, and the public at large can benefit from integrated reviews such as this one to experiment with groundwater strategies and approaches that best suit their particular conditions, needs, and expectations.

Acknowledgments: The authors gratefully acknowledge the support of the International Water Security Network, funded by Lloyd's Register Foundation (LRF), a charitable foundation in the United Kingdom helping to protect life and property by supporting engineering-related education, public engagement, and the application of research. The work could not have been done without the initial backing of the UNESCO International Hydrological Program (IHP), as part of the project "Groundwater Governance—A Global Framework for Action", sponsored by the Global Environment Facility (GEF), the Food and Agriculture Organization of the United Nations, UNESCO-IHP, the International Association of Hydrogeologists, and the World Bank. We further acknowledge the Inter-American Institute for Global Change Research (IAI), for Project SGP-CRA005, supported by U.S. National Science Foundation (NSF) Grant No. GEO-1138881; and for Research Project CRN3056, supported by NSF Grant No. GEO-1128040. The paper also benefited from support by the Technology and Research Initiative Fund, through the University of Arizona (UA) Water Sustainability Program and Water Resources Research Center (WRRC); the Morris K. Udall and Stewart L. Udall Foundation in Tucson, Arizona; and the USGS Water Resources Research 104b Program, administered by the UA WRRC.

Author Contributions: All authors have contributed to the writing of this paper. Andrea K. Gerlak, Robert G. Varady and Sharon B. Megdal designed the methodology and created the new framework for analyzing groundwater governance based on their previous work (Thematic Paper #5). Adriana A. Zuniga-Teran reviewed the papers and developed the map. Each author analyzed one of the pillars of the framework and wrote their own sections. All authors contributed to the editing of the paper as a whole and its revisions.

Conflicts of Interest: The authors declare no conflict of interest.

Abbreviations

The following abbreviations are used in this manuscript:

ADWR	Arizona Department of Water Resources
AMA	Active Management Areas
CAP	Central Arizona Project
CBM	Community Based Management
GEF	Global Environment Facility
GMA	Groundwater Management Act
OECD	Organization for Economic Co-operation and Development
TBS	Tarun Bharat Sangh

References

1. Groundwater Governance—A Global Framework for Action. Available online: http://www.groundwatergovernance.org/home/en/ (accessed on 19 September 2016).

2. Organisation for Economic Co-Operation and Development (OECD). Principles on Water Governance. Available online: http://www.oecd.org/environment/watergovernanceprogramme.htm (accessed on 23 June 2016).
3. Varady, R.G.; van Weert, F.; Megdal, S.B.; Gerlak, A.K.; Iskandar, C.A.; House-Peters, L. Groundwater Governance: A Global Framework for Country Action. Available online: http://www.yemenwater.org/wp-content/uploads/2015/04/GWG_Thematic5_8June2012.pdf (accessed on 19 September 2016).
4. Megdal, S.B.; Gerlak, A.K.; Varady, R.G.; Huang, L.-Y. Groundwater governance in the United States: Common priorities and challenges. *Groundwater* **2015**, *52*, 677–684. [CrossRef] [PubMed]
5. Garfin, G.M.; Romero-Lankao, P.; Varady, R.G. Editorial: Rethinking integrated assessments and management projects in the Americas. *Environ. Sci. Policy* **2013**, *26*, 1–5. [CrossRef]
6. Lemos, M.C.; Agrawal, A. Environmental governance. *Annu. Rev. Environ. Nat. Resour.* **2006**, *31*, 297–325. [CrossRef]
7. Lemos, M.C.; Morehouse, B.J. The co-production of science and policy in integrated climate assessments. *Glob. Environ. Chang.* **2005**, *15*, 57–68. [CrossRef]
8. Scott, C.A.; Varady, R.G.; Meza, F.; Montaña, E.; de Raga, G.B.; Luckman, B.; Martius, C. Science-policy dialogues for water security: Addressing vulnerability and adaptation to global change in the Arid Americas. *Environ. Sci. Policy Sustain. Dev.* **2012**, *54*, 30–42. [CrossRef]
9. Gupta, J. An essay on global water governance and research challenges. In *Principles of Good Governance at Different Water Governance Levels*; van der Valk, M., Keenan, P., Eds.; The Netherlands National Committee IHP-HWRP: Delft, The Netherlands, 2011.
10. Water Governance in OECD Countries. Available online: http://www.oecd-ilibrary.org/environment/water-governance-in-oecd-countries_9789264119284-en (accessed on 19 September 2016).
11. Hoff, H. Global water resources and their management. *Curr. Opin. Environ. Sustain.* **2009**, *1*, 141–147. [CrossRef]
12. Garduño, H.; van Steenbergen, F.; Foster, S. Stakeholder Participation in Groundwater Management. Available online: http://siteresources.worldbank.org/EXTWAT/Resources/4602122-1210186362590/GWM_Briefing_6new.pdf (accessed on 19 September 2016).
13. Solanes, M.; Jouravlev, A. *Water Governance for Development and Sustainability*; United Nations CEPAL: Santiago, Chile, 2006.
14. Mukherji, A.; Shah, T. Groundwater socio-ecology and governance: A review of institutions and policies in selected countries. *Hydrogeol. J.* **2005**, *13*, 328–345. [CrossRef]
15. Scott, C.A.; Kurian, M.; Westcoat, J.L., Jr. The Water-energy-food nexus: Adaptive capacity to complex global challenges. In *Governing the Nexus: Water, Soil and Waste Resources Considering Global Change*; Kurian, M., Ardakanian, R., Eds.; Springer: Berlin, Germany, 2015; pp. 15–38.
16. Scott, C.A.; Sugg, Z.P. Global energy development and climate-induced water scarcity—Physical limits, sectoral constraints, and policy imperatives. *Energies* **2015**, *8*, 8211–8225. [CrossRef]
17. Gerlak, A.K.; Varady, R.G.; Petit, O.; Haverland, A. Hydrosolidarity and beyond: Can ethics and equity find a place in today's water resource management? *Water Int.* **2011**, *36*, 251–265. [CrossRef]
18. Priscoli, J.D.; Wolf, A.T. *Managing and Transforming Water Conflicts*; Cambridge University Press: Cambridge, UK, 2009.
19. Moench, M.; Burke, J.J.; Moench, Y. *Rethinking the Approach to Groundwater and Food Security*; Food & Agriculture Organization: Rome, Italy, 2003.
20. Knüppe, K. The challenges facing sustainable and adaptive groundwater management in South Africa. *Water SA* **2011**, *37*, 67–79. [CrossRef]
21. Cash, D.W.; Borck, J.C.; Patt, A.G. Countering the loading-dock approach to linking science and decision making comparative analysis of El Niño/Southern Oscillation (ENSO) forecasting systems. *Sci. Technol. Hum. Values* **2006**, *31*, 465–494. [CrossRef]
22. Mott Lacroix, K.E.; Megdal, S.B. Explore, synthesize, and repeat: Unraveling complex water management issues through the stakeholder engagement wheel. *Water* **2016**, *8*, 118. [CrossRef]
23. Villholth, K. Groundwater assessment and management: Implications and opportunities of globalization. *Hydrogeol. J.* **2006**, *14*, 330–339. [CrossRef]
24. Food and Agriculture Organization of the United Nations (FAO). Definition of Policy. Available online: http://www.fao.org/wairdocs/ILRI/x5499E/x5499e03.htm (accessed on 9 May 2016).

25. Megdal, S.B.; Scott, C.A. The importance of institutional asymmetries to the development of binational aquifer assessment programs. *Water* **2011**, *3*, 949–963. [CrossRef]
26. Ocampo-Melgar, A.; Vicuña, S.; Gironas-Leon, J.; Varady, R.G.; Scott, C.A. Science-policy co-production of climate-change-adaptation indicators: A prototype approach based on the Maipo River basin, Chile. *Environ. Sci. Policy Sustain. Dev.* **2016**, *58*, 24–37. [CrossRef]
27. Megdal, S.B.; Dillon, P.; Seasholes, K. Water Banks: Using Managed Aquifer Recharge to Meet Water Policy Objectives. *Water* **2014**, *6*, 1500–1514. [CrossRef]
28. Shah, T. Mobilising social energy against environmental challenge: Understanding the groundwater recharge movement in Western India. *Nat. Resour. Forum* **2000**, *24*, 197–210. [CrossRef]
29. Llamas, R.M.; Garrido, A. Chapter 13: Lessons from intensive groundwater use in Spain: Economic and social benefits and conflicts. In *The Agricultural Groundwater Revolution: Opportunities and Threats to Development*; Giordano, M., Villholth, K.G., Eds.; International Water Management Institute (IWMI): Colombo, Sri Lanka, 2007; pp. 266–298.
30. Henriques, C.; Holman, I.P.; Audsley, E.; Pearn, K. An interactive multi-scale integrated assessment of future regional water availability for agricultural irrigation in East Anglia and North West England. *Clim. Chang.* **2008**, *90*, 89–111. [CrossRef]
31. Van den Broek, M.; Brown, J. Blueprint for breakdown? Community Based Management of rural groundwater in Uganda. *Geoforum* **2015**, *67*, 51–63. [CrossRef]
32. Everard, M. Community-based groundwater and ecosystem restoration in semi-arid north Rajasthan (1): Socio-economic progress and lessons for groundwater-dependent areas. *Ecosyst. Serv.* **2015**, *16*, 125–135. [CrossRef]
33. Figureau, A.-G.; Montginoul, M.; Rinaudo, J.-D. Policy instruments for decentralized management of agricultural groundwater abstraction: A participatory evaluation. *Ecol. Econ.* **2015**, *119*, 147–157. [CrossRef]
34. Kuzdas, C.; Warner, B.P.; Wiek, A.; Vignola, R.; Yglesias, M.; Childers, D.L. Sustainability assessment of water governance alternatives: the case of Guanacaste Costa Rica. *Sustain. Sci.* **2016**, *11*, 231–247. [CrossRef]
35. Megdal, S.B. The role of the public and private sectors in water provision in Arizona, USA. *Water Int.* **2012**, *37*, 156–168. [CrossRef]
36. Ingram, H.M. Beyond universal remedies for good water governance: A political and contextual approach. In *Water for Food in a Changing World*; Garrido, A., Ingram, H.M., Eds.; Routledge: Abingdon, UK, 2011; pp. 241–261.
37. De Boer, C.; Vinke-de Kruijf, J.; Ozerol, G.; Bressers, H. *Water Governance, Policy and Knowledge Transfer: International Studies on Contextual Water Management*; Routledge: Abingdon, UK, 2013.
38. Mechlem, K. Groundwater governance: The role of legal frameowrks at the local and national level—Established practices and emerging trends. *Water* **2016**, *12*, 347. [CrossRef]
39. Lemos, M.C. Usable climate knowledge for adaptive and co-managed water governance. *Curr. Opin. Environ. Sustain.* **2015**, *12*, 48–52. [CrossRef]
40. Burton, M.; Molden, D. Making sound decisions: Information needs for basin water management. In *Irrigation and River Basin Management: Options for Governance and Institutions*; Svendsen, M., Ed.; CABI Publishing: Boston, MA, USA, 2005; pp. 51–74.
41. Goulden, M.; Conway, D.; Persechino, A. Adaptation to climate change in international river basins in Africa: A review. *Hydrol. Sci. J.* **2008**, *54*, 805–827. [CrossRef]
42. Lejano, R.; Ingram, H. Collaborative networks and new ways of knowing. *Environ. Sci. Policy* **2009**, *12*, 653–662. [CrossRef]
43. Mostert, E. Conflict and cooperation in international freshwater management: A global review. *Int. J. River Basin Manag.* **2003**, *1*, 267–278. [CrossRef]
44. Newig, J.; Haberl, H.; Pahl-Wostl, C.; Rothman, D. Formalised and non-formalised methods in resource management—Knowledge and social learning in participatory processes: An introduction. *Syst. Prac. Act. Res.* **2008**, *21*, 381–387. [CrossRef]
45. Edelenbos, J.; van Buren, A.; van Schie, N. Co-producing knowledge: Joint knowledge production between experts, bureaucrats and stakeholders in Dutch water management projects. *Environ. Sci. Policy* **2011**, *14*, 675–684. [CrossRef]

46. Armitage, D.; de Loë, R.; Morris, M.; Edwards, T.D.; Gerlak, A.; Hall, R.; Huitema, D.; Ison, R.; Livingstone, D.; MacDonald, G.; et al. Science-policy processes for transboundary water governance. *Ambio* **2015**, *44*, 353–366. [CrossRef] [PubMed]
47. Groenfeldt, D. *Water Ethics: A Values Approach to Solving the Water Crisis*; Earthscan from Routledge: London, UK; New York, NY, USA, 2013.
48. Groenfeldt, D.; Schmidt, J. Ethics and water governance. *Ecol. Soc.* **2013**, *18*, 14. [CrossRef]
49. Lu, F.; Ocampo-Raeder, C.; Crow, B. Equitable water governance: Future directions in the understanding and analysis of water inequities in the global South. *Water Int.* **2014**, *39*, 129–142. [CrossRef]
50. Perreault, T. What kind of governance for what kind of equity? Towards a theorization of justice in water governance. *Water Int.* **2014**, *39*, 233–245.
51. Roa-Garcia, M. Equity, efficiency and sustainability in water allocation in the Andes: Trade-offs in a full world. *Water Altern.* **2014**, *7*, 298–319.
52. Wilder, M.; Ingram, H. Knowing equity when we see it: Water equity in contemporary global contexts. In *Oxford Handbook on Water Policy and Politics*; Conca, E., Weinthal, E., Eds.; Oxford University Press: Oxford, UK, 2016.
53. Emerson, K.; Gerlak, A.K. Adaptation in Collaborative Governance Regimes. *Environ. Manag.* **2014**, *54*, 768–781. [CrossRef] [PubMed]
54. Morinville, C.; Harris, L.M. Participation, politics, and panaceas: Exploring the possibilities and limits of participatory urban water governance in Accra, Ghana. *Ecol. Soc.* **2014**, *19*, 36. [CrossRef]
55. Wester, P.; Merrey, D.J.; de Lange, M. Boundaries of Consent: Stakeholder Representation in River Basin Management in Mexico and South Africa. *World Dev.* **2003**, *31*, 797–812. [CrossRef]
56. Blomquist, W. *Dividing the Waters: Governing Groundwater in Southern California*; Institute for Contemporary Studies (ICS) Press: New York, NY, USA, 1992.
57. Brunner, R.; Steelman, T. Beyond Scientific Management. In *Adaptive Governance: Integrating Science, Policy, and Decision Making*; Brunner, R., Steelman, T., Coe-Juell, L., Cromley, C., Edwards, C., Tucker, D., Eds.; Columbia University Press: New York, NY, USA, 2005; pp. 1–46.
58. Holmes, J.; Scott, A. Bridging the Gaps between Science and Policy: A Review of the Evidence and some Principles for Effective Action. In *Water System Science and Policy Interfacing*; Quevauviller, P., Ed.; Royal Society of Chemistry: Cambridge, UK, 2010; pp. 15–35.

water

MDPI

Review

Enhancing Groundwater Governance by Making the Linkage with Multiple Uses of the Subsurface Space and Other Subsurface Resources

Jac van der Gun [1,*], Alice Aureli [2] and Andrea Merla [3]

1 Consultant, Ganzeboomsweg 3 C, Schalkhaar 7433ES, The Netherlands
2 UNESCO International Hydrological Programme (UNESCO–IHP), 7, place Fontenoy, Paris 07 SP 75352, France; a.aureli@unesco.org
3 Consultant, Via Ponte dei Galli 2, Assisi 06081, Italy; merla.andrea@gmail.org
* Correspondence: j.vandergun@home.nl; Tel.: +31-570-854-611

Academic Editors: Sharon B. Megdal, Susanna Eden and Eylon Shamir
Received: 18 February 2016; Accepted: 17 May 2016; Published: 25 May 2016

Abstract: One of the aspects highlighted in the Framework for Action and other key documents produced by the *Groundwater Governance Project* (funded by GEF and implemented by UNESCO, FAO, World Bank and IAH) is the interdependence between groundwater and human activities related to other physical components of the real world. Consequently, it is important in groundwater governance to make essential linkages with other components of the water cycle (IWRM), with sanitation and wastewater management, with land use and land use practices, with energy and with the uses of subsurface space and other subsurface resources. This paper presents an overall description of the multiple uses of the subsurface space and of the exploitation and management of subsurface resources. It attempts to give an impression of intensities and trends in use and exploitation, of the possible interactions and of current and potential efforts to control negative impacts of such interactions. It concludes by briefly summarizing in three simple steps how to improve groundwater governance by making appropriate linkages with uses of the subsurface space and subsurface resources.

Keywords: groundwater governance; linkages; use of subsurface space; use of subsurface resources

---

1. Introduction

Groundwater governance is a complex concept that can be viewed from different angles. Hence, several definitions have been proposed, each with its own merits. We present here a definition adopted during the final stage of the *Groundwater Governance Project*, funded by GEF and implemented by UNESCO, FAO, World Bank and IAH:

> *"Groundwater governance comprises the promotion of responsible collective action to ensure socially-sustainable utilisation, control and protection of groundwater resources for the benefit of humankind and dependent ecosystems"* [1]

This definition refers implicitly to the main components of groundwater governance: (1) actors; (2) information on the groundwater resources and their beneficial use; (3) a vision on goals to be pursued regarding groundwater and on how to achieve them (policy and planning); and (4) tools for regulation, control and protection (laws, regulations, incentives, *etc.*).

"Collective action" implies the existence of several actors that have to cooperate *unisono* in order to achieve the goals set. In relation to groundwater there is a large diversity of actors (governmental organizations, non-governmental organizations, research institutes, the private sector, individual citizens) who may act at different levels (ranging from local to national, and in some cases even supra-national levels). Many of these actors focus explicitly on groundwater, in their endeavors for using, draining, augmenting or protecting it, while the activities of many others address something else, but, at the same time, cause unintended impacts on groundwater ('side effects'). It goes without saying that information on and knowledge of the local groundwater systems and their state and context is indispensable for making informed decisions on groundwater. In many cases, however, there is much to be desired in this respect, with the result of considerable uncertainty: on current conditions and even more on how the future will look, either with or without the implementation of measures for control and protection. However, inadequate information is not the only reason for poor groundwater policies, planning and management. A myopic view may be another reason. Groundwater is not an isolated part of the world, but it is interacting with its surroundings, e.g., with surface water, with sanitation and with waste and wastewater management, with land use and land use practices, with energy and with the use of subsurface space and subsurface resources. Good governance requires that essential linkages are made, tailor-made to each particular case.

This paper focuses on only one of the mentioned categories of interactions: that between groundwater and the uses of the subsurface space and other subsurface resources. These uses will be briefly reviewed, while paying attention to their possible interaction with groundwater and to current and potential efforts to control negative impacts of such interactions. This may contribute to enhanced awareness of these interactions and to being alert to taking them duly into account in policies, planning and operational groundwater management. This is followed by a few suggestions on how to make these linkages in practice. The paper relies to a large extent on the outcomes of the Groundwater Governance Project, in particular on Thematic Paper No. 10 [2].

2. Panorama of the Multiple Uses of the Subsurface Space and Subsurface Resources

The most common human activities that take place in the subsurface are listed in Table 1. They fall into two main categories: (a) extraction of natural resources (water, energy, minerals, building materials, *etc.*) and (b) use of the underground space (for construction, transport and temporal or permanent storage of substances). Shallow groundwater abstraction and some forms of mining have been practiced already from time immemorial. Many other activities started only in more recent times. In general, one may say that human activities are nowadays encroaching rapidly into the subsurface, especially in densely populated areas, where the demand for natural resources is high and space is becoming scarce. These are the areas where conflicts between the different uses of the subsurface space and of its resources are most likely to occur.

Table 1. Main subsurface human activities.

Category	Type of Activity	Geographic Distribution
1. Groundwater development and management	Groundwater withdrawal for different uses	Widespread across the globe (withdrawal is still rare at greater depths, beyond 600 m)
	Drainage of excess shallow groundwater	In selected flat areas, mostly lowlands (USA, Europe, China, Pakistan, India, Mexico)
	Managed aquifer recharge	Mostly in water-scarce areas, and in coastal zones (to counterbalance seawater intrusion)
2. Mining	Extraction of minerals, coal, lignite, building materials, *etc.*	Scattered over the globe (where such profitable resources have been identified)
3. Geo-energy development (fluids as carrier of energy)	Oil and gas development	In major geological basins on- and off-shore
	High-enthalpy geothermal energy development *	Areas of favorable temperature anomalies (e.g., in USA, Japan, Iceland, Italy, Central America, Indonesia, Philippines, Kenya, China (Tibet), *etc.*)
	Low-enthalpy geothermal resources development *	Major sedimentary basins (e.g., France, Germany, Brazil, *etc.*)

Table 1. *Cont.*

Category	Type of Activity	Geographic Distribution
4. Disposal and storage of hazardous waste	Waste disposal by deep well injection	Often associated with mining, or with the oil and gas industry
	Carbon capture and sequestration	Major projects in the North Sea, in Canada and in Algeria
	Subsurface storage of radioactive waste	Mainly in selected countries, e.g., USA, France, Russia, Japan and India
	Nuclear weapons testing and nuclear power accidents	e.g., in Western USA and French Polynesia (tests); Russia and Japan (accidents)
5. Injection and recovery	Solution mining (e.g., using acids and other lixiviants)	In selected mining areas
	Injecting residual geothermal fluids	In geothermal energy development areas
	Temporary storage of heat	Summer heat surplus used during winter
	Storage of hydrocarbons and fluids associated with oil and natural gas production	In or near oil and gas production areas
	Hydraulic fracturing or "fracking"	In zones where shale gas is explored or exploited
6. Construction into the underground space	Pipelines, sewerage systems and cables	Very general around the world, in particular in developed countries
	Tunnels and underground railways	In particular in many urban centers around the world
	Underground car parks and other underground constructions	In particular in urban areas of industrialized countries where space is scarce

Note: * Enthalpy, which can be considered more or less proportional to temperature, is used to express the heat (thermal energy) content of the fluids that act as the carrier transporting heat from the deep hot rocks to the surface. According to the enthalpy criterion, geothermal resources can be divided into low-, medium- and high-enthalpy (or -temperature) resources, which gives a general indication of their potential forms of utilization. The classification is, in practice, mostly based on temperature, but standardization of the temperature limits to be applied has not yet been agreed upon in the scientific community, thus leaving some room for confusion and ambiguity. Upper temperature limits for low-enthalpy resources, as proposed by selected authors, vary between 90 °C and 200 °C, while lower temperature limits for high-enthalpy resources are in the range of 150 °C to 250 °C [3].

2.1. Groundwater Development and Management

Groundwater withdrawal is by far the most widespread and most common human activity in relation to the subsurface. The volumes of groundwater abstracted annually on Earth—during 2010, some 982 km^3 [4]—very significantly exceed the volumes of any other substance extracted from the subsurface. In contrast to most other subsurface activities that require highly specialized technical expertise, groundwater abstraction is extremely decentralized, since the resource and conventional withdrawal technology are within relatively easy reach of anyone who owns a piece of land. Groundwater therefore is sometimes called a social or democratic resource, but the drawback is competition and interference between uncoordinated groundwater pumpers. Scarcity of groundwater is pushing the groundwater frontier to greater depths, which requires more technological expertise and larger investments. At greater depths (beyond 600 m), significant reservoirs filled with fresh groundwater (so called "deep-seated aquifers") have been identified in several regions, but often these are relatively isolated from the currently active hydrological cycle and thus are not or only poorly recharged. This makes such aquifers more suitable as a strategic resource for emergency supplies than for regular use. An additional problem related to deep-seated aquifers is that in many countries, their depth has the implication that mining law is applicable, not water law; this may also have consequences for institutional jurisdictions and mandates.

The ubiquitous presence of groundwater causes it to be exposed and vulnerable to almost any human subsurface activity, often with negative impacts. Given the undisputed huge importance of groundwater for domestic, environmental and economic purposes, due attention should be paid to these interactions in order to prevent or minimize negative impacts.

Groundwater, in turn, can also have negative impacts on the human living environment and on human activities above or below the ground surface. The most common corrective intervention in this context is artificial drainage of excess groundwater, which has a long tradition in several countries. The reverse action—artificial replenishment of groundwater in order to enhance water availability during dry periods—is nowadays known under the term Managed Aquifer Recharge (MAR). Its worldwide implementation is still modest, but growing rapidly to enhance and secure vulnerable supplies through water banking.

2.2. Mining

The extraction of building materials from the subsurface and the mining of coal and of metallic and non-metallic minerals have a millennia-long history, or even date back to pre-historic times. Thousands of years ago, shafts were already constructed to reach depths of several tens of meters below the surface. The mining industry expanded particularly during the 18th and 20th centuries, and its production is still increasing. In 2013, the world's annual mining production, excluding building materials, reached approximately 9.95 billion metric tons: 1.65 for metals, 0.78 for industrial minerals and 7.52 for mineral fuels (without oil and gas) [5]. Mining generates enormous economic benefits around the world, but it also produces considerable risk to groundwater and the environment. First, because most mining operations, both underground and open-pit or surface mining, require drainage. This modifies the groundwater regime locally or sometimes even regionally, such as in the Ruhr area in Nordrhein-Westfalen (Germany) where approximately 1.2 billion cubic meters of groundwater annually is drained in order to enable open-pit mining of lignite [6]. Furthermore, mining often produces significant environmental pollution, leading to degraded groundwater and surface-water quality. These negative impacts of mining do not stop immediately when the mining operations are discontinued, but may still continue for a long time afterwards. An example is the Cornish Wheal Jane Mine in southwest England where tin was extracted from the mid-18th century until 1992. Following abandonment, the mine flooded and acid drainage water escaped into the surface drainage. The subsequent remedial works were demanded under environmental legislation at a cost of over 300 million USD [2]. Countries with weak legislation and poor enforcement capabilities related to mining are likely to be faced with a growing legacy of resource and environmental problems.

2.3. Geo-Energy Development Based on Fluids as a Carrier of Energy

The oil and gas sector ranks under the world's largest industries. In 2014, the world production of oil was 1659 billion barrels and that of natural gas was 201,771 billion cubic meters [7]. Oil and gas wells started to be drilled systematically during the 19th century, and currently they form the second-widest intrusion into deep underground space, after water wells. In spite of advanced technology and strict environmental regulations, all phases of hydrocarbon exploration, production and abandonment can pose a risk to groundwater resources, because of accidents that are usually the result of human errors. During drilling, loss of control over high-pressure zones can lead to a severe pressure kick or full-scale well blow-out; depending on the drilling phase, shutting in a high-pressure kick can force the reservoir fluids and gases into overlying permeable geological horizons, including aquifers. Furthermore, any deficits on the well construction related to the integrity of the casing and/or grouting can lead to problems during the operation of production-phase oil and gas wells. Holes in the casing or leakages between the casing and the well wall can allow formation fluids to migrate into overlying aquifer formations. The disposal of oil field brines by injection wells can be another source of pollution if leaks occur.

Geothermal energy development generally is based on groundwater as a carrier of heat. On average, the temperature of groundwater increases by 30 °C per km of depth, but this temperature gradient is steeper in zones of thermal anomalies, thus causing groundwater of 400 °C or more to be found at economically viable depths. Global use of geothermal energy is annually growing at a rate of 14%, but it still remains an underdeveloped energy resource, compared to the vast amount of thermal energy stored within the Earth's crust. Access to this huge, low-cost, clean energy resource which could be considered infinite on a human time scale, is possible, but requires a number of technical and financial constraints to be overcome. The potential effects of geothermal energy development on groundwater vary according to the applied technology. Table 2 provides an impression, with differentiation between low-enthalpy and high-enthalpy geothermal resources.

Table 2. Potential effects of geothermal energy development on groundwater (adapted from [8]).

Actions **and Impacts**	**Low-Enthalpy Systems—HSA ***	**High-Enthalpy Systems—EGS ****	
		(a) Vapour-Dominated	**(b) Liquid-Dominated**
Drilling operations:			
Contamination by drilling fluids	Little	Moderate	Moderate
Mass withdrawal:			
Depletion of groundwater	No effect	Little	Moderate
Hydrothermal eruptions:			
Ground temperature changes	No/Little	Little	Moderate
Waste liquid disposal:			
Infiltration of surface disposal	Little	Little	Moderate
Re-injection contamination of groundwater	Little	Little	Little

Notes: * HSA = Hot Sedimentary Aquifer; ** EGS = Enhanced Geothermal Systems.

2.4. *Disposal and Storage of Hazardous Waste*

Apart from unintentional encroachment of solid and liquid waste into the subsurface, e.g., by leaking sewerage systems and septic tanks, or as a side effect of drilling, mining and tunneling, the subsurface is progressively being used intentionally for storing hazardous solid or liquid waste. Deep-well injection is a commonly used technique for the disposal of liquid waste or solid waste that can be reworked to a slurry. To apply this technique successfully, a permeable formation with good storage capacity is needed to receive the liquid waste, covered by an effectively impermeable rock unit to ensure that the stored waste remains permanently isolated from upper layers and the biosphere. It is clear that the risks of this technique include leakage towards other compartments of the underground and the permanent disqualification of the storage reservoir for other purposes. Typical waste products disposed by deep-well injection include waste produced in the oil and gas industry (oil field brines, cuttings, drilling mud, sulphides, mercury compounds, arsenic, cadmium), liquid waste from mining operations and all kinds of industrial and municipal liquid waste.

Also under this category comes Carbon Capture and Sequestration (CCS), which consists of storing carbon underground in order to curb accumulation of carbon dioxide in the atmosphere. It is undertaken because natural sinks of carbon (forests, oceans and soils) are considered unable to accommodate the increasing quantities of carbon dioxide emitted by humans, with consequences in terms of climate change. Major storage of CO_2 is ongoing in large projects in the North Sea, in Canada and in Algeria.

Radioactive waste ranks highest within the category of hazardous waste. Its potential subsurface disposal is still controversial and unresolved, although there is wide agreement that storing nuclear waste in a geologically stable and isolated location underground is a far more promising option than dumping it in sealed barrels into the ocean, which was practiced for some time during the period 1940–1960 for low-level radioactive waste. The problem of nuclear waste lies not only in the extremely devastating impact of potentially released radiation on the exposed biosphere (including humans), but also in the enormous persistence of its hazardous properties. Radiation half-lives of important nuclear waste components are in the order of thousands to millions of years, which means, in practice, that a safe repository should isolate the waste from the human environment for indefinite time. Various approaches to the disposal or containment of radioactive waste can be observed. Controlled sanitary landfills are used for low-level radioactive waste, sometimes after on-site containment for a number of years as to allow decay to a safe level. Current practices in managing high-level radioactive waste are a combination of temporary storage and investigating options for permanent storage. Examples of potential sites selected and studied in detail are in volcanic rock high above the water table (e.g., Yucca Mountain, Nevada, USA), in crystalline rock (e.g., Finland, Sweden and India), in thick impermeable clay (e.g., Boom Clay Formation, Belgium) and in salt domes (e.g., Gorleben site, Germany).

A special category of radioactive pollution hazards is formed by nuclear weapons testing and nuclear power accidents. They form a significant and rather unpredictable risk to groundwater and the environment.

2.5. Injection and Recovery

Unlike permanent storage as discussed above, human activities also may produce the temporary storage of substances into the subsurface, to be followed by their recovery or removal. This includes, firstly, injection and recovery for mineral extraction ('solution mining'), which developed rapidly during the second half of the 19th century as a result of progress in drilling and pump technology. Until the 1950s this technique was carried out largely using water or steam, but later the use of chemical leaching solutions (lixiviants) came into vogue.

Second is the reinjection of residual geothermal fluids. It started purely as a disposal method, but has more recently been recognized as an essential and important part of reservoir management. Reinjection serves not only to maintain reservoir pressure, but also to increase energy extraction efficiency over the life of the resource.

Third is the temporal storage in shallow aquifers of the surplus heat available during low-demand periods of the year in order to recover it during periods when the energy demand is higher. A common application of such stored energy is heating of buildings during winter.

Fourth is the injection and recovery of crude oil and liquid hydrocarbons in underground caverns—natural or man-made—and in reservoirs/aquifers. The hydrocarbons are injected into the formation (in caverns or an aquifer) for storage and later pumped back out for processing and use. The underground reinjection of produced liquid hydrocarbons and natural gas can have different purposes: (i) provide the industry with short-term deliverability during peak demand and/or set aside long-term strategic reserves; (ii) increase production and prolong the life of oil-producing fields.

Fifth is 'hydraulic fracturing', a controversial technique used to increase the rate at which oil, water or natural gas can be produced from natural subterranean reservoirs, including unconventional reservoirs, such as shale rock or coal beds. Hydraulic fractures are formed by pumping a fracturing fluid under pressure into the reservoir formation. The injected fluid is typically a slurry of water, 'proppants' (usually sieved round sand) and chemical additives.

All these applications of injection and recovery present risks to aquifers and to the environment in general; effective management thus is required in order to ensure an acceptable level of risk.

2.6. Construction into the Underground Space

This use of subsurface space is not a new phenomenon, but it has expanded very rapidly during the last few decades. Compared to the subsurface human activities described above, underground constructions are limited to a very shallow upper zone of the subsurface.

Included are, firstly, pipelines (for water, gas and oil), sewerage systems and cables (for electricity and communication, including radio, television and Internet). In densely populated areas these are usually buried, at shallow depth and above the groundwater table. Leaks in sewerage systems and pipelines for oil and gas may constitute potential pollution risks for groundwater.

Tunnels and underground railways also belong to this category. If they are located below the groundwater table, which is often the case, then the groundwater regime may be significantly disturbed during construction (by artificial drainage) and compaction of compressible layers may occur. Modified hydraulic properties of these layers, in combination with constructed artificial obstacles, may, after construction, have a permanent influence on the shallow groundwater regimes. Some tunnels may form preferential entry paths for pollutants into groundwater systems.

In urban areas there is a tendency to locate new car parks underground in order to save space. In addition, plans are being developed in several countries for a much more intensive use of the shallow subsurface for buildings with commercial or public functions, especially under cities where extraction of building materials for centuries had created impressive subsurface open spaces. The potential risks of these car parks and underground buildings regarding groundwater are similar to those of tunnels and subways.

3. Diagnostic

3.1. Interactions and Their Impacts

Use of the subsurface and its resources is steadily becoming more intensive, both by the proliferation of the conventional uses and by the emergence of relatively new, non-conventional uses. From what has been presented above, it may be concluded that most of the related subsurface activities are potentially interacting and are likely to produce negative side effects, such as pollution or increased pollution risks, undesired modification of the hydrogeological regimes, or increasing competition for limited space or water. Although the interactions act in all directions, groundwater systems are particularly vulnerable and threatened by pollution and depletion. The potential impacts of these changes in state include health risks—for instance, if subsurface storage would fail to isolate hazardous substances (in particular radioactive waste or carbon dioxide) from the human environment, loss of valuable sources of water, the increasing cost of water and various types of environmental degradation (land subsidence, degradation of ecosystems, *etc.*). The exploitation of groundwater, on the other hand, may affect other uses of the subsurface and its resources. For instance, groundwater level declines resulting from groundwater abstraction may lead to differential subsidence of the land surface, causing damage to buried pipelines, sewerage systems and various types of subsurface constructions. Groundwater level declines, however, also open up the space required for the application of managed aquifer recharge, similar to the way exploited gas and oil fields offer repositories for storing hazardous waste.

3.2. Expertise, Regulation and Governance in Each of the Sectors

Technical expertise deployed in human activities in the subsurface varies from case to case, and from sector to sector. Except for some types of small-scale and simple interventions, a high level of technical expertise is generally present, especially if the activities take place at great depths and are carried out by highly specialized or large commercial companies (oil companies, mining companies, *etc.*). Usually the expertise allows to do more than focus on the immediate purpose of the activities (such as how to exploit a specific resource) and enables to assess the associated risks, constraints and direct impacts on other activities or interests (economic externalities) or figure out how to mitigate the latter. This technological knowledge and expertise is subject to continuous innovation.

Most countries have laws and regulations on the most important categories of human activities in the subsurface, but their quality and comprehensiveness are variable, as is the effectiveness of their implementation. These legal tools have to ensure that risks of poor concepts or designs, inadequate practices and harmful human errors are minimized, and that the industries put safety and environmental protocols duly in place and comply with these. The laws and regulations—and also the government agencies in charge—tend to lack coordination and tend to be fragmented over the separate subsurface use sectors.

3.3. *The Need for Coordination and Control*

The steadily more intensive multiple use of the subsurface space and subsurface resources leads undeniably to a progressively stronger interaction between the different uses, usually with negative externalities. The risks are very substantial in areas of intensive use or where hazardous projects are implemented, while any degradation of the subsurface and its resources (in particular groundwater) is virtually irreversible due to the inertia of subsurface processes. Those who produce the externalities may not be in the position or may not be motivated to avoid them, either due to lack of information and awareness, or because more priority is given to saving costs and to maximizing profits from their activities. Government agencies usually still have very fragmented involvement and mandates in the various sectors of subsurface activity. Hence, where multiple use of the subsurface is already present or is expected to come into existence in the foreseeable future, provisions should be made for coordination and control, in order to ensure sustainable use of the space and resources offered by the subsurface.

3.4. *Options and Constraints*

At first glance, it seems clear what should be aimed for: integrated management of the subsurface space and all subsurface resources. However, is this a feasible option and is it compatible with other aspirations for integration? Alternatively, should—for the time being—more modest steps towards broader governance of the subsurface domain be pursued?

To answer these questions, it is useful to have a look first at the most common constraints to achieving 'good groundwater governance', as observed in many countries [1]:

- Insufficient and/or unreliable information on the local factual situation (including geology), the predicted futures and the associated benefits, costs, risks and external impacts;
- Lack of awareness at all levels of what is at stake;
- Weak implementation of legal and regulatory frameworks;
- Often poor policies or even absence of policies;
- Disappointing performance of government agencies designated to play a leading role in groundwater governance, due to limitations in capacity, mandate and/or finances;
- Little or no effective involvement of relevant stakeholders.

Next, the interdependence between groundwater and surface water pleads strongly for adopting an integrated water resources management approach (IWRM), while other interdependencies (such as with land use, with sanitation, with energy, *etc.*) also call for linkages in policy, planning and management. Already in conventional groundwater management, the number and diversity of actors do form a significant challenge at the level of implementation. Adding those actors involved in all other sectors of subsurface activities will indisputably lead to much more complexity, certainly also due to the fragmentation of their goals, mandates and legal frameworks. The most likely result then will be a subsurface governance set-up too complex to function.

Therefore, an incremental approach to linking groundwater and other sectors of subsurface uses is considered more appropriate. Smart incremental steps should be designed, in terms of ambition, content and pace of implementation compatible with the local conditions. Examples of such steps include: (a) creating transparency on all subsurface activities in the area concerned, e.g., by raising awareness and public information programs, web portals and publicly accessible databases; (b) announcing to the general public all significant planned new projects related to the subsurface and permitting stakeholders to appeal against these or propose amendments; (c) introducing obligatory 'groundwater impact assessments' for new projects aimed at any use of subsurface space or subsurface resources. First of all, however, efforts are needed to get the subject included on the political agenda and to negotiate the allocation of the means required for implementation of planned action. To this end, awareness should be raised among politicians, key stakeholders and other important decision-makers, in order to gain political support and willingness to cooperate.

4. Conclusions

Multiple uses of the subsurface space and subsurface resources are rapidly encroaching in many areas around the world, especially in densely populated areas. Their interactions are becoming steadily more intensive and often create negative externalities or have the potential to do so. Groundwater and aquifers are, in particular, at risk, or have been affected already by different human subsurface activities, causing degradation of the groundwater quality or depletion of the resources. Traditionally, the different types of subsurface activities are, in most countries, carried out and governed in isolation from each other, subject to separate policies and separate legal and institutional regimes. From the point of view of groundwater governance, the need is felt to abandon this isolation and to make a linkage between groundwater and uses of the subsurface space and of other subsurface resources, in order to protect and manage the groundwater resources optimally.

Decision-makers and stakeholders involved in decision-making on the use of the subsurface need to be made aware both of the many current and potential uses of the subsurface and its resources in their area of concern, and of the geological characteristics of the area's subsurface. Furthermore, they need to have a reliable picture of the benefits, costs and risks of the individual subsurface activities (in comparison with those of non-subsurface alternatives), but they also need to understand that the different uses may interfere and produce externalities. In addition, there are two more key lessons for them to learn about the subsurface domain: First, it is difficult and expensive to explore the subsurface adequately, which results, in practice, in large uncertainties on the assumed properties and potential of the underground. Second, many processes that produce change in the subsurface are virtually irreversible, on a human time scale—or are reversible only at a very high cost.

Professionals responsible for groundwater resources management should be active and creative in finding out how to link their policies and practices to other subsurface use sectors. They should make sure that they are well-informed on all new initiatives for subsurface activities; that sufficient efforts are made to assess external risks and other external impacts; that regulations exist to promote orderly behavior; that all projects comply with these regulations; and that changes over time of the state of the subsurface and related environment are duly monitored.

Acknowledgments: The authors would like to thank all colleagues who have been contributing to the GEF funded Groundwater Governance Project (*Groundwater Governance: a Framework for Action*), in particular Michael Jones and Jake Burke as co-authors of Thematic Paper No. 10, and the members of the project's Steering Committee. The useful comments and suggestions of two anonymous reviewers are also gratefully acknowledged.

Conflicts of Interest: The authors declare no conflict of interest.

Abbreviations

The following abbreviations are used in this manuscript:

CCS	Carbon Capture and Sequestration
EGS	Enhanced Geothermal Systems
FAO	Food and Agriculture Organization of the United Nations
GEF	Global Environment Facility
HSA	Hot Sedimentary Aquifer
IAH	International Association of Hydrogeologists
IHP	International Hydrological Programme
IWRM	Integrated Water Resources Management
MAR	Managed Aquifer Recharge
NGO	Non-Governmental Organisation
UNESCO	United Nations Educational, Scientific and Cultural Organization

References

1. GEF Groundwater Governance Project. Global Diagnostic on Groundwater Governance. Groundwater Governance—A Global Framework for Action. 2015. Available online: http://www.groundwatergovernance.org (accessed on 1 February 2016).
2. Van der Gun, J.; Merla, A.; Jones, M.; Burke, J. Governance of the subsurface and groundwater frontier. In *Groundwater Governance, A Global Framework for Action*; Thematic Paper No. 10; Available online: http://www.groundwatergovernance.org/resources/thematic-papers/en/ (accessed on 1 February 2016).
3. Dickson, M.H.; Fanelli, M. What is Geothermal Energy? International Geothermal Association. February 2004. Available online: http://www.geothermal-energy.org/what_is_geothermal_energy.html#c317 (accessed on 9 May 2016).
4. Margat, J.; Van der Gun, J. *Groundwater around the World: A Geographic Synopsis*; CRC Press/Balkema: Leiden, The Netherlands, 2013.
5. Reichl, C.; Schatz, M.; Zsak, G. *World Mining Data*; International Organizing Committee for the World Mining Congress: Vienna, Austria, 2015.
6. Stuurman, R. *Schetsen van Het Nederlands Grondwatersysteem in 2050*; TNO-Report 2007-U-R0225/B; TNO: Utrecht, The Netherlands, 2007.
7. Eni Spa. *World Oil and Gas Review*; Eni Spa: Rome, Italy, 2015.
8. Hunt, T.M. *Five Lectures on Environmental Effects of Geothermal Utilization*; Report 1-2000; United Nations University—Geothermal Training Program: Reykjavik, Iceland, 2001; p. 109.

water

MDPI

Article

Water Governance Decentralisation and River Basin Management Reforms in Hierarchical Systems: Do They Work for Water Treatment Policy in Mexico's Tlaxcala Atoyac Sub-Basin?

Cesar Casiano Flores *, Vera Vikolainen † and Hans Bressers †

Department of Governance and Technology for Sustainability (CSTM), University of Twente, Enschede 7500AE, The Netherlands; v.vikolainen@utwente.nl (V.V.); j.t.a.bressers@utwente.nl (H.B.)

* Correspondence: c.a.casianoflores@utwente.nl; Tel.: +31-68-174-6250

† These authors contributed equally to this work.

Academic Editors: Sharon B. Megdal, Susanna Eden and Eylon Shamir

Received: 22 February 2016; Accepted: 11 May 2016; Published: 19 May 2016

Abstract: In the last decades, policy reforms, new instruments development, and economic resources investment have taken place in water sanitation in Mexico; however, the intended goals have not been accomplished. The percentage of treated wastewater as intended in the last two federal water plans has not been achieved. The creation of River Basin Commissions and the decentralisation process have also faced challenges. In the case of Tlaxcala, the River Basin Commission exists only on paper and the municipalities do not have the resources to fulfil the water treatment responsibilities transferred to them. This lack of results poses the question whether the context was sufficiently considered when the reforms were enacted. In this research, we will study the Tlaxcala Atoyac sub-basin, where water treatment policy reforms have taken place recently with a more context sensitive approach. We will apply the Governance Assessment Tool in order to find out whether the last reforms are indeed apt for the context. The Governance Assessment Tool includes four qualities, namely extent, coherence, flexibility, and intensity. The assessment allows deeper understanding of the governance context. Data collection involved semi-structured in-depth interviews with stakeholders. The research concludes that the observed combination of qualities creates a governance context that partially supports the implementation of the policy. This has helped to increase the percentage of wastewater treated, but the water quality goals set by the River Classification have not been achieved. With the last reforms, in this hierarchical context, decreasing the participation of municipal government levels has been shown to be instrumental for improving water treatment plants implementation policy, although many challenges remain to be addressed.

Keywords: Mexico; water governance; governance assessment; sanitation; water treatment plants policy; Tlaxcala sub-basin

---

1. Introduction

During the last decades of policy reforms, new instruments have been developed and substantial economic resources have been invested to improve water sanitation quality in Mexico. An important apparatus of institutions at river basin level has been established, such as river basin organisations, river basin councils and auxiliary bodies. "Since 1992, in addition to the 13 river basin organisations implementing the National Commission of Water (CONAGUA) policies in each hydrographic region, 26 river basin councils have been created as consultative bodies, working closely with 32 river basin commissions [...]" [1] (p. 107). However, the creation of River Basin Organisations has been described as a mere offices' decentralisation instead of a decentralisation of the policy. The case of Tlaxcala exemplifies this national situation. "Decision-making power is still centralised within the hands of CONAGUA and its regional/local offices" [1] (p. 17).

Despite the governmental efforts, pollution levels have increased in different parts of the country and only a few studies have addressed this [2]. Water pollution contributes to ecological damage and health issues of the population with economical consequences [3]. "Over half of the world's hospital beds are occupied with people suffering from illnesses linked with contaminated water and more people die as a result of polluted water than are killed by all forms of violence including wars" [4]. In this paper, we focus on Tlaxcala's Alto Atoyac sub-basin, where almost 80% of the state population lives and whose main river Zahuapan, a tributary of the Atoyac River, is considered the third most polluted in Mexico. The Alto Atoyac sub-basin has called national attention as well as media coverage, since the river pollution affects more than two million people in the states of Tlaxcala and Puebla [5], The NGO *Fray Julian Garces* from Tlaxcala, produced a documentary about the health impacts on the population that lives close to the river, and the media have also reported cases of leukaemia and respiratory diseases [6].

In order to solve water issues, the Mexican government has promoted a more integrated water management at the basin level. In the case of Tlaxcala, the federal government created the Atoyac-Zahuapan River Basin Commission on 26 November 2009. The Commission was expected to facilitate the implementation of river basin councils' strategy; inter-governmental co-ordination; and social participation at the sub-basin level [1] (p. 111). However, in the following six years the Commission never had a session and most of the state and municipal actors as well as NGOs were unaware of its existence. Governmental actors commonly perceive that stakeholders' participation increases complexity without positive results. The inoperability of the Atoyac-Zahuapan River Basin Commission shows that following the worldwide trend towards decentralisation in water management could not produce the expected results in a centralised policy making context. Water policy is still implemented through the national programs of CONAGUA. Hence it is necessary to consider the importance of the context and investigate whether the reform was apt for the situation, because central government could still play an important role [7] (p. 22).

Governance as a normative concept has been highly promoted by international organisations as an answer to solve water problems. Contextual factors are also mentioned as an important element to consider when policies are created and implemented. Although literature provides examples of how context matters, it generally presents "water governance" as the recipe for all water management problems. In this study, we give a more contextual answer and demonstrate the importance of context in a systematic manner. The central question we ask in this article is: how apt have the water treatment policy reforms been for the governance context in Tlaxcala's Alto Atoyac sub-basin? We assess the governance context of the Water Treatment Plant (WTP) policy in Tlaxcala, including planning, construction, operation, and monitoring. The assessment is conducted using the Governance Assessment Tool (GAT), which includes descriptive-analytical and semi-normative elements.

2. Theoretical Framework and Methodology

As Ostrom explains "[t]here is no panacea where a single type of governance system applies to all environmental problems" [8] (p. 15176). In this sense, one of the main concerns from the academic perspective is to propose frameworks that consider contextual factors [8–15]. "[W]e observe a gradual shift from the promotion of simplistic panaceas for water governance reform towards more context-sensitive approaches" [12] (p. 11), as well as the necessity of developing frameworks and methods that allow comparison across cases [10,12,16–19]. In addition to this, "[t]he field of water governance lacks both a systematic empirical base and a theoretical understanding of governance system" [12] (p. 20). Furthermore, "there is a lack of attention to implementation of agreements or regulations from a governance perspective" [15] (p. 736), while "methods to deal with the complexity of governance systems are missing in general" [12] (p. 17). The Tlaxcala case presented here contributes in different degrees to these academic concerns from an institutional perspective. We believe that our framework and method can clarify the complexity when analysing water governance implementation. The methodology used in our study is called the Governance Assessment Tool (GAT).

The GAT defines governance as:

"the combination of the relevant multiplicity of responsibilities and resources, instrumental strategies, goals, actor-networks and scales that forms a context that, to some degree, restricts and, to some degree, enables actions and interactions" [20] (p. 6).

The GAT is based on Contextual Interaction Theory (CIT) [21–24]. CIT focuses on the context in which people work as being pivotal to the outcome of their interactions and divides this context into descriptive-analytical and semi-normative categories. By semi-normative we imply that the normative content of these qualities is derived from and thus dependent on the importance and urgency of the implementation of policies and projects under assessment. The GAT analyses what governance conditions are especially supportive or restrictive for the effective realisation of such interventions.

The assessment of these categories allows deeper understanding of the governance context and how it impacts the policy implementation. The categories are separated in two sets: the five dimensions (multi-level, multi-actor, multi-faceted, multi-instrument and multi-resource based) and the four semi-normative qualities (coherence, extent, flexibility, and intensity) that are employed to analyse the governance interactions [14]. The GAT is made up of a 'matrix' model consisting of these five dimensions and four qualities [25]. See Table 1 below.

Table 1. Water governance matrix.

Governance Dimension	Extent	Coherence	Flexibility	Intensity
Levels & Scales	How many levels are involved and dealing with an issue? Are there any important gaps or missing levels?	Do these levels work together and do they trust each other between levels? To what degree is the mutual dependence recognised?	Is it possible to move up and down levels (upscaling and downscaling) given the issue at stake?	Is there a strong impact from a certain level towards behavioural change or management reform?
Actors & Networks	Are all relevant stakeholders involved? Who are excluded?	What is the strength of interactions between stakeholders? In what way are these interactions institutionalised in stable structures? Do the stakeholders have experience in working together? Do they trust and respect each other?	Is it possible that new actors are included or even that lead shifts from one actor to another when there are pragmatic reasons for this? Do the actors share in social capital allowing them to support each other's tasks?	Is there a strong impact from an actor or actor coalition towards behavioural change or management reform?
Problem Perspectives & Goal Ambitions	To what extent are the various problem perspectives taken into account?	To what extent do the various goals support each other, or are they in competition or conflict?	Are there opportunities to re-assess goals?	How different are the goal ambitions from the *status quo*?
Strategies & Instruments	What types of instruments are included in the policy strategy and are implemented; and which are excluded?	To what extent is the resulting incentive system based on synergy? Are there any overlaps or conflicts of incentives created by the included policy instruments?	Are there opportunities to combine or make use of different types of instruments? Is there a choice?	What is the implied behavioural deviation from current practice and how strongly do the instruments require and enforce this?
Responsibilities & Resources	Are responsibilities clearly assigned and sufficiently facilitated with resources?	To what extent do the assigned responsibilities create competence struggles or cooperation within or across institutions?	To what extent is it possible to pool the assigned responsibilities and resources as long as accountability and transparency are not compromised?	Is the amount of allocated resources sufficient to implement the measures needed for the intended change?

By analysing the five dimensions of governance according to the four qualities of the governance regime, one can attain a very pragmatic understanding of how different elements of governance interact and hence influence a particular implementation setting. Extent refers to the completeness of the regime in terms of relevant aspects, such as actors or instruments. Coherence relates to how the various elements of the regime strengthen or weaken each other. Flexibility refers to the existence of different roads according to the opportunities or threats that arise during the implementation. And finally, intensity is "the degree to which the regime elements urge changes in the *status quo* or in current developments" [23] (p. 93).

As demonstrated in previous research, the GAT framework can be applied when there is a multi-level setting with interdependency among the actors. This means that different levels should act as semi-autonomous units and power must be diversified [26]. This interdependence must at least be classified as a "legislatively initiated coordination" [26] even if it is not fully implemented. The GAT has shown important strengths in the analysis of water projects implementation in the Netherlands [23], Canada [24], in North Western Europe [27], Romania [28], and Mexico [29]. The methodology has proven to be useful for comparison across governance structures. This is important, since this study is part of a broader comparative project regarding different institutional arrangements at the sub-basin level. The Governance Assessment Tool (GAT) applied in this research is related to the UN's (United Nations) fifth methodology [30] and it is part of the 25 assessment tools compiled by OECD (Organisation for Economic Co-operation and Development) [31] (p. 31).

Data collection consisted of 21 semi-structured in-depth interviews that took place between May 2014 and July 2015. The questions in Table 1 were used as an interview guideline. The interviewees included representatives of three government levels, the private sector and non-governmental organisations. In most of the organisations, the directors and heads of the departments were interviewed. Table 2 shows the affiliations of the interviewed actors.

Table 2. Actors interviewed per stakeholder category in the Tlaxcala Alto Atoyac sub-basin.

Water Utility (WU)	State Level	Federal Level	Industry Sector	Organisations of Civil Society
CAEM Tlaxcala (WU of Tlaxcala municipality)	Coordinacion General de Ecologia	CONAGUA Balsas Organization	Textile Industry	Colegio de Tlaxcala
CAEM Chiautempan (WU of Chiautempan municipality)	SECODUVIT (Ministry of Infrastructure)	CONAGUA Delegation in Tlaxcala		Fray Julian Garcés NGO
CAEM Huamantla (WU of Huamantla municipality)	CSITARET (Centre for Integral Water Treatment)			PRONATURA NGO
CAEM Apizaco (WU of Apizaco municipality)	CEAT (State Water Commission)			
CAEM Tlaxco (WU of Tlaxco municipality)				

The document for the creation of the Atoyac-Zahuapan Commission was the starting point to select the actors to be interviewed at the sub-basin level. The fact that the last federal administration focused its water sanitation policy on building Water Treatment Plants (WTPs) in the municipalities of San Pablo del Monte, Chiautempan, Huamantla, Contla de Juan Cuamatzi, and the optimisation of Tlaxcala, Tlaxco, Apizaco and Ixtacuixtla de Mariano Matamoros WTPs [32] (p. 99) was also considered in sampling the interviewees.

The quality of the water governance regime is assessed inductively based on the interviewees' answers. To support this primary research, official international and national documents, and electronic newspapers were also reviewed. Each response was first assessed individually and then compared with the rest of the actors in order to reach an assessment per cell. The result was then contrasted with the other four categories. When the quality was ranked as moderate or high in most of the categories, the quality was graded as supportive. Otherwise it was ranked as restrictive.

3. Location of the Study

The Alto Atoyac sub-basin covers 69 municipalities in the states of Puebla and Tlaxcala. Its main rivers are the Zahuapan, Atoyac, and Alseseca. The sub-basin belongs to the Balsas River Basin formed by 12 sub-basins in eight states [32]. The state of Tlaxcala has a population of 1,169,936 inhabitants according to the last census in 2010. It is the smallest federal entity in size just after Mexico city. It represents 0.2% of the Mexican territory and generates 0.56% of the National Gross Domestic Product [33]. The municipalities with the largest populations are Tlaxcala (89,795), Huamantla (84,979), and Apizaco (76,492) [34]. The municipalities of Apizaco, Tlaxcala, and Chignahuapan alone concentrate 30% of the economic activities, 35% of employees, and 54% of the economic production of the state [33]. The main economic activities of the population are: 53% services, 32% industry, and 15% agriculture. The manufacturing industry employs 34% of the workforce [33]. The part of the Atoyac-Zahuapan sub-basin that belongs to Tlaxcala hosts 79.5% of the state inhabitants [35] (p. 20). Among the most important municipalities are the Tlaxcala municipality, Villa Vicente Guerrero, Apizaco, and Chiautempan [35] (p. 18). The sub-basin covers a territory of 1485 km^2 [36]. The Zahuapan River is the most important and crosses 25 of the 60 municipalities that comprise the state, while other 20 municipalities discharge into it indirectly [37] (pp. 18–19). Figure 1 shows the location of the study.

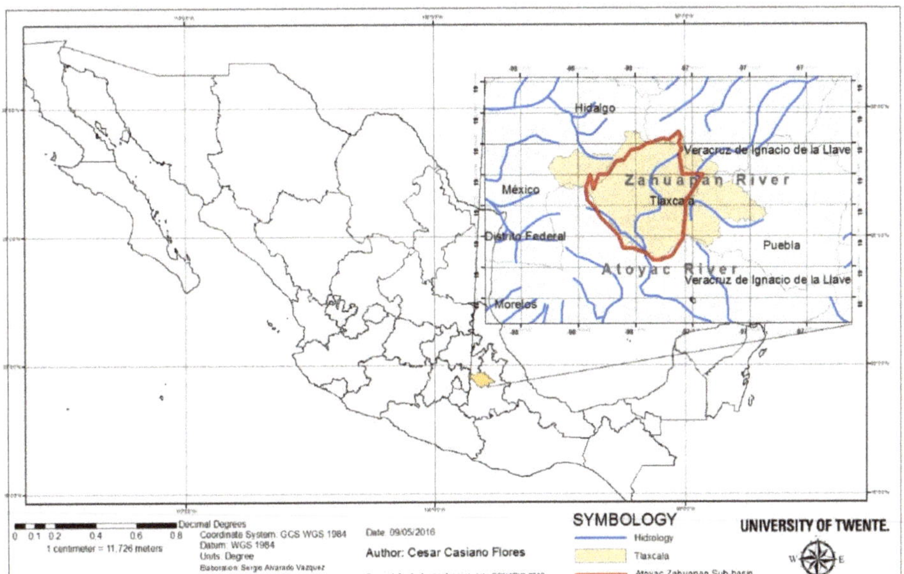

Figure 1. Location of the Atoyac-Zahuapan sub-basin in the state of Tlaxcala. Source: Authors' elaboration.

4. Background of the Tlaxcala Case Study

The federal government is the actor with the most decision-making power in the Mexican water sanitation policy and the institution in charge is CONAGUA. The CONAGUA is organised in three levels: Central Headquarters, Basin Organisations and Local Delegations [38] (p. 12). Between 1999 and 2010 the budget from the federal government for water programs increased from 17.64 to 360.2 USD millions [39] (p. 6). In 2016 the CONAGUA budget is more than 2000 USD millions for direct investments [40] and the water sanitation capacity in the country in 2015 was estimated to be 55% [41].

Although the federal government transferred its responsibilities to lower governmental levels decades ago, CONAGUA remains the main water management actor. The delegation of water related services to the municipal level started in the 1980s [42] (p. 33). In 1983, the reform established water related services as an exclusive function of the municipalities and the state government as transitional entities. In most of the cases the state government transferred the new obligation to the municipalities immediately [43] (p. 84). However, the 115 Article of the Mexican Constitution also mentioned that municipalities could coordinate and associate their efforts with other municipalities or the state level to provide the water services [44] (p. 17). The policy of delegating water related services to water utilities (WUs) at the municipal level has been criticised during the last years. The municipal government has two characteristics: it is the closest to the population, but has a short-term government of three years [42] (p. 37).

In the state of Tlaxcala the interest in water sanitation started in 1985 with the creation of [35] (p. 16):

- The Company for Control of Polluted Water in the State of Tlaxcala (ECCAET)
- The State Drinking Water and Sewage Commission (CAPAET)

In 1994, the Coordination of Ecology (CGE) office took over the functions from the ECCAET, and the CAPAET's responsibilities were conferred to the municipalities [35] (p. 16). In 2009, the State Water Law created two organisations whose operations started in 2011 with the newly elected government:

- The State Water Commission (CEAT): the maximum organisation for water related matters with technical, normative and consultation capacity;
- The Centre for Integral Water Treatment (CSITARET), in charge of the water sanitation policy, with the operation and the monitoring of water discharges quality of some WTPs' among its functions.

The sanitation of the Alto Atoyac basin has been among the main projects since the last federal government 2006–2012 [45] (p. 64). Between 2007 and 2011, CONAGUA with the support of the state and municipal level conducted an important number of studies and WTPs were built as well as collectors. In the last decade, NGOs and the media from the states of Puebla and Tlaxcala pressured the government by stating the high levels of pollution and its impacts. Also the green political party in the national Congress and the Senate pointed out this situation [46]. The results of the Alto Atoyac basin sanitation have been mixed. The state of Tlaxcala reports more improvements than the state of Puebla. The last study found that the main polluters in the Alto Atoyac sub-basin are the textile industry, slaughterhouses, the food and construction industries, and the municipal discharges [47]. Only 400 out of 3675 industries have water discharge permits [48], and the pollution levels are eight times higher than the national norm [49].

In the state of Tlaxcala, the amount of discharges treated between 2010 and 2013 remained around 40%–43%. In September 2014, the CSITARET implemented a program to increase water treatment capacity by rehabilitating 15 WTPs. These plants do not require energy to work. The state expects to treat 63% of water discharges with this investment [50]. However, industrial discharges are seen as one of the most important problems. The industrial park Quetzalcoatl has been highly criticised and linked to health problems by different NGOs.

In 2011, only two out of 128 water treatment plants were operating and in May 2014, the state government announced that 61 WTPs were working, representing 61.08% of the water treatment capacity. The state government directly operates nine plants, which are located in the municipalities of Apetatitlan, Apizaco, Tetla de la Solidaridad, Huamantla, Tlaxco, Ixtacuixtla, Tepeyanco, and Tlaxcala. This is shown in Figure 2 below.

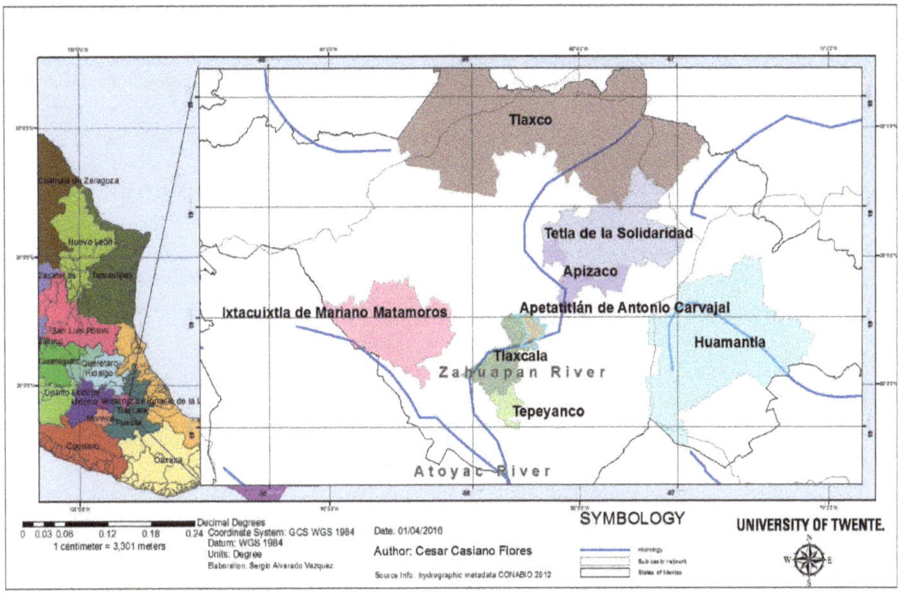

Figure 2. Location of the municipalities where WTPs are operated by the state level. Source: Authors' elaboration.

Three main legal reforms also took place during this period: the River Classification (2012), the sanctions to water polluters in the State Penal Code (2013), and changes to the Federal Fund for Strengthening the Municipalities (2014). River Classification is very important, because it is the legal instrument that sets water quality discharge limits, goals and stages for the basin [51].

Along with the governmental efforts mentioned, civil society has been demanding an improvement in the water treatment quality. One of the most important situations occurred in 2006, when NGOs sued the government for health damages caused by the Atoyac River pollution in the *Tribunal Latinoamericano del Agua*. The tribunal exhorted the governments from Tlaxcala and Puebla to solve the pollution problem [52]. During that year, the state government signed an agreement with the *Instituto Politécnico Nacional* (IPN) to diagnose the situation. However, the study did not include industrial discharges nor heavy metal discharges from the municipal water utilities and was not open to the citizens. This brought another legal process from the NGOs to ask for transparency. The NGO *Centro Fray Julián Garcés Derechos Humanos y Desarrollo Local A.C* has continued very actively in its demands for improvement of the water quality. In the light of its wide coverage in the national media in 2014, communities in the Atoyac-Zahuapan sub-basin required the intervention of the Congress for health issues. The Congress demanded the intervention of the Ministry of the Environment and Natural Resources, CONAGUA and the Attorney General's Office for Environmental Protection to monitor that wastewater discharges met the norm in the sub-basin. Some months later, CONAGUA answered this demand by providing information about the actions they took to monitor wastewater discharges in the last year as well as the actions they were developing.

5. Governance Assessment of the Context

The governance assessment results are presented in the following sub-sections.

5.1. *Levels & Scales—The Three Governmental Levels Are Involved; However, the Federal and State Government Are the Main Drivers of the Policy*

While three levels of government (municipal, state, and federal) are involved in water sanitation policy, the state and federal play the main role. Their relation is institutionalised by the Rules of Operation for Water Programs from CONAGUA and the regionalisation policy executed by the state government. The first aspect, the Rules of Operation are applied in the Wastewater Treatment Program (PROTAR) launched in 2009, the Urban Potable Water and Sewerage Program (APAZU) from 1990 and the Rural Waterworks Development Program (PROSSAPYS) from 1996. The Rules include the creation of the Commission of Regulation and Follow up (CORESE). This Commission is established between the federal government through CONAGUA and the state government, excluding the municipal level. The two actors have monthly meetings, where the state level can present proposals to CONAGUA about programs and actions. The second aspect is the regionalisation policy for water sanitation. The Tlaxcala Water Law allows this process through agreements between the state government and the WUs at the municipal level. Currently the state government is in charge of nine WTPs. The state and federal government are studying the possibility of strengthening this policy. Most of the WUs agree with the policy.

The state and the federal government collaborate in the planning and the construction of federal projects, while the operation is the responsibility of the municipal level. The state government has also developed its own sanitation programs in collaboration with the WU at the municipal level. One example is the construction of the WTP for the slaughterhouse in the municipality of Apizaco between 2012 and 2013, its operation being the responsibility of the WU.

There is a certain trust among the levels, but lack of trust is perceived within each level. For example at the municipal level there is a lack of information sharing between the infrastructure department and the WU. The Basin Council and CONAGUA's Delegation are not working as close as in 2013 due to political changes after elections and because the Council mentioned that other sub-basins had priority. However, the Council experiences commitment from the state government. The mutual dependence is recognised by the three governmental levels since each one has specific responsibilities and resources. For example there is a necessity to co-invest, to build and operate the required infrastructure for the sanitation of the sub-basin. On many occasions the WU mentioned the importance of the federal government resources.

There is no possibility to move responsibilities up and down levels in a pragmatic manner given the issue at stake. The state and federal law as well as the Rules of Operation are clear about the role of each level. The order is respected from top to down: federal, state, and municipal. However, the agreements signed between the municipal and the state level, have permitted to consolidate the state level as the implementer of the water sanitation policy.

The strongest impulse towards behavioural change comes from the federal level followed by the state level. CONAGUA has the most important budget and main programs; it is also the actor that participates the most and prepares the national water legislation. The actors interviewed experience that the state and federal level work to strengthen the sanitation and regionalisation policy and to rehabilitate different plants already constructed.

5.2. *Actors & Networks—The Network to Implement the Policy Is Composed of Governmental Actors Who Do Not Allow the Participation of Social Actors*

Not all the interviewed stakeholders are included in the policy implementation. Exclusion of the municipality is common in planning and construction. The social actors that are mentioned in the document establishing the Zahuapan Commission do not participate actively in the water policy. Some very active organisations, such as NGO *Fray Julian Garces*, were even excluded. The industry sector as well as civil society is not involved in the WTP policy. NGOs are not commonly taken into consideration, neither is academia and when they are, it is because the political actors feel comfortable with that specific group. One state actor said, "society is merely political groups who complicate the decision-making process". From the government perspective, society is not interested and the industry is limited to meeting the requirements. Civil society participates by pushing the government.

The main strength of interaction is found in the CONAGUA's programs, they are institutionalised through the Rules of Operation, the most important action being the CORESE monthly meetings between the federal and state government. The state government has seen benefits from the CORESE: they have made proposals there, which have received positive answers from the federal government. The Zahuapan River Basin Commission should be the auxiliary organisation where different stakeholders meet and make proposals for basin sanitation policy. However, the Commission has never had any session.

As in other parts of the country, government elections affect continuity of the policies, their implementation and the trust in the network. Many interviewees at the federal level mention the lack of interest from the municipal government to operate the plants. Nationally, "[m]unicipalities change government every three years, and water service providers change General Directors every 18 months on average" [1] (p. 29). The federal government has been more stable, many of the directors and heads of the departments have continued after the 2012 election at the federal government. But the high profile actors such as the federal delegate and the Balsas Basin director were changed. There is no tradition of different stakeholders working together. Since the policy of regionalisation, cooperation between the federal and state level has improved and they have been working closer.

The relation between the federal and the state government with the NGOs is generally characterised by a lack of trust. Besides the legal demand in 2006, in 2008 the NGO *Fray Julian Garces* started a legal process to oblige the government to publish the 2008 results about water pollution in Tlaxcala. After many allegations the government accepted this, but the results were highly criticised since they were general and did not include the industrial sector. The NGO has also produced a documentary about the health impacts of the river pollution. There is a higher degree of trust from the government to the industry, but the feeling is not mutual. One of the state level actors said "we have been explaining the situation to the industry sector and their response has been positive". The industry sector, on the other hand, distrusts the government mainly for their discretional law implementation and lack of support for the industry.

The policy process takes place only among government actors, without including new actors. It is not a common practice that leadership shifts from one actor to another. The main impact seems to come from the coalition of the state and federal government. The social capital sharing is limited to the governmental actors. The impulse towards change comes from two actors, the government and the NGOs, but there is no coalition among them. The NGO has only managed to set the importance of water sanitation on the agenda.

5.3. *Problem Perspectives and Goal Ambitions—The Main Perspectives Considered for the Policy Implementation Are Those of the Federal and State Government, and the Achievements Are Far from the Goals Set*

The main perspectives considered are of the federal and state level. Participation from industry, society and the municipal level is not really promoted and their perspectives are not considered. This has had implications in the past with the failure of WTPs' construction, when for example the state and federal government built a plant for a municipality with Israeli technology and the municipality did not have the resources to operate it. The building of the water treatment plants has just started, considering the lack of resources from the municipalities to operate them. The problem perspectives of NGOs are not considered by the government.

The system was created to support the governments' goals. For example, one of the WUs did not have debts with CONAGUA and this helped them to invest in the municipality via payback from CONAGUA. However, most of the WUs have debts with CONAGUA due to lack of payments for water extraction and wastewater discharges; therefore, they cannot access these benefits.

There is a competition between the goals of economy and the environment. On some occasions the government does not enforce the law because industry threatens to close factories, provoking unemployment. One of the government actors said that some industries have been permitted to operate without water treatment for 10 years. The NGOs comment that the Mexican government has been easy with companies in order to promote jobs. Except for the operation of some WUs, the perspective and goal ambitions of the state and federation are more aligned than before. Both have a similar time perspective and a more balanced relation. As for the industry sector, each actor has its own perspective due to its particular interest and vision of the problem. NGOs perceive lack of openness from the government to their critique of economic and political interests.

It is not a common practise to re-assess the goals. Opportunities for this are not really there, and goals are set by the central offices. In the case of water sanitation, the government considers the reports that show different advances related to the WTPs, such as the number of plants built, rehabilitated and operating, to be an assessment. The 2014–2018 National Water Plan is the first that includes a biannual policy assessment [53] (p. 57).

The goals and standards of the discharges established by the River Classification are still far from being achieved compared with the current situation, although the governmental levels report some improvements. Now the federal government is concerned because they know that the standards they set are not going to be achieved. The industry sector is more inclined to meet the requirements, but the main challenge is with the municipalities. One of the interviewees even said "the municipalities do not do anything". Only 20% of the municipalities in the state have discharge permits. The co-investment programs have allowed the WUs to create sanitation infrastructure. The government recognises that problems such as the inoperability of the WTPs due to inadequate technology, legal problems or lack of capacity from the municipalities are still present.

5.4. Strategies & Instruments—New Instruments at the State Level and the Strategy of Regionalisation Have Supported the Policy Implementation. However, Important Reforms Are Still Required

The main documents establishing water policy with a long-term vision, including sanitation, were the 2030 Water Agenda created in 2011 and now the 2014–2018 National Water Plan. The most important legislative framework for water sanitation consists of: chapter three of the General Law on Environmental Protection, the National Water Law, the State Water Law, the River Classification and the norms. The River Classification is the latest legal instrument that sets water quality discharge limits, goals, and stages for the Atoyac river [51]. NOM-001-SEMARNAT-1996 (establishes the pollution levels for wastewater discharges in national waters), NOM-002-SEMARNAT-1996 (establishes the pollution levels for wastewater discharges in municipal infrastructure) and NOM-003-SEMARNAT-1997 (establishes the pollution levels for treated water and its reuse for public services). Currently, some state reforms to automatically discount the costs of the water treatment plants operation have been applied. This is the case with the Federal Fund for strengthening the municipalities (FORTAMUN), as published in the Official State Gazette on 31 January 2014 and the 2009 reform of article 523 of the Financial Code for the State of Tlaxcala and its Municipalities.

The main programs come from CONAGUA's implementation of the National Water Plan. The master document at the state level is the 2011–2016 State Development Plan. It enables the CEAT with economic resources to create the Program for the State Water Infrastructure [54] (p. 2). Different state actors interviewed referred to the commitment from the governor and the importance of the topic in the state plan. Currently, the state and federal government monitor the industrial sector. However, state level monitoring has no legal consequences.

A new diagnosis of the regionalisation policy is being developed. This study is between the state and federal levels. The monitoring network and improved laboratories are also being built. One of the WUs that is not in charge of WTPs has decided to support the policy with education campaigns in schools to make people aware of the pollution problems. However, the WU mentioned that more efforts are required to make people aware of the importance of water service payment, because "society is not used to paying for water".

Among the non-implemented actions is the Atoyac-Zahuapan Commission. The industry sector says that educational programs, technical support or even co-investment policies are missing. Officials at the state level consider that real time monitoring in the river could help them to have more accurate information on the water quality and to find the sources of pollution easier.

Since people have complained that the WTPs are not operating, the government is inviting schools to visit the plants. The state government also keeps its website updated with information about state government actions in favour of water treatment.

The system is based on incentives and it is supposed to create synergy, therefore co-investment for projects is an essential part of the programs. The programs establish clear limits to participation, reducing it in most cases to those with economic contributions. Different actors comment that politicisation and lack of will from various levels of government complicate correct implementation, especially by the municipalities, in part because the state government is an intermediary between the municipalities and the federal programs. Politicisation of the WUs contributes to the lack of enforcement of payments. WUs try to avoid political costs from obliging the citizens to pay. There are no incentives to promote the joint participation between the government and the NGOs. The CORESE meetings do not facilitate this social participation. The government and NGOs do not work together, diminishing the possibilities of synergy created by their cooperation, and the programs working separately lack integral vision.

There are no opportunities to combine or make use of the different types of instruments. Permits can only be adapted to new regulations after five years. Each instrument has its own rules and works separately. The only opportunity to bring the strategies of different stakeholders together is the Atoyac-Zahuapan Commission. It might help to improve coordination, bring efforts together, and provide an integral vision. At the moment, cooperation is the result of obeying the upper levels.

Considering the current practice, important behavioural changes are still required, including stronger enforcement. "For example, wastewater discharge-related sanctions established by the 1981 Federal Duties Law are rarely enforced, and inspections to ensure their compliance are conducted infrequently [...]" [1] (p. 57).

No measures have been taken to enforce WTPs operation by the WUs. Very few municipalities have discharge permits or operate their WTPs and the federal government finds enforcement complicated. One interviewee said "when an industry is not meeting the norm we can go and seal the company to oblige them, but in the case of the municipalities we cannot do that, reforms need to take place so we can have a way to really force the municipalities to comply with the norm". Enforcement by municipalities in the industry sector is also lacking. NGOs report lack of enforcement in the textile sector and the Industrial Park Quetzalcoatl is among the most criticised.

The industry sector experiences mostly "uneven enforcement". They commented on a case where one company was closed down and apparently due to political contacts it was opened immediately, and also noted that municipalities were not being punished. However, they recognise that there is more pressure from current state and federal administration.

There are also gaps in the current legislation. As mentioned before, the most important is the lack of secondary legislation of the National Water Law, making it difficult to implement [1] (p. 41). The national norms as well as the River Classification need to be updated. NGOs have claimed that there is no regulation on the impact of the mix of various wastes from different industries. They have demanded the creation of a law that stipulates norms depending on the kind of industry and they insist on an integral policy. The Latin American Tribunal of Water notes the lack of law enforcement from all three levels of government.

Currently, the state-owned water quality laboratory monitors the nine water treatments plants operated by the state government as well as 60 municipal water treatment plants. However, the laboratory is not certified, so its results do not have any legal authority. There is still lack of interest from the municipalities. The state government complains that they rehabilitate the water treatment plants and the municipalities do not operate them, and they do not have the capacity to sanction them. They are now signing agreements with municipalities to oblige them to operate the plants. According to the CSITARET, all the WTPs employ biological processes, such as: activated sludge, aerated lagoon, anaerobic digester, facultative lagoon, percolator filter, Imhoff tank or artificial wetlands.

5.5. Responsibilities & Resources—Responsibilities Are Clearly Assigned, However a Considerable Amount of Economic Resources Is Still Required for Almost All the Phases of Policy Implementation

According to the interviewed actors, responsibilities are clearly assigned by the law. However, important differences are found when it comes to economic resources. While the federal and the state government seem to have resources to continue implementing the water sanitation policy, the municipal government is struggling. For this reason, co-investment in water sanitation programs has been implemented. One study showed that Mexico's water utilities income represents 75% of their expenses [44] (p. 20). In this regard, some of the interviewed WUs mentioned that the income goes to daily payments and there are no resources left to invest.

The state government has been able to provide economic support to the municipal level; however, more resources are required. The policy of regionalisation has helped to improve this situation, but more resources are needed for rehabilitation programs, to build more laboratories and to maintain the monitoring network. In contrast, the implementation of the Zahuapan Commission has not been facilitated with resources. Cases of corruption have also been found in the application of the programs. A report by the Superior Auditor of Mexico found that there are "limitations in CONAGUA's capacity to document how states manage their resources. It also signalled irregularities in the use of federal programmes' funds such as [...] APAZU" [1] (p. 161).

The three government levels require more staff to monitor the different levels and the industry sector. The government levels that operate WTPs also require more staff for the operation of them.

With the reforms at the state level, more resources for water treatment plants operation have been secured. Also since some plants fulfil the norm, they are receiving economic support from the federal level for their operation. At the same time, money is being invested in the rehabilitation of water treatment plants, as in the case of the plants in Apizaco and Tlaxcala city.

Lack of cooperation is more common at the same government level, than across different levels. At the state level, lack of coordination between Water, Infrastructure and Rural Development was mentioned, however the situation has been improving. The respondents describe the relationship among the Ministries of Finance, Infrastructure, and Water (CEAT) as close. For example, the Ministry of Finance retains the money of the municipalities and make it available for the CEAT, so they can finance the plants operation. Within each government entity a high degree of communication was reported.

One of the most important incoherencies related to the resources is that water tariffs are not established by the WUs according to their necessities, but by the State Congress [44] (p. 21). This leads to politicisation of the costs. Complains about the lack of fulfilment of the municipal responsibilities was common. CONAGUA civil servants commented that only 12 out of 60 are working on their water discharges.

Resources cannot be pooled outside CONAGUA or state programs and they are only available to government actors. The industry and the NGOs work on their own and there is no collaboration among them. Actually, the industry sector perceives NGOs as only criticising without an interest in collaboration.

There is no flexibility in the application of the resources and programs created by the federal government. It is expected that they will be implemented as programmed without changes. Once the program is approved, there is no possibility to make changes until the following year. The state government only has one year to plan and execute programs, so bad planning and poor execution is common. Some projects require 6–8 months to be planned and if so, the government only has the next few months to implement them.

The inflexibility in the programs has caused implementation problems and has not contributed to promote accountability. Institutional arrangements also limit accountability. For example, according to the Federal Audit Law, if the Superior Auditor of Mexico finds corruption in a state tender process, he cannot act immediately: the revision process is annual and the Auditor can act just after its completion. Then, he can demand actions from the State Comptroller, appointed by the state governor and therefore part of the same government cabinet. As a result, it is very unlikely that the Comptroller will apply sanctions to the same state government. In March 2013, the Comptroller started an administrative process against state government employees related with PROTAR for the mismanagement of four million USD in 2012 [55]. However, no important actions have been undertaken to address this situation.

Even when there are economic resources, they have not been sufficient to achieve the intended change. Nevertheless, the state government reports progress and the state and federal government feel that there has been sufficient support. Lately, the co-investment budget presented by the federation has been matched by the state government for the water treatment plants construction. CONAGUA "is the biggest spender in the water sector" [1] (p. 27).

In 2013, with the continuous operation of the nine WTPs, 42% of total discharges in the state are treated meeting the norm NOM-001-SEMARNAT-1996 [54] (p. 9). To achieve this percentage 1.5 million USD were required by the state government, the CONAGUA, the industry sector and the 14 municipalities. Eighty five percent of the total resources was self-generated [54] (p. 10). Monitoring is one of the main challenges, but some efforts are being made. For example, to monitor the accomplishment of the norm NOM-001-SEMARNAT-1996, the state government did 1700 laboratory analyses last year [54] (p. 11). However, there are not enough resources for the inspections to make the industry comply with the norm. When the federal government has tried to enforce the law, it has reported that some industries play around the rules. For example, in some cases textile companies ask their employees to take fabric to their home to be tinted there and the employees accept because they are afraid they will be fired. This makes monitoring more complicated. WUs lack resources to operate the water treatment plants and to monitor the industry sector.

6. Summary of Findings

After a systematic analysis of each governance dimension and quality, we assess extent and flexibility as restrictive and coherence and intensity as moderate supportive. In general terms, these qualities create a context that still poses important restrictions, although improvements have been made with the recent reforms at the state level. In the section below we explain this situation in more detail.

Extent was assessed as low, thus restrictive. Extent has been reduced by the regionalisation policy. The state government participates more actively now in the policy implementation through the plants operation and monitoring. There is also a more balanced relationship among the two governmental actors. But the policy is highly governmental-actor centric, there is no social or industry participation and municipal involvement is partial. Various stakeholders' perspectives are not considered. There is a lack of important instruments such as secondary legislation. While both the state reforms and programs have improved the implementation, there is an important lack of resources for monitoring.

Coherence was assessed as moderate, therefore supportive. There is moderate coherence among both relationships: the state and federal level, and the state and municipal government. The regionalisation policy is increasing the coherence between the state and the federal level. The CORESE has had a positive impact, although the Atoyac-Zahuapan Commission has never functioned. There is trust among governmental actors, but they are not trusted by the society. By decreasing the participation of the municipality through the regionalisation policy, the coherence among the state and federal actors' perspectives has increased. However, social actors' perspectives are not considered. Coherence of instruments has also increased with the federal and state reforms, although they still work separately. Now that the state has played a more active role in providing more legal and economic resources, coherence with tasks has increased in this regard. However, municipalities are still facing important challenges in operating WTPs.

Flexibility was assessed as low, thus restrictive. There is no degree of adaptation, the rules must be followed as established. The government actors' network is not inclusive and does not create social capital. It is not possible to recombine goals during implementation; until a new administration starts. It is not possible to combine different instruments, each one has its own rules that must be followed. Pooling resources is only possible through the programs' Rules of Operation, and there is no flexibility to do it outside established programs.

Intensity was assessed as low to moderate supportive (improving). Intensity towards the policy implementation is driven by the federal and state government. The industry sector and NGOs do not work together. Goals are still far from being achieved. Many changes still need to take place and laws require enforcement. For example, (1) increasing the number of discharge permits for municipalities; (2) allowing the municipalities to establish their own tariffs; (3) if municipalities have to operate the water treatment plants they should participate from the start of the planning process; (4) more resources should be allocated for monitoring. Table 3 below summarises these findings per governance dimension and quality.

Table 3. Assessment results.

	Qualities of the Governance Regime			
Governance Dimension	**Extent**	**Coherence**	**Flexibility**	**Intensity**
Levels & Scales	Moderate	Moderate	Low	Moderate
Actors & Networks	Low	Low	Low	Low
Problem Perspectives & Goal Ambitions	Low	Moderate	Low	Low
Strategies & Instruments	Moderate	Moderate	Low	Moderate
Responsibilities & Resources	Low	Moderate	Low	Moderate
Assessed as	**Low/Restrictive**	**Moderate supportive**	**Low/Restrictive**	**Low to Moderate**

Source: Authors' elaboration.

7. Discussion

In this paper we asked the following question: how apt have the water treatment policy reforms been for the governance context in Tlaxcala's Alto Atoyac sub-basin?

We found that the reforms to promote a more integrated water management and decentralisation in Tlaxcala have failed to provide sufficient results. Examples of cases where bottom-up processes took place are found in common-pool resources literature. Fewer cases have analysed these in a context-sensitive perspective. When examples of water governance improvement are given by international organisations (OECD, UN, CEPAL, World Bank), they correspond to a normative meaning of water governance with "successful" examples of transition from centralised to decentralised systems. At the end such organisations "[...] are increasingly playing a role in the spread of ideas, programs and institutions around the globe" [56] (p. 11). However, decentralisation processes present many challenges and, as research shows, are not a panacea. Our case study provides an example of a more context-sensitive analysis. As we can see from Tlaxcala's case, state reforms have decreased the participation of the municipal government and the Atoyac-Zahuapan Commission has never had a session. This sounds negative, but in this hierarchical context with low capacity at the municipal level, these changes have actually increased coherence in the implementation and have allowed a more balanced relationship among the governmental actors.

State and federal governments have the main economic resources and are more stable since they stay in office for six years. Some Mexican states have already decided to manage water at the state level instead of strengthening the decentralisation process, for example in Baja California, Nuevo León, Tabasco, Oaxaca, and Querétaro [44] (p. 17). While only few water utilities among hundreds operate at the state level, five of them are in the top-ten performance [44] (p. 33). However, low stakeholder participation limits the ability to increase resources and makes policies less effective [57] (p. 162). Limited participation lacks the stock of social capital "that is created when a group or organisation develops the ability to work together for mutual productive gain" [58] (p. 302). If more instruments and actors are included, the discretionary decisions can be decreased [59,60]. The Zahuapan River Basin Commission was created, but never enforced. River basin and integrated water resources management are seen as the natural approach for water management [61] (p. 12). However, they have been difficult to implement even in countries that promote them [62] (p. 2). There is an important gap between promise and practice [62] (p. 1). River basin institutions have not been a magic bullet [62] (p. 13). For example, due to the institutional arrangement the Presa Guadalupe Basin Commission in the Estado de Mexico "can only encourage others to take measures and their own capacity to act is very limited" [63] (p. 197). The decentralisation policy strengthens the state level capacities. However, the municipal level is still highly dependent on the upper levels, in economic and politic terms, despite the decentralisation efforts [64]. Federal resources are "the main sources of financing for public water investments" [65]. There has been no improvement in the outcomes of the WUs since the decentralisation process started in the 1980s [66]; in the end, the water treatment plant policy has been implemented through the CONAGUA's programs in a multi-level fashion. Monitoring is still one of the most important challenges at all governmental levels and this creates a serious threat for implementation since "regulations and agreements that cannot be enforced will suffer from a lack of credibility and, in the end, legitimacy" [15] (p. 737).

8. Conclusions

We can conclude that the reforms toward decentralisation and integrated water management were less apt for the Tlaxcala sub-basin than the contextual reforms implemented by the state government. This is because the municipality lacks the capacity for policy implementation and because the hierarchical context as well as the institutional arrangement complicates integrated water management implementation. To answer the research question, we applied the Governance Assessment Tool, and the results were as follows: extent and flexibility were assessed as restrictive, while coherence and intensity were found to be moderate supportive for WTP policy implementation. Based on previous research about the qualities of the governance context, we can determine that implementation in the Tlaxcala context is "[...] very strict as it tries to force change through pre-dictating the processes that are followed to achieve the desired results of the upper governmental levels" [24] (p. 57). This situation is supported by the lack of flexibility in the hierarchical context. Therefore, in this context decreasing extent through the regionalisation policy, supports coherence. Increasing coherence has been supported by the most recent reforms at the state level. In this hierarchical context, it actually helps that decisions are taken by a limited number of actors, an outcome that is counter-intuitive when dealing with decentralisation efforts.

The state and federal actors are more stable and have more resources. Municipalities are not equipped and decentralisation reforms of decision making have not had the intended results. The municipalities are highly dependent on the upper levels for resources. In this particular context, state government actors are willing to be involved, but lack legal obligations, so long-term implementation depends on state policy-makers' will. Nonetheless, it delivers better results than leaving the responsibility to the municipalities with their low priority and skills for operating the WTPs. The dependence on the state level creates a challenge, because this level does not have the main legal responsibility and new administrations take over every six years, often with contrasting policy priorities. Without the willingness of the state level, the upper level expectations are more prone to be dashed.

The contextual limitations of social actors' participation decreases trust, accountability and long-term implementation. For example, none of the water utility directors knew about the 2030 Water Agenda. In order to improve this situation, higher levels of transparency are still required. Therefore, legal reforms that give the state level more responsibilities and capacities and strengthen transparency, might be a step towards a more effective implementation.

Acknowledgments: We express our gratitude to CONACYT and CONCYTEP for funding this research and to Larry O'Toole, Claudia Pahl-Wostl and the anonymous reviewers from the journal, whose helpful comments improved the quality of this paper.

Author Contributions: The first author conducted this research during his doctoral program under the supervision of the third author. The second author supervised the writing of this article and also reviewed it.

Conflicts of Interest: The authors declare no conflict of interest.

Abbreviations

The following abbreviations are used in this manuscript:

APAZU	the Urban Potable Water and Sewerage Program
CAPAET	the State Drinking Water and Sewage Commission
CEAT	the State Water Commission
CEPAL	Economic Commission for Latin America and the Carribean
CGE	the Coordination of Ecology
CIT	Contextual Interaction Theory
CONAGUA	National Commission of Water
CORESE	Commission of Regulation and Follow up
COTAS	Technical groundwater committees
CSITARET	the Centre for Integral Water Treatment
ECCAET	the Company for Control of Polluted Water in the State of Tlaxcala
FORTAMUN	Federal Fund for Strengthening the municipalities
GAT	Governance Assessment Tool
IPN	National Polytechnic Institute
NGO	Non-governmental organization
OECD	Organisation for Economic Co-operation and Development
PROSSAPYS	the Rural Waterworks Development Program
PROTAR	the Wastewater Treatment Program
UN	United Nations
WTP	Water Treatment Plant
WTPs	Water Treatment Plants
WU	Water utility
WUs	water utilities

References

1. The Organisation for Economic Co-Operation and Development (OECD). *Making Water Reform Happen in Mexico*; OECD Publishing: Paris, France, 2013. Available online: http://dx.doi.org/10.1787/9789264187894-en (accessed on 1 December 2015).
2. Pacheco-Vega, R. Arreglos institucionales para el saneamiento de aguas residuales en México. Un caso de estudio en la cuenca Lerma-Chapala. In *Gestión del Agua: Una Visión Comparativa entre México y Brasil*; Osorio, I.S., do Carmo, R.L., Velázquez, S.V., Guzmán, N.B., Eds.; Instituto Mexicano de Tecnología del Agua (IMTA): Jiutepec, Mexico, 2009; pp. 97–106.

3. UN-WATER. Wastewater Management, A UN-WATER Analytical Brief. *UN-WATER*. 2014. Available online: http://www.unwater.org/fileadmin/user_upload/unwater_new/docs/UN-Water_Analytical_Brief_Wastewater_Management.pdf (accessed on 26 March 2016).
4. Corcoran, E.; Nellemann, C.; Baker, E.; Bos, R.; Osborn, D.; Savelli, H. *Sick Water? The Central Role of Waste-Water Management in Sustainable Development. A Rapid Response Assessment*; United Nations Environment Programme; UN-HABITAT, GRID-Arendal: Arendal, Norway, 2010. Available online: http://www.unep.org/pdf/SickWater_screen.pdf (accessed on 5 April 2016).
5. Damián, F. PVEM pide sancionar a quien contaminó el río Atoyac (Green party asks sanctions for the Atoyac river polluters). Milenio. 20 March 2016. Available online: http://www.milenio.com/politica/Atoyac_contaminacion_PVME_atoyac-rio_atoyac_contaminacion_partido_verde_0_704329723.html (accessed on 22 March 2016).
6. Servicios Informativos y Publicitarios del Sureste. Contaminación en el río Atoyac causa enfermedades graves (Pollution in the Atoyac river provokes serious deseases). Milenio. 20 March 2016. Available online: http://sipse.com/mexico/contaminacion-en-el-rio-atoyac-causa-enfermedades-196718.html (accessed on 22 March 2016).
7. Havekes, H.; Hofstra, M.; van der Kerk, A.; Teeuwen, B. *Building with Blocks for Good Water Governance*; Water Governance Centre: The Hague, The Netherlands, 2013; Available online: http://watergovernance.s3.amazonaws.com/files/WGC_building_blocks_for_good_governance_def.pdf (accessed on 1 December 2015).
8. Ostrom, E.; Janssen, M.; Anderies, J. Going beyond panaceas. *Proc. Natl. Acad. Sci. USA* **2007**, *104*, 15176–15178. [CrossRef] [PubMed]
9. Ansell, C.; Gash, A. Collaborative Governance in Theory and Practice. *J. Public Adm. Res. Theory* **2007**, *18*, 543–571. [CrossRef]
10. Pahl-Wostl, C. A conceptual framework for analysing adaptive capacity and multi-level learning processes in resource governance regimes. *Glob. Environ. Chang.* **2009**, *18*, 354–365. [CrossRef]
11. Pahl-Wostl, C.; Holtz, G.; Kastens, B.; Knieper, C. Analyzing Complex Water Governance Regimes: The Management and Transition Framework. *Environ. Sci. Policy* **2010**, *13*, 571–581. [CrossRef]
12. Pahl-Wostl, C. *Water Governance in the Face of Global Change: From Understanding to Transformation*; Springer: London, UK, 2015.
13. Thiel, A.; Egerton, C. Re-scaling of Resource Governance as institutional change: The case of water governance in Portugal. *J. Environ. Plan. Manag.* **2011**, *54*, 383–402. [CrossRef]
14. Bressers, H.; Kuks, S. Water governance regimes: Dimensions and dynamics. *Int. J. Water Gov.* **2013**, *1*, 133–156. [CrossRef]
15. Van Rijswick, M.; Edelenbos, J.; Hellegers, P.; Kok, M.; Kuks, S. Ten building blocks for sustainable water governance: An integrated method to assess the governance of water. *Water Int.* **2014**, *39*, 725–742. [CrossRef]
16. Ostrom, E. A General Framework for Analyzing Sustainability of Social-Ecological Systems. *Science* **2009**, *325*, 419–422. [CrossRef] [PubMed]
17. Hinkel, J.; Cox, M.E.; Schülter, M.; Binder, C.R.; Falk, T. A Diagnostic Procedure for Applying the Social-Ecological Systems Framework in Diverse Cases. *Ecol. Soc.* **2015**, *20*. [CrossRef]
18. Binder, C.; Hinkel, J.; Bots, P.; Pahl-Wostl, C. Comparison of Frameworks for Analyzing Social-Ecological Systems. *Ecol. Soc.* **2013**, *18*. [CrossRef]
19. Mollinga, P.; Gondhalekar, D. Finding Structure in Diversity: A Stepwise Small-N Qualitative Comparative Analysis Approach for Water Resources Management Research. *Water Altern.* **2014**, *7*, 178–198.
20. Bressers, H.; de Boer, C.; Lordkipanidze, M.; Özerol, G.; Vinke-De Kruijf, J.; Furusho, C.; Lajeunesse, I.; Larrue, C.; Ramos, M.; Kampa, E.; *et al.* Water Governance Assessment Tool: With an Elaboration for Drought Resilience. 2013. Available online: http://doc.utwente.nl/86879/1/Governance-Assessment-Tool-DROP-final-for-online.pdf (accessed on 31 July 2014).
21. Bressers, H.; Kuks, S. *Integrated Governance and Water Basin Management: Conditions for Regime Change towards Sustainability*; Kluwer Academic Publishers: Dordrecht, The Netherlands; Boston, MA, USA; London, UK, 2004.
22. Bressers, H. From public administration to policy networks: Contextual interaction analysis. In *Rediscovering Public Law and Public Administration in Comparative Policy Analysis: A Tribute to Peter Knoepfel*; Narath, S., Varone, F., Eds.; Presses Polytechniques: Lausanne, Switzerland, 2009; pp. 123–142.

23. De Boer, C.; Bressers, H. *Complex and Dynamic Implementation Processes: The Renaturalization of the Dutch Regge River*; University of Twente, in Collaboration with the Dutch Water Governance Centre: Enschede, The Netherlands, 2011.
24. De Boer, C. Contextual Water Management: A Study of Governance and Implementation Processes in Local Stream Restoration Projects. Ph.D. Thesis, Universiteit Twente, Enschede, The Netherlands, 2012.
25. Bressers, H.; Bressers, N. *Governance Assessment Guide*; DROP Project: Almelo, The Netherlands, 2015; Available online: http://www.dropproject.eu/publications (accessed on 1 December 2015).
26. Gage, R.; Mandell, M. *Strategies for Managing Intergovernmental Policies and Networks*; Praeger: New York, NY, USA, 1990.
27. Bressers, N.; DROP Partners. *Benefit of Governance in Drought Adaptation. A Handbook for Regional Water Authorities*; DROP Project: Almelo, The Netherlands, 2015; Available online: http://www.dropproject.eu/publications (accessed on 1 December 2015).
28. Vinke de Kruijf, J.; Kuks, S.; Augustijn, D. Governance in support of integrated flood risk management? The case of Romania. *Environ. Dev.* **2015**, *16*, 104–118. [CrossRef]
29. Casiano, C.; de Boer, C. Symbolic implementation: Governance assessment of the water treatment plant policy in the Puebla's Alto Atoyac sub-basin. *Int. J. Water Gov.* **2015**, *4*, 1–22.
30. Jacobson, M.; Meyer, F.; Oia, I.; Reddy, P.; Tropp, H. User's Guide on Assessing Water Governance. 2013. Available online: http://www.undp.org/content/undp/en/home/librarypage/democratic-governance/oslo_governance_centre/user-s-guide-on-assessing-water-governance/ (accessed on 1 December 2015).
31. The Organisation for Economic Co-operation and Development (OECD). *OECD Inventory. Water Governance Indicators and Measurement Frameworks*; OECD Publishing: Paris, France, 2015. Available online: http://www.oecd.org/gov/regional-policy/Inventory_Indicators.pdf (accessed on 1 December 2015).
32. Comisión Nacional del Agua (CONAGUA). *Programa Hídrico Regional Visión 2030, Región Hidrológico-Administrativa IV Balsas*; Secretaría de Medio Ambiente y Recursos Naturales: México D.F., Mexico, 2012.
33. Instituto Nacional de Estadística y Geografía (INEGI). *ConociendoTlaxcala*; INEGI: Aguascalientes, Mexico, 2013. Available online: http://www.inegi.org.mx/prod_serv/contenidos/espanol/bvinegi/productos/estudios/conociendo/TLAXCALA.pdf (accessed on 25 March 2016).
34. Instituto Nacional de Estadística y Geografía (INEGI). *Información por Entidad, Tlaxcala*; INEGI: Aguascalientes, Mexico, 2010. Available online: http://cuentame.inegi.org.mx/monografias/informacion/tlax/territorio/div_municipal.aspx?tema=me&e=29 (accessed on 25 March 2016).
35. Rodríguez, E. Gobernanza del Saneamiento en la Cuenca Atoyac-Zahuapan del Estado de Tlaxcala. Master's Thesis, Instituto Mexicano de Tecnología del Agua (IMTA), Jiutepec, Mexico, 2010. Available online: http://www.imta.edu.mx/images/pdf/tesis_eva%20rodriguez%20rodriguez_2.pdf (accessed on 1 December 2015).
36. Alvarado, M.; Martínez, M.; Rodríguez, R. *Evaluación y Monitoreo Ambiental "Cuenca del Río Zahuapan, Estado de Tlaxcala"*; SEMARNAT: México D.F., Mexico, 2011. Available online: http://www.inecc.gob.mx/descargas/cuencas/2011_cnch2_mon_malvarado.pdf (accessed on 26 March 2016).
37. UNAM-Red del Agua and Instituto de Ingeniería. Programa de Apoyo al Desarrollo Hidráulico de los Estados de Puebla, Oaxaca y Tlaxcala. *Coordinación Técnica de la Red del Agua UNAM*. 2012. Available online: http://www.agua.unam.mx/padhpot/assets/cdh/generales/Reportefinal_formulacion_200812.pdf (accessed on 1 December 2015).
38. Noria, D. *Guía para Legisladores en Recursos Hídricos*; Fondo para la Comunicación y la Educación Ambiental, A.C.: Mexico D.F., Mexico, 2012. Available online: http://www.agua.org.mx/h2o/index.php?option=com_content&view=article&id=24529:guia-para-legisladores-en-recursos-hidricos&catid=1185:documentos&Itemid=100117 (accessed on 1 December 2015).
39. Comisión Nacional del Agua (CONAGUA). *Política Pública de Mejoramiento de Eficiencias en los Sistemas Urbanos de Agua Potable y Saneamiento en Mexico*; SEMARNAT: México D.F., Mexico, 2012.
40. Secretaría de Hacienda y Crédito Público. Presupuesto de egresos de la federación 2016 programas y proyectos de inversion. 2016. Available online: http://pef.hacienda.gob.mx/work/models/PEF/2016/docs/16/r16_pir.pdf (accessed on 25 March 2016).
41. Comisión Nacional del Agua (CONAGUA). Avance en metas relevantes. 2015. Available online: http://www.conagua.gob.mx/CONAGUA07/Contenido/Documentos/Metas_Relevantes.pdf (accessed on 26 March 2016).

42. Rodríguez, E. Agua y saneamiento en México: Avances, errores y alternativas. In *Agua Potable en Mexico*; Olivares, R., Sandoval, R., Eds.; Asociación Nacional de Empresas de Agua y Sanamiento, A.C.: Mexico D.F., Mexico, 2008; pp. 29–47. Available online: http://www.aneas.com.mx/contenido/EL%20AGUA%20POTABLE%20EN%20MEXICO.pdf (accessed on 16 August 2013).
43. Jardines, J. La Infraestructura hidráulica Municipal y la participación en su creación y operación de los Gobiernos Federal y Estatal. In *Agua Potable en Mexico*; Olivares, R., Sandoval, R., Eds.; Asociación Nacional de Empresas de Agua y Sanamiento, A.C.: Mexico D.F., Mexico, 2008; pp. 77–96. Available online: http://www.aneas.com.mx/contenido/EL%20AGUA%20POTABLE%20EN%20MEXICO.pdf (accessed on 30 August 2013).
44. Instituto Mexicano para la Competividad. Guía para la creación de organismos metropolitanos de agua potable y saneamiento en México. Instituto Mexicano para la Competitividad. 2014. Available online: http://imco.org.mx/wp-content/uploads/2014/03/AguaPotable.pdf (accessed on 1 December 2015).
45. Comisión Nacional del Agua (CONAGUA). *Estadísticas del Agua en Mexico, Edición 2013*; SEMARNAT: México D.F., Mexico, 2013; Available online: http://www.agua.org.mx/index.php/biblioteca-tematica/estadisticas/30681-estadisticas-del-agua-en-mexico-edicion-2013 (accessed on 1 December 2015).
46. Mercado, A.; Arellano, S. Senadores exigen rescate del río Atoyac y acciones inmediatas (Senators demand the Atoyac river recovery and inmediate accions). Milenio. 15 March 2016. Available online: http://www.milenio.com/politica/Senadores_Rio_Atoyac-Rio_Atoyac-Rio_Atoyac_sanciones-Rio_Atoyac_Senado_0_701330102.html (accessed on 25 March 2016).
47. Staff Puebla on Line. El Atoyac, un río muerto por las descargas de 3 mil 500 industrias (The Atoyac, a dead river due to 3500 wastewater discharges from the industry sector). Staff Puebla on Line. 15 March 2016. Available online: http://pueblaonline.com.mx/2015/portal/index.php/estado/item/36617-el-atoyac-un-rio-muerto-por-las-descargas-de-3-mil-500-industrias#.VvWYi2PeLaY (accessed on 25 March 2016).
48. Damián, F.; Zambrano, J. Rio Atoyac: pide el verde sanciones por la contaminación (Atoyac river: The green party requests sanctions for polluters). Milenio. 21 March 2016. Available online: http://www.milenio.com/estados/rio_Atoyac-partido_Verde-sanciones-contaminacion_de_rios_0_704929546.html (accessed on 27 March 2016).
49. Castillo, K.; Viñas, J. Organización demanda nueva ley de agua y servicio municipalizado (NGO demands new water law and municipal water service). E-Consulta.com. 22 March 2016. Available online: http://www.e-consulta.com/nota/2016-03-22/sociedad/organizacion-demanda-nueva-ley-de-agua-y-servicio-municipalizado (accessed on 25 March 2016).
50. La Redacción. Inicia CEAT la rehabilitación de 15 plantas tratadoras de aguas residuales (CEAT starts the rehabilitation of 15 water treatment plants). La Jornada de Oriente. 1 April 2014. Available online: http://www.lajornadadeoriente.com.mx/2014/04/01/inicia-ceat-la-rehabilitacion-de-15-plantas-tratadoras-de-aguas-residuales/ (accessed on 1 December 2015).
51. Secretaría de Gobernación. Declaratoria de clasificación de los ríos Atoyac y Xochiac o Hueyapan, y sus afluentes. Diario Oficial de la Federación. 6 July 2011. Available online: http://dof.gob.mx/nota_detalle.php?codigo=5199672&fecha=06/07/2011 (accessed on 1 December 2015).
52. Tribunal Latinoamericano del Agua. Veredictos de la Audiencia Pública Regional, México Tribunal Latinoamericano del Agua. Tribunal Latinoamericano del Agua. 2006. Available online: http://tragua.com/wp-content/uploads/2012/04/R%C3%ADo-Atoyac-final.pdf (accessed on 1 December 2015).
53. Comisión Nacional del Agua (CONAGUA). Programa Nacional Hídrico 2014–2018. Available online: http://www.conagua.gob.mx/conagua07/contenido/documentos/PNH2014-2018.pdf (accessed on 1 December 2015).
54. Comisión Estatal de Agua de Tlaxcala and Centro de Servicios Integrales para el Tratamiento de Aguas Residuales del Estado de Tlaxcala. Informe Annual 2013. Available online: http://platrans.tlaxcala.gob.mx/sistemas/transparencia/view_docs.php?recno=7287 (accessed on 1 December 2015).
55. Ruiz, M. Andan bailando 53 millones, el gobierno federal ya investiga. (Doubts about 53 millions, the federal government is already investigating). E-Consulta Tlaxcala.com. 12 August 2014. Available online: http://archivo.e-consulta.com/blogs/senoriotlaxcalteca/?p=3955 (accessed on 1 December 2015).
56. Dolowitz, D.; Marsh, D. Learning from Abroad: The Role of Policy Transfer in Contemporary Policy-Making. *Governance* **2000**, *13*, 5–23. [CrossRef]

57. Sandfort, J.; Milward, B. Collaborative Service Provision in the Public Sector. In *Handbook of Inter-Organizational Relations*; Cropper, S., Ebers, M., Huxham, C., Smith, P., Eds.; Oxford University Press: Oxford, UK, 2008; pp. 147–174.
58. Agranoff, R.; McGuire, M. Big questions in public network management research. *J. Public Adm. Res. Theory* **2001**, *11*, 295–326. [CrossRef]
59. Bryson, J. What to do when stakeholders matter: Stakeholder identification and analysis techniques. *Public Manag. Rev.* **2004**, *6*, 41–53. [CrossRef]
60. Klijn, H.; Skelcher, C. Democracy and Governance Networks: Compatible or Not? *Public Adm.* **2007**, *85*, 587–608. [CrossRef]
61. Vinke-de Kruijf, J.; Özerol, G. Water management solutions. On panaceas and policy transfer. In *Water Governance, Policy and Knowledge Transfer: International Studies on Contextual Water Management*; de Boer, C., Vinke-de Kruijf, J., Ozerol, G., Bressers, H.T.A., Eds.; Earthscan Studies in Water Resource Management; Routledge: Aubingdon, Oxon, UK, 2013; pp. 12–35.
62. Ingram, H. Beyond Universal Remedies for Good Water Governance: A Political and Contextual Approach. 2008. Available online: https://www.researchgate.net/publication/238658113_Beyond_Universal_Remedies_for_Good_Water_Governance_A_Political_and_Contextual_Approach (accessed on 18 March 2016).
63. Franco-Garcia, M.L.; Hendrawati-Tan, L.; Gutierrez-Diaz, C.; Casiano, C.; Bressers, J.T.A. Institutional innovation of water governance in Mexico: The case of Guadalupe Basin, near Mexico City. In *Water Governance, Policy and Knowledge Transfer: International Studies on Contextual Water Management*; de Boer, C., Vinke-de Kruijf, J., Ozerol, G., Bressers, H.T.A., Eds.; Earthscan Studies in Water Resource Management; Routledge: Aubingdon, Oxon, UK, 2013; pp. 188–204.
64. Rodríguez, V. *Decentralization in Mexico*; Westview Press: Boulder, CO, USA, 1997.
65. Wilder, M.; Romero, P. Paradoxes of decentralization: Water reform and social implications in Mexico. *World Dev.* **2006**, *34*, 1977–1995. [CrossRef]
66. Comisión Nacional del Agua (CONAGUA). Identificacion de Costos para la Sostenibilidad de los Organismos Operadores de Agua. 2015. Available online: http://www.agua.org.mx/biblioteca-tematica/gestion-del-agua/1258-transparencia-y-rendicion-de-cuentas/37007-identificacion-de-costos-para-la-sostenibilidad-de-los-organismos-operadores-de-agua (accessed on 20 March 2016).

water

MDPI

Article

The Fall and Rise of the Kishon River

Tal Golan

Department of History, University of California, San Diego, La Jolla, CA 92093-0104, USA; tgolan@ucsd.edu

Academic Editors: Sharon B. Megdal and Susanna Eden
Received: 16 March 2016; Accepted: 13 June 2016; Published: 9 July 2016

Abstract: This paper recounts the environmental history of a main waterway in Northern Israel—the Kishon, and deploys this history to examine the evolution of Israel water policy as it struggled to bridge the growing gap between its ambitions of development and the realities of its limited water supply. The first part of the paper describes the decay of the Kishon since the early 1950s, and the multiple scientific, political and legal attempts to alleviate its misfortunes, and discusses the reasons for their failings. Some of these reasons were administrative by nature, but the paper suggests a deeper reason, rooted in the ideological core of the infant state that was overwhelmingly concerned with the development of its infrastructure, and invited the pioneering Israeli society to consider the demise of the Kishon as a necessary sacrifice for progress. The second part of the paper describes the late-20th century developments that allowed for the recovery of the ailing river. Changing social mores, the growing importance of environmental politics, the advance of Israel's water technologies, and an environmental scandal that endowed the rehabilitation of the Kishon with a new political and moral meaning, have all contributed to the rehabilitation of the river. Once a testament for the sacrifices involved in a struggle to create a viable state, the Kishon has become a theater for a confident society that has triumphed in its struggle against nature.

Keywords: Kishon River; Israel water policy; environmental politics; water technologies

1. Introduction

What drives the ubiquitous conflict between national interests of development and environmental protection, and how is it best managed? This paper describes how Israel's water policy makers have struggled to answer these questions, and analyzes how their answers were shaped by the evolving relations between nature, society, ideology, technology, and the young state. Israel deserves our attention because its subjects its water to intense regulation [1]. Well aware of the country's limited supply, Israeli legislators developed, early on, a powerful legal framework to administrate Israel's water resources [2]. However, this powerful machinery could not prevent the destruction of Israel's natural waterways system [3]. Their headwaters redirected to quench the thirst of the growing population and agricultural-based economy, Israel rivers and streams languished and their flow became dominated by the wastewater of the thriving modern nation: Agricultural runoff, industrial waste and urban sewage [4].

The paper focuses on the environmental history of a main waterway in Northern Israel—the Kishon. The first part of the paper describes the decay of the river since the early 1950s, and the multiple scientific, political and legal attempts made to alleviate its misfortunes, and discusses the reasons for their failings. Some of the obstacles were administrative by nature, but the paper suggests a deeper reason, rooted in the ideological core of the infant state that was overwhelmingly concerned with the development of its infrastructure [5]. The Zionist policy of aggressive development, the paper suggests, paid little attention to water it could not use, and invited the pioneering Israeli society to consider, and even celebrate, the demise of the Kishon as a necessary sacrifice for progress [6].

The second part of the paper describes the late-20th century developments that finally allowed for the recovery of the ailing river [7]. By the 1990s, Israel was a developed country that had reached a per capita income level approaching that of the long-independent Western nations. Changing social mores and the growing importance of environmental politics have improved regulation and reduced pollution. Meanwhile, the advance of Israel's water technologies—dripping, recycling, and lately desalination—have increased Israel's stock of good-quality water, reduced wastewater, as well as the state dependency on fresh water, and allowed Israel to return some of its old flow to the river, even in hot summer months. The recovery was further expedited by a scandal extraordinaire that erupted in 2000, when veterans of one of Israel's elite military units claimed that they contracted cancer from their trainings in the polluted river. The military denied the claim and the dispute escalated into a bitter public dispute that endowed the rehabilitation of the Kishon with a new political and moral meaning [8].

2. The 1950s and 1960s: A Necessary Price for Progress

The Kishon rises in the northern part of the Samaria mountain ridge and flows north-westerly for about 70 km through the Jezareel Valley and into the Haifa Bay. On its way to the sea, the river first runs through the agricultural Valley of Jezareel, where it collects the runoff water, heavy with fertilizers and pesticides, from the intensely-cultivated fields. It then cuts through the northern metropolitan area, where it collects the ever-growing urban load. Finally, on the last leg of its journey to the sea, the river rubs backs with Israel's largest industrial port, at the Haifa Bay, where it is greeted by a dense cluster of heavy industries–shipyard, refineries, petrochemical plants, etc.—all of which have been discharging their wastewater directly into its course [9].

Perennial for centuries, the Kishon might have been able to carry its heavy load to the open sea. However, in 1953, as part of the its National Water Carrier project, Israel's national water company, Mekorot, dammed the Kishon's upper watershed, in the Jezareel Valley, and redirected its flow into an artificial lake, to be stored for irrigation during the long dry summer season. The capturing of the headwater transformed the lower Kishon into an ephemeral stream that occasionally receives natural flow during wet winters [10]. After that, the lower Kishon quickly deteriorated into an outright sewage conduit. The river continued to languish for half a century, until the late 1990s when the state finally decided to rehabilitate the river.

The first warnings about the deteriorating condition of the Kishon were sounded in 1953, just a few months after the river was robbed of its headwater. Scientists from the Sea Fisheries Research Station (SFRS), a research unit within the Ministry of Agriculture (MoA) that studied fisheries of commercial importance, and monitored physical, chemical and biological parameters of the Israeli coastal water, detected the pollution in the lower Kishon and warned of the "great physical and financial damages" it could cause to submerged structures and vessels porting the river [11]. By 1956, the warnings led to the creation of a technical committee, the first in a long lineage, to inquire into the problems of the Kishon. The committee brought together technical, scientific and administrative personnel from the Ministries of Transportation and Agriculture, the City of Haifa, Navy, Technion (Haifa's Technological Institute), and the SFRS. Lacking political clout, the committee succeeded, after repeated requests, to secure a small budget that allowed the SFRS to initiate a research program that monitored the pollution in the Kishon and looked into its sources [12–14]. Seven, and later, eleven, monitoring stations were established along the Kishon, and SFRS scientists began to collect weekly samples and analyze them. By the end of the 1950s, enough scientific evidence had accumulated to portray a clear picture of an ailing watercourse, dominated by toxic waste and raw sewage [15].

In 1961, a regional sewage treatment plant was erected on the banks of the lower Kishon, to treat the urban wastewater deposited into the Kishon by the growing population in the region. The first to operate in the country, the Haifa Sewage Treatment Plant was designed to serve a population of 255,000. However, its capacity was quickly outstripped by the growing population, and, as the overload built up, the plants' output deteriorated and malfunctions multiplied. Ending up in the

Kishon, the plant's hefty discharge soon equaled the entire natural flow of the river, turning it into an outright sewage conduit, especially in the summer months when natural flow was minimal [10]. The results, once again, were quickly noticed. By the end of the decade, the SRFS's yearly reports documented, not only increase of industrial wastes, but also a dramatic increase of organic wastes, accompanied by enormous levels of pathogens and massive fish kills [16].

In January 1962, off-the-chart concentrations of *E. coli* bacteria and other pathogens in the water of the Kishon prompted the Health Ministry to assemble a high-level administrative committee to address the troubles of the ailing river [17]. Chaired by the Health Ministry's chief engineer, the committee included representatives from the two bodies charged with national water management—the Water Commissioner and the National Water Planning Authority—and from local stakeholder, such as Haifa's Port Administration, City Engineer, and Health Services. Ordered to present their report within three months, the committee bravely took a boat trip down the Kishon, visited some of the implicated plants and met with their respective management; assembled reports from the Haifa city engineering department, SFRS, Haifa port administration, and Haifa health services; consulted hydrological and sanitation experts; heard the representatives from six of the accused plants, and digested reports submitted by other plants. Finally, in May 1962, two months behind schedule, the committee presented its official report [18].

The committee framed the problem in administrative terms. The pollution at the Kishon was cast as a public nuisance, to be controlled by the local authorities via the licensing process. The 1958 *Statute of Arts and Industries* empowered the local authorities to permit licensed businesses in their jurisdiction to discharge their wastewater into the municipal sewer system or natural waterways, provided their wastewater satisfied certain criteria that delimited their injurious effects [19]. However, the local authorities had neither the technical facilities, nor the trained personnel, needed to monitor the various sources of wastewater. Moreover, local authorities were often among the main polluters and had little incentive to abate it. The committee identified twenty-nine plants that discharged their wastewater directly into the Kishon, and all of them were properly licensed to do so, even though none of them met their license criteria. The local authorities never bothered to monitor the plants, let along take action to force the plants to comply with the license terms. This feeble political will, the committee further noted, was further weakened by an incoherent political structure. The lower part of the Kishon region was not controlled by any single authority. The oil refineries, by the terms of their old British license, constituted a special zone, free from municipal jurisdiction; Kishon harbor was also out of municipal control, while other parts of the Kishon were controlled by various local municipalities, each operating independently and uncoordinated with the others [10].

To vitalize the political will, if not the river, the committee prescribed the creation of a permanent council that will consist of representatives from all relevant government ministries and local authorities, and would coordinate the scientific, administrative and legal efforts needed for the proper functioning of the Kishon [10]. Similar statutory bodies—composed of representatives from a wide range of stakeholders, and equipped with sufficient power to prevent pollution and promote restoration along their designated streams—had been proven useful in other countries [20]. In 1965, the Israeli Parliament did its part and passed the *Stream and Springs Authority Bill*, which allowed for the creation of such Israeli boards [21]. Unfortunately, the territory of this new administrative species overlapped with that of another, well-established authority—the Drainage Authorities, which oversaw the proper drainage of these same streams. The Drainage Authorities were controlled by the powerful MoA, which saw no reason to share its powers and budgets. Consequently, the ministries of agriculture and the interior, which were given control over the new law, did not implement it. Not a single stream authority was created for the next 30 years.

Still, it would be a mistake to reduce this leniency to political paralysis or administrative turf wars. Well aware of the country's limited water supply, Israeli legislators had developed, early on, a powerful legal framework to administrate Israel's water resources. Topping these efforts was the 1959 Water Law, which, as one commentator out it, "established a water management program for other countries to envy [22]." The Water Law had a vested ownership of each and all water resources in the state, and provided a strong administrative framework, supported by a comprehensive water code, for a centralized state control over the development, regulation, conservation, supply, and protection of all water resources. The Water Law further fashioned a Water Commissioner (WC) executive position and equipped it with draconian powers to oversee the execution of the law. The WC was authorized to allocate water to all users, set conditions for the protection of all water sources, take measures to prevent any damage to water sources, and restore polluted water resources to their original state at the expanse of the polluter [23].

The Israeli water market was, therefore, subjected to intense government regulation, and plenty of legal and administrative means existed to protect Israel's water resources from pollution. However, neither the powerful WC, nor any other authority, made any attempt to force industries to treat their effluent. However, what the 1962 committee diagnosed as a lack of political will could be considered as the potent manifestation of a different political will; one deeply rooted in the ideological core of the infant state, which was overwhelmingly concerned with the development of its infrastructure. The millions of Jews who heeded the Zionist call to settle in Israel needed housing, food, jobs, and security; and the development of the infrastructure that would meet these challenges was given a carte blanche by political decision-makers. "We shall plant for you and build for you," promised one of Israel's most prominent poets. "We shall beautify you greatly. We shall cloth you in a robe of cement and concrete [24]".

The establishment, in 1966, of a new government-owned plant, Haifa Chemicals (HC), on the bank of the lower Kishon, illustrates the dictates of the ideology of aggressive development and their environmental implications. Using originally-patented processes, HC converted Israeli phosphate rock, mined and shipped from Southern Israel, into products of high added-value for use in agriculture, industry and food production. Nothing came closer to the Zionist dream of a flourishing and sophisticated industry built upon Jewish brilliance and local resources [25,26]. Still, producing fertilizers and industrial chemicals, HC quickly emerged as one of the Kishon's worst polluters, discharging large volumes of toxic waste, daily, into the river [27].

3. The 1970s: New Expertise, Old Ideology

Their resources overextended by the massive activities involved in the National Water Carrier project, Israeli planning authorities paid little attention during the 1960s to the decay of the natural waterways system. That began to change in 1970, when a cholera epidemic erupted in Jerusalem, killing a few children, and was quickly traced to the illegal irrigation of crops with Jerusalem's free-flowing wastewater [28]. The outbreak disrupted agricultural export and tourism, and moved policy makers to pay attention to the potential dangers flowing down Israel's riverbeds [29]. Prompt by the epidemic and alarmed by the extreme levels of pathogens measured in the Kishon, the Haifa regional health authorities launched a comprehensive public-health study of the Kishon and its tributaries. Put together over two years by a team of sanitation engineers, the ensuing report carried a no-nonsense tone. "Many governmental agencies have shown their concern", it noted, "but nothing was done, and the condition of the Kishon has deteriorated from year to year. The political view that the Kishon could serve the industry without causing damages to the shipping and fishing facilities located at its mouth, as well as becoming a place of recreation and leisure, is far detached from reality". In reality, the report dryly stated, the lower Kishon could only be referred to as a severe public nuisance; a body of mostly standing water "so heavily contaminated with industrial wastes that it wiped out any sign of biological life [10] p. 1".

The report stayed away from political and legal analyses. Instead, the authors analyzed four possible strategies to rehabilitate the Kishon: (1) Reduce the pollution to a volume the Kishon could handle; (2) force each plant to treat its own wastewater before dumping into the Kishon; (3) collect and treat all industrial effluents at the Haifa's Sewage Treatment Plant; (4) bypass the Kishon all together by conveying the effluents directly to the sea with a dedicated pipeline. What is urgently needed next, the authors concluded, is a feasibility study that would carefully examine these options and determine the best one. It would then be the politicians' turn to force the solution on industry [10] pp. 90–92.

The Haifa Regional Union of Cities and the local industries jointly hired Balasha-Jalon Infrastructure Systems, a private consulting company that specialized in hydraulic and agricultural planning to perform the feasibility study. Published in 1975, the Balasha-Jalon report clarified early on that, if the Kishon was to be rehabilitated, the industry must cease dumping its effluents into its waters. That is because no matter how well the industrial wastes were treated, some of the toxic constituents were simply too difficult to remove [30]. Instead, Balasha-Jalon recommended the construction of a dedicated pipeline that would collect and carry the industrial effluents far away from the coast into the deep sea. The report developed two alternative schemes for the hypothetical pipeline. In the first scheme, effluents would first be sorted, and the more toxic effluents would undergo a primary treatment and then be conveyed to the sea via a dedicated pipeline. The less-toxic effluents would be treated to a level that could be reused for irrigation, or be discharged back into the Kishon. In the second scheme, all industrial effluents were to be collected in a central facility, where they would undergo a primary treatment before being conveyed to the sea via a three-kilometer-long pipeline. Balasha-Jalon recommended the second scheme. Dumping all industrial effluents into the deep sea, the report concluded, would be cheaper to maintain, and produce the best results for the Kishon. Still, expectations should be kept low. For even the best option would not restore the Kishon to its former glory. It would cease to be a public nuisance, and its basin may be able to support life again. However, the Kishon's water will remain dangerous for many years to come, and direct human contact with the water would have to be strictly forbidden [30].

This last bleak warning hinted at a disturbing finding published a year earlier, while the Balasha-Jalon study was still in the making. In 1974, two young researchers, Joel Kronfeld and Jerzy Navrot, decided to take a close look, not at the water—the focus of all previous attention—but at the thick sludge that had steadily accumulated at the riverbed of the lower Kishon. Using advanced spectro-photometrical techniques, the two detected substantial amounts of various heavy metal constituents in the sludge. The finding was somewhat perplexing since the Kishon's water showed almost no trace of these toxic metals. This, Kronfeld and Navrot noted, was fools' luck. The high pH levels of the Sewage Treatment Plant's effluent prevented the heavy metals from dissolving in the water. Consequently, the heavy metals settled and got trapped in the riverbed sediment. Ironically, Kronfeld and Navrot concluded, that any significant reduction in Treatment Plant's output may actually add to the "biological menace" of the Kishon, and further endanger the entire Haifa Bay [31,32].

The authors of the Balasha-Jalon report were aware of these new findings [33]. They were also aware of the impending plan for a major upgrade of Haifa's Sewage Treatment Plant. According to this plan, the treated effluent would be collected in a seasonal impound some 20 km away, where it would be diluted with agricultural runoff and be used for summer irrigation. Devoid of the large wastewater discharge, the Balasha-Jalon report warned, the flow in the Kishon would be significantly reduced, which would worsen its condition and heighten the perils [30], p. 2.

The improvement plan of the Haifa's Sewage Treatment Plant was part of an overall reorientation of the national water policy that gained momentum through the 1970s and gave priority to the development of wastewater recycling projects [34]. The impetus for this reorientation was the growing realization in the late 1960s, among Israeli water planners, that the country was approaching its limits in developing freshwater resources. The demand for water, nevertheless, continued to grow, and predictions indicated that the gap between demand and supply could reach 300 million cubic meters by 1990 [35,36]. The gap was temporarily covered by over-pumping water from the main aquifers, but a long-term solution was still in need. Seawater desalination was first considered as the preferred strategy to increase water supply, but the available technology was found to be immature and too expensive [37]. A consensus evolved that, until the arrival of cheap desalination, any major addition to the water supply would have to come from the reclamation of wastewater. Meanwhile, the activities involved in the National Water Carrier and the associated regional water development projects had come to their conclusion, and the massive water resources development program had almost ground to a halt. Consequently, the entire water community threw their full weight behind the development of a wastewater-recycling plan. The design of a national plan began in 1970, and by 1973 it came into effect as the Israel Sewerage Project (ISP) [38].

ISP dictated a comprehensive change. The national water company had used a single supply system for both domestic and agricultural use. However, a more complex system had to be built, and conventional thinking among water planners had to change, from using a single high standard of water quality to a hierarchical system that classifies, transports, stores, distributes and monitors water of varying qualities. High-quality water was to be protected and utilized for higher quality purposes, such as drinking, while lower quality water should be used for all other purposes, including agriculture. The 1970s saw Israel entering a new era of innovative research and development concerning all aspects of wastewater reclamation and reuse. By the end of the decade, Israel had already become a world leader in wastewater recycling, having developed sizable technoscientific expertise and an elaborate administrative capacity for water quality management and conservation [39]. However, the new program was driven by the same old ideology. That is, not by environmental concerns, but by the growing gap between the ideological ambition of development and the realities of inadequate water supply. The goal was not to clean the polluted streams, but to have treated wastewater supplying the greater part of agricultural irrigation by the 1990s.

The new water recycling policy made it possible to look for solutions for the ongoing contamination of the country's waterways with sewage flows. Seasonal effluents impoundments, like the one planned for the Haifa's Sewage Treatment Plant, provided a feasible solution to the recovered wastewater for agricultural use, coupled with the reduction of the pollution discharged into streams, such as the Kishon. Still, as the Balsha-Yalon report made clear, the industrial effluents were freighted with toxic materials that were difficult to filter out. The planning authorities preferred however to focuse on the bulk of the wastewater that could be recycled for agricultural purposes and refrained from forcing the polluting plant to introduce expensive technologies for the flows lost to the sea. Without political wind in its sails, the Balasha-Jalon report was ignored and soon forgotten. The dedicated pipe that would divert the industrial effluent from the Kishon and transport it into the deep sea did not materialize. The government did not want to finance it; and, in spite of the powerful administrative and legal mechanisms available, no attempt was made to force the industry to do so.

4. 1980: Delays and Evasions

Considered an acceptable price, necessary perhaps, for progress, the ailing Kishon had little choice but patiently wait for a change in values. That change started in the late 1970s, with the fall from power of the ruling socialist coalition that had been at the helm since the early days of the century. The newly-elected liberal-nationalist coalition, known as the Likud, carried no special attachment to the socialists-dominated agriculture sector. Some of the implications of this change were clarified early by Joseph Tamir, an urban, liberal politician, who played a central role in the 1975 establishment of *Life and Environment*, an umbrella organization that coordinated activities among Israel's fledgling non-governmental environmental organizations (NGOs). When the Likud rose to power in late 1977, Tamir became the new Chair of the Knesset's powerful *Internal Affairs Committee*. He quickly added the "Environment" to the committee's title and dominion, held a series of hearings concerning the Kishon, reprimanded the WC for its neglect of the Kishon, which "has turned during the years to a lifeless sewage conduit that constitutes a severe ecological and environmental nuisance", and ordered the WC to take immediate actions, using all the means in its possession, to reduce the severe pollution in the Kishon [40].

Ordered to report back every six months to the *Internal Affairs and the Environment Committee* (IAEC), the WC reluctantly moved into action. In early 1978, the commissioner issued a series of edicts to about fifteen small plants located along the Gedura stream, the Kishon's main tributary. The edicts demanded that the plants comply with the terms of their license, cease dumping their effluent into the stream, and to connect to the local sewage system. The bigger plants located along the Kishon's Banks, posed a more difficult problem, as their waste contained toxic ingredients that could not be treated by the municipal sewage treatment system. A different set of edicts was therefore issued to ten of these plants, which included a new standard that specified maximum thresholds for six commonly used water quality indicators: Acidity, suspended solids, oxygen demand, ammonia, detergents, and oils. The edicts demanded that all industrial effluents dumped into the Kishon must comply with the new standard [41].

The edicts presented, for the first time, a clear set of official parameters for the plants to comply with [42]. The demands were anything but harsh. The standards mentioned neither the hard-to-remove heavy metals, nor other significant pollutants, such as phosphates, fluorides or nitrates. The maximum thresholds permitted for the six mentioned indicators were also relatively lenient. Nevertheless, compliance was not an easy task for the plants, which had gotten used to having their way for so long. To comply, significant technological changes had to be made. New filtering technologies had to be installed, and, possibly, even fundamental changes in the production processes were needed. All these involved a financial burden that the plants were reluctant to carry alone without significant government help, which was not forthcoming.

On his part, the powerful WC, was hesitant to enforce the new standard on the plants, many of which were owned directly by the government and were central to Israel's economy. The refineries, for example, provided crucial energy to Israel's rapidly growing population and its modernizing economy. The petrochemical industries were among Israel major exporters, providing it with much needed foreign currency. Moreover, the plants provided livelihood to thousands in the Haifa area, and were backed by the *Histadrut*, Israel's powerful umbrella labor union, which controlled much of local politics. In this context, any attempt to enforce regulation that added to the cost of production, was conceived as a threat, not only to the viability of the plants, but also to the nation's economy, and would be met with resistance, not only from the plants' managements, who resisted any change in the status quo, but also from the powerful labor unions, as well as local and national politicians. Neither the WC, nor the plants, were, therefore, ready to make the difficult decisions needed to meet the challenges set by the edicts.

The edicts required the plants to present a preliminary plan within six months from the issuance date; a final plan within twelve month from the date of the approval of the preliminary plan; and full compliance within three years from the date of the approval of the final plan. In July 1979, a year after the issuance of the edicts, the WC reported back to the IAEC that most of the smaller factories along the Gedura tributary have complied with the edicts and began to deliver their waste to the municipal sewage treatment facility. That was not the case, however, with the bigger plants along the Kishon, "where real progress was yet to be made". Tellingly, instead of pressing the plants, the WC chose to plead their case before the committee. "These are large plants", the WC, Ben Meir, explained to the IAEC, "each unique in kind in Israel, facing difficult problems in treating their effluents. Taking note of this objective situation, we gave them long period (up to five years) to solve their problems and stop dumping their wastes into the Kishon". Even then, predicted Ben Meir, only few of the plants will comply with the prescribed timetable [43].

The plants took advantage of a loophole in the edicts, which did not set absolute dates for compliance. Instead, each phase in the time-table was conditional on the successful completion of the previous one. That opened the door for repeated delays, as each stage was put off until the successful completion of the previous stage. Consequently, four and half years after the issuance of the edicts, at the theoretical date of final compliance, all the big polluters were still far from compliance [44]. Facing noncompliance, the WC chose to avoid confrontation and abstained from taking punitive actions against any of the polluting plants. Instead, it sought to cultivate cooperation through ongoing negotiation. The WC's main argument against the plants was its control over their water allocation. The plants' water allocation had to be annually approved by the WC, and the WC refused to approve it unless the plants showed some advancement in their treatment plans. The plants responded by doing just enough to persuade the WC not to cut off their water supply, but never enough to comply with the demands of the 1978 edicts.

The actions of the biggest polluter, Haifa Chemicals (HC), may serve as an example. In October 1978, the date set by the edicts for the completion of the preliminary plans, HC asked for an extension. The WC extended the deadline by three months to December 1979. By January 1980, HC still failed to present the WC with the requested plans, and the WC agreed to give HC another three-month extension, until March 1980. In March, HC finally presented the WC with a treatment plan that fell short of the required standard. In the accompanying letter, HC explained that full compliance would necessitate too large of an investment and intimated that it could jeopardize the existence of the plant and the livelihood of its 580 workers. The WC rejected the alternative plan and the two sides negotiated the hiring of an external arbitrator. A company specializing in environmental technology was contracted, which after a great deal of research concluded what everybody had already suspected. An adequate solution would necessitate great expenditure and fundamental changes in the production processes themselves. HC rejected this option as impractical and resubmitted its previous plan. Unless a practical solution could be found, it reiterated, it will have to consider the possibility of shutting down its operation and dismissing its 580 workers [45–48].

A solution was finally unearthed from the long-forgotten recommendations of the 1975 Balasha-Jalon study—a dedicated pipeline that would divert HC's highly toxic effluent away from the Kishon and into the Mediterranean Sea. Both sides seemed content with this elegant solution that promised to hide everybody's problems deep in the sea. Best of all, there was plenty of work to be done before the pipeline could materialize. Detailed surveys of the Haifa Bay had to be performed, followed by studies of potential influences on the marine environment. Plans had to be made, and data be collected to satisfy the demands of the various planning and municipal authorities. The physical course of the pipeline had to be negotiated, and objections from the various authorities had to be considered. All this was to take a long time and HC was in no hurry. By the early 1990s it was still knee deep in the preliminary stages of the pipeline planning. As long as the bureaucratic process was in motion, temporal solutions could be legitimated. These fell far short from satisfying the standard demanded by the 1978 edicts. However, HC could always pull out its ace, the marine pipeline, and promise that the final solution was well on its way [49,50].

Other plants followed similar strategies. Many of them investing a substantial amount of money and implemented various apparatuses to reduce their pollution. Still, by 1993, fifteen years after the issuance of the edicts, none of the plants along the Kishon had met the required standard. Moreover, while the plants may have somewhat improved the quality of their effluents, they were also generating much more of it, as both industry and population continued to grow. The condition of the Kishon continued, therefore, to deteriorate during the 1980s, with no clear solution in sight [51].

5. 1990s: New Champions for the Old River

Some things, nevertheless, had changed by the 1990s, not the least of which was the global rise of environmental politics. American President Richard Nixon declared the 1970s to be "the decade of the environment", and established the Environmental Protection Agency [52]. The 1970s saw also the appearance on the political stage of NGOs, such as Friends of the Earth and Greenpeace, which kept steady pressure on authorities to keep good on their environmental commitments. A defining moment of this growing international environmentalism took place in 1972, in Stockholm, Sweden, where 113 countries attended the first global conference on the environment, organized by United Nations. Israel sent a high-level delegation to Stockholm, headed by the foreign minister, Abba Eban, who assured the assembly of Israel's commitment to sound environmental policy [53]. Inspired by the historic event, the Israeli delegation members lobbied, upon their return, to Israel for the creation of a full-fledged environmental ministry, but had to settle for a small department within the Prime Minister's office, equipped with an amorphous mandate to advise the government on environmental issues and coordinate relevant activities among the various ministries. It took another decade and a half for Israel to finally establish a Ministry of the Environment (MoE), which remained poorly funded and enjoyed neither political cooperation nor public support. Eight ministers came and went by the end of the century and none of them stayed long enough to make a significant mark on Israeli politics [54].

The municipalities along the Kishon and the industrial complex at its basin, continued to dump their effluents into the Kishon with little interruption. By the end of the 1980s, scientists from the *National Oceanographic and Limnological Research Institute*, who have been monitoring the river since the 1950s, began to warn the authorities that the toxic sedimentation accumulating at the Kishon's bottom was slowly chocking it and diminishing its capacity to channel large flows to the sea [55]. No one paid attention until the lower Kishon overflew during the exceptionally wet winter of 1992, and covered extensive residential and industrial areas with thick toxic sludge and transported massive quantities of it into the bay [56]. The flooding and its high cleaning costs increased the pressures on the political system to find a solution to the Kishon ills [57]. In 1994, thirty-two years after it was originally recommended, the Kishon Authority (KA) was finally created to administer the political, legal, scientific, and administrative efforts needed for the rehabilitation of the Kishon [58].

Facing formidable resistance, the KA continued to avoid confrontation, abstained from taking punitive actions against any of the big polluting plants, and sought instead to cultivate cooperation by ongoing negotiation. The plants, almost all of them privatized by now, responded with delay and evasion, doing just enough to avoid sanctions, but never enough to fully comply. The stage was set, therefore, for the continuation of the status quo between government and industry, were it not for the surprising appearance of a new kid in town—a young NGO called the *Israeli Union for Environmental Defense* (IUED).

Founded in 1991 by an American immigrant, Alon Tal, and funded by private, largely American donors, the new NGO struggled to take root in Israeli society, which lacked a tradition of grassroots civic activism, constantly occupied itself with existential threats, and preferred to negotiate its collective problems through state channels and consensual politics [59]. The IUED looked for a dramatic case that will allow it to raise Israeli environmental consciousness by using the legal courts as a public stage for dramatizing Israel's growing environmental problems. The Kishon came up early and often on the IUED agenda as a possible candidate [60]. It was the most polluted waterway in Israel, the polluters were known, and plenty of evidence was available. Alas, IUED did not have the legal standing required to prosecute the polluters. Israeli law provided powerful tools to prosecute water pollution, but these tools were all concentrated in the hands of the state. Some of them were deposited with the WC, who refused to go to battle with powerful rivals over water lost to the sea. Others were in the hands of the local municipalities, which were among the largest polluters and did little to stop it.

The government's grip over water regulation began to loosen up in the late 1980s, under the pressure of growing international environmental law. During the previous decade, Israel was privy to the negotiations that culminated in the 1976 Barcelona Convention for the Protection of Mediterranean Sea against Pollution, and the following years saw Israel's executive and legislative branches ratifying the protocols of the Barcelona Convention [61]. One of these was the *Protocol for the Protection of the Mediterranean Sea against Pollution from Land-Based Sources* (LBS Protocol), which was adopted in 1980, legislated by the Israeli Parliament in 1988, and ratified in 1990 [62]. The LBS Statute is a criminal statute that forbids the discharge or disposal of any substances into the sea without a permit. Most importantly, among its various provisions, the new legislation allowed for the first time for interest groups, properly certified as such by the Ministry of the Interior, to enter complaints in the name of the public.

In 1993, the Ministry of the Interior certified IUED to represent the public under the LBS Statue. Shortly after, another passage, this time to the powerful Water Law, was opened when IUED learned of a group of fishermen who ported at the Kishon and complained that the polluted water had caused great physical and financial damages to their fishing boats. The revised Water Law allowed the fishermen to pursue legal action to protect their property and livelihood. Still, taking on the mighty plants was not an easy task, and none of the lawyers the fishermen turned to thought that the amount of damage that could be proven and compensated for, justified the expense and effort required to fight the powerful plants. The fishermen were therefore delighted when IUED contacted them and offered a barter deal: Free legal representation in return for incorporating environmental concerns into the suit. The fishermen agreed and the young NGO finally found the legal showdown it was looking for.

In December 1994, IUED entered criminal charges against the two biggest polluters at the Kishon Basis: Haifa Chemicals and Deshanim, for violating the Water Law and the LBS Statue by dumping dangerous wastewater into to the Kishon, and from there to the Mediterranean [63]. Caught off-guard by the aggressive suits, the plants hired some of the top Israeli law firms in attempt to shake them off. A long and stormy battle ensued. IUED provided the courts with samples taken from the plants' outpour to the Kishon, which showed grave violations of the 1978 industrial wastewater standard [64]. Deshanim's lawyers successfully shook off the suit by arguing that that IUED test results, which found detergents 676 times the maximum permitted level, indicated that the sample was mistakenly taken from a stream that belonged to their next-door neighbor, a detergents plant [65]. HC was not so lucky. Unable to shake off the suit, its lawyers conducted a campaign of attrition designed to run down the fledgling IUED through legal costs. They insisted on detailed and lengthy discoveries, asked for repeated delays, and altogether launched a steady barrage of procedural missiles at the suit. However, IUED enjoyed the good fortune of a sympathetic Magistrate Judge, Yitzhak Dar, who ruled time and again in IUED's favor.

HC tried also to dismiss the suit through its political connections. It pleaded with the Attorney General for a delay in the case until a proper solution could be found [66]. It emphasized the plant's importance to Israel's economy and hinted to the possibility that the suit may force it to close shop, bringing ruin to hundreds of families in the Haifa Metropolitan area. However, the General Attorney refused to intervene in the proceedings. The LBS statue mandated that every discharge into the Mediterranean must be registered and authorized by an inter-ministerial committee headed by a representative of the Minister of the Environment [62]. Haifa Chemicals had never bothered to ask for such a permit, but now it desperately tried to get one, as a legal shield against the IUED suit. It summoned its best political allies to plead with MoE to provide it with a temporary permit, until a solution for their wastewater could be found. However, this time, it was MoE's turn to turn a deaf ear. Permits, the ministry insisted, could only be given in return for concrete plans and guarantees for future improvement. Haifa Chemicals had no such plans to present, apart from its hypothetical marine pipeline proposal, for which they had no tangible plan or timetable, and which was strongly objected to by the MoE. The requested permit was not given [67].

Finally, after two more years, in November 1996, the court approved a consent agreement in which Haifa Chemicals agreed to cover the full costs of the damage to fishermen's boats, fully cover IUED's legal expenses, and create a $250,000 environmental protection fund that would finance the monitoring of the Kishon and relevant education activities. More importantly, the legal settlement set, for the first time, a detailed industrial-waste standard for HC to abide by, with a clear time-table to do so, and a crowded list of sanctions if it failed to do so [68]. Thus, after more than two decades of firmly insisting that it had no other financially-feasible solutions for its industrial wastes, except for the dedicated marine pipeline, HC moved quickly to introduce the necessary changes to its production processes. HC began to import raw materials that were more expensive but with less impurities and invested more than ten million dollars into new treatment facilities. Other plants at the Kishon basin followed course and upgraded their treatment facilities in order to secure their LBS permit. The local municipalities along the river's banks also combined forces to improve their sewage infrastructure.

Meanwhile, in 1999, after a severe drought, the Israeli government finally decided that the cost of desalination had come down enough, and approved a massive desalination program that would meet the growing urban, domestic, and industrial demands, and perhaps some of its natural waterways' needs as well [69]. Thus, by the final years of the second millennium, through the combined efforts of a polite administration, aggressive civil activism, and technological advances, the Kishon finally had something to look forward to. Aquatic life and recreational parks slowly began to appear along its banks. Only the warning signs that were spread along the banks, bluntly forbidding any bodily contact with water, served as a reminder that not all past sins can be easily washed to the sea, and that under the façade of the lazy clean water still lay toxic sludge.

6. The 2000s: A Proving Ground for the Neo-Liberal Regime

In the spring of 2000, a public scandal erupted that brought the gloomy Kishon out from the obscurity of scientific journals and political backrooms and endowed its rehabilitation with new moral meanings and political urgency. It all started when veterans of the Israeli Navy Seals, an elite military unit, disclosed to the media that their unit has secretly trained in the Kishon's port for half a century [70]. The veterans claimed that the trainings in the polluted water caused a rash of cancers among them, and demanded that the Ministry of Defense (MoD) take responsibility for their illness, finance their medical treatments, and support their families if they should die. Worried about opening a floodgate of similar claims from soldiers in other units, the military denied any causal connection between the training in the polluted river and the veterans' cancers, and rejected the veterans' demands [8].

The dispute between the military and some of its elite soldiers quickly escalated into a bitter public controversy. Military service constitutes a central unifying narrative within Israeli society. Most young Jewish citizens are required to serve in the military, during which time they must follow orders, sometimes to the extent of risking their lives. In return, the State, by means of the MoD, guarantees full care of them and their families in case of injury or death in the line of duty. Trust is at the core of this vital contract between the State, the soldiers and their families. The elite military units, the Navy Seals prime among them, embodied the very essence of this contract, as their soldiers took the greatest risks and in return received society's highest appreciation. The notion, therefore, that this trust had been compromised—and, of all soldiers, with the elite Navy Seals—mesmerized the public. Newspapers, radio stations, and TV channels alike raced for the personal stories of the distressed soldiers and their families, and dug out the long history of disregard and neglect of the Kishon, and highlighted the refusal of the government to take responsibility for the misfortunes of both her young soldiers and old river.

A high-rank committee led by President Emeritus of the Israeli Supreme Court Meir Shamgar, and two prominent scientists, toxicologist Meir Wilchek and epidemiologist Gad Rennert, were called upon to investigate the disputed causal connection between the polluted water and divers' cancers. It did not take long for the committee's members to realize the scale of the calamity they were dealing with. The evidence quickly mounted and left little doubt that the lower Kishon River had been heavily polluted for many years with impunity. The Commission decided, therefore, not to wait for the completion of its entire inquiry, which included a large-scale retrospective risk study that would take almost two years, but to go ahead and publish an interim report that would address the urgent needs of both the sick soldiers and the sick river.

Published in July 2001, the preliminary report spared no party involved: The plants that took the river hostage and pumped their toxic wastes into it; the municipalities that did the same with their domestic sewage; the administrators and politicians who knew but looked the other way; the naval commanders who ignored the warnings and failed to guard the safety of their soldiers. All of them were privy to this scandalous affair, and each of them could and should have taken action to stop it. Indeed, the Commission found the neglect to be so widespread and systematic that it saw no purpose in pointing fingers. Instead it concluded its preliminary report with two sets of recommendations. The first set addressed the military: All activities involving physical contact with the Kishon must be immediately suspended; the sick soldiers and their families should receive interim medical and financial help, pending final resolution; and future training sites should be tested for environmental safety before soldiers be allowed to train there. The second set addressed the disturbing state of public affairs the committee encountered during its investigation. The culture of disregard towards the environment must not be tolerated anymore. The civil authorities must exercise the legal and technical means available to prevent further pollution of the environment at large, and the Kishon in particular [71].

In November 2001, the government officially adopted the recommendations of the preliminary report and instructed the MoD to implement the first set of the military recommendations, and the MoE to follow up on the second set [72]. The military ceased operations in the Kishon and appointed professional teams to study the environmental safety of its extensive network of training sites [73], while an inter-ministerial committee headed by the MoE embraced an ambitious master program proposed by the KA to stop all wastewater discharge into the Kishon, increase its share of good-quality water, dredge and remove the toxic sludge accumulated at its bottom, and vastly expand the parks around it [74].

In April 2003, the Commission published its final report. To the dismay of all, after nearly three years of intense inquiries the committee failed to resolve the main issue of contention. Having found no statistically-significant correlation between the cancer among the veterans and the water they trained in, the two scientists rejected the veterans' claim. The jurist, on the other hand, found the accumulated evidence, which included also direct testimonies and other non-quantifiable data, strong enough to support a causal connection. Conflicted, the committee's final report failed to put a closure to the painful rapture between the state and its elite soldiers. Once a public reminder for the sacrifices that the Zionists had to endure in their struggles to reinvent Jewish society, the polluted Kishon turned into a monument for the state's neglect and betrayal of its soldiers and rivers [75].

The political reaction came quickly. Within days of the publication of the final report, the Minister of Defense announced that the government decided to embrace the jurist's minority opinion and give the veterans and their families all the help they needed [76]. The ailing river received its share as well. Three months earlier, in January 2003, Israel opened the largest reverse-osmosis desalination facility in the world, the first in a series of five plants erected in Israel in the next decade, which produce about 600 million cubic meters of desalinated seawater per year, at a reasonable cost of under $0.60 per cubic meter—enough, not only to meet the growing urban demand, but also to make it economically feasible to start rehabilitating the depleted water aquifers and replenish the natural waterways with good–quality water. The dredging of the toxic sludge proved to be more challenging, but in April 2011, the government finally launched a larger-scale project to clean up the Kishon's riverbed. In June 2012 the Kishon Drainage Authority issued an international tender for the excavation of 400,000 cubic meters of contaminated sediment along a seven-kilometer route of the river [77]. In January 2013, the Canadian firm, EnGlobe Corp, was selected to perform the dredging, and, in February 2015, the project began, with a 20-month time-table [78].

So stands at the moment the story about the fall and rise of the Kishon. The old river has been domesticated by the grace of the modern state. Its flow is strong again and its water and banks show new signs of life. Visitors who happen upon the lower Kishon these days are treated with lavish recreational parks, fully furnished with lawns, BBQs, swings, and walking, running and biking paths. The long-range plan includes also an upscale residential zone that will look over the rehabilitated waterway, which will enjoy a public amphitheater, as well as a tourist-oriented nursery, fruit orchards and a produce market [79]. Once a testament for the sacrifices involved in a struggle to overcome a hostile nature and create a viable state, the rehabilitated Kishon has now become a theater for a confident society that has triumphed in its struggle against nature.

Acknowledgments: I would like to thank the U.S.-Israel Center on Innovation and Economic Sustainability, at the University of California, San Diego, for supporting some of the research for this paper.

Conflicts of Interest: The author declares no conflict of interest.

Abbreviations

KCF	Kishon Commission Files
MoA	Ministry of Agriculture
MoD	Ministry of Defense
MoE	Ministry of the Environment

References

1. Galnoor, I. Water policymaking in Israel. *Policy Anal.* **1978**, *4*, 339–367.
2. Lester, R. Legal Aspects of Water Quality Management in Israel. In *Water Quality Management under Conditions of Scarcity: Israel as a Case Study*; Shuval, H., Ed.; Academic Press: New York, NY, USA, 1980.
3. Tal, A. *Pollution in the Promise Land: An Environmental History of Israel*; University of California Press: Berkeley, CA, USA, 2002.
4. Zaslavski, D. *Below the Red Line, or Water as an Allegory*; Technion Press: Haifa, Israel, 2002.
5. Blass, S. *Water in Strife and Action*; Massada: Ramat Gan, Israel, 1973. (In Hebrew)
6. Alatout, S. States of scarcity: Water, space, and identity politics in Israel, 1948–1959. *Environ. Plan. D* **2008**, *26*, 959–982. [CrossRef]
7. Feitelson, E. The Four Eras of Israeli Water Policies. In *Water Policy in Israel: Context, Issues and Options*; Becker, N., Ed.; Springer: Berlin, Germany, 2013; pp. 15–32.
8. Golan, T. The Kishon Affair: Science, Law, and the Politics of Causation. *Sci. Context* **2012**, *23*, 535–569. [CrossRef]
9. For Maps and Additional Information (in Hebrew) about the Kishon, Visit the Kishon River Authority Site. Available online: http://www.kishon.org.il (accessed on 14 March 2016).
10. Avital, Y.M.; Shelef, G.; Braskin, A.; Kazer, M.; Shemesh, D.; Noy, Y.; Yuls, D. *Public Health Survey of the Kishon and Its Brooks*; Ministry of Health: Haifa, Israel, 1972.
11. Oren, H. Letter to the Director of the Fishing Department. In *KCF*; Military Archives: Jerusalem, Israel, 29 November 1953.
12. Oren, H. Letter to the Director of the Fishing Department and the Ministry of Agriculture. In *KCF*; Military Archives: Jerusalem, Israel, 3 February 1957.
13. Oren, H. Letter from the Navy's headquarters to the Fishing Department and the Ministry of Agriculture. In *KCF*; Military Archives: Jerusalem, Israel, 13 February 1957.
14. Oren, H. The Technical Committee on the Kishon Pollution, Protocol. In *KCF*; Military Archives: Jerusalem, Israel, 14 October 1956.
15. Oren, H. The Kishon's water around the Harbor. In *Alon Miktzo'ee La'Dayag (Fishermen 's Professional Bulletin)*; MoA: Haifa, Israel, 27 March 1959.
16. Komarovsky, B.; Litay, E. *Survey of Polluted Waterways: Summery Report for the Period 30.9.1970–1.4.1971*; MoA: Haifa, Israel, 1971; pp. 10–14.
17. Peper, A. *Letter to Engineer Watson*; MoH: Jerusalem, Israel, 1961.
18. The Kishon Pollution Committee. *Intermediary Report*; The Kishon Pollution Committee: Haifa, Israel, 1962.
19. *Local Municipalities Statue*; State of Israel: Jerusalem, Israel, 1958.
20. Porter, E. *Water Management in England and Wales*; Cambridge University Press: Cambridge, UK, 1978.
21. Kensset. Rivers and Stream Authorities Statue. 1965. Available online: http://fs.knesset.gov.il//5/law/5_lsr_210348.PDF (accessed on 27 June 2016).
22. Lester, R. Israel. In *International Encyclopedia of Laws Environmental Law*; Blanpin, R., Boes, M., Eds.; Kluwer: Cambridge, MA, USA, 1973; p. 52.
23. Water Law. In *Kovetz Takanot*; State of Israel: Jerusalem, Israel, 1991; p. 470.
24. Alterman, N. A Morning Song to the Homeland. Available online: http://www.zemereshet.co.il/song.asp?id=3292 (accessed on 14 March 2016).
25. Herzl, T. *Altneuland*; Hermann Seemann Nachfolger: Leipzig, Germany, 1902.
26. Barell, A. *Engineer-King: David Ben Gurion, Science, and National Building*; Ben Gurion University Press: Sede Boker, Israel, 2014.

27. The Kishon Enterprises. Available online: http://www.sviva.gov.il/subjectsEnv/SeaAndShore/MarinePollutionLand/ProjectionSources/Pages/KishonPlants.aspx (accessed on 14 March 2016).
28. Haaretz. 1970. Available online: http://www.haaretz.co.il/news/education/1.1278551 (accessed on 14 March 2016).
29. Shuval, H. Quality Management Aspects of Wastewater Reuse in Israel. In *Water Quality Management under Conditions of Scarcity: Israel as a Case Study*; Shuval, H., Ed.; Academic Press: New York, NY, USA, 1980; p. 222.
30. Balasha-Jalon Infrastructure Systems. *Prevention of the Kishon's Pollution by Industrial Wastes: A Preliminary Report*; Balasha-Jalon Infrastructure Systems: Haifa, Israel, 1975.
31. Kronfeld, J.; Navarot, J. Transitional Metal Contamination in the Qishon River System, Israel. *Environ. Pollut.* **1974**, *6*, 281–288. [CrossRef]
32. Kronfeld, J.; Navarot, J. Aspects of Trace Metal Contamination in the Coastal Rivers in Israel. *Water Air Soil Pollut.* **1974**, *4*, 127–134. [CrossRef]
33. Kronfeld's and Navarot's paper was included in the Balasha-Jalon bibliography.
34. Selbst, N. Economic, Social and Administrative Considerations in Reuse of Wastewater. In *Water Quality Management under Conditions of Scarcity: Israel as a Case Study*; Shuval, H., Ed.; Academic Press: New York, NY, USA, 1980; pp. 243–262.
35. Vardi, Y. National Water Resources Planning and Development in Israel—The Endangered Resource. In *Water Quality Management under Conditions of Scarcity: Israel as a Case Study*; Shuval, H., Ed.; Academic Press: New York, NY, USA, 1980.
36. Bone, Y. The Historical Development of Underground Water Production. In *Water in Israel*; Grinvald, Z., Ed.; Ministry of Agriculture: Jerusalem, Israel, 1973; Part A.
37. Hofman, D. *Desalination and Its Application in Israel*; National Water Authority: Jerusalem, Israel, 2011.
38. Gabbay, S. *The Environment in Israel*; Ministry of the Environment: Jerusalem, Israel, 1992.
39. Shuval, H. *Water Quality Management under Conditions of Scarcity: Israel as a Case Study*; Academic Press: New York, NY, USA, 1980.
40. Internal Affairs and Environment Committee. *Conclusions of the Knesset's Internal Affairs and Environment Committee Concerning the Kishon's Pollution*; Knesset Records: Jerusalem, Israel, 4 July 1978.
41. Ben Meir, M. *Treatment of the Kishon's Pollution Problem: Half-Annual Report by the Water Commissioner to the Knesset's Internal Affairs Committee*; Knesset Records: Jerusalem, Israel, 10 July 1979.
42. The indices seemed to be based on the 1976 Water Quality Standards published by the American Environmental Protection Agency, which included more than fifty indicators. The Water commissioner edicts chose six of the most commonly used indicators and specified a maximum threshold for each. The indices chosen were easy and cheap to implement and provided a good picture on the overall conditions. (1) Biological Oxygen Demand (BOD < 30 mg/L). Moderately polluted rivers may have BODs in the range 2–8 mg/L. Efficiently treated municipal sewage will have a BOD value of about 20 mg/L. The threshold set by the edicts, 30 mg/L, was a sad reminder that about half of the Kishon's water came from Haifa's sewage treatment plant; (2) Total Suspended Solids (TSS< 60 mg/L) The threshold specified by the edicts, 60 mg/L, was very lenient. In heavily urbanized and industrialized areas, such as in Haifa, a river may be fishless with TSS concentration as law as 15 mg/L; (3) Ammonia (<12 mg/L). Ammonia nitrogen (N) is a product of microbiological activity, as well as a key ingredient in fertilizers. When found in natural water it is indicative of sanitary pollution. However; (4) Detergents (<1 mg/L). Detergents were environmentally targeted in the 1970s and 80s because they contained large amounts of phosphates, which can promote excessive growth in algae and aquatic plants that disrupts normal functioning of the ecosystem and cause variety of problems, from odors to killing fish and increasing pathogenic populations; (5) Oils and Mineral Oils (<1 mg/L). Concentrations as law as 0.1 ml/L were shown to be lethal to marine organisms, so the threshold of 1 mg/L, specified in the edicts is, again, quite lenient; (6) pH levels (between 6 and 9). The pH levels specified in the edicts were a little more lenient that EPA's (between 6.5 and 9), probably to accommodate the prevalence of HC's effluent.
43. Ben Meir, M. *Report of the Water Commissioner to Mr. Joshua Barazani, Secretary of the Internal Affairs and Environment Committee Concerning the Kishon Pollution*; Knesset Records: Jerusalem, Israel, 4 July 1979.
44. Environmental Protection Agency. *1981 Annual Report*; Environmental Protection Agency: Jerusalem, Israel, 1982; p. 151.

45. Water Commissioner to Haifa Chemicals LTD. In *KCF*; Military Archives: Jerusalem, Israel, 30 October 1978 & 15 January 1978.
46. Haifa Chemicals to the Water Commissioner. In *KCF*; Military Archives: Jerusalem, Israel, 12 March 1980 & 4 May 1981.
47. Forrer, Z. Letter to Haifa Chemicals LTD. In *KCF*; Military Archives: Jerusalem, Israel, 4 April 1984.
48. Forrer, Z. Letters to the Water Commissioner. In *KCF*; Military Archives: Jerusalem, Israel, 8 February 1989 & 9 February 1990.
49. Letters from Haifa Chemicals to Forrer, Z. In *KCF*; Military Archives: Jerusalem, Israel, 21 October 1990 & 1 November 1990.
50. Letter from B. Flikstein, Vice to the Ministry of the Environment. In *KCF*; Military Archives: Jerusalem, Israel, 6 August 1991.
51. Reports of the Israel Oceanographic and Limnological Research from the Late 1980s to Mid 1990s. In *KCF*; Military Archives: Jerusalem, Israel, 2001.
52. Reorganization Plan No. 3 of 1970. In *U.S. Code, Congressional and Administrative News*; 91st Congress—2nd Session, 1970; Volume 3.
53. Eban in Stockholm. *Yediot Achronot*, 7 June 1972.
54. Tal, A. *Pollution in the Promise Land: An Environmental History of Israel*; University of California Press: Berkeley, CA, USA, 2002; pp. 259–262.
55. Hornung, H.; Kres, N.; Krumgalz, M. *The Presence of Heavy Metals in Sediments Taken from the Kishon's Bed and the Docks*; Israel Oceanographic and Limnological Research Reports; Jerusalem, Israel, October 1990.
56. *Kishon Authority Annual Report*; Ministry of the Environment: Jerusalem, Israel, 2000; p. 90.
57. See for the protracted litigation between insurance companies and the various local authorities regarding liability, in Haifa District Court, see cases 404/93 and 307/99. The litigation later reached the Israeli Supreme Court. See cases 2906/01, 3049/01, and 3139/01.
58. Kishon River Authority. Available online: http://www.kishon.org.il/ (accessed on 14 March 2016).
59. Morag-Levine, N. The Politics of Imported Rights: Transplantation and Transformation in an Israeli Environmental Cause Lawyering Organization. In *Cause Lawyering and the State in a Global Era*; Sarat, A., Scheingold, S., Eds.; Oxford University Press: Oxford, UK, 2001; pp. 334–352.
60. Interview with Alon Tal. 11 March 2014.
61. Convention for the Protection of Mediterranean Sea against Pollution. Available online: http://195.97.36.231/dbases/webdocs/BCP/BC76_Eng.pdf (accessed on 27 June 2016).
62. Regulations for the Prevention of Marine Pollution from Land Based Sources. Available online: http://www.sviva.gov.il/English/Legislation/Documents/Seas%20and%20Coasts%20Laws%20and%20Regulations/PreventionOfSeaPollutionFromLand-basedSourcesLaw1988.pdf (accessed on 27 June 2016).
63. Israeli Union for Environmental Defense v. Haifa Chemicals. Haifa Peace Court records, criminal complaint # 5790/94. 1994.
64. Mona Nuffi. *Expert Report*, Haifa Peace Court records, criminal complaint #5790/94. 1994.
65. Protocols; Haifa Peace Court records, criminal complaint # 5790/94. 1994.
66. Request for delay of procedures, Haifa Chemicals' letter to the Attorney General. In *KCF*; Military Archives: Jerusalem, Israel, 11 November 1995.
67. Adler, E. Testimony before the Shamgar Committee of Inquiry. In *KCF*; Military Archives: Jerusalem, Israel, 2001.
68. The agreement is attached to the protocol of the 25 November 1996 meeting; Haifa Peace Court records: criminal complaint # 5790/94. 1994.
69. *Government Decision #4895*; State of Israel: Jerusalem, Israel, 1999. Available online: https://www.knesset.gov.il/mmm/data/pdf/m02176.pdf (accessed on 14 March 2016).
70. Tal-Shir, A.; Yechezkely, T. Cancerous Diving; Investigative Report. *Yediot Aharonoth*, 25 May 2000.
71. Kishon Commission. *Interim Report*; State of Israel: Jerusalem, Israel, 2001; pp. 235–242.
72. *Government Decision #969*; State of Israel: Jerusalem, Israel, 2001.
73. Almog, S.; Amitai, Y. Diving in Polluted Water: Proposed Guidelines Report of the Israeli Expert Committee. In Proceedings of the Conference on Diving in Polluted/Contaminated Water, Washington, DC, USA, 25 July 2006; pp. 25–26.
74. *Protocols*; Government decision #1509; State of Israel: Jerusalem, Israel, 2002.

75. Kishon Commission. *Final Report*; State of Israel: Jerusalem, Israel, 2003.
76. Protocols. In *Proceedings of Government Meetings*; State of Israel: Jerusalem, Israel, 27 April 2003.
77. Dredging and Sediment Remediation in the Kishon River. Available online: http://www.sviva.gov.il/English/env_topics/RiversAndStreams/Documents/EnvBulletin-DredgingAndSedimentRemediationInTheKishonRiver-July2013.pdf (accessed on 14 March 2016).
78. Israel Begins Sediment Dredging in One of Its Most Notorious Waterways. Available online: http://www.imsdredge.com/blog/river-dredging/israel-begins-sediment-dredging.html (accessed on 14 March 2016).
79. Kishon River: From Poison to Pristine. Available online: http://www.israel21c.org/kishon-river-from-poison-to-pristine/ (accessed on 14 March 2016).

water

MDPI

Article

Water and Agriculture in a Mediterranean Region: The Search for a Sustainable Water Policy Strategy

Maria Llop * and Xavier Ponce-Alifonso

Research Centre on Industrial and Public Economics (CREIP) and Department of Economics, Universitat Rovira i Virgili, Avda Universitat 1, Reus 43204, Spain; xavier.ponce@urv.cat

* Correspondence: maria.llop@urv.cat; Tel.: +34-977-759898; Fax: +34-977-758907

Academic Editors: Sharon B. Megdal, Susanna Eden and Eylon Shamir

Received: 17 November 2015; Accepted: 5 February 2016; Published: 19 February 2016

Abstract: This paper analyzes two of the main challenges facing agriculture in Europe: technological changes and the application of the principle of cost recovery to water services. Our study takes into account the economic, social, and ecological consequences associated with these measures. Specifically, we consider the effects of these two situations not only on water consumption, but also on environmental, social, and economic indicators. Our study also includes two institutional scenarios involving the possibility or impossibility of performing transactions in formal water markets. By using a computable general equilibrium (CGE) model for the economy of Catalonia, a region located in Northeastern Spain, our results suggest that institutions related with water markets matter in terms of the effects that agricultural policies cause on water resources. They also suggest that greater economic efficiency is not necessarily optimal if we consider social or environmental criteria.

Keywords: agricultural technological change; computable general equilibrium model; economic impact; water policy

1. Introduction

Water has historically been considered an abundant resource in Spain. However, there have always been considerable differences in the temporal and spatial distribution of water across the country.

In the late nineteenth century, an intellectual and political movement called regenerationism established a link between the problem of water and the Spanish economy's backwardness compared to other European countries [1]. According to this movement, water infrastructure would enable the expansion of irrigated agriculture and, thus, increased productivity, which in turn would lead to an increase in agricultural exports and an improvement in the trade balance. Furthermore, these necessary infrastructures would not only facilitate the settlement of the population throughout the territory, but would also generate the hydropower necessary to drive industrial development. Given the difficulties in obtaining the private funding needed, the idea that the public sector should take charge of planning and financing these waterworks slowly began to gain momentum [2]. Public funding would enable users to have access to water at a price below its real cost, thereby encouraging the expansion of irrigation and the modernization of the country. This new water supply policy, therefore, became a win-win game, ensuring profits for farmers, builders, hydropower companies, and financial institutions [3]. It was also seen by the political class as a way to gain legitimacy, support, and prestige. This convergence of interests explains the effort made throughout the twentieth century to transform Spain into the European country with the largest number of reservoirs.

However, over time this water supply policy has become a problem rather than a solution. In a scenario of rapid growth of water demand for non-agricultural uses, and a slowdown of investment to expand supply, the water policy has not prevented an increasing water shortage in some parts of

Spain, such as Catalonia. The main criticism of this water supply policy is that the low price of water has not created incentives to use it efficiently. As a result, despite the large volume of water used by farmers, the price they traditionally paid for this resource did not reflect its real cost, and hid a cross-subsidization between different users. This situation hindered the ability of prices to transmit shortage signals and led to mismanagement of the resource, comprising huge losses in distribution channels, the use of out-dated irrigation techniques, and the production of low-yielding crops, among other problems.

As water scarcity increased, a change in water policy became a matter of urgency. The new water policy implemented in Spain in recent decades encourages a more efficient and sustainable use of water, but also considers releasing water from agriculture to other activities with a higher economic or social value [4]. However, a more efficient use of water in agriculture does not necessarily save water but, instead, as noted by the Jevons' Paradox, may eventually lead to an increase in the amount of water consumed [5,6]. Given this possibility, the literature has considered various institutional scenarios that would help to mitigate these adverse effects.

In this paper, we analyze the impacts of technological changes in agriculture, considering different institutional frameworks. First, we consider the existence of formal water markets; second, we analyze the application of the full cost recovery principle for water services contained in the European Union Water Framework Directive [7]. These two factors are among the most important measures that have been applied in the Spanish water policy in the recent years. Our analysis considers not only how these measures contribute to saving water, but also their economic impact both at an aggregated and disaggregated level.

In order to analyze the effects of technological change in agriculture in Catalonia and to discuss the issues mentioned above, we define a static computable general equilibrium (CGE) model using a social accounting matrix (SAM) database with the most recent data available, which is for 2001. We chose Catalonia because, as a Mediterranean region, it can be regarded as a typical case in which Spanish water problems have been even more intensive than at the national level (dependency on rainfall, predominance of agriculture in water uses, high population density, high levels of economic activity, *etc.*).

Over the last 20 years, CGE models have largely been used to analyze the effects of agricultural productivity gains in areas such as poverty, food production, and trade. For instance, Lofgren and Robinson [8] presented some modifications to the specified standard CGE models to incorporate a more realistic technology in the agricultural sector. Arndt *et al.* [9] used the CGE approach to analyze both improvements in agricultural productivity and reductions in marketing costs in Mozambique. Prasada [10] studied the general equilibrium impacts of technological changes in Canadian agriculture, which were modeled as productivity rises in the use of intermediate inputs and primary factors. More recently, Belhaj *et al.* [11] investigated the influence of trade openness on both agricultural technological change and poverty in the Tunisian economy.

The CGE framework has also been extensively used to investigate water issues. In general, the CGE literature has treated water as a commodity subject to the market and, accordingly, there is a price for that commodity that is used to obtain the corresponding agents' demands. Among other contributions, Berck *et al.* [12], Seung *et al.* [13] and Goodman [14] used CGE analysis for water issues in various regions of the United States. Diao and Roe [15] studied the general equilibrium effects of some possible reforms in the water management in Morocco. Gómez *et al.* [16] presented a CGE analysis of water issues in the Balearic Islands; Letsoalo *et al.* [17] applied the CGE analysis to South Africa; Strzepek *et al.* [18] constructed a CGE model to analyze the effects of the construction of the Aswan dam in the Egyptian economy; and Lennox and Diukanova [19] applied a general equilibrium model to water reallocation in Canterbury. At the global level, Berrittella *et al.* [20] showed the potential of CGE analysis by providing a multi-regional model of the world economy that evaluated sustainable water supply uses. Most of the CGE water contributions treat water as an input used in the production system and define both the intermediate and the final usages of water. Moreover, these contributions usually focus on the impacts of an exogenously defined water reallocation and water policy in the economic system.

The structure of this paper is as follows. Section 2 describes the main features of the regional CGE model and Section 3 shows the simulation analysis undertaken and the main results. In the last section, we make some concluding remarks.

2. The Model

Computable general equilibrium techniques have advantages over other partial equilibrium models, as they provide a complete representation of the economic agents and their optimization behavior [21]. CGE models also take into account all the interactions between economic agents by completing the representation of the circular flow of income and accordingly, are not limited to the production side of the economy. Moreover, CGE models allow consideration of nonlinearities in the equations that define the optimization rules of economic agents and are, therefore, a useful method for capturing the complex interrelations within an economy.

The definition of equilibrium used in our model is based on the Walrasian notion, which has been extended to include not only producers and consumers, but also government and foreign agents. Analytically, the model is a set of equations containing the equilibrium conditions of all economic agents and all markets. The solution of the model consists of a set of endogenous variables (a vector of prices, a vector of activity levels, and other macroeconomic indicators) that clear all markets and allow all agents to reach their optimization plans.

2.1. Production

The structure of production assumes perfect competition in all markets. It shows 16 differentiated sectors (j = 1,...,16), with one representing the agricultural activity (j = 1) and one representing the production and distribution of water (j = 3).

Each sector is assumed to produce a homogeneous final good and has a nested technology that shows constant returns-to-scale in production. The first level of the production function defines the total output in each sector (Q_j), following the Armington specification [22], using a Cobb–Douglas aggregator of domestic production (X_{dj}) and imports from abroad (X_{Mj}):

$$Q_j = \delta_j X_{dj}^{\gamma_j} X_{Mj}^{1-\gamma_j} \quad j = 1,\ldots,16 \tag{1}$$

The second level of the production function defines the domestic production. Our analysis assumes that domestic production complies with a Cobb–Douglas combination of intermediate inputs and value added, as follows:

$$X_{dj} = \lambda_j X_{1j}^{\varphi_{1j}} X_{2j}^{\varphi_{2j}} \ldots X_{16j}^{\varphi_{16j}} VA_j^{\varphi_{vj}} \quad j = 1,\ldots,16 \quad \textstyle\sum_{j=1}^{16} \varphi_{kj} + \varphi_{vj} = 1 \tag{2}$$

where j = 1,...,16 represent the production activities, X_{dj} is the domestic production in j, λ_j is a scale parameter, φ_{kj} are parameters that show the response of X_{dj} when there are changes in X_{kj}, and VA_j is the value added of j.

The production function in Equation (2) allows us to analyze different institutional frameworks related to water policy. Since a Cobb–Douglas function allows a certain degree of substitutability between productive factors, this situation can be interpreted as a scenario in which there is a water market that facilitates substitutability through the buying and selling of water use rights. On the contrary, the absence of water markets is represented in the model by fixing the level of activity in the sector of water distribution (j = 3). This situation could be interpreted as a scenario of null changes in the quantities of water sold to other sectors and this would, therefore, implicitly mean that no water market to exchange water exists.

The sectoral value added, which is defined in the last (third) level of the production function, is obtained by combining labour and capital using a Cobb–Douglas function:

$$VA_j = \beta_j L_j^{1-\alpha_j} K_j^{\alpha_j} \qquad j = 1, \ldots, 16 \tag{3}$$

where L_j and K_j are the labor and capital, respectively, in sector j, and α_j and β_j are the parameters or exogenous variables obtained through calibration.

2.2. Consumers

Our CGE model includes a private consumer that maximizes a Cobb–Douglas utility function (in logarithms) according to Equation (4):

$$U = \sum_{h=1}^{10} \gamma_h \ln c_h + \gamma_s \ln c_s \qquad \gamma_h, \gamma_s > 0; \qquad \sum_{h=1}^{10} \gamma_h + \gamma_s = 1 \tag{4}$$

The model differentiates two types of goods: production goods (j = 1,...,16) and consumption goods (h = 1,...,16). Consumption goods are a combination of production goods obtained through a conversion matrix of fixed and linear coefficients. In Equation (4), C_h represents the consumption of good "h" and C_S is the private saving or future consumption. The model identifies the final demand for water (h = 3) that corresponds to the amounts of water production used by the final demand of the economy.

The objective of consumers is to maximize their utility subject to the budget constraint. This constraint establishes that households' expenditure (arising from consumption and saving) cannot exceed their disposable income (arising from labor income, capital income, and public transfers net of taxation). We obtain the demand functions for consumption goods and saving from the optimization of consumers.

2.3. Public Administration

The model also includes a public agent that is assumed to produce public goods and public services. This agent also demands public services and investment goods. The public administration has a Leontief [23] utility function of public consumption (C_{16}^G) and public investment (C_I^G) in a fixed relation, which is determined by the parameter $\gamma_G > 0$:

$$U^G = \min\left[C_{16}^G, \gamma^G C_I^G\right] \tag{5}$$

The public agent maximizes Equation (5) subject to a budget constraint. In specific terms, this constraint establishes that public consumption and public investment must be equal to public revenues, which come from taxation once social transfers have been subtracted from public revenues. The model also contains a stock of public borrowing or bonds that the public agent can issue in the event of deficit. The public budget is defined as:

$$P_{16}C_{16}^G + P_1 C_I^G \leqslant I^G + \omega_I^G P_I \tag{6}$$

where ω_I^G is the debt that this agent issues if there is a situation of public deficit and I_G is the income from taxation which comes from an indirect tax on consumption, a direct tax on income, taxation on domestic production, and social security taxation. Additionally, the public administration transfers income to consumers and sectors, which have to be subtracted from taxation revenues.

The objective of the public agent is to maximize the utility function (Equation (5)) subject to the public budget constraint (Equation (6)). We obtain both the public consumption demand and the public investment from the solution to this optimization problem.

2.4. Foreign Agent

The foreign sector is assumed to produce a traded good by using regional exports through a fixed coefficients technology. The model assumes that the regional economy may receive income from abroad and may simultaneously make transfers to the external agents.

Additionally, to guarantee the macroeconomic equilibrium between investment and saving at the aggregated level, the model allows a situation of external deficit that must be used as savings by the foreign agent.

2.5. Ecological Sector

As our aim is to show the effects of agricultural technological changes on water resources, our model distinguishes between the water used by economic agents and the total water resources. By taking into account that there is an amount of water not used by the economic activity, usually referred to as the environmental or ecological flow of water, we can approximate the effects of any new situation on the water ecosystems.

In particular, to show the trade-off between the water used by the economy and the environmental flow of water, we define a level of activity in the ecological sector (*Ye*) which is residually calculated by taking into account the natural restriction between the total water endowments and total water uses in the regional economy, as follows:

$$1 = w_3 Y_3 + w_e Y_e \tag{7}$$

where w_3 is the percentage of water resources used by the economy (both by production and consumption activities), w_e is the percentage of ecological water not used in the economy and returned to nature), and, finally, Y_3 represents the level of activity in sector 3 (production and distribution of water). As *Ye* is endogenously determined by the model, Equation (7) assumes that there is no limitation in the amount of water because inter-basin transfers and seawater desalination automatically allow to cover any increase in the uses of water.

2.6. Database

Following the standard procedure in CGE modeling, all the model's parameters are obtained by applying the calibration procedure, which assumes that the initial situation of the economy is a starting equilibrium (benchmark situation). In this situation, the prices and activity levels are equal to one and consequently, the benchmark solution exactly reproduces the empirical information shown in the social accounting matrix database used to calibrate the parameters of the model.

The empirical application is based on a 2001 social accounting matrix of the Catalan economy. We use 2001 data because the regional water statistics available do not provide more recent data.

According to the agents and goods described in the model, the regional SAM shows 16 differentiated activities: one agricultural sector, one sector for the production and distribution of water, nine industrial activities, and five service sectors. The SAM also distinguishes between ten consumption goods, which are those used in the consumers' optimization problem. Our database also shows two factors of production (labor and capital), an account for the private income and expenditures of households, and a capital account containing all the saving and investment in the regional economy. The public agent's accounts involve four different taxes (on domestic production, on household income, on consumption, and on firms' social security contributions) and an account for the income flows and expenditures related to the public sector. Finally, the SAM includes a consolidated account for the foreign sector that simultaneously contains imports, exports, and foreign income transactions in the regional economy.

To calibrate the parameters of the ecological sector, we use additional information on water uses [24]. This information comprises the water resources in the natural regime, which are regulated through dams, groundwater pumping, and water reuse. Our model, therefore, accounts for the water that is used in the economy by sectors and households, and by residual; it also accounts for the ecological water that it is not used by economic activities and is returned to ecosystems.

The first solution in the model consists of calculating the reference equilibrium (the benchmark situation). We then use the model to analyze the effects of technological changes in agriculture on both the economic indicators and the ecological variables (e.g. the ecological flow of water).

As the analytical framework is based on a computable general equilibrium model, it is important to bear in mind some important features of the CGE in order to correctly understand the outcomes of the model. First, one of the model's equilibrium conditions is always redundant under CGE analysis, according to Walras's law [25]. Accordingly, we can take one price as being unitary in all of the perturbations. We assume that this fixed unitary price (the *numéraire*) corresponds to the wage, which will be always equal to one. The implication of this is that prices in all the new equilibriums are not absolute prices, but relative prices with respect to the *numéraire*. Second, the level of activity of the public agent is variable and the public deficit is fixed, whereas the foreign sector has a variable activity level and a fixed deficit.

3. Simulation Analysis

The belief that water was abundant in Spain, but irregularly distributed, meant that redistributing the water throughout the territory could easily solve water issues. However, this situation also led the public administration to pay little attention to inefficient uses of water that was considered to be a good in public ownership.

Nevertheless, the discourse in Spain has begun to change in recent years. Conflicts between territories and between water users have intensified in a context in which water is an increasingly scarce resource. There is, therefore, an increasing need to modernize irrigation and to foster technological changes in order to achieve a more efficient use of water in agriculture. However, the important questions are: (1) what will the environmental effects of this modernization process be; and (2) what institutional arrangements can help to mitigate those potential adverse environmental effects?

3.1. Technological Change in Agriculture Without Water Markets

In recent years, Spanish governments have established a framework of incentives to promote modernization in agriculture irrigation systems: the National Irrigation Plan [26], the A.G.U.A. Programme [27], the Emergency Plan for the Modernization of Irrigation [28], and the National Strategy for the Sustainable Modernization of Irrigation [29]. These supports for technological changes in agriculture are justified by the relative inability of private enterprises to undertake modernization processes, but these grants can also be understood as a compensation for the social benefit generated from the water savings obtained by using more efficient technologies. They can, therefore, be interpreted as compensation paid to farmers for creating positive externalities that contribute to the preservation of the environment.

For example, the National Strategy for the Sustainable Modernization of Irrigation [29] requires that each project for improving the distribution infrastructure of irrigation must include, along with the planned investment, an estimation of the amount of water that the project aims to save. It must also propose training activities that will expand the knowledge and professional skills of farmers, the diversification of activities and the implementation of environmentally-friendly production systems. However, this political support for technological change is not free of controversy because these subsidies are not always sufficient for farmers wishing to implement technological change: the uncertainty about the potential earnings of some innovations and the presence of high sunk costs often lead to a delay in farmers' investments in new technologies [30,31].

Our first simulation analyzes the impact of a 25% increase in the total factor productivity of agriculture by, for instance, modernizing transportation systems and the distribution and application of water. Following the idea that the government should subsidize the required investments to overcome farmers' reluctance to technological change [32], our simulations do not assume any increase in the producers' costs. In other words, technological changes are exogenously applied as a mere modification in TFP, without changing any other variable in the model.

The main results of Simulation 1 are shown in Table 1. Our results show that an increase in TFP reduces agricultural prices (−4.5%), although this is not translated into significant changes in the quantity produced (2.5%). As noted by King's Law, a reduction in agricultural prices only leads to a small variation in the quantity produced. As a result, the fact that agricultural prices plummet in this simulation means that farmers could see a negative impact on their profits.

Table 1. Simulation 1: 25% increase in TPF of agriculture (without water market).

Sectors	Changes in Prices (%)	Changes in Production (%)
1. Agriculture	−4.55%	2.50%
2. Energy	−0.19%	−0.06%
3. Water distribution	−0.28%	0.00%
4. Chemistry	−0.19%	−0.16%
5. Metals and electric equipment	−0.17%	0.04%
6. Automobiles	−0.17%	0.02%
7. Food production	−1.18%	0.07%
8. Textiles	−0.27%	0.03%
9. Paper	−0.15%	−0.02%
10. Other industries	−0.19%	0.03%
11. Construction	−0.03%	0.14%
12. Commerce	−0.10%	0.22%
13. Transports and communications	−0.03%	−0.02%
14. Finance	0.02%	−0.03%
15. Private services	0.00%	0.01%
16. Public services	−0.02%	0.11%
Water variables	**Changes in water variables (%)**	
Final water demand	0.32%	
Intermediate water demand	−0.29%	
Ecological water	0.00%	
Gross domestic production	**Changes in GDP (%)**	
Real GDP	0.31%	

At the regional level, technological change based on growth in TFP leads to a growth in real GDP (0.31%). However, the controversy surrounding incentives to technological change in agriculture is reinforced by doubts about the effect such a change would have on water consumption at the aggregate level. The results suggest that it only leads to a reduction in the level of intermediate water consumption of around 0.3%.

Since measures to improve water efficiency are not effective in reducing the economic pressure exerted on water resources they, therefore, need to be accompanied by other policy measures. We address this issue in the following simulations. Specifically, we analyze the impact of the EU Water Framework Directive and the establishment of formal water markets.

3.2. Technological Change in Agriculture and the EU Water Directive Framework

Water policy must provide suitable external incentives to farmers to use water more efficiently, but water policy must also encourage water savings in agriculture. As the European Union Water Framework Directive points out, water management should promote the sustainable exploitation of water resources so as to meet present needs without endangering the supply for future generations. To achieve this goal, the EU Water Framework Directive requires that water prices in the member countries of the European Union reflect their cost. Although applying the principle of cost recovery to water services can generate an additional cost that causes farmers to lose competitiveness, it can be also a stimulus to start the process of technological change that the state subsidizes.

However, this proposal leads to resistance in the agricultural sector. In fact, the water price for agriculture has been kept low by governments. For example, the average price paid for water in Catalonia is 1.7€/m^3. Nevertheless, if the EU Water Framework Directive's principle of cost recovery for water services was strictly applied, the price of water would increase to 3.30 €/m^3 [33]. The reason for this is that farmers would bear most of the cost of a significant increase in water price and, as the logic of collective action suggests, small distributional coalitions like farmers would have an incentive to form lobbies and influence policies in their favor and could, thus, hurt economic growth. Governments have, therefore, usually avoided applying the principle of cost recovery to water services in agriculture. This hidden subsidy has traditionally been justified by the need for food security and the need to establish equality between farmers with an unequal ability to pay.

In any case, we use our CGE model to simulate technological changes in agriculture in conjunction with a rise in water prices. In specific terms, we apply a 25% tax on the water price to the previous simulation of agricultural technological change. The results of Simulation 2 are shown in Table 2. As in the previous case, the agricultural price decrease (−4.6%) is greater than the increase in the quantity produced (3.2%). Furthermore, real GDP growth is also similar (0.3%). However, unlike the previous simulation, the increase in water price leads to a significant decrease in the level of intermediate water consumption (−8.3%) which is completely offset by an increase in final water demand (11%).

Table 2. Simulation 2: 25% increase in TPF of agriculture and 25% tax on water price (without water market).

Sectors	Changes in Prices (%)	Changes in Production (%)
1. Agriculture	−4.61%	3.22%
2. Energy	−0.44%	0.09%
3. Water distribution	−9.96%	0.00%
4. Chemistry	−0.24%	0.77%
5. Metals and electric equipment	−0.22%	0.71%
6. Automobiles	−0.21%	1.02%
7. Food production	−1.23%	0.88%
8. Textiles	−0.32%	1.03%
9. Paper	−0.20%	0.93%
10. Other industries	−0.24%	0.79%
11. Construction	−0.07%	0.06%
12. Commerce	−0.14%	0.58%
13. Transports and communications	−0.09%	0.55%
14. Finance	0.00%	0.27%
15. Private services	−0.04%	0.13%
16. Public services	−0.07%	−0.04%
Water variables	**Changes in water variables (%)**	
Final water demand	11.09%	
Intermediate water demand	−8.38%	
Ecological water	0.00%	
Gross domestic production	**Changes in GDP (%)**	
Real GDP	0.32%	

3.3. Technological Change in Agriculture and Water Markets

The water market is an institutional arrangement to allocate water among competing users that could enhance the economic efficiency of water uses. The pros and cons of water markets have been discussed widely in the literature [34–37]. Under certain circumstances, water markets could make it easier for farmers to internalize the opportunity cost of using an increasingly scarce resource and this could, therefore, encourage farmers to save water. To that end, the Spanish Water Act was reformed in 1999 by introducing the possibility of transferring water-use rights and, thus, enabling the creation of formal water markets. Additionally, the Spanish Water Act also introduced significant restrictions, such as who can participate in those formal water markets, which use is given to water, how long water rights may be transferred, how much water can be sold, *etc.* Therefore, the effect of water market in terms of improving allocative efficiency has been very limited.

Simulation 3 again shows the results of a 25% increase in the total factor productivity of agriculture, but now we allow water transactions between intersectoral users, as occurs in formal water markets. As shown in Table 3, allowing users to exchange water does not substantially affect the results compared to Simulation 1, in which water markets were not considered; the decline in agricultural prices (−4.5%) are greater than changes in the quantity produced (2.5%). However, the results between simulation 1 and 3 differ slightly at an aggregated level: although both lead to an increase in real GDP (0.3%), the existence of water markets lead to a greater reduction in the level of intermediate water consumption (−0.6%). These results could indicate that the impact of water markets is reduced. Indeed, as discussed below, what is questionable is the impact of subsidies for technological change.

Table 3. Simulation 3: 25% increase in TPF of agriculture (with water market).

SECTORS	Changes in Prices (%)	Changes in Production (%)
1. Agriculture	−4.55%	2.50%
2. Energy	−0.19%	−0.07%
3. Water distribution	0.01%	−0.30%
4. Chemistry	−0.18%	−0.16%
5. Metals and electric equipment	−0.17%	0.04%
6. Automobiles	−0.17%	0.02%
7. Food production	−1.18%	0.07%
8. Textiles	−0.27%	0.03%
9. Paper	−0.15%	−0.02%
10. Other industries	−0.19%	0.03%
11. Construction	−0.03%	0.14%
12. Commerce	−0.10%	0.22%
13. Transports and communications	−0.03%	−0.02%
14. Finance	0.02%	−0.03%
15. Private services	0.00%	0.01%
16. Public services	−0.02%	0.11%
Water variables	**Changes in water variables (%)**	
Final water demand	0.03%	
Intermediate water demand	−0.59%	
Ecological water	0.11%	
Gross domestic production	**Changes in GDP (%)**	
Real GDP	0.30%	

3.4. Designing a Win-Win Water Strategy

The challenge of water policy is to design a win-win strategy that encourages water efficiency and savings. In other words, water policy must generate better outcomes in terms of economic efficiency and environmental sustainability at the aggregated level, whilst also ensuring that this does not lead to increased political and social costs.

As we have seen before, applying the principle of cost recovery of water services may not be enough to improve economic efficiency or environment sustainability. Furthermore, farmers are the biggest consumers of water, so they will probably be reluctant to accept an increase in the price of water. The problem of political feasibility, therefore, continues to present itself, and raises the question of how to overcome farmers' reluctance and how to simultaneously improve environmental sustainability. This issue is not trivial.

However, this result may be offset. In Simulation 4, where we analyze a situation in which there is a 25% tax on the water price with the sale of water between users and uses (*i.e.*, water markets), the results differ. Table 4 shows these new results: both the price and quantity of the agricultural sector remains almost unchanged, and the real GDP decreases slightly (−0.03%). However, the results show a major impact at an aggregated level: a reduction in the level of intermediate and final water consumption (−24% and −21%, respectively). The main result is, therefore, an increase in water resources for maintenance of the environmental flow (around 8%).

Table 4. Simulation 4: 25% tax on water price (with water market).

Sectors	Changes in Prices (%)	Changes in Production (%)
1. Agriculture	0.11%	−0.07%
2. Energy	0.56%	−0.45%
3. Water distribution	27.89%	−23.43%
4. Chemistry	0.11%	−0.06%
5. Metals and electric equipment	0.07%	−0.01%
6. Automobiles	0.07%	−0.02%
7. Food production	0.08%	−0.05%
8. Textiles	0.09%	−0.04%
9. Paper	0.07%	−0.03%
10. Other industries	0.09%	−0.02%
11. Construction	0.06%	0.13%
12. Commerce	0.07%	−0.05%
13. Transports and communications	0.09%	−0.07%
14. Finance	0.02%	−0.04%
15. Private services	0.05%	0.00%
16. Public services	0.09%	0.36%
Water variables	**Changes in water variables (%)**	
Final water demand	−21.87%	
Intermediate water demand	−24.63%	
Ecological water	7.80%	
Gross domestic production	**Changes in GDP (%)**	
Real GDP	−0.03%	

The rising price of water may help to transmit signals that water is a scarce commodity. However, if buying or selling water is impossible, the incentives to internalize the opportunity cost of water use are smaller. This is why the establishment of water markets is crucial. As long as water use rights cannot be exchanged, water consumption will focus mainly on agriculture and, specifically, on farms that have historically had a larger water supply regardless of their efficiency. Without water markets, the rising price of water implies a double jeopardy for farmers: a drop in the prices of agricultural products and in the marginal value of the allocated water. This situation can be offset by the creation of water markets, allowing farmers to obtain an income from the sale of water.

4. Conclusions

In this paper, we have analyzed some challenges facing the agricultural sector, such as the process of technological change and the implementation of the EU Water Framework Directive. Furthermore, we consider how the institutional framework, such as the existence of formal water markets, modifies the effects of these measures. Our analysis involves a computable general equilibrium model that reflects all the connections and interactions between the economic agents. A unique feature of our model is that it shows the effects on the environmental flow of water, and provides information about the ecological consequences of each measure in terms of water resources. We, therefore, consider both economic variables and water uses.

The first result we obtain is that any technological change reduces farmers' expected revenue, which helps to explain their reluctance to undertake technological changes. The same results are obtained when we apply the principle of cost recovery in water services for agriculture. As suggested by the literature on public policy making, if a new policy is to be socially accepted, it not only needs to be economically rational, but also politically sensitive to social an environmental conditions during its implementation. However, in our simulations there is a divergence between farmers' preferences and the optimal social choice. In other words, there is a trade-off between economic efficiency and political viability.

The second conclusion we draw is that the debate about formal water markets not only matters in terms of efficiency, but also in terms of environmental sustainability. Our analysis shows that technological change or higher water prices are not sufficient conditions to ensure water savings. Our results suggest the need to clearly differentiate two types of measures and strategies that, as stated by Jevons' Paradox, are not always consistent: increasing technical efficiency and reducing water withdrawals.

The third conclusion is that environmental improvement depends on the institutional framework. Water markets enhance the political feasibility and environmental impact of the EU Water Framework Directive. Water markets provide an indication of the opportunity cost of water. Water markets should, therefore, be judged not only in terms of economic efficiency but also based on environmental sustainability.

Finally, there appears to be a trade-off between economic efficiency, environmental sustainability, and political viability in agricultural policies. In other words, a policy that leads to greater economic efficiency does not necessarily lead to environmental improvement, and nor is it the most likely to be accepted by farmers. In this context, the choice of an economic second-best improves the environmental impact and also creates greater consensus regarding its application. This conclusion leads us to the following future line of research: the challenge for water policy is to design other win-win strategies that encourage savings and efficient use of water and which, at the same time, have reduced political and social costs when they are implemented. Put another way, there needs to be a policy that generates a certain degree of consensus among those who believe that water should be treated as a commodity, and those who view water as a social asset that should be allocated outside the market. The European Union faces a new challenge in this respect: while the reform of the Common Agricultural Policy (CAP) and the implementation of the Water Framework Directive aim for greater economic efficiency accompanied by environmental improvement, these policies may also lead to a reduction in farmers' incomes. Our objective will be to evaluate the joint impact of the rise in the price of water proposed by the European Union Water Framework Directive and the reduction in agricultural prices resulting from the reform of the CAP. On the basis of this analysis, we aim to determine what kind of water policy would help turn this threat into a win-win situation.

We must bear in mind that, given that our CGE model is static, we can only observe the final impacts on endogenous variables without being able to reflect the temporal adaptation of those variables. Although this limitation makes it difficult to have a complete perspective of the effects of a win-win water strategy, our analysis helps to clarify its final economic and ecological consequences.

Another feature of the model that should be borne into account is that it implicitly assumes that any rise in water demand is automatically covered by the corresponding water supply. A future research, beyond the scope of this paper, will focus on the introduction of mechanisms able to reflect possible technological restrictions (e.g., the existing infrastructures and storage capacities) and risks of failures (e.g., floods and droughts) in water provision.

Acknowledgments: Acknowledgments: The authors acknowledge financial support from the Spanish Ministry of Culture (Grant ECO2013-41917-P), and the Government of Catalonia (Grant SGR2014-299, "Xarxa de Referencia d'R+D+I en Economia i Polítiques Públiques" and "Xarxa de Referencia d'R+D+I en Economia Aplicada"). Useful comments and suggestions by two anonymous referees and by the editor of the journal have substantially improved an earlier version of the manuscript.

Author Contributions: Author Contributions: Maria Llop constructed the database, designed the CGE model and performed the simulations. Xavier Ponce-Alifonso analysed the results and interpreted the outcomes with the related literature. Both authors wrote the paper.

Conflicts of Interest: Conflicts of Interest: The authors declare no conflict of interest.

References

1. Costa, J. *Política Hidráulica;* Biblioteca Joaquín Costa: Madrid, Spain, 1911.
2. Embid, A. The Evolution of Water Law and Policy in Spain. *Int. J. Water Resour. Dev.* **2002**, *18*, 261–283. [CrossRef]
3. Ramos, J.L. La formulación de la política hidrológica en el siglo XX: Ideas e intereses, actores y proceso politico. *Ekonomiaz* **2001**, *47*, 126–151. (In Spanish).
4. Llop, M. Water reallocation in the input–output model. *Ecol. Econ.* **2013**, *86*, 21–27. [CrossRef]
5. Batchelora, C.H.; Ratna Reddyb, V.; Linsteadc, C.; Dhard, M.; Royd, S.; Mayc, R. Do water-saving technologies improve environmental flows? *J. Hydrol.* **2014**, *518*, 140–149. [CrossRef]
6. Berbel, J.; Gutiérrez-Martín, C.; Rodríguez-Díaz, J.A.; Camacho, E.; Montesinos, P. Literature Review on Rebound Effect of Water Saving Measures and Analysis of a Spanish Case Study. *Water Resour. Manag.* **2015**, *29*, 663–678. [CrossRef]
7. European Parliament. Directive 2000/60/EC of the European Parliament and of the Council of 23 October 2000, establishing a framework for the community action in the field of water policy. Available online: http://eur-lex.europa.eu/legal-content/EN/TXT/?uri=celex%3A32000L0060 (accessed on 5 February 2016).
8. Lofgren, H.; Robinson, S. *The Mixed Complementary Approach To Specifying Agricultural Supply in Computable General Equilibrium Models;* International Food Policy Research Institute: Washington, DC, USA, 1997.
9. Arndt, C.; Jensen, H.T.; Robinson, S. Marketing margins and agricultural technology in Mozambique. *J. Dev. Stud.* **2000**, *37*, 121–137. [CrossRef]
10. Prasada, P. General equilibrium impacts of technological change under different market structures: A comparison of supply managed and other primary agricultural markets in Canada. *Sri Lankan J. Agric. Econ.* **2007**, *9*, 1–21. [CrossRef]
11. Belhaj Hassine, N.; Robichaud, V.; Decaluwé, B. Agricultural Trade Liberalization, Productivity Gain and Poverty Alleviation: A General Equilibrium Analysis. 2010. Available online: http://papers.ssrn.com/sol3/papers.cfm?abstract_id=1687135 (accessed on 5 February 2016).
12. Berck, P.; Robinson, S.; Goldman, G. The use of computable general equilibrium models to assess water policies. In *The Economics and Management of Water and Drainage in Agriculture;* Dinar, A., Zilberman, D., Eds.; Kluwer Academic Publishers: Boston, MA, USA, 1991; pp. 489–509.
13. Seung, C.K.; Harris, T.R.; MacDiarmid, T.R.; Shaw, W.D. Economic impacts of water reallocation: A CGE analysis for the walker river basin of Nevada and California. *J. Reg. Anal. Policy* **1998**, *28*, 13–34.
14. Goodman, D.J. More reservoirs or transfers? A computable general equilibrium analysis of projected water shortages in the Arkansas River Basin. *J. Agric. Resour. Econ.* **2000**, *25*, 698–713.
15. Diao, X.; Roe, T. The win-win effect of joint and trade reform on interest groups in irrigated agriculture in Morocco. In *The Political Economy of Water Pricing Reforms;* Dinar, A., Ed.; Oxford University Press: Oxford, UK, 2000.
16. Gómez, C.M.; Tirado, D.; Rey-Maquieira, J. Water exchanges *versus* water works: Insights from a CGE model for the Balearic Islands. *Water Resour. Res.* **2004**, *4*, 1–11.

17. Letsoalo, A.; Blignaut, J.; de Wet, T.; de Wit, M.; Hess, S.; Tol, R.; van Heerden, J. Triple dividends of water consumption charges in South Africa. *Water Resour. Res.* **2007**, *43*, 1–11. [CrossRef]
18. Strzepek, K.M.; Yohe, G.W.; Tol, R.J.; Rosegrant, M.R. General equilibrium modelling of the value of the high Aswan dam to the Egyptian economy. *Ecol. Econ.* **2008**, *66*, 117–126. [CrossRef]
19. Lennox, J.; Diukanova, O. Modelling the regional general equilibrium effects of water allocation in Canterbury. *Water Policy* **2011**, *13*, 250–264. [CrossRef]
20. Berrittella, M.; Hoekstra, A.Y.; Rehdanz, K.; Roson, R.; Tol, R. The economic impact of restricted water supply: A computable general equilibrium analysis. *Water Res.* **2007**, *41*, 1799–1813. [CrossRef] [PubMed]
21. Cardenete, M.A.; Guerra, A.I.; Sancho, F. *Applied General Equilibrium. An Introduction*; Springer: Berlin/Heidelberg, Germany, 2012.
22. Armington, P. A theory of demand for products distinguished by place of production. *Int. Monet. Fund Staff Papers* **1969**, *16*, 159–178. [CrossRef]
23. Leontief, W. *Input-Output Economics*; Oxford University Press: New York, NY, USA, 1966.
24. Termes, M.; Guiu, R. Anàlisi de la tendència del consum d'aigua a Catalunya i marges d'estalvi. *Nota d'Economia* **2009**, *93–94*, 21–33.
25. Patinkin, D. *The New Palgrave: A Dictionary of Economics*, 2nd ed.; Palgrave Macmillan: Hampshire, UK, 2008.
26. Ministerio de Agricultura, Pesca y Alimentación. Real Decreto 329/2002, de 5 de abril, por el que se aprueba el Plan Nacional de Regadíos. Available online: https://www.boe.es/diario_boe/txt.php?id=BOE-A-2002-8129 (accessed on 15 February 2016).
27. Ministerio de Agricultura, Alimentación y Medio Ambiente. Programa Actuaciones para la Gestión y la Utilización del Agua (A.G.U.A.). Available online: http://www.magrama.gob.es/es/agua/planes-y-estrategias/informes-de-viabilidad-de-obras-hidraulicas/actuaciones-y-proyectos/index_agua.aspx (accessed on 15 February 2016).
28. Ministerio de la Presidencia. Real Decreto 287/2006, de 10 de marzo, por el que se regulan las obras urgentes de mejora y consolidación de regadíos, con objeto de obtener un adecuado ahorro de agua que palie los daños producidos por la sequía. Available online: https://www.boe.es/diario_boe/txt.php?id=BOE-A-2006-4415 (accessed on 15 February 2016).
29. Ministerio de Agricultura, Alimentación y Medio Ambiente. Estrategia para la Modernización Sostenible de los Regadíos. Available online: http://www.magrama.gob.es/es/calidad-y-evaluacion-ambiental/participacion-publica/PP_2009_p_019.aspx (accessed on 15 February 2016).
30. Carey, J.M.; Zilberman, D. A model of investment under uncertainty: Modern irrigation technology and emerging markets in water. *Am. J. Agric. Econ.* **2002**, *84*, 171–183. [CrossRef]
31. Koundouri, P.; Nauges, C.; Tzouvelekas, V. Technology adoption under production uncertainty: Theory and application to irrigation technology. *Am. J. Agric. Econ.* **2006**, *88*, 657–670. [CrossRef]
32. Bjornlund, H.; Nicol, L.; Klein, K.K. The adoption of improved irrigation technology and management practices. A study of two irrigation districts in Alberta, Canada. *Agric. Water Manag.* **2009**, *96*, 121–131. [CrossRef]
33. Catalan Water Agency (ACA). *Catalan Water Management*; Catalan Water Agency: Barcelona, Spain, 2010.
34. Easter, K.W.; Rosegrant, M.W.; Dinar, A. *Markets for Water: Potential and Performance*; Kluwer Academic Publishers: Boston, MA, USA, 1998.
35. Garrido, A. Water markets design and evidence from experimental economics. *Environ. Resour. Econ.* **2007**, *38*, 311–330. [CrossRef]
36. Grafton, R.Q.; Libecap, G.; McGlennon, S.; Landry, C.; O'Brien, B. An Integrated Assessment of Water Markets: A Cross-Country Comparison. *Rev. Environ. Econ. Policy* **2011**, *5*, 219–239. [CrossRef]
37. Palomo-Hierro, S.; Gómez-Limón, J.A.; Riesgo, L. Water Markets in Spain: Performance and Challenges. *Water* **2015**, *7*, 652–678. [CrossRef]

water

MDPI

Article

Re-Linking Governance of Energy with Livelihoods and Irrigation in Uttarakhand, India

Stephanie Buechler [1,2,*], Debashish Sen [3], Neha Khandekar [3,4] and Christopher A. Scott [1,2]

1 School of Geography and Development, University of Arizona, Tucson, AZ 85721, USA; cascott@email.arizona.edu
2 Udall Center for Studies in Public Policy, University of Arizona, Tucson, AZ 85721, USA
3 People's Science Institute, Dehra Dun, Uttarakhand 248006, India; debu_manu@yahoo.co.in (D.S.); khandekar05@gmail.com (N.K.)
4 The Energy and Resources Institute (TERI), New Delhi, Delhi 110003, India
* Correspondence: buechler@email.arizona.edu; Tel.: +1-520-626-5314

Academic Editor: Arjen Y. Hoekstra
Received: 1 May 2016; Accepted: 21 September 2016; Published: 8 October 2016

Abstract: Hydropower is often termed "green energy" and proffered as an alternative to polluting coal-generated electricity for burgeoning cities and energy-insecure rural areas. India is the third largest coal producer in the world; it is projected to be the largest coal consumer by 2050. In the Himalayan state of Uttarakhand, India, over 450 hydroelectric power schemes are proposed or are under development. Hydropower projects ranging from micro hydro (run-of-the-river systems with generating capacity up to 100 kW) to large reservoirs (storage systems up to 2000 MW) such as the Tehri Dam are in various stages of planning, construction or implementation. Run-of-the-river hydropower projects are being developed in Uttarakhand in order to avoid some of the costs to local communities created by large dams. Stakeholders in this rapid hydropower expansion include multiple actors with often diverging sets of interests. The resulting governance challenges are centered on tradeoffs between local electricity and revenue from the sale of hydropower, on the one hand, and the impacts on small-scale irrigation systems, riparian-corridor ecosystem services, and other natural resource-based livelihoods, on the other. We focus on the Bhilangana river basin, where water dependent livelihoods differentiated by gender include farming, fishing, livestock rearing and fodder collection. We examine the contradictions inherent in hydropower governance based on the interests of local residents and other stakeholders including hydropower developers, urban and other regional electricity users, and state-level policymakers. We use a social justice approach applied to hydropower projects to examine some of the negative impacts, especially by location and gender, of these projects on local communities and then identify strategies that can safeguard or enhance livelihoods of women, youth, and men in areas with hydropower projects, while also maintaining critical ecosystem services. By assessing the Bhilangana basin case, we also offer hydropower–livelihoods–irrigation nexus lessons for headwater regions across the Himalayas and globally.

Keywords: hydropower; governance; livelihoods; gender; irrigation; ecosystem services

1. Introduction

Rampant urban and industrial growth and an agricultural sector increasingly dependent on groundwater pumping for irrigation have placed rapidly mounting demands on electrical power generation across South Asia. The Gangetic Plain and Himalayan headwaters in India and Nepal are witnessing rapid social–ecological change. Hydropower development is occurring on rivers where irrigation, livestock rearing, and other natural resource-based activities are already stretched in their ability to meet local residents' livelihood needs. Water is the central thread that interweaves energy (e.g., mechanically powered mills, electricity generation), agriculture (irrigated and rainfed crops, fodder for livestock), fishing, and ecosystems (provisioning of water, regulation of micro-climates)—a prime example of the water–energy–food nexus [1]. Yet, the integration of hydropower as the newcomer into local and basin-scale resource-use practices poses very significant governance challenges. Our purpose here is to identify strategies in the Bhilangana River basin in Uttarakhand state, in northern India, that safeguard or enhance the livelihoods of women, youth, and men in areas with hydropower projects, while also maintaining critical ecosystem services. By assessing the Bhilangana basin case in Uttarakhand, we also offer hydropower–livelihoods–irrigation nexus lessons for headwater regions across the Himalayas and globally.

Hydropower supplied about 16% of energy globally in 2014, on the rise since 2004 among countries in the Global South with the rising percentages of their growing populations demanding access to energy [2,3]. Hydropower is one component of the energy portfolio of India. It was estimated that in 2014, India used only 21% of its potential energy from hydropower [2]. Rural and urban electricity is highly subsidized by the central government. Electricity demand is increasing steadily. Most of the electricity consumers are either located in urban areas or in breadbasket regions where extensive groundwater pumping exerts electricity demand [4,5].

Low-income rural people must often make do with inferior sources of energy such as fuel wood, kerosene, animal waste and coal. They also do without power for oil expellers, rice hulling and wheat grinding, as well as for uses such as electrical tools, and dairy processing. Household needs that are often neglected are the blending of food, a fan for hot weather and lighting for use at nighttime for children's homework and to extend women's and men's productive hours.

The social impacts of the extraction of natural resources for energy production are often borne by rural communities. The urban populace, however, also pays for the current dependence on coal for energy production through very poor air quality; the capital city of Delhi has received the dubious distinction of the city with the worst air quality in the world [6]. Thus, urban and rural built environments, as well as ecosystems, are impacted by the water–energy–food nexus. As Rasul has argued [7] (p. 9):

> *Greater policy coherence among the three sectors (water, energy and agriculture) is critical for decoupling increased food production from water and energy intensity and moving to a sustainable and efficient use of resources. The nexus approach can enhance understanding of the interconnectedness of the sectors and strengthen coordination among them.*

The water–energy–food nexus is impacted by localized special interests that include hydropower developers, construction companies/contractors, landowners, village leaders, and others, each with their own agendas. A paradox related to the push towards hydropower for the production of energy in India is that it is a federal program, yet it is being executed in an ad hoc manner by usually private actors. Some of these private actors are based in areas distant from the villages impacted by the hydropower projects.

This paper examines the governance challenges of integrating hydropower with irrigation and other livelihoods by considering nexus dynamics in a headwater basin. This is set against the backdrop of rural–urban exchanges of energy through transmission–grid extension; agricultural production through market penetration; labor migration as a result of stark differences in wages and services; and a range of digital, financial, and other "divides" between plains cities and more remote headwaters regions. Based on this brief introductory framing, we proceed to discuss the case-study location, as well as its features of more generic relevance; present a social–justice conceptual approach centered on governance and institutions; analyze hydropower-impacted livelihoods; and conclude with a discussion of strategies to better balance livelihoods and ecosystem services.

2. Research Sites and Methods

Uttarakhand is a state in northern India bordering Nepal to the east and China (Tibet) to the north. Two major rivers of north India, the Ganges and the Yamuna, originate in Uttarakhand. These rivers and their tributaries provide the opportunities for hydropower generation. Uttarakhand has four ecological zones which, from South to North are: Shivalik, Lesser Himalaya, Great Himalaya, and Trans Himalaya. There are two administrative divisions: the Kumaon and Garhwal regions. This study was conducted in the Lesser Himalaya zone in the northwestern Garhwal division within the Tehri Garhwal district (refer to Figure 1). Eighty-nine per cent of the state of Uttarakhand is mountainous, which significantly restricts the area capable of sustaining agricultural production. In the mountainous terrain, factors such as deforestation due to climate change and heavy felling of trees for construction, other development projects and livestock grazing, have increased the incidence of landslides. Uttarakhand is also prone to erosion; 53 per cent of the state is categorized as severely or very severely eroded and dams in the state have experienced serious challenges with sedimentation [8]. Seventy per cent of the population in Uttarakhand is involved in agriculture, primarily subsistence agriculture [9]. The cropping intensity has been calculated at 160 per cent (i.e., all arable land is cultivated for at least one crop season, and an additional 60 per cent of this has a second crop). The average landholding in the hills per household is 0.68 acre (often divided into patches). The Tehri Garhwal district has a high poverty rate: more than 45 per cent of the population lives below the poverty line [8].

Figure 1. Map of Uttarakhand showing locations of Garhwal and Kumaon regions. Source: https://en.wikipedia.org/wiki/Garhwal_division.

Technocrats and policymakers in India in general and in the water-rich Himalayan states of Arunachal Pradesh, Himachal Pradesh, Jammu and Kashmir and Uttarakhand in particular, overwhelmingly believe that hydropower generation will have the least negative environmental impacts of any energy option, and it can also bring in state revenue. These states have been termed the "Power Banks" of India for their capacity to generate power for their own demand and that of other states [2]. As Reddy et al. have argued with respect to hydropower in Himachal Pradesh, the state northwest of Uttarakhand: *"the main winners are the state and commercial developers who have gained access to water for producing electricity"* [10]. Approximately 450 new hydropower projects are currently being developed in Uttarakhand to partially "fill the gap" between utilization and the state government's estimate of 27,039 MW of hydropower potential. The state and central (federal) governments, in joint ventures between public and private entities, or with private companies (nationally and internationally based) are coordinating these power projects.

Experiences with large and medium scale hydropower projects, the types that have largely been promoted to date in Uttarakhand, have been mixed [11]. The recent shift from large reservoir-based hydropower generation to run-of-the-river (ROR) projects, are seen by proponents to have tackled displacement and rehabilitation issues, as well as the environmental impacts related to the construction and management of dams. In India, there are 22 large dams with a height of 100 m or higher. In Uttarakhand, some of these include the Tehri dam (1000 MW capacity), Koteshwar dam (400 MW capacity) and Ramganga dam (198 MW capacity). The Tehri dam, completed in 2006, has a height of 260 m (855 feet). It is the highest dam in India (and only one other dam is over 200 m high) and fifth highest globally. This dam caused the displacement of 110 villages and over 100,000 people [12] and the severe disruption of livelihood systems [13]. In terms of climate change, large dams, especially those located in tropical regions, have been implicated in large greenhouse gas emissions largely from decaying organic matter [14], putting into question the green credentials of hydroelectric power production [15].

ROR hydroelectricity, unlike hydropower via large dams, requires little to no water storage. Instead, the natural flow or elevation of the river is used to generate electricity; the water is diverted from the river with a diversion weir or small dam and then channeled into the powerhouse to turn turbines (see Figure 2). Sometimes a small dam is created called pondage where a small area is flooded but this area is much smaller than with conventional dams. These systems produce little or no greenhouse gasses except during the construction stages and are thus considered to be carbon-neutral.

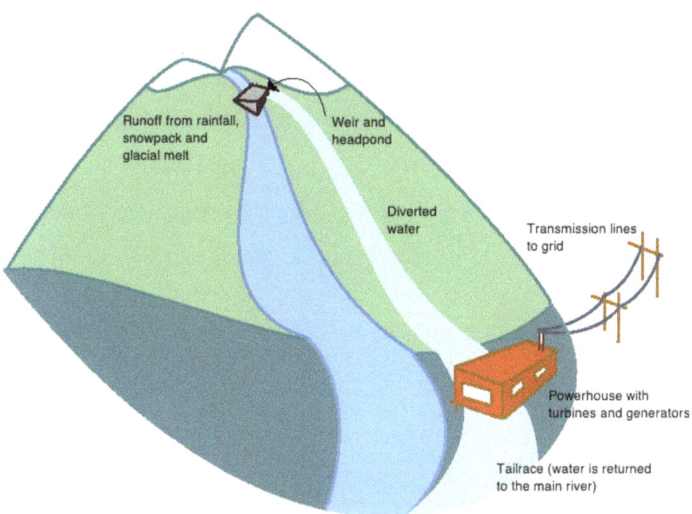

Figure 2. A small hydro, or run-of-river facility, drawing power from a mountain stream. Source: http://www.energybc.ca/profiles/runofriver.html.

ROR projects have not been as effective in the redress of local environmental concerns. ROR can affect the health of a watershed. Partly this is because the abstraction of water from the river leaves an area downstream, between the weir and the powerhouse, without normal river flows. Sufficient environmental flows of water (in this case river water made available via management regimes) are critical for ecosystems and the environmental, economic and social benefits these ecosystems bring. A healthy watershed provides multiple ecosystem goods and services for human, animal and plant populations within its boundaries. Postel and Thompson summarize these benefits as [16] (p. 98):

> *... water supplies for agricultural, industrial, and urban–domestic uses; water filtration/ purification; flow regulation; flood control; erosion and sedimentation control; fisheries, timber and other forest products; recreation/tourism; habitat for biodiversity preservation; aesthetic enjoyment; climate stabilization; and cultural, religious and inspirational values.*

Altering river flow leads to sedimentation, more turbidity, warmer water, and a change in nutrients that harms riparian biodiversity and the aquatic and animal populations dependent on the river [17]. Also, in Uttarakhand, water for ROR is often channeled through large pipes to achieve greater hydrostatic head (height difference between water upstream and downstream) and water flows to improve power production potential. The blasting techniques followed for constructing the tunnels through the mountains for diversions often leads to destabilization of the hillsides as well as detrimental effects on homes, cattle sheds and other infrastructure in the area.

Figure 3 indicates the location of Uttarakhand, the Bhilangana river basin, and the Tehri dam. The Bhilangana river is a tributary of the Bhagirathi river, which in turn is a tributary of the very large glacier-fed Ganges (known in India as the Ganga) river. The Bhilangana river basin is 2000 square kilometers in size. In April 2015 there were six operational projects, two under construction, six proposed, and nine identified sites for hydropower development in this basin [18] (fieldwork 2015). The hydropower projects are located in the narrow river valleys that are densely populated with villages and agricultural smallholdings on terraced land.

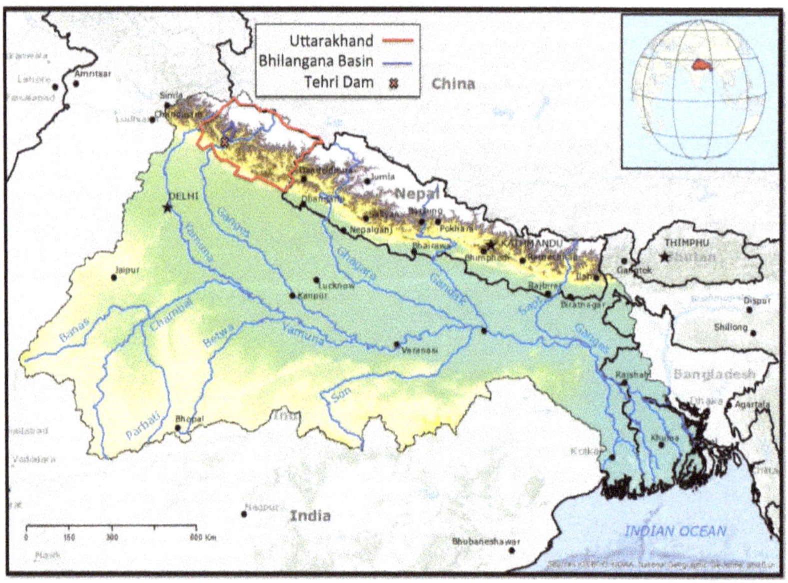

Figure 3. Study location in Uttarakhand. Source: Base map from Challenge Program for Water and Food (CPWF). http://waterandfood.org/river-basins/ganges-2/.

The research sites that include villages and their respective agricultural fields are located along the Bhilangana river and its tributaries including the Balganga river. These sites include villages impacted by the three operational run-of-the-river hydropower projects: Phalenda village (Bhilangana hydropower project with 22.5 MW capacity); Chakar Gaon and Devling villages (Bhilangana III with 24 MW capacity) and Chani village (Agunda–Thati with 3 MW capacity) (refer to Figure 4).

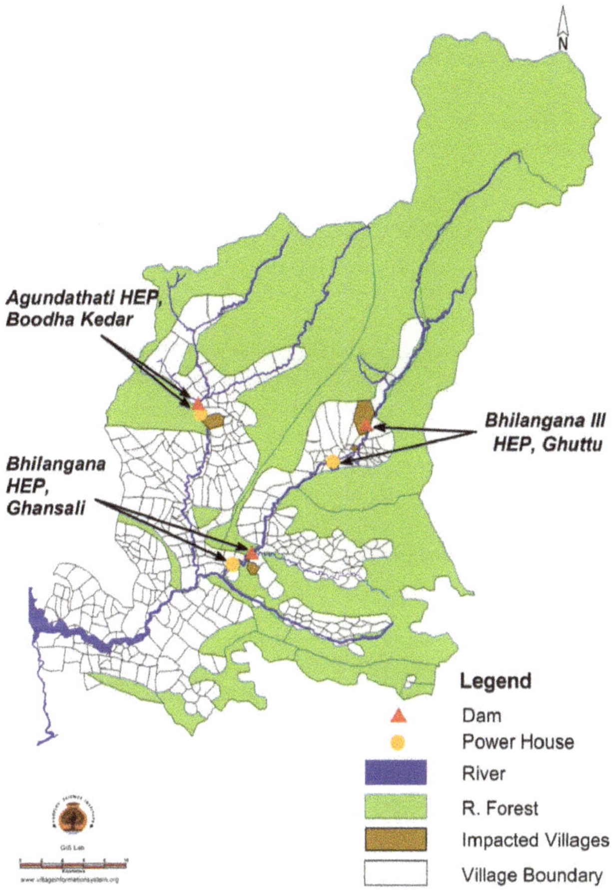

Figure 4. Research sites (impacted villages) and associated hydropower projects in the Bhilangana River Basin within Uttarakhand, India. Source: People's Science Institute, Dehra Dun.

Research methods utilized for this ongoing study include rapid rural appraisal (RRA), focus group discussions and household surveys (25 per cent stratified sampling) conducted between April 2015 and May 2016. Secondary research included a review of hydropower project related documents such as some of the detailed project reports (DPRs) on the hydropower projects, village agreements and writ petitions.

Urban-centered decision-making processes predominate in hydropower development. Pearce (2005) has argued, with respect to large dams in northeastern India, that [19] (p. 99): "Politically, they (large dams) have become a weapon for the rich, urban and powerful to take control of water resources away from the poor, rural and dispossessed". D'Souza's argument about the effects of large dams and other river diversions in India is also applicable [20] (p. 71):

> ... *though announced as neutral technological artefacts, (large dams and diversions) are deeply implicated in several processes that have been integral to affecting types of enclosure, hydraulic transfer, the expropriation and elimination of other water management skills and traditions and inevitably the externalisation of real costs through displacement onto the most marginal and impoverished communities.*

In India, a large power gap exists between rural and urban residents. In 2012, those in rural areas consumed 7 kWh per capita per month whereas their urban counterparts consumed 24 kWh per capita per month [21]. Urban dwellers are thus benefitting more from hydropower development than rural peoples. However, previously energy-deprived rural populations are also deriving benefits in some cases. A range of actors including locally elected politicians, private developers, non-governmental organizations (NGOs), and others are actively working to secure funding for energy projects from Central and State government sources [22]. This is part of a general scramble for development projects in rural areas, a process that gained significant momentum with the National Rural Employment Guarantee Act, 2005 (later renamed the Mahatma Gandhi National Rural Employment Guarantee Act). Rural political leaders use the promise of such projects to bring development to their region. The desire to attract employment is an important component of such development projects. There is a rural employment crisis that has led to labor out-migration from rural areas. Gender is an important factor in employment and migration issues [23]. In the region studied in Uttarakhand, male migration is higher than female migration. Younger men migrate first for educational opportunities then for employment. The migration of women tends to occur more often after they marry to join their husbands. Ironically, migration to cities swells urban population, thus increasing urban demand for energy that contributes to a vicious cycle of rural expropriation (in this case of water) for largely urban energy use. It also has another societal cost: diminished food production in rural areas.

We take a social justice approach, especially as applied to energy projects [24], by contending that rural peoples should be given a stronger voice [25] than urban-based developers and political power brokers. This approach also leads to questions over centralized projects for resource provision and management, including hydropower in the case study we consider; however, centralized photovoltaic and wind-power projects are also limited in their capacity to meet social–justice criteria. Young [26] has characterized social justice as a condition in which the institutional context makes it possible for everyone to be included in governance processes and entails distributive results such as the satisfaction of basic needs. As Gross [27] (p. 2727) has argued with respect to fairness or justice in renewable energy projects (specifically, wind energy):

> *Questions of justice are ... at the heart of many environmental disputes, such as ... infrastructure development projects... While broad in scope these environmental disagreements share common characteristics which include how decisions are made and how public goods, such as power generation infrastructure, and environmental burdens ... are distributed.*

Questions of fairness in a social justice framework incorporate issues pertaining to project benefits and costs and how these are distributed by gender, social class/caste, and geographical location [28]. In India, these distributional issues have so far been largely studied as they are related to the effects of large dams [29–31]. An overarching distributional issue is that which is mentioned above, that is, urban elites' expropriation of river water from rural populations living and farming in particular locations in riparian areas. In India, no water law exists that confers water rights to local people; only land rights exist [32].

Innovative technologies and social and legal processes and rules can help rural stakeholders generate energy for their own benefit rather than for urban, centralized, politically powerful actors. Rural populations are demanding regular power supply from the government. Electricity in rural Uttarakhand is used for lighting (including energy-saving compact fluorescent lights (CFL) bulbs), charging cell phones, and for blenders, irons, televisions, and refrigerators (usually only a few per village). Some of the benefits they seek from connection to small hydropower generation include longer hours of power and/or more reliable power, as the centralized grid supply in this area is still very poor. Smaller ROR projects that use the water alternately during the day and night for power generation and irrigation can help ensure power and livelihoods for local needs.

A more equitable distribution of environmental burdens can also be achieved in more centralized ROR power projects. Governance related measures that would help ensure this include the following: compensation by the developer for damage done to local populations' property during the construction phase. Minimization of the environmental burden on local livelihoods during the lifetime of the ROR project via government regulations applied to power developers' operations. Such regulations would incorporate sufficient environmental flows. Finally, governance measures would include more equitable distribution of benefits through profit sharing.

Social justice issues related to water resources have emphasized that water for multiple different uses must be protected in any watershed. The concept described above of environmental flows helps bring to light the multiple human and non-human water needs in a particular, managed, area. Chowdhury and Rasul's argument [25] (p. 46) with respect to irrigation projects also holds true for hydropower's effects on agricultural and domestic water resource distribution and access across different socio-economic groups: *"(a project) ... may affect their initial water access and use rights, entitlements and affect their well-being differently"*. As competition grows between different water users (agricultural, fishing and hydropower in this case study) due to increasingly scarce water resources, social justice issues will become an increasingly pressing area for policymakers to address.

3. Results

3.1. Main Actors and Governance Linkages in the Study Area

The main local actors in the Bhilangana basin involved in hydropower projects include: Village Councils (*Panchayats*), Farmers' Groups, Irrigation Users' Groups, Non Governmental Organizations, Trusts, Cooperative Societies, Producer cooperatives or businesses, Village Energy Committees, Forest Councils (*Van Panchayats*), Self Help Groups, Government Departments, village women's welfare groups (*Mahila Mangal Dals*, MMDs), and village youth welfare groups (*Yuvak Mangal Dals*, YMDs). According to the villagers in the rural communities studied, the hydropower developers and state energy department are the least significant and accessible in terms of services and benefits accrued; in Figure 5 below, this is indicated by the smaller size of the circles for the developers, state energy department and courts, and the distance of the circles from the hydropower impacted villages.

An environmental impact assessment (EIA) must be undertaken for hydropower projects by Indian law (the Environmental Protection Act passed in 1986). The Ministry of Environment, Forestry and Climate Change, through environmental appraisal committees executes the EIA for hydropower projects before they decide on whether or not to approve them. The assessment is used as a tool to identify the environmental, social and economic impacts of a hydropower project. Recently, however, the National Institute of Disaster Management (NIDM) recommended in a report that disaster impact studies should be part of the EIA. This recommendation stemmed from the common perception that hydropower projects have increased the damage in Uttarakhand caused by floods, especially flash floods and lake bursts from glacier-fed lakes, such as the catastrophic one that caused large loss of life in 2013. The NIDM study also pointed out that: "hydro-power projects are blamed for disturbing ecological balance in the sensitive fragile zone of Himalayas, leading to more landslides and other associated risks" (NIDM in [33]). Small hydropower projects of up to 25 MW with an outlay of less than Rs. one thousand million (i.e., one billion) are exempted from the EIAs but they do need to obtain all clearances from the corresponding state government.

Women are the cultural and economic backbone of life in the Garhwal Himalayas (as well as in the neighboring Kumaon Himalayas of Uttarakhand [34]). They do most of the farming and household work, which involves collection of drinking water, fodder, fuel wood and attending to the cattle (mostly cows and water buffaloes for their milk), besides all the daily household chores. Girls often work alongside their mothers. Male out-migration from the area has further heightened women's roles in agriculture such as in paddy (rice) cultivation, the cultivation of other grains and pulses such as lentils and the production of vegetables. Similar patterns have been documented in other areas of Uttarakhand state such as in the Nanda Devi biosphere reserve area [35], which is located to the northeast of the Bhilangana river basin research sites we studied. Women's welfare groups i.e., MMDs are therefore recognized as one of the most important institutions by the local communities and are therefore represented by the biggest circle. In all the impacted villages studied, MMDs were also reported to have actively participated in the protests against hydropower developers to fight for their rights related to land, forest and water. Therefore, these are perceived to be closer to the village in terms of their relationship with the villagers.

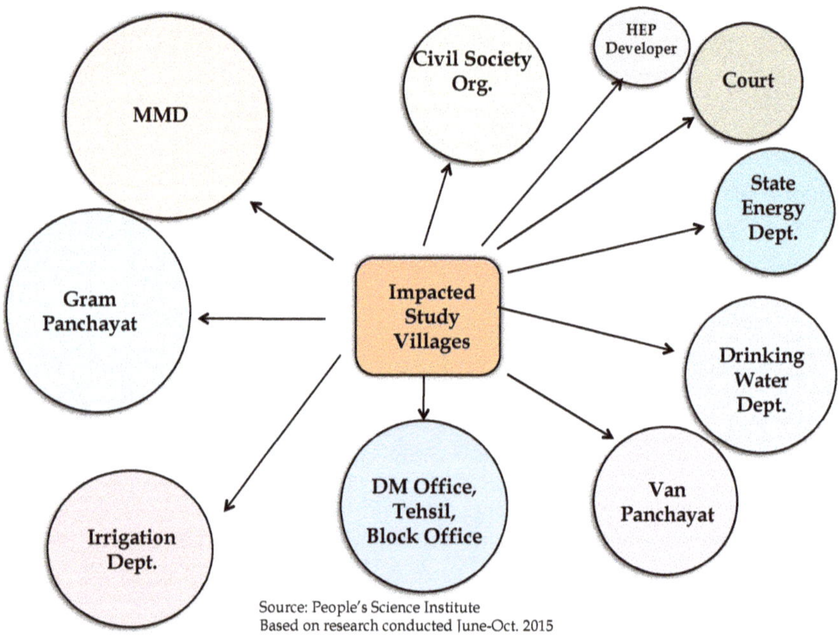

Figure 5. Stakeholder mapping: Bhilangana River Basin study sites.

An example of MMD's brave, active and prolonged (therefore costly) participation is reflected in the words of a woman who is a former member of the MMD of Phalenda village and was a leader in the protest:

> *We came to know about the ongoing construction plans only when we saw the machines. We led the group and reached the construction site. The moment villagers went to stop the construction, police were employed. The police would not even let us access our fields. We did not want them to take away the water used for irrigating our fields. So we started the protest that lasted for about a year. We would take turns going to our fields for work, attending to the needs of our children and also participating in the protest. For hours, groups of men and women would sit determined by the side of the road and would get up only when surrounded by the police. We participated in frequent hunger strikes. In my memory we stayed hungry for four days continuously one time and eight days another time. We were sent to jail at three different places three times. A settlement was reached after one year of tiring agitation, when we got water for irrigating our fields. However, if you ask me, the most important impact has been on our mental peace of mind and well-being, which can't be reimbursed (translation from Hindi ours).*

Due to the prolonged agitation by the MMD, farmers of Phalenda managed to get a better share of irrigation water from the hydropower developers in comparison to other affected villages.

Gram Panchayat as an institution in the three-tier local governance system is the lowest institution in the hierarchy through which communities participate in the planning and execution of all village development activities. So, it is perceived as the second most important and still close to the village in terms of accessibility and usefulness to the villagers.

Yuvak Mangal Dal or the youth welfare's group is still active but heavy out-migration from the area has diminished its level of activity to a considerable extent. Among all the village level institutions, this is perceived to be least active.

The next set of primarily government institutions are mainly service providers and are approached only in case of need. According to the villagers' own words: "As and when these institutions are approached, our work gets done. However, we have no additional expectations from them". The irrigation and water departments are shown separately, as per their roles, but according to the villagers, the majority of the repair and maintenance work for the pipelines and *guhls* (irrigation canals) is performed by the villagers themselves via voluntary labor contributions. Therefore, the villagers' perceptions of their need for these organizations and these departments' efficacy is lower (represented by smaller circles) and they are placed further away.

Lastly, hydropower developers and the district courts are the two institutions that are perceived to be least important. The developers' main interest appears to be earning revenues from energy generation; local communities' interests do not come first within their agenda and often get sidelined. Similarly, the communities have had largely negative experiences with the courts in terms of resolving conflicts over benefits sharing and compensation related to hydropower projects. However, as we discuss below, the state and federal governments could play a role in mandating, through legal instruments, more equitable benefit sharing between the community and hydropower developers. This would also translate into a better balance between meeting the needs of rural and urban populations. The courts would then be responsible for upholding these legal measures aimed at achieving social justice and environmental sustainability. In addition, a stronger, more socially just water–energy–food nexus would be ensured because of the critical role rural populations continue to play in the production of food for both rural and urban areas.

3.2. *Case Studies*

Developers of ROR projects have frequently overlooked the energy and irrigation needs of the local communities. They have not allowed for sufficient environmental flows downstream of the ROR projects. This lack of transparency and coordination with local communities has translated into negative impacts on the livelihoods of female and male villagers neighboring the project sites, as highlighted below in the case studies of four villages in the Bhilangana basin in Uttarakhand.

3.2.1. Observed Common Social Impacts in All Studied Hydroelectric Project (HEP) Areas

The negative effects of hydropower development reported by villagers living and farming in the vicinity of these ROR projects include land destabilization, leading to infrastructure damage including cracks/seepage in houses and/or community buildings; and destabilization of slopes with greater chance of landslides, fissures and land subsidence (some of these effects are due to dynamite blasting of tunnels for the HEP project); as well as an accumulation of muck (rock and debris created by tunneling during the construction of hydropower projects) on grazing/agriculture land. There are also myriad water-related effects such as reduced water resource availability including defunct water mills (for grinding grains) due to less water in the river in locations where water has been extracted for the HEP; reductions in domestic and irrigation water supply due to spring sources disturbed especially by tunnel blasting; irrigation canals destroyed due to landslides or abandoned due to too little water available; water pipelines carrying spring water for domestic water use and home gardens destroyed; dry riverbeds during part of the year; more wild animal attacks on crops and livestock due to deforestation affecting food supply for these animals; and less water in the river; reduction in irrigated area; reduction in availability of fodder and fuelwood due to less vegetation along the river banks; reduction in manure due to less livestock kept because of decreased availability of fodder; loss of cremation ghats (riverside cremation grounds); and a reduction in fish population due to less water in the river and poorer water quality. These effects are not limited to the study area. Dry springs and increased vulnerability to landslides and deforestation, for example, were similarly identified as effects of hydropower projects in the Nanda Devi biosphere reserve in Uttarakhand to the northeast of the Bhilangana study sites [34]. Damage to and lack of water for water mills and irrigation canals was also reported in a study conducted on the effects of small ROR projects in Himachal Pradesh state, India [36]. These environmental burdens are experienced by agricultural communities that depend on natural resources for their livelihoods and reside in the ROR project areas. As will be seen, some local residents are affected even more than others.

3.2.2. Impacts Observed in the Case of Devling and Chakar Village, Bhilangana III HEP Area

Impacts of blasting for the construction of the tunnels for the HEP can be clearly seen at the Bhilangana III HEP areas. Some of the respondents expressed that: "The land beneath our houses would shake, as if an earthquake had hit the area. We couldn't sleep during those nights". In Devling village located near Bhilangana III HEP (24 MW), as many as 86 out of the 110 resident households observed visible cracks in their houses after the construction of the HEP started. In terms of their springs (also a source for irrigation), seven out of eight are reported to have subsequently partially or completely dried up. Due to the construction of tunnels, irrigation channels in the neighboring villages have been affected. The agricultural fields have suffered a loss of productivity due to reduced irrigation water availability, which also led to a reduced amount of fodder (straw) from paddy and wheat crops. Approximately 95% of the households reported a loss of productivity in agriculture. The reduction in irrigated area has affected food availability from farm and agricultural income. Agricultural land has also been affected through the dumping of construction material from the hydropower projects on this land. Dependency on the market for food grains has therefore increased.

Debris from the blasting of the tunnels for the hydropower projects has been dumped on grazing lands accessed by all households in the village. When asked how the hydropower project has impacted their daily life, a respondent immediately reacted by forcefully stating that:

> *They caused damage to our agricultural fields and did not compensate us for it. Our harvest has been reduced due to less water available for irrigation and a lot of dust from the construction. The springs feeding our guhls (irrigation channels) have disappeared because of the blasting. This pollution in the area has also impacted our grazing land. It's just not the same grass anymore and animals won't consume it. Many of us have therefore sold off fifty per cent of our milch (dairy) cattle.*

Households that earlier ran water mills from the mountain streams for additional monetary income and for food security, have now been affected by a loss of income due to greatly reduced water flows. As many as ten water mills have become defunct in Devling and Chakar villages.

The People's Science Institute's (PSI) field research team, together with residents of Devling village, carried out mapping of impacts on a village map of Devling, indicating present land use and land cover, which was available with the revenue official at the site (see Figure 6). A range of features including impacted springs, impacted irrigated/unirrigated area, houses that developed cracks, defunct water mills, land deposited with muck, etc. were marked after visual verification by the field research team.

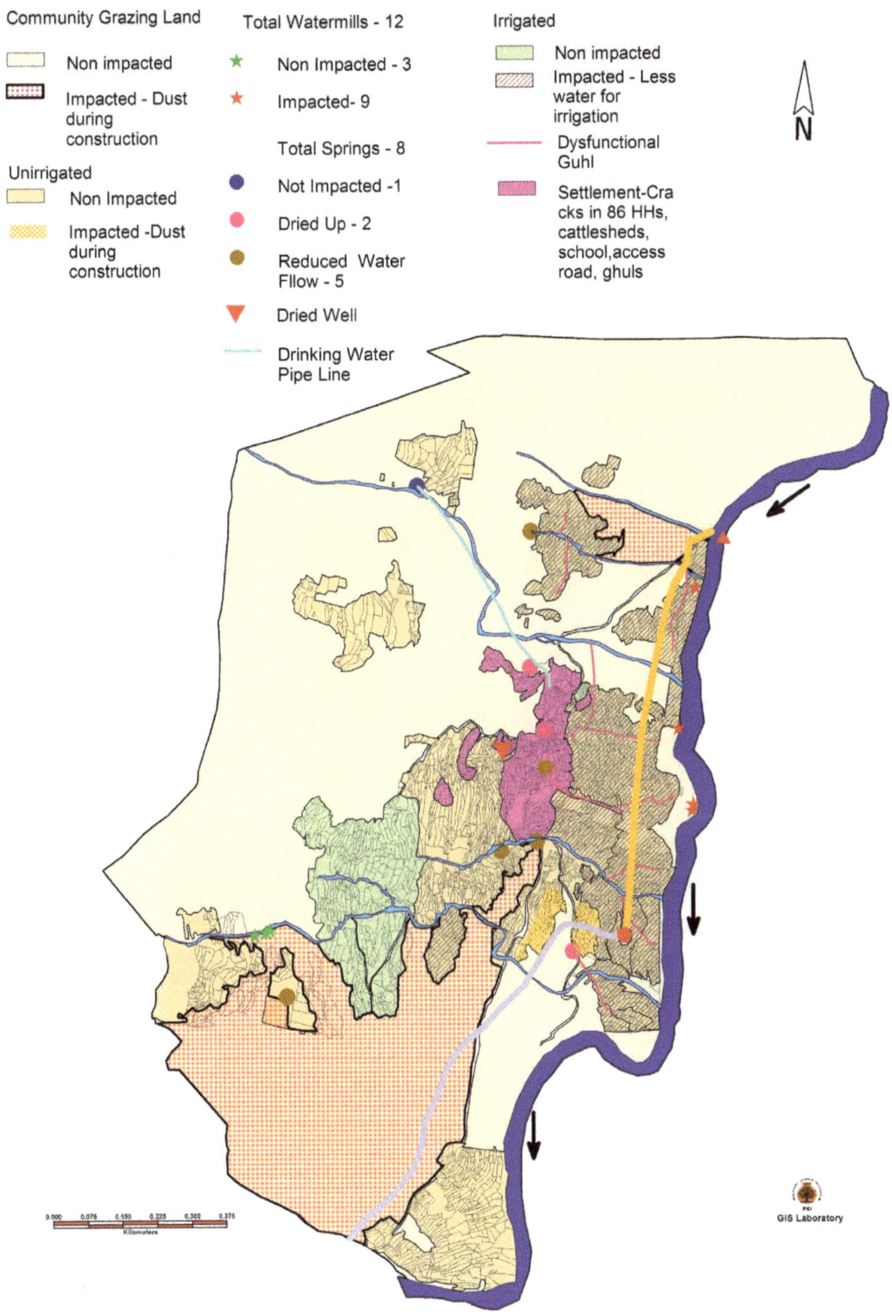

Figure 6. Map of Run-of-the-River (ROR) Project Impacted Areas in Devling Village.

The map shows the extent of the damage exacted on important livelihood-related infrastructure, privately-held and common lands and on water resources for domestic and agricultural use.

Gendered Impacts

All hydropower projects have had gendered impacts; in the study area, in the case of all women interviewed, their workload related to the collection of water, fodder and fuelwood increased. Loss of fodder and fuelwood sources has forced them to go deeper into the forests for collection. More men were reported to have migrated in the post hydropower project period to make up for livelihoods-related losses from the hydropower project. Women have been left alone to tend to animals and to farms with decreased access to the means of production.

Villagers have responded by reducing their herds of cows, bulls and buffaloes, and/or by renting the use of bulls for ploughing. In Devling village, the average annual loss in income per household from milk has been around Rs. 14,000 since the HEP started functioning. In Chakar village near the Bhilangana III hydroelectric plant, women told our research team:

> *The forest is far away. Lower availability from other sources has increased our time and effort. There is a fear of wild animals in deeper nooks and crannies in the forest. We have no help from the men. So we have reduced the number of animals we keep.*

Thus, hydropower projects in this area have clear social–justice related implications for rural populations, revealed in gendered impacts on labor allocations, income and access to food.

Compensation for Losses

Some livelihood compensation has been offered by the hydropower projects but these vary across and within the affected villages causing a great deal of unhappiness. Very few households from the affected villages received employment at the hydropower projects, for example, only three from Devling. Only some households in the villages received monetary compensation and this compensation was reported to be inadequate. In Devling, only four out of nine households received compensation for defunct water mills. Land compensation rates for land taken over by a hydropower developer ranged from Rs. 40,000 to Rs. 95,000 per *nali* (50 *nalis* = 1 ha) for 20 households. No compensation was given for damaged houses and cattle sheds in any of the affected villages. A few households reported investing Rs. 15,000 to Rs. 20,000 for reconstructing cattle sheds. In some villages, pipelines have been built by developers from the sources of springs to the village. However, in Devling village, due to disputes over labor charges for villagers volunteering to help, pipelines were not installed. An ad hoc compensation of Rs. 1 lakh (Rs. 100,000) was given to the *Gram Panchayat* of Chakar Gaon village.

3.2.3. Impacts Observed in Chani Village, Agunda–Thati HEP Area Despite Signed Agreement

In Chani, a village affected by the 3 MW Agunda–Thati hydropower project, the villagers stated that they were warned by nearby villages in the area about the negative effects that they had endured. Villagers interviewed in adjoining Agunda and Thati villages said that they had turned down the hydropower developers' requests to construct a hydropower project in their location. As a result, the project was moved to Chani. The signed agreement stated that, aside from the land handed over on paper to the developers by the villagers, no other land would be affected. The signed agreement also assured that any individual in the *gram sabha* (i.e., all villagers over age 18) who suffered any kind of damage to their property during the HEP construction phase would be compensated by the company. Muck disposal would be done in the depositing yard, not on farmers' land. They were promised that any destruction to the canals, crops, paths and fields would be repaired. However, none of these promises were subsequently met.

The villagers explained that:

> *We have been living here for generations and now they have taken everything away. The developer belongs to the same district (but is now a large industrialist living in the large city of Dehradun). We told others: learn a lesson from our village.*

Most of the landowners (127 out of 135 households) reported having irrigated land along the riverbank. This land of the village had become barren downstream of the power project due to the diversion of river water for power generation. In Chani village, families lived for six months near the river and farmed there and then moved up to higher elevation lands and lived and farmed there for six months. Now they must live up in the higher elevations year-round.

A reduction in the number and depth of pools in the river has caused fish populations to diminish, affecting male fishermen primarily and their household food and income security. A fisherman farmer from Chani village explained that fishing has been harmed by the Agunda–Thati hydropower project. He decried that:

> *I have seen three varieties of fish in the river. We call them as machibag, machhi, gadiyal locally. One of them is red and yellow in color. Earlier we would go for fishing in the summers. In the monsoon season, we would catch fish from standing water in the paddy crop (coming up from the river). Now, since most of the water is diverted in the hydropower channel, I do not see any fish in these waters. My fellow men and I from the village do not visit the river for fishing anymore.*

Nine fishermen households from Chani village reported an average annual economic loss of over Rs. 10,000 due to the fact that there are almost no fish in the river anymore due to inadequate flows in the dry season. These households attempted to intensify their agricultural production, but several factors have limited their opportunities to do so. Thus, fisherfolk are particularly vulnerable to the environmental burdens of hydropower projects, with clear social justice implications for their livelihoods and for household food security.

Wild vegetation used to grow on land along the riverbank in Chani but now it has been affected by the hydropower project construction debris and the lower volume of water in the river. This reduces the availability of proximate fodder for livestock. Since women and children (mainly girls) collect fuelwood and fodder, the intensity of women and children's fodder-collection labor has increased.

The residents of Chani realized that they were fooled and demanded that the developers should have at least abided by the signed contract. The villagers shared with us that: "our lawsuit focuses on compensation for our land, forest and water ... How do we keep the youth here when what we have are land and water based which also has been taken away?" They contended that their skills as farmers and their farm-based livelihoods provide a strong degree of self-reliance and as a result are more respectable than urban-based menial jobs as laborers or hotel employees.

The villagers now obtain most of their food grains at subsidized rates from the ration store that is reserved for families living below the poverty level, or, for families above the poverty line, from the market. Even the hydropower project operator from the village complained about receiving low wages for his work from the developer. He is also pressured by the developer to tell the villagers not to protest otherwise he will lose his job. He thinks that the village should take over the project and he would work for them instead. Fear of retribution by developers or the government was also noted in a study of HEP projects in Sikkim state in Northeast India [37].

3.2.4. Impacts on Environmental Flows Observed in Phalenda Village, Bhilangana HEP Area

In a hamlet of Phalenda village there is a small hydel project (Bhilangana) of 22.5 MW. Villagers had protested the construction of the HEP for a long period, demanding benefits such as the construction of an access road and a school, in addition to a fair share of water for irrigation. Women and even children went to jail due to their active participation in the protests. Though the villagers here got a better share of irrigation water compared to villages affected by the Bhilangana III and Agunda–Thati projects, they reported a loss of productivity from their fields in the lowlands especially in the winter season (wheat crop) when HEP developers did not come through with a timely and adequate release of irrigation water. This community could possibly have benefitted from power generated for lift irrigation for their fields in the upland areas but due to a lack of community awareness no such demand was made during the protests or during the signing of the agreement between the HEP developer and the impacted residents.

Fifty to sixty houses of Phalenda, especially in the upper slopes, developed cracks in them from the blasting for the tunnel construction. Although in Phalenda there was a distribution of compensation for infrastructure-related damage, according to villagers, the distribution was not done properly and not all families got the same sum. In the words of a respondent: "the sum is not enough to get the repair work done. I had to take a loan from a neighbor to get it done". About seven springs were reported to be affected by the blasting operations, four of which were partially affected earlier by the 1991 earthquake. Only eight people from the village derived employment from the project.

The women and men in the villages have a keen understanding of the political, social, economic and environmental dynamics associated with the abstraction of natural resources from rural areas for the provision of power to urban areas. The diversion of water from the river for the hydropower project did not allow for sufficient environmental flows to sustain local livelihoods. A woman from Gawani hamlet (in Phalenda village) told us: "Food (i.e., power) is being prepared in our village but someone else is getting to eat it". In other words, their lived experiences with hydropower production have enabled them to understand the true costs of the expropriation of river water near their villages in terms of its effects on their access to water for irrigation and household use.

4. Discussion

4.1. Governance of Run-Of-The-River Hydropower

Studies conducted on the social and environmental impacts of centralized hydropower schemes in India have revealed negative gender impacts related to access and control over resources [30,38]. The risks to which women of affected communities have been exposed, through such projects, include loss of land, loss of houses, resettlement, loss of fisheries, loss of access to forests, influx of (male) workers, etc. Even the more recently commissioned Tehri (Uttarakhand) and Sardar Sarovar (Gujarat) projects seem not to have learned lessons from the past experiences of Bhakra–Nangal (Himachal Pradesh) and Hirakud (Odisha) dams [39–43]. This situation is not restricted to India, as the report of the World Commission on Dams [44] (p. 114) highlights how large hydropower projects all over the world have *"widened gender disparities either by imposing a disproportionate share of social costs on women or through an inequitable allocation of the benefits generated"*. There have been no strategic efforts by the developers, either private or governmental, to involve women right from the planning process. A recent gender analysis of renewable energy projects in India states that there has been a *"lack of gender mainstreaming in the energy sector and a lack of understanding of how to incorporate gender concerns in a sector that has primarily been technology-driven"* [45] (p. 93). In other words, within the water–energy–food nexus, women's roles are not recognized even though they are the principal agricultural producers in rural communities in Uttarakhand.

Hydropower projects are also subjected to risks and damage from natural disasters including earthquakes and floods. The disastrous Koyna earthquake of 10 December 1967, is often attributed to reservoir-induced seismicity [46,47]. Questions are often raised about the earthquake resistance ability of the Tehri dam, the tallest dam in India, which has been built in a seismic zone [48]. The 2013 floods of Uttarakhand have highlighted that the design of most of the 24 existing and under construction hydropower projects in the Upper Ganges basin have not accounted for events such as cloud bursts, suffering not only damage but also aggravating the proportion of the disaster [49,50]. This has further raised concerns over the viability of such projects, especially in the context of a changing climate. Thus, one can argue that there are hardly any examples of centralized hydropower schemes in India that have been informed by more progressive gender and risk-based impact assessment.

4.2. Benefits Sharing

The concept of benefits sharing in hydropower projects has evolved over time, starting from a notion that local communities would derive trickle down benefits of hydropower development initiated by a regional or national entity and then encompassing compensation to local communities for harm done to their property or livelihoods. Currently, though, as Shrestha et al. (2016) explain, there is greater potential for local communities to gain *"less conflict and greater voice as well as greater ecosystems services"* because *"benefit sharing is increasingly defined as going beyond the mitigation of project impacts and beyond compliance to a situation where the local affected population directly benefits from the project."* [51] (p. 7). A benefits' sharing partnership enables the local population to have an equity stake in the project and to strive for more socially just project outcomes for villagers. Hydropower developers in the Indian Himalayan region need to be encouraged to invest in livelihood enhancing activities of local villages. According to hydropower developers in the study area in Uttarakhand, they allocate 1 per cent of project costs towards a local area development fund (LADF). Similarly, 3 per cent of generated power is to be given to the concerned Panchayti Raj Institutions of the impacted villages. However, discussions with local communities and members of Panchayati Raj Institutions (PRIs) revealed multiple discrepancies between such policy measures, signed agreements and ground realities (fieldwork, 2015). Village communities in the study area are largely unaware of such policies and contributions made to PRIs. This calls for revisions in institutional mechanisms and ensuring transparency in the process. It also requires enforcement of rules and legislation.

It is worthwhile to mention here that the neighboring country of Nepal is moving toward a new electricity act that would mandate that ten per cent of the shares of the hydropower project would go to the impacted local community, with the smallest bids considered first in a market-based bid allocation system [51,52]. Private companies are creating public companies to distribute shares in hydropower projects. Chilime, a 22.1 MW hydropower project in Nepal, distributed shares to the affected local people in 2010 for the first time in the history of Nepal. Since then, other ROR hydropower projects have also allocated 10% shares to affected communities [51] (p. 16). This is particularly salient in the context of the continued expansion of hydropower in Nepal, where the government has been actively promoting hydropower in order to reduce the country's power load shedding to zero. Nevertheless, the April 2015 earthquake had serious impacts on hydropower facilities and infrastructure of all kinds. With the urgent need for urban housing and commercial reconstruction, however, these delays in hydropower expansion may provide the opportunity to enhance integrated governance, including benefit sharing [51].

4.3. Decentralization

Decentralized Distributed Generation (DDG) through community managed small hydro projects can be a suitable method for the provision of electricity to the forest fringe and scattered villages in the Indian Himalayan region [53,54]. Operation and maintenance of these projects can be undertaken by the local community after formal training. Such projects have a short gestation period with minimum environmental impacts. In the mountain state of Uttarakhand, the Uttarakhand Renewable Energy Development Agency (UREDA) is a state agency, established in 2001 as a registered society under the Department of Alternate Energy, Government of Uttarakhand. Its primary purpose is to promote the development of renewable energy resources in Uttarakhand. It has constructed 44 off-grid and on-grid micro-hydel projects with a total installed capacity of 4.29 MW. Another 19 projects (2.315 MW) are under construction [18]. The off-grid projects are managed by committees of the beneficiary villages. These projects are meant for electricity generation for feeding into the central grid or for local domestic household consumption, in cases where they are not already electrified by the central grid. Small hydropower schemes with lower head can incorporate fish ladders, for example, that can protect fish migration routes and thus livelihoods derived from fishing.

Jansamarth, a Delhi-based voluntary organization with the goal of promoting local livelihoods, has helped in the installation of 17 micro-hydro power projects in the two Himalayan states of Jammu and Kashmir and Uttarakhand. One such example is that of a community-managed 50 KW capacity hydropower plant established in 2006 in Agunda village located in Balganga valley of the Bhilangana study area in Uttarakhand [52]. This multipurpose hydropower unit mechanically powers an oil press, rice huller and flourmill, creating livelihood opportunities for the village community consisting of 65 families. The same water is used to irrigate 6–7 hectares of land. The villagers are proud owners of the project, because they put in the labor for its construction and because it is now operated by trained community members. An older man from this village stated: "This project is ours—we brought it here. While we are living we won't allow any other project".

River basins can be glacial or spring-fed. The performance and life of hydropower projects located in the basins of spring-fed rivers depend on sustained discharge of springs contributing to the river flows. Over the last two decades a substantial decrease in spring discharge has been reported to be affecting flows in spring-fed rivers in the Himalayas [55,56]. This is mostly due to environmental degradation caused in the catchments of the springs due to anthropogenic activities and more recently due to climate change. Spring-shed development can help revive springs. Sustained flows are essential to sustain hydropower projects, especially those with a smaller capacity and those run by spring-fed rivers. Spring development therefore needs to be incorporated into hydropower planning in Himalayan states. Springshed development involves hydro-geological mapping, delineation of the concerned aquifer, seasonal monitoring of spring discharge, identifying the potential recharge area, formation and capacity building of spring user group, planning and implementation of treatment measures in the recharge area with community's participation, and subsequent monitoring and impact assessment. Treatment measures include a combination of tree plantation, staggered trenches, brush wood check dams, etc. which reduce surface run off and increase infiltration leading to spring recharge [57].

People's Science Institute (PSI), a research and development organization based in Dehradun in Uttarakhand, together with other non-governmental organizations such as WWF-India, ACWADAM-Pune, and Arghyam–Bangaluru, assisted the government of the state of Sikkim in the eastern Himalayas in India to successfully launch the Dhara Vikas (Springs Development) initiative in 2008. This spring-shed development initiative was launched under the centrally sponsored Mahatma Gandhi National Rural Employment Guarantee Act (MGNREGA) scheme, covering a total area of 400 hectares and resulting in an annual groundwater recharge of 900 million·L. The scheme led to the revival of five lakes and fifty springs. PSI is in dialogue with Uttarakhand and other Himalayan states concerning similar initiatives to link hydropower with livelihoods, taking into special consideration ecosystems and climate change.

4.4. Planning for a Basin with Decentralized and Centralized Distributed Generation Solutions

Decentralized power projects cater to energy requirements that transcend local needs. However, these projects have their own limitations in terms of accessibility (high cost of transmission lines and maintenance) and quality (reduced supply hours and poor load), especially in remote areas as evident in the Himalayan region of India. In such cases, there is a clear need for integrating decentralized projects with centralized schemes. This would result in a more cost effective delivery of power in addition to a decrease in undesirable environmental and social impacts. Apart from the above, such decentralized projects could open up new and enhance existing economic activities as demonstrated by the small-scale, community-based hydropower projects established by Jan Samarth in India. Integration of centralized and decentralized projects would involve (1) proper estimation of energy requirements at the basin level and even beyond considering different sectorial needs such as household lighting, potable and irrigation water supply, health and sanitation, education, communication and economic activities; (2) identification of vulnerable areas within the basin which cannot be reached by centralized supply; (3) identification of alternative energy sources which could be owned and run by vulnerable communities; (4) participation of local communities especially women in the planning/decision-making of centralized solutions to minimize detrimental effects and ensure access to natural resources; (5) strengthening of local institutions (such as *panchayats* i.e., local governance bodies) for ensuring just flow and sharing of benefits for affected communities; and (6) robust institutional mechanisms at various levels within the basin involving various stakeholders to monitor and evaluate benefit sharing activities in the impacted areas. This also calls for the establishment of a grievance redressal mechanism for negotiated settlements of disputes benefiting local communities as well as developers. Thus, integration of centralized and decentralized power schemes needs "adaptive, multilevel, and collaborative institutional arrangements" [58] that are decentralized and account for local conditions [59].

5. Conclusions

Multipurpose, run-of-the-river hydropower development can contribute to a healthier water–energy–food nexus in the Himalayas. However, at present, the benefits are skewed in favor of urban populations and hydropower developers. The study, based on a social justice approach, illuminated imbalances in decision-making power between these urban interests and rural villagers who are dependent on agriculture and from whom river water has been expropriated. Mountain communities are caretakers of ecosystem services. Run-of-the-river hydropower, especially larger projects, has been causing environmental harm (albeit less harm than large dams) that in turn has gendered livelihood impacts. A social justice approach and the inclusion of the exact words of villagers interviewed shed light on the desires of local populations with respect to hydropower projects and the areas where social equity in hydropower governance could be enhanced. Rural populations desire local hydropower for livelihood activities, irrigation benefits, household chores and the education of children. Locally controlled, smaller HEP projects in Uttarakhand have enabled rural peoples to be prime decision-makers in the control over power generation and water allocations.

Where the HEP project will negatively impact environmental flows for agriculture and fishing, fuelwood and fodder, and will damage village infrastructure, populations in Uttarakhand are demanding compensation and, increasingly, benefits sharing. Women are strong actors in these protest movements. This is largely due to the fact that women bear an unequal burden within their rural communities when irrigation and household water, land, and forest resources are reduced due to hydropower, climate and land intensification-induced changes. Rural men and youth are migrating to urban areas at a higher rate than before, partly because of diminished income from agriculture and livestock rearing due to HEP impacts; this migration is adding to urban energy demand and adding to rural women's work. These HEP impacts on labor reduce rural food production; they thus also exact a cost for the central government in terms of increased urban demand for services such as energy, and declining rural food production at a time when it is necessary to feed growing urban populations. Perhaps, if more women were involved at the global level in renewable energy issues, these regional water–energy–food nexus challenges could be addressed. At present, at the global scale in energy governance institutions, only 4% of the top World Energy Council (WEC) chair positions are women and only 18% of the secretary positions in WEC are women [60].

Participatory governance and clear guidelines and mechanisms for benefits sharing are essential at the local, basin, and transboundary scales. Following the example of neighboring country Nepal, higher rates for benefit sharing between hydropower developers and local communities near the hydropower project could be instituted and enforced to ensure greater equity. State and federal governments could play a role in mandating, through legal instruments, more equitable benefit sharing between the community and hydropower developers that would be enforced by lower courts. If disaster impact assessments were mandated as part of environmental impact assessments carried out by the Ministry of Environment, Forests and Climate Change, disaster could be redefined to include the negative effects found in the villages in this study of the Bhilangana river basin. Such assessments would ideally contain a greater emphasis on gender-related assessments to avoid both greater workloads for women and girls and negative impacts on household food security and income.

These measures would also translate into a better balance between meeting the needs of rural and urban populations. Springshed development has shown promise in other regions in India and could help prevent some of the harm to water sources for irrigation and household use for riparian communities in Uttarakhand. Small hydropower schemes, built and operated by local populations and that meet hydropower generation and irrigation needs, can reduce the negative impacts and increase the benefits of hydropower for rural communities, as can a combination of decentralized and centralized schemes.

Future research needs include an examination of how local, state and central governments can support the local management of hydropower schemes and how springshed development and other initiatives shown to be successful in one area can be adapted by communities in other states to meet their needs. Other research topics that need to be studied include analyses of hydropower–agriculture tradeoffs and an examination of HEP effects on female and male youth's labor-related decision-making to gain a clearer vision of future scenarios for these rural communities and possible policy responses.

Acknowledgments: The authors gratefully acknowledge the support of the Water, Land, and Ecosystems Program of the CGIAR, and specifically funding for the project "The Irrigation–Hydropower Nexus in the Ganges Headwaters" provided by the International Water Management Institute.

Author Contributions: Stephanie Buechler conceived the analysis and took primary responsibility for analysis and writing. Debashish Sen contributed through analysis of hydropower governance and extensive expertise in Uttarakhand. Neha Khandekar conducted primary field research in the Bhilangana basin, aided and supervised by Stephanie Buechler, Debashish Sen, and Christopher Scott. Christopher Scott contributed conceptual analysis and writing to link hydropower governance with the water–energy–food nexus.

Conflicts of Interest: The authors declare no conflict of interest.

Abbreviations

The following abbreviations are used in this manuscript:

DPR	detailed project report
EIA	environmental impact assessment
HEP	hydroelectric project
LADF	local area development fund
MMD	*Mahila Mangal Dal* (women's village welfare groups)
NIDM	National Institute of Disaster Management
PRI	Panchayati Raj Institution
ROR	Run-of-the-river
YMD	*Yuvak Mangal Dal* (youth welfare groups)

References

1. Scott, C.A.; Crootof, A.; Thapa, B.; Shrestha, R.K. The water-energy-food nexus in the Ganges Basin: Challenges and opportunities. In *Water Management in the Ganges Basin*; Bharati, L., Smakhtin, V., Sharma, B.R., Eds.; Earthscan: London, UK, 2016.
2. Kumar, D.; Katoch, S.S. Sustainability indicators for run of the river (RoR) hydropower projects in hydro rich regions of India. *Renew. Sustain. Energy Rev.* **2014**, *35*, 101–108. [CrossRef]
3. REN21. Renewables Global Status Report (GSR). Paris, REN21 Secretariat. 2015. Available online: http://www.ren21.net/status-of-renewables/global-status-report/ (accessed on 12 February 2016).
4. Mukherji, A.; Molden, D.; Rasul, S.G.; Wagnon, P. Himalayan waters at the crossroads: Issues and challenges: Editorial. *Int. J. Water Resour. Dev.* **2015**, *31*, 151–160. [CrossRef]
5. Shah, T. Climate change and groundwater: India's opportunities for mitigation and adaptation. *Environ. Res. Lett.* **2009**, *4*, 035005. [CrossRef]
6. World Health Organization. Ambient (Outdoor) Air Pollution in Cities Database 2014. Available online: http://www.who.int/phe/health_topics/outdoorair/databases/AAP_database_results_2014.pdf (accessed on 5 April 2016).
7. Rasul, G. Managing the food, water, and energy nexus for achieving the Sustainable Development Goals in South Asia. *Environ. Dev.* **2015**. [CrossRef]

8. Watershed Management Directorate, Dehra Dun. Uttarakhand State Perspective and Strategic Plan 2009–2027. Dehra Dun, Uttarakhand, 2010; p. 288. Available online: http://wmduk.gov.in/Perspective_Plan_2009--2027.pdf (accessed on 1 March 2016).
9. Joshi, K.; Bhardwaj, N. Women and natural resource management: A study of 'communities of practice' prevailing in women farmers' community management of water and forests of lesser Himalayan region in India. *Int. J. Adv. Res.* **2015**, *3*, 363–374.
10. Reddy, V.R.; Uitto, J.I.; Frans, D.R.; Matin, N. Achieving global environmental benefits through local development of clean energy? The case of small hilly hydel in India. *Energy Policy* **2006**, *34*, 4069–4080. [CrossRef]
11. Ministry of Environment and Forests. Assessment of Environmental Degradation and Impact of Hydroelectric Projects during the June 2013 Disaster in Uttarakhand, Main Report, MoEF. Government of India, April 2014. Chapter 2, ToR 2.1 a. p. 34. Available online: http://www.indiaenvironmentportal.org.in/files/file/environmental%20degradation%20&%20hydroelectric%20projects.pdf (accessed on 10 January 2016).
12. Gopalakrishnan, M. Resettlement and Rehabilitation: Lessons from India. In *Impacts of Large Dams: A Global Assessment*; Dogan, A., Biswas, A., Tortajada, C., Eds.; Springer: Berlin/Heidelberg, Germany, 2012.
13. Asthana, V.; Cheney, W.A. Forced Displacement: A Gendered Analysis of the Tehri Dam Project in India. *Econ. Political Wkly.* **2012**, *47*, 96–102.
14. Mäkinen, K.; Khan, S. Policy considerations for greenhouse gas emissions from freshwater reservoirs. *Water Altern.* **2010**, *3*, 91–105.
15. Ahlers, R.; Budds, J.; Joshi, D.; Merme, V.; Zwarteveen, M. Framing hydropower as green energy: Assessing drivers, risks and tensions in the Eastern Himalayas. *Earth Syst. Dyn. Discuss.* **2014**, *5*, 1521–1541. [CrossRef]
16. Postel, S.L.; Thompson, B.H. Watershed protection: Capturing the benefits of nature's water supply services. *Nat. Resour. Forum* **2005**, *29*, 98–108. [CrossRef]
17. ESHA. The European Small Hydropower Association, 2009. Environmental Barometer on Small Hydro Power. Brussels, Belgium. Available online: http://www.esha.be/fileadmin/esha_files/documents/SHERPA/Environmental_Barometer_SHP.pdf (accessed on 29 February 2016).
18. Uttarakhand Jal Vidyut Nigam, Ltd.; Uttarakand, India. Personal communication, 2013.
19. Pearce, F. Dams and floods. In *Large Dams for Hydropower in Northeast India: A Dossier*; Menon, M., Kohli, K., Eds.; SANDRP: Pune, India, 2005.
20. D'Souza, R. Filling Multipurpose Reservoirs with Politics: Displacing the Modern Large Dam in India. In *Large Dams in Asia*; Nüsser, M., Ed.; Springer: Dordrecht, The Netherlands, 2014.
21. Bhattacharyya, S.C. Energy access programmes and sustainable development: A critical review and analysis. *Energy Sustain. Dev.* **2012**, *16*, 260–271. [CrossRef]
22. Urpelainen, J. Grid and off-grid electrification: An integrated model with applications to India. *Energy Sustain. Dev.* **2014**, *19*, 66–71. [CrossRef]
23. Zepeda, E.; McDonald, S.; Panda, M.; Kumar, G.; Sapkota, C. Employing India: Guaranteeing jobs for the rural poor. A Report Published by the Carnegie Endowment for International Peace and the UNDP. p. 86. Available online: http://issuu.com/carnegie_endowment/docs/india_rural_employment/1?e=3035200/5270418 (accessed on 26 September 2016).
24. Wüstenhagen, R.; Wolsink, M.; Bürer, M.J. Social acceptance of renewable energy innovation: An introduction to the concept. *Energy Policy* **2007**, *35*, 2683–2691. [CrossRef]
25. Chowdhury, A.J.U.; Rasul, G. Equity and social justice in water resource governance: The case of Bangladesh. *South Asian Water Stud.* **2011**, *2*, 44–58.
26. Young, I.M. *Justice and the Politics of Difference*; Princeton University Press: Princeton, NJ, USA, 2011.
27. Gross, C. Community perspectives of wind energy in Australia: The application of a justice and community fairness framework to increase social acceptance. *Energy Policy* **2007**, *35*, 2727–2736. [CrossRef]
28. Braun, Y.A. Interrogating large-scale development and inequality in Lesotho. In *A Political Ecology of Women, Water and Global Environmental Change*; Buechler, S., Hanson, A., Eds.; Routledge: Abingdon-on-Thames, UK, 2015.
29. Rai, K. *Dam Development: The Dynamics of Social Inequality in a Hydropower Project in Nepal*; Cuvillier Verlag: Göttingen, Germany, 2005.

30. Mathur, H.M. *Displacement and Resettlement in India: The Human Cost of Development*; Routledge: London, UK, 2013.
31. Baviskar, A. Written on the body, written on the land: Violence and environmental struggles in central India. In *Nancy Peluso and Michael Watts*; Cornell University Press: Ithaca, NY, USA, 2001.
32. Asthana, V.; Shukla, A.C. *Water Security in India: Hope, Despair and the Challenges of Human Development*; Bloomsbury Publishing: New York, NY, USA, 2014.
33. Mohan, V.; Thakur, P. Make Disaster Study a Must for Uttarakhand Hydel Projects. Times of India, 17 August 2015. 07.09 AM IST. Available online: http://timesofindia.indiatimes.com/india/Make-disaster-study-must-for-Uttarakhand-hydel-projects/articleshow/48507584.cms (accessed on 20 April 2016).
34. Tiwari, P.; Joshi, B. Gender processes in rural out-migration and socio-economic development in the Himalaya. *Migr. Dev.* **2016**, *5*, 330–350. [CrossRef]
35. Ogra, M.V.; Badola, R. Gender and climate change in the Indian Himalayas: Global threats, local vulnerabilities, and livelihood diversification at the Nanda Devi Biosphere Reserve. *Earth Syst. Dyn.* **2015**, *6*, 505–523. [CrossRef]
36. Baker, J.M. Small hydropower development in Himachal Pradesh. An analysis of socioecological effects. *Econ. Political Wkly. (EPW)* **2014**, *49*, 21.
37. Huber, A.; Joshi, D. Hydropower, Anti-Politics, and the Opening of New Political Spaces in the Eastern Himalayas. *World Dev.* **2015**, *76*, 13–25. [CrossRef]
38. Lahiri-Dutt, K. Large dams and changes in an agrarian society: Gendering the impacts of Damodar Valley Corporation in eastern India. *Water Altern.* **2012**, *5*, 529–542.
39. Asthana, V. Women and Forced Displacement in the Tehri Dam Project. Available online: http://refugeewatchonline.blogspot.com/2011/02/women-and-forced-displacementin-tehri.html (accessed on 28 February 2011).
40. Baruah, B. The Narmada Valley Project: Displacement of local populations and impact on women. *Nat. Resour. Forum* **1999**, *23*, 81–84. [CrossRef]
41. Barve, N.S. Economic, Social, and Environmental Impacts of Sardar Sarovar Dam Resettlement. Available online: http://scholarworks.sjsu.edu/etd_theses/3846 (accessed on 26 September 2016).
42. Rangachari, R. *Bhakra-Nangal Project: Socio-Economic and Environmental Impacts*; Oxford University Press: Oxford, UK, 2006.
43. Nayak, A.K. Development, Displacement and Justice in India: Study of Hirakud Dam. *Soc. Chang.* **2013**, *43*, 397–419. [CrossRef]
44. World Commission on Dams. *Dams and Development: A New Framework for Decision-Making, the Report of the World Commission on Dams*; Earthscan Publications Ltd.: London, UK; Sterling, VA, USA, 2000.
45. Integrated Research and Action for Development (IRADe). *Gender Analysis of Renewable Energy in India: Present Status, Issues, Approaches and New Initiatives*; Integrated Research and Action for Development: New Delhi, India, 2009.
46. Chopra, A.K.; Chakrabarti, P. The Koyna earthquake and the damage to Koyna Dam. In *Bulletin of the Seismological Society of America*; Seismological Society of America: Albany, CA, USA, 1973.
47. Bhatia, S. The Danger of Reservoir-Induced Seismicity. The Hindu. Available online: http://www.thehindu.com/todays-paper/tp-features/tp-openpage/the-danger-of-reservoirinduced-seismicity/article900414.ece (accessed on 10 January 2016).
48. Down to Earth. *Is the Tehri Dam Safe?*; Centre for Science and Environment: New Delhi, India, 1998.
49. South Asia Network on Dams, Rivers and People. *Two Years of Uttarakhand Flood Disaster of June 2013: Why is State & Centre Gambling with the Himalayas, the Ganga & Lives of Millions?*; South Asia Network on Dams, Rivers and People (SANDRP): Delhi, India, 2015.
50. Chopra, R. Uttarakhand: Development and Ecological Sustainability. Report produced for Oxfam India, New Delhi. 2014, p. 53. Available online: http://www.environmentportal.in/files/file/UttarakhandDevpEcoSustainabiity.pdf (accessed on 15 January 2016).
51. Shrestha, P.; Lord, A.; Mukherji, A.; Shrestha, R.K.; Yadav, L.; Rai, N. *Benefit Sharing and Sustainable Hydropower: Lessons from Nepal*; ICIMOD: Kathmandu, Nepal, 2016.
52. Securities Board of Nepal. *Securities Registration and Issue Regulations 2008*; First Amendment: Kathmandu, Nepal, 2010.

53. Chopra, R. *Hydropower Development in Uttarakhand: A Situation Analysis Report*; People's Science Institute: Uttarakhand, India, 2012.
54. Chandra, K.K. Electricity Is a Beginning in the Uplift of Villages and Not the End. Available online: http://www.theweekendleader.com/Causes/1339/beyond-power.html#sthash.yjjCSfG4.dpuf) (accessed on 20 January 2016).
55. Mehta, G.S. *Development of Uttarakhand: Issues and Perspectives*; APH Publishing Corporation: New Delhi, India, 1999.
56. Valdiya, K.S. *Environmental Geology: Ecology, Resource and Hazard Management*; McGraw-Hill Education: New York, NY, USA, 2013.
57. Tambe, S.; Kharel, G.; Arrawatia, M.L.; Kulkarni, H.; Mahamuni, K.; Ganeriwala, A.K. Reviving Dying Springs: Climate Change Adaptation Experiments from the Sikkim Himalaya. *Mount. Res. Dev.* **2012**, *32*, 62–72. [CrossRef]
58. Molle, F.; Wester, P. *Developing and Managing River Basins: The Need for Adaptive, Multilevel, Collaborative Institutional Arrangements, Comprehensive Assessment of Water Management in Agriculture*; International Water Management Institute: Colombo, Sri Lanka, 2007.
59. Kemper, K.; Blomquist, W.; Dinar, A. *Integrated River Basin Management through Decentralization*; Springer: Cham, Switzerland, 2007.
60. Environment and Gender Index. Women's Participation in Global Environmental Decision Making. Available online: https://portals.iucn.org/union/sites/union/files/doc/egi_factsheet_desicion_making_web_sept2015.pdf (accessed on 28 April 2016).

MDPI AG
St. Alban-Anlage 66
4052 Basel, Switzerland
Tel. +41 61 683 77 34
Fax +41 61 302 89 18
http://www.mdpi.com

Water Editorial Office
E-mail: water@mdpi.com
http://www.mdpi.com/journal/water

www.ingramcontent.com/pod-product-compliance
Lightning Source LLC
Chambersburg PA
CBHW041751220926
51847CB00014B/167
9783038424468